APPLIED POLYMER SCIENCE

Applied Polymer Science

Editor:

Asad Rahman

Delve Publishing

www.delvepublishing.com

Applied Polymer Science

Editor: **Asad Rahman**

708 3rd Avenue, 6th Floor
New York, NY 10017
United States of America
www.delvepublishing.com

ISBN: 978-1-68095-895-9

Library of Congress Control Number: 2016951756

About the Editor

Asad Rahman

Asad Rahman obtained MSc Chemical Engineering from University of Nottingham in 2012. Currently He is working towards PhD in Energetic Material Synthesis and its Performance Analysis in Micro Thrusters. His research interests are in Ultrasound based particle treatment, Energetic material synthesis and Micro thruster applications.

Contents

List of Contributors

Liqing Wei
Department of Forest, Rangeland and Fire Sciences, University of Idaho, Moscow, ID 83844-1132, USA

Armando G. McDonald
Department of Forest, Rangeland and Fire Sciences, University of Idaho, Moscow, ID 83844-1132, USA

Ralf Zimmermann
Leibniz Institute of Polymer Research Dresden, Max Bergmann Center of Biomaterials Dresden, Hohe Strasse 6, 01069 Dresden, Germany

Carsten Werner
Leibniz Institute of Polymer Research Dresden, Max Bergmann Center of Biomaterials Dresden, Hohe Strasse 6, 01069 Dresden, Germany
Technische Universität Dresden, Center for Regenerative Therapies Dresden, Tatzberg 47, 01307 Dresden, Germany

Jérôme F. L. Duval
Laboratoire Interdisciplinaire des Environnements Continentaux (LIEC), CNRS UMR 7360, 15 avenue du Charmois, B.P. 40, F-54501 Vandoeuvre-lès-Nancy cedex, France

Kristin Trommer
Research Institute of Leather and Plastic Sheeting (FILK) Freiberg, Meißner Ring 1–5, 09599 Freiberg, Germany

Carina Petzold
Research Institute of Leather and Plastic Sheeting (FILK) Freiberg, Meißner Ring 1–5, 09599 Freiberg, Germany

Bernd Morgenstern
Research Institute of Leather and Plastic Sheeting (FILK) Freiberg, Meißner Ring 1–5, 09599 Freiberg, Germany

Brendon A. Parsons
Wilkes Honors College of Florida Atlantic University, 5353 Parkside Drive, Jupiter, FL 33458, USA
Department of Chemistry, University of Washington, Box 351700, Seattle, WA 98195-1700, USA

Olivia Lin Smith
Wilkes Honors College of Florida Atlantic University, 5353 Parkside Drive, Jupiter, FL 33458, USA

Myeong Chae
Wilkes Honors College of Florida Atlantic University, 5353 Parkside Drive, Jupiter, FL 33458, USA

Veljko Dragojlovic
Wilkes Honors College of Florida Atlantic University, 5353 Parkside Drive, Jupiter, FL 33458, USA

Roberto Scipioni
Dipartimento di Chimica, Sapienza Università di Roma, Piazzale Aldo Moro 5, 00185 Roma, Italy

Delia Gazzoli
Dipartimento di Chimica, Sapienza Università di Roma, Piazzale Aldo Moro 5, 00185 Roma, Italy

Francesca Teocoli
Centro di Ricerca Hydro-Eco, Sapienza Università di Roma, Piazzale Aldo Moro 5, 00185 Roma, Italy

Oriele Palumbo
Istituto dei Sistemi Complessi-Consiglio Nazionale delle Ricerche–ISC-CNR U.O.S. Sapienza, Piazzale A. Moro 5, 00185 Roma, Italy

Annalisa Paolone
Istituto dei Sistemi Complessi-Consiglio Nazionale delle Ricerche–ISC-CNR U.O.S. Sapienza, Piazzale A. Moro 5, 00185 Roma, Italy

Neluta Ibris
Dipartimento di Scienze, Università della Basilicata, Viale dell'Ateneo Lucano 10, 85100 Potenza, Italy

Sergio Brutti
Istituto dei Sistemi Complessi-Consiglio Nazionale delle Ricerche–ISC-CNR U.O.S. Sapienza, Piazzale A. Moro 5, 00185 Roma, Italy
Dipartimento di Scienze, Università della Basilicata, Viale dell'Ateneo Lucano 10, 85100 Potenza, Italy

Maria Assunta Navarra
Dipartimento di Chimica, Sapienza Università di Roma, Piazzale Aldo Moro 5, 00185 Roma, Italy

Ichiko Yamada
Department of Frontier Materials, Graduate School of Engineering, Nagoya Institute of Technology, Gokiso, Showa-ku, Nagoya 466-8555, Japan

Koji Takagi
Department of Materials Science and Engineering, Graduate School of Engineering, Nagoya Institute of Technology, Gokiso, Showa-ku, Nagoya 466-8555, Japan

Yasuhiko Hayashi
Department of Frontier Materials, Graduate School of Engineering, Nagoya Institute of Technology, Gokiso, Showa-ku, Nagoya 466-8555, Japan

Tetsuo Soga
Department of Frontier Materials, Graduate School of Engineering, Nagoya Institute of Technology, Gokiso, Showa-ku, Nagoya 466-8555, Japan

Norio Shibata
Department of Frontier Materials, Graduate School of Engineering, Nagoya Institute of Technology, Gokiso, Showa-ku, Nagoya 466-8555, Japan

Takeshi Toru
Department of Research, Nagoya Industrial Science Research Institute, 1-13 Yotsuyadori, Chikusa-ku, Nagoya 464-0819, Japan

Fan Li
Department of Chemistry, State Key Laboratory of New Ceramics and Fine Processing, Tsinghua University, Beijing 100084, China
Department of Material Science and Engineering, Tsinghua University, Beijing 100084, China

Chong Geng
Department of Chemistry, State Key Laboratory of New Ceramics and Fine Processing, Tsinghua University, Beijing 100084, China

Qingfeng Yan
Department of Chemistry, State Key Laboratory of New Ceramics and Fine Processing, Tsinghua University, Beijing 100084, China

Ting Huang
State Key Laboratory Cultivation Base for Nonmetal Composites and Functional Materials, Southwest University of Science and Technology, Mianyang 621010, Sichuan, China

Bo Jin
State Key Laboratory Cultivation Base for Nonmetal Composites and Functional Materials, Southwest University of Science and Technology, Mianyang 621010, Sichuan, China
School of National Defence Science and Technology, Southwest University of Science and Technology, Mianyang 621010, Sichuan, China

Ru Fang Peng
State Key Laboratory Cultivation Base for Nonmetal Composites and Functional Materials, Southwest University of Science and Technology, Mianyang 621010, Sichuan, China

Cong Di Chen
State Key Laboratory Cultivation Base for Nonmetal Composites and Functional Materials, Southwest University of Science and Technology, Mianyang 621010, Sichuan, China

Rong Zong Zheng
State Key Laboratory Cultivation Base for Nonmetal Composites and Functional Materials, Southwest University of Science and Technology, Mianyang 621010, Sichuan, China

Yi He
School of National Defence Science and Technology, Southwest University of Science and Technology, Mianyang 621010, Sichuan, China

Shi Jin Chu
State Key Laboratory Cultivation Base for Nonmetal Composites and Functional Materials, Southwest University of Science and Technology, Mianyang 621010, Sichuan, China

Layth Mohammed
Centre for Advance Materials, Department of Mechanical Engineering, Universiti Tenaga Nasional, 43000 Kajang, Selangor, Malaysia

M. N. M. Ansari
Centre for Advance Materials, Department of Mechanical Engineering, Universiti Tenaga Nasional, 43000 Kajang, Selangor, Malaysia

Grace Pua
Centre for Advance Materials, Department of Mechanical Engineering, Universiti Tenaga Nasional, 43000 Kajang, Selangor, Malaysia

Mohammad Jawaid
Laboratory of Biocomposite Technology, Institute of Tropical Forestry and Forest Products (INTROP), Universiti Putra Malaysia (UPM), 43400 Serdang, Selangor, Malaysia

M. Saiful Islam
Department of Chemistry, Faculty of Science, Universiti Putra Malaysia (UPM), 43400 Serdang, Selangor, Malaysia

Tsukuru Yaglhara
Department of Applied Chemistry, Faculty of Science and Technology, Keio University, 3-14-1 Hiyoshi, Kohoku-ku, Yokohama 223-8522, Japan;

Shuichi Matsumura
Department of Applied Chemistry, Faculty of Science and Technology, Keio University, 3-14-1 Hiyoshi, Kohoku-ku, Yokohama 223-8522, Japan;

Zinia Nasreen
Department of Chemistry, Dhaka University of Engineering and Technology, Gazipur 1700, Bangladesh

Mubarak A. Khan
Radiation and Polymer Chemistry Laboratory, Institute of Nuclear Science and Technology, Bangladesh Atomic Energy Commission, P.O. Box 3787, Dhaka 1000, Bangladesh

A. I. Mustafa
Department of Applied Chemistry and Chemical Engineering, University of Dhaka, Dhaka 1000, Bangladesh

Leif P. Jentoft
Harvard School of Engineering and Applied Sciences, Cambridge, MA 02138, USA

Aaron M. Dollar
Department of Mechanical Engineering and Materials Science, Yale School of Engineering and Applied Science, New Haven, CT 06511, USA

Christopher R. Wagner
Formerly with the Harvard School of Engineering and Applied Sciences, Cambridge, MA 02138, USA

Robert D. Howe
Harvard School of Engineering and Applied Sciences, Cambridge, MA 02138, USA

Julieta Ledesma
Área Tecnología Química, Facultad de Ciencias Bioquímicas y Farmacéuticas, Universidad Nacional de Rosario, Suipacha 531, S2002LRK Rosario, Argentina

Santiago A. Bortolato
Instituto de Desarrollo Tecnológico para la Industria Química (INTEC), Güemes 3450, S3000GLN Santa Fe, Argentina

Carlos E. Boschetti
Área Tecnología Química, Facultad de Ciencias Bioquímicas y Farmacéuticas, Universidad Nacional de Rosario, Suipacha 531, S2002LRK Rosario, Argentina

Débora M. Martino
Instituto de Desarrollo Tecnológico para la Industria Química (INTEC), Güemes 3450, S3000GLN Santa Fe, Argentina
Facultad de Humanidades y Ciencias, Universidad Nacional del Litoral, Paraje El Pozo, S3080HOF Santa Fe, Argentina

Yi-Chen Lin
Graduate Institute of Photonics and Optoelectronics, National Taiwan University, No. 1, Sec. 4, Roosevelt Rd., Taipei 106, Taiwan

Chung-Yi Yu
Graduate Institute of Photonics and Optoelectronics, National Taiwan University, No. 1, Sec. 4, Roosevelt Rd., Taipei 106, Taiwan

Chung-Min Li
Graduate Institute of Photonics and Optoelectronics, National Taiwan University, No. 1, Sec. 4, Roosevelt Rd., Taipei 106, Taiwan

Chin-Heng Liu
Graduate Institute of Photonics and Optoelectronics, National Taiwan University, No. 1, Sec. 4, Roosevelt Rd., Taipei 106, Taiwan

Jiun-Peng Chen
Graduate Institute of Communication Engineering, National Taiwan University, No. 1, Sec. 4, Roosevelt Rd., Taipei 106, Taiwan

Tah-Hsiung Chu
Graduate Institute of Communication Engineering, National Taiwan University, No. 1, Sec. 4, Roosevelt Rd., Taipei 106, Taiwan

Guo-Dung John Su
Graduate Institute of Photonics and Optoelectronics, National Taiwan University, No. 1, Sec. 4, Roosevelt Rd., Taipei 106, Taiwan

Shuai Liang
Key Laboratory of Fine Petrochemical Engineering, Changzhou University, Changzhou 213164, China

Geng Zhang
Department of Chemistry, College of Science, Huazhong Agricultural University, Wuhan 430070, China

Jianzhong Min
Key Laboratory of Fine Petrochemical Engineering, Changzhou University, Changzhou 213164, China

Junqiao Ding
State Key Laboratory of Polymer Physics and Chemistry, Changchun Institute of Applied Chemistry, Chinese Academy of Sciences, Changchun 130022, China

Xingmao Jiang
Key Laboratory of Fine Petrochemical Engineering, Changzhou University, Changzhou 213164, China
Changzhou Yingzhong Nano Technology Company Limited, Changzhou 213023, China

Yoshikuni Teramoto
Course of Applied Life Science, Faculty of Applied Biological Sciences, Gifu University, 1-1 Yanagido, Gifu 501-1193, Japan

Marek Krehel
Empa, Swiss Federal Laboratories for Materials Science and Technology, Laboratory for Protection and Physiology, Lerchenfeldstrasse 5, St. Gallen 9014, Switzerland
ETH Zurich, Swiss Federal Institute of Technology, Department of Information Technology and Electrical Engineering, Gloriastrasse 35, Zurich 8092, Switzerland

René M. Rossi
Empa, Swiss Federal Laboratories for Materials Science and Technology, Laboratory for Protection and Physiology, Lerchenfeldstrasse 5, St. Gallen 9014, Switzerland

Gian-Luca Bona
Empa, Swiss Federal Laboratories for Materials Science and Technology, Laboratory for Protection and Physiology, Lerchenfeldstrasse 5, St. Gallen 9014, Switzerland
ETH Zurich, Swiss Federal Institute of Technology, Department of Information Technology and Electrical Engineering, Gloriastrasse 35, Zurich 8092, Switzerland

Lukas J. Scherer
Empa, Swiss Federal Laboratories for Materials Science and Technology, Laboratory for Protection and Physiology, Lerchenfeldstrasse 5, St. Gallen 9014, Switzerland

Devesh Shukla
Department of Polymer and Process Engineering, Indian Institute of Technology, Roorkee 247001, India

Yuvraj Singh Negi
Department of Polymer and Process Engineering, Indian Institute of Technology, Roorkee 247001, India

Vijai Kuma
Central Institute of Plastic Engineering and Technology, Lucknow 226008, India

Julius I. Osayi
School of Chemical and Metallurgical Engineering, University of the Witwatersrand, 1 Jan Smut Avenue, Braamfontein, Private Bag 3, Johannesburg 2050, South Africa

Sunny Iyuke
School of Chemical and Metallurgical Engineering, University of the Witwatersrand, 1 Jan Smut Avenue, Braamfontein, Private Bag 3, Johannesburg 2050, South Africa

Samuel E. Ogbeide
Department of Chemical Engineering, Faculty of Engineering, University of Benin, P.M.B. 1154, Benin City, Edo State, Nigeria

Magdalèna Couty
Univ. Paris Sud, Laboratoire IEF (Institut d'Electronique Fondamentale), UMR-8622, Bat. 220, F91405 Orsay, France;
CNRS (Centre National de Recherche Scientifique), F-91405 Orsay, France;
LGPPTS (Laboratoire du Génie des Procédés Plasmas et Traitements de Surface), Chimie ParisTech/UPMC EA3492, 11 rue Pierre et Marie Curie, F-75005 Paris, France;

Marion Woytasik,
Univ. Paris Sud, Laboratoire IEF (Institut d'Electronique Fondamentale), UMR-8622, Bat. 220, F91405 Orsay, France;
CNRS (Centre National de Recherche Scientifique), F-91405 Orsay, France;

Jean-Christophe Ginefri
CNRS (Centre National de Recherche Scientifique), F-91405 Orsay, France;
Univ. Paris Sud, Laboratoire IR4M (Imagerie par Résonance Magnétique Médicale et MultiModalités), UMR-8081, Bat. 220, F-91405, Orsay, France

Anne Rubin
Univ. Paris Sud, Laboratoire IEF (Institut d'Electronique Fondamentale), UMR-8622, Bat. 220, F91405 Orsay, France;
CNRS (Centre National de Recherche Scientifique), F-91405 Orsay, France;
I2BM (Institut d'Imagerie BioMédicale), Neurospin, CEA Saclay, Bat. 145, F-91191 Gif-sur-Yvette, France;

Emile Martincic
Univ. Paris Sud, Laboratoire IEF (Institut d'Electronique Fondamentale), UMR-8622, Bat. 220, F91405 Orsay, France;
CNRS (Centre National de Recherche Scientifique), F-91405 Orsay, France;

Marie Poirier-Quinot
CNRS (Centre National de Recherche Scientifique), F-91405 Orsay, France;
Univ. Paris Sud, Laboratoire IR4M (Imagerie par Résonance Magnétique Médicale et MultiModalités), UMR-8081, Bat. 220, F-91405, Orsay, France

Luc Darrasse
CNRS (Centre National de Recherche Scientifique), F-91405 Orsay, France;
Univ. Paris Sud, Laboratoire IR4M (Imagerie par Résonance Magnétique Médicale et MultiModalités), UMR-8081, Bat. 220, F-91405, Orsay, France

Fawzi Boumezbeur
I2BM (Institut d'Imagerie BioMédicale), Neurospin, CEA Saclay, Bat. 145, F-91191 Gif-sur-Yvette, France;

Franck Lethimonnier
I2BM (Institut d'Imagerie BioMédicale), Neurospin, CEA Saclay, Bat. 145, F-91191 Gif-sur-Yvette, France;

Michael Tatoulian
LGPPTS (Laboratoire du Génie des Procédés Plasmas et Traitements de Surface), Chimie ParisTech/UPMC EA3492, 11 rue Pierre et Marie Curie, F-75005 Paris, France;

Elisabeth Dufour-Gergam
Univ. Paris Sud, Laboratoire IEF (Institut d'Electronique Fondamentale), UMR-8622, Bat. 220, F91405 Orsay, France;
CNRS (Centre National de Recherche Scientifique), F-91405 Orsay, France;

Ahmed Alfadhel
[1]Computer, Electrical and Mathematical Sciences and Engineering Division (CEMSE), King Abdullah University of Science and Technology (KAUST), Thuwal 23955-6900, Saudi Arabia

Mohammed Asadullah Khan
Computer, Electrical and Mathematical Sciences and Engineering Division (CEMSE), King Abdullah University of Science and Technology (KAUST), Thuwal 23955-6900, Saudi Arabia

Susana Cardoso
INESC-Microsystems and Nanotechnologies (INESC-MN), Rua Alves Redol, 9, Lisbon 1000-029, Portugal
Instituto Superior Técnico IST, Physics Department, Universidade de Lisboa, Lisbon 1049-001, Portugal

Diana Leitao
INESC-Microsystems and Nanotechnologies (INESC-MN), Rua Alves Redol, 9, Lisbon 1000-029, Portugal
Instituto Superior Técnico IST, Physics Department, Universidade de Lisboa, Lisbon 1049-001, Portugal

Jürgen Kosel
Computer, Electrical and Mathematical Sciences and Engineering Division (CEMSE), King Abdullah University of Science and Technology (KAUST), Thuwal 23955-6900, Saudi Arabia

Preface

Applied Polymer Science is the subfield of material science and organic chemistry, which focuses on polymeric materials and their applications. It includes polymer synthesis, recycling, bulk production and applications in biomedical, material science, plastic industry, fiber and composite material industry and many more. Polymers are the backbone of current manufacturing industry, and they are found everywhere around us including both natural and synthetic polymers.

Polymers are used for making household items, protective coatings, bodies of electronic gadgets, solar panels, and they are a big part of our food chain as well. Straight chain polymers and branched polymers are the two basic types of polymers. Among branched polymers, there are several sub categories, such as Isotactic, Atactic, Syndiotactic and Eutactic polymers. The type of branching determines the physical properties of the polymer, such as their melting point, tensile strength, flexibility etc. Therefore, in this book, several articles covering different aspects of applied polymer science are presented. It includes latest applications of polymeric materials, recycling of polymers, and niche polymers, which are used in unique applications

Editor
Asad Rahman

CHAPTER 1

A REVIEW ON GRAFTING OF BIOFIBERS FOR BIOCOMPOSITES

Liqing Wei and Armando G. McDonald

Department of Forest, Rangeland and Fire Sciences, University of Idaho, Moscow, ID 83844-1132, USA

ABSTRACT

A recent increase in the use of biofibers as low-cost and renewable reinforcement for the polymer biocomposites has been seen globally. Biofibers are classified into: lignocellulosic fibers (*i.e.*, cellulose, wood and natural fibers), nanocellulose (*i.e.*, cellulose nanocrystals and cellulose nanofibrils), and bacterial cellulose, while polymer matrix materials can be petroleum based or bio-based. Green biocomposites can be produced using both biobased fibers and polymers. Incompatibility between the hydrophilic biofibers and hydrophobic polymer matrix can cause performance failure of resulting biocomposites. Diverse efforts have focused on the modification of biofibers in order to improve the performances of biocomposites. "Grafting" copolymerization strategy can render the advantages of biofiber and impart polymer properties onto it and the performance of biocomposites can be tuned through changing grafting parameters. This review presents a short overview of various "grafting" methods which can be directly or potentially employed to enhance the interaction between biofibers and a polymer matrix for biocomposites. Major grafting techniques, including ring opening polymerization, grafting via coupling agent and free radical induced grafting, have been discussed. Improved properties such as mechanical, thermal, and water resistance have provided grafted biocomposites with new opportunities for applications in specific industries.

INTRODUCTION

A composite is usually defined as the combination of two or more different components, in which one plays the role of filler or reinforcement (such as biofibers), while the other performs as a resin or polymer matrix material. One advantage of biofibers-based composites is their low density yet higher mechanical strength and stiffness compared to glass fibers, lower impact on the environment, lower manufacturing cost, and great biodegradability. To date, biofibers-filled petroleum derived thermoplastics and thermosetting polymers have been extensively used for composites manufacturing [1,2,3,4,5,6,7]. Both the polymer matrix and reinforcement determine the overall physicochemical properties of the resulting composites.

According to market data reported by Bioplastic Magazine, 352,000 tons of the wood—(74% of the market share) and natural fibers (26% of the market share) based composites were produced in 2012 in the European Union, and the wood plastic composites (WPC) will increase to 350,000 tons by 2015 [8]. About 38% of WPC was used for the automotive industry in 2012, primarily for rear shelves and trims for trunks

Citation: Wei, L.; McDonald, A.G. A Review on Grafting of Biofibers for Biocomposites. Materials 2016, 9, 303. doi:10.3390/ma9040303

and spare wheels, as well as in interior trims for doors; while natural fibers-based composites are mainly used for interior trims for high value doors and dashboards. Asia-Pacific is estimated to be the largest biocomposites market in the coming years due to the increasing concerns of environment and the market of construction, manufacturing and automotive sectors [9]. Although biofibers are good candidates for composites production, they also bring about challenges. The high hydrophilicity due to a large number of hydroxyl groups (Figure 1 and Figure 2) on fiber surfaces, and this can result in poor interfacial bonding/adhesion between fibers and the, generally hydrophobic, polymer matrix. Poor adhesion will make the composite vulnerable to environmental attacks (weathering, water absorption, and biodeterioration), mechanical failure, and therefore reduce service time. Composite properties can be improved via exchanging biofibers with cellulose because of its improved thermophysical properties. Such biocomposites will have great potential to be used in the automobile industry with increasing demands for recyclable new cars [10].

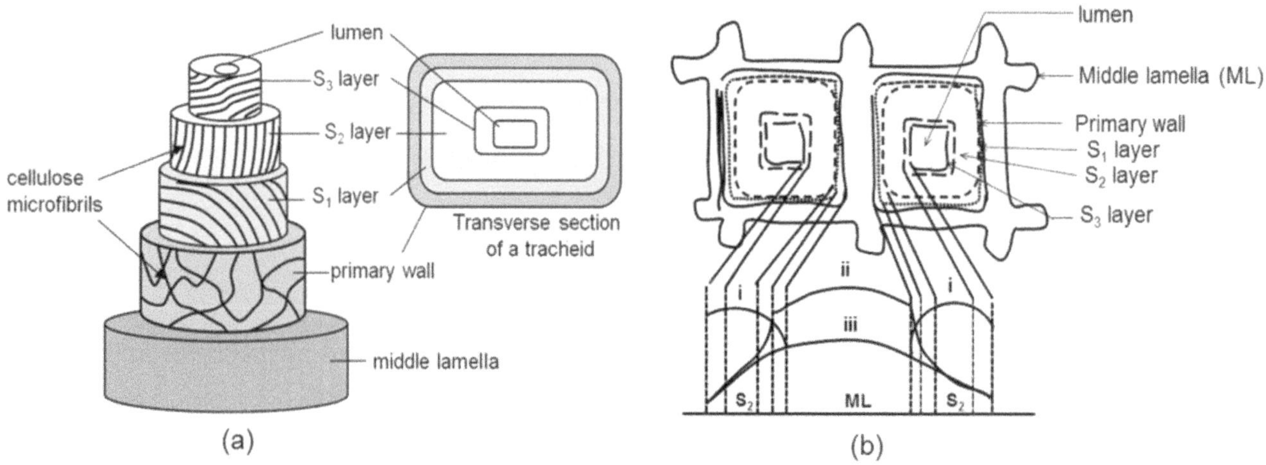

Figure 1: (**a**) Three-dimensional structure of the secondary cell wall of a xylem cell and (**b**) the relative amounts of cellulose, hemicellulose, and lignin across a cross-section of two wood cells (i: cellulose; ii: lignin; and iii: hemicelluloses).

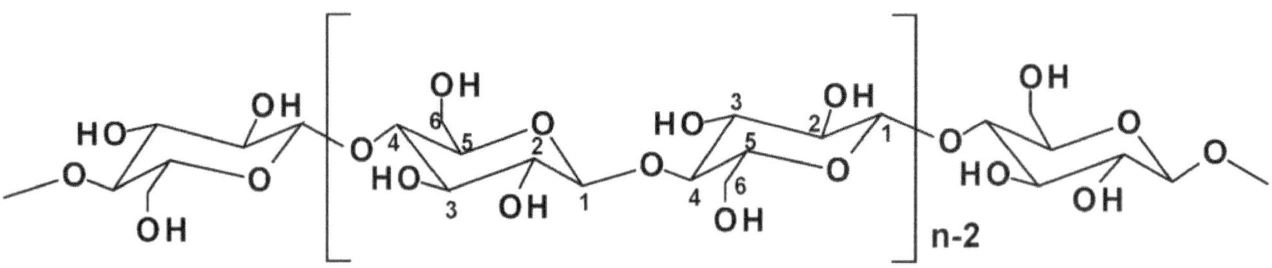

Figure 2: The molecular structure of cellulose (*n* = DP; the carbon positions of the ring structure are marked).

To reduce the hydrophilicity of cellulose fiber surface this can be achieved either through a physical or chemical treatment, and therefore to improve the interfacial adhesion between the two phases. Although these modifications result in a decrease in moisture absorption and an increase in mechanical properties, biodurability and weatherability, the processes used for cellulose modification are costly and involve toxic chemicals which could be a deterrent to its use [11].

Very recently, through the method, "grafting copolymerization", modifying the properties of biofibers by imparting the desired and targeted properties of a polymer has attracted great interest. This method can significantly improve the stress transfer between the two phases. Review articles were published on cellulose fiber grafted to polymers via ring opening polymerization (ROP) [11,12]. However, inadequate information is available in the existing literature which summarizes other types of biofibers as well as other grafting techniques for biocomposites production.

This review intends to show the feasibilities of employing recently developed copolymerization via grafting of various biofibers as reinforcement to polymer matrices for biocomposites applications. The overall characteristics of biofibers and polymer matrix materials used in biocomposites and different modification methods to improve the biocomposites properties will be reviewed. Different grafting copolymerization techniques based on grafting reaction mechanisms are focused upon. The overall comparison of performance changes, structure manipulation, and potential applications due to grafting is discussed in detail. Finally, the review will conclude with future trends of "grafting" used for applications in biocomposites.

BIOFIBERS

Lignocellulosic Biofibers

Lignocellulosic fibers (or natural fibers) are used as reinforcement in biocomposites and have been recently reviewed [2,13]. Fibers were categorized into six types: bast fibers (jute, flax, hemp, kenaf and ramie), leaf fibers (abaca, sisal and pineapple), seed fibers (coir, cotton and kapok), core fibers (kenaf, hemp and jute), grass and reed fibers (wheat, corn and rice) and other types (wood, food/agricultural residues, recycled newspaper/magazine/office paper fibers, bamboo/rattan, and roots). Natural fibers have a hollow structure, providing insulation against noise and heat. Each fibril has a complex and layered structure as shown in Figure 1 [14]. The middle lamella performs like a glue between two adjacent cells within the plant tissue. The primary wall is the first layer deposited during the formation of cell. Three layers, the outer layer (S1), middle layer (S2), and inner layer (S3), compose the secondary cell wall [15]. The distribution of chemical compositions (cellulose, hemicelluloses, and lignin) across middle lamella and wood cell walls of hardwood is shown Figure 1b [15].

In the primary wall, the orientation of cellulose microfibrils is random from 0° to 90° relative to long axial direction of the cell. For wood fibers, the angle (microfibrillar angle) between microfibrils direction and the long axis is between 50° and 89° for S1 layer [16]. These microfibrils have a diameter in the range of 10–30 nm and are typically made up of 30–100 cellulose molecules in an extended chain conformation, which provides the mechanical strength of the fibers [14]. In addition, the mechanical properties of natural fibers are also influenced by their moisture content and voids between fibers [13]. The properties of natural fiber, such as the fiber morphology, defects, crystallinity and structures, have great impacts if fibers are used as a reinforcement of the polymer matrix for composite materials.

The chemical composition of lignocellulosic biofibers varied with source and preparation methods. However, the main components of natural fibers generally include cellulose, hemicellulose, lignin, and a small amount of pectin and wax [17]. The combined effect of these components determines the overall thermal and mechanical properties of fibers. The major chemical components of selected natural fibers are summarized in Table 1.

Table 1: Major chemical composition of selected natural fibers (%)

Type of Fiber	Cellulose	Hemicellulose	Lignin	References
Black locust	41.61	17.66	26.70	[18]
Hybrid poplar	44.70	18.55	26.44	[18]
Eucalyptus	49.50	13.07	27.71	[18]
Pine	44.55	21.90	27.67	[18]
Switchgrass	31.98	25.19	18.13	[18]
Bagasse	54.3–55.2	16.8–29.7	24.3–25.3	[19]
Bamboo	34.50	20.50	26.00	[20,21]
Rattan	35.6–52.9	22.8–34.7	21.0–22.0	[22]
Flax	70.50	16.50	2.50	[20,21]
Hemp	70–92	18-22	3–5	[23]
Kenaf	53.50	21.00	17.00	[20,21]
Jute	67.00	16.00	9.00	[20,21]
Oil palm	14.3–65.2	12.5–38.7	17.3–26.5	[22]

Cellulose

All the plant-based fibers contain a high proportion of cellulose (40%–90%) (Figure 2) [24]. Cellulose is a hydrophilic polysaccharide that consists of a linear macromolecular chain of 1 → 4 linked β-d-glucopyranosyl units to form linear chains in the cell wall [25]. All three hydroxyl groups in each monomeric unit are in the equatorial plane. The numerous hydroxyl groups on the glucose ring contribute to extensive intra- and inter-molecular C-H···O hydrogen bonds forming crystalline cellulose microfibrils (Figure 3a) [11,25,26]. The hydrogen bonding within and between cellulose chains are responsible for its high strength, stiffness, crystallinity, durability, and biocompatibility [12,26]. There are six polymorphic forms of cellulose identified, including cellulose I, II, III_I, III_{II}, IV_I and IV_{II}. Among them, the cellulose I is found in nature and has two suballomorphs, termed I_α and I_β. For both I_α and I_β, cellulose chains have "parallel-up" configurations, but different hydrogen bonding patterns and therefore different crystalline structure. It was reported by Nishiyama *et al.* [27,28], that I_α is comprised of alternating glucose units in each chain with slightly different conformation but all molecular chains are the same, while I_β has two types of parallel chains. The authors used Fourier-difference analysis to demonstrate that hydrogen bonding within the I_β allomorph is complex and disordered.

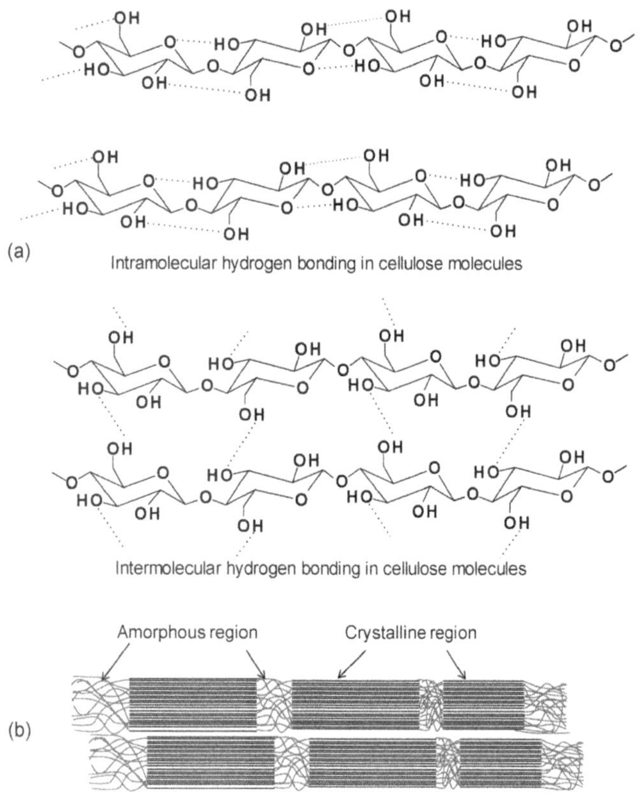

Figure 3: (**a**) Intra- and inter-molecular hydrogen bonding and (**b**) amorphous and crystalline region in cellulose molecules [11].

The degree of polymerization (DP) for cellulose is about 8000 in the primary cell wall while the DP in the secondary wall is >15,000 [29]. Microfibrils are formed from bundles of cellulose chains and form larger fibrils. Cellulose microfibrils have both crystalline (highly ordered) and amorphous (less ordered) regions (Figure 3b). Generally, wood cellulose is about 60%–70% crystalline [15]. The cellulose amorphous region can absorb water more easily due to its greater accessibility. Hence, the crystallinity of cellulose is one of the key characteristics affecting its final properties, including its reactivity, mechanical performances, water absorption, and biodurability.

Cellulose is reactive due to the presence of hydroxyl groups and hence can be chemically modified to form cellulose derivatives such as cellulose esters and ethers (e.g., cellulose acetate/propionate/butyrate, methyl/ethyl cellulose, and hydroxypropyl cellulose). The reactivity of cellulose also depends on the positions of hydroxyl groups, for example, during esterification, the OH groups on the C_6 position is 10 times more reactive than those attached on C_2 and C_3 [11]. The overall reactivity of the three types of OH groups are $OH\text{-}C_6 >> OH\text{-}C_2 > OH\text{-}C_3$ [10]. The properties of cellulose derivatives are altered to native cellulose and also determined by the degree of substitutions.

Nanocellulose

Cellulose nanomaterials have been known since the 1950s and have gained extraordinarily renewed attention recently. Generally, there are two forms of nanocellulose, the cellulose nanocrystals (CNCs) and cellulose nanofibrils (CNFs). The CNCs are also known as cellulose whiskers, nanowhiskers, or nanorods. The CNCs can be often obtained by a controlled acid hydrolysis in which the amorphous or crystalline regions of the cellulose, hemicellulose and lignin are removed. Thus, the rod-like acid-resistant CNCs are produced with reduced DP as compared to the native cellulose, and the degree of drop-off depends on the cellulose origins (e.g., cotton, ramie fibers, algae, bleached wood pulp, bacterial cellulose, or tunicate) [30,31,32]. The dimensions of CNCs vary upon the cellulose origins as well and range from 3 to 20 nm in diameter and 100 nm to several micrometers (from celluloses of tunicates, algae, bacteria) in length (aspect ratio, length/diameter, was from 10 to 100) [26,33,34,35,36]; CNCs from animal tunicate have a large aspect ratio close to 100 [37]. The Young's modulus of CNCs can reach to the range of 120 to 170 GPa [33,37,38,39,40]. CNCs have a high specific surface area (150–170 m^2/g) and high crystallinity (54%–90%) with reactive surfaces due to hydroxyl groups [33,41,42].

CNFs are string-like particles which can be produced via different methods, such as TEMPO-mediated oxidation (2,2,6,6,-tetrame thylpipelidine-1-oxyl radical), mechanical fibrillation using high-pressure homogenization, ultrasonic methods, and enzymatic hydrolysis [30,43]. CNFs have a high aspect ratio (4–20 nm in diameter and several micrometers in length). The CNFs generated by TEMPO oxidation have the dimensions of 3–10 nm in diameter and a few microns in length [44]. CNFs contain both crystalline and amorphous regions.

Lignin

In addition to cellulose, lignocellulosic fibers also contain hemicellulose (branched polysaccharide) and lignin. Lignin is a complex three dimensional hydrocarbon polymer with both aliphatic and aromatic constituents. Lignin is biosynthesized from three major monomeric units, coniferyl alcohol, synapyl alcohol, and p-coumaryl alcohol (Figure 4). The composition of monolignols is dependent on plant classes and species (*i.e.*, softwood, hardwood, grass). The empirical formula for lignin is $C_9H_{10}O_2$ $(OCH_3)_n$, with n being the ratio of MeO to C_9 groups: $n = 1.4$, 0.94 and 1.18 for hardwood, softwood and grasses, respectively. The main functional groups are phenolic hydroxyl (ArOH), aliphatic hydroxyl (AlOH), carboxylic acid (COOH), and methoxy (OMe) in hardwood kraft lignin with corresponding concentration of 4.3, 1.7, 0.5, and 580 mmol/g, respectively [45]. Lignin is usually a waste byproduct of the cellulosic bioethanol and pulp and paper industry with low cost material. Hence, it has become a potential candidate as renewable materials used for sustainable polymer biocomposite as a matrix, coupling agent, coatings, and reinforcement in resin, fibers and aerogels [46,47,48,49,50,51,52,53,54,55]. With an increasing awareness of building a sustainable society, the incorporation of lignin into a biopolymer matrix as the reinforcement for biocomposites has attracted more interest. Recently, the feasibility of this idea was reviewed elsewhere [56,57].

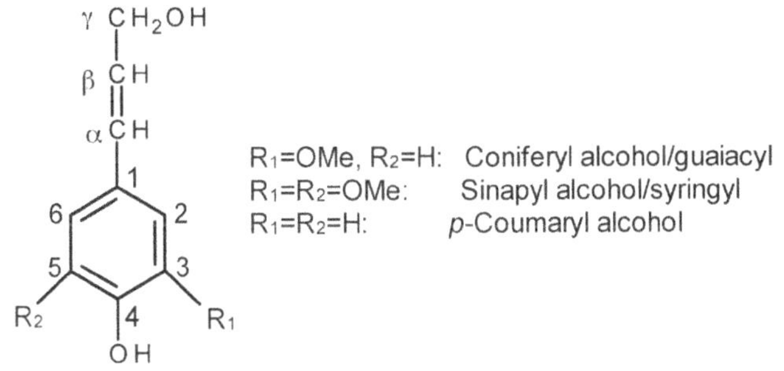

Figure 4: The three basic building blocks of lignin.

Bacterial Cellulose

The species of bacteria which produces cellulose is generally called *Acetobacter xylinum*. The microbial or bacterial cellulose (BC) shows great mechanical properties due to its crystalline structure. BC has triclinic (I_α) form crystalline structure, while wood and other higher plant cell wall celluloses are majorly comprised of monoclinic (I_β) polymorph. In addition, BC has a higher degree of crystallinity (60%–90%) than plant cellulose and forms characteristic ribbon-like microfibrils, which gives BC excellent mechanical strength. The Young's modulus of BCs was reported to be 15 to 35 GPa isotropically across the surface of a plane [26,58], while that of the bacterial nanocellulose (BNC) fibers can reach to 138 GPa [59]. The tensile strength of BCs can reach 450 MPa [60]. The length ofBNCs ranges from 100 nm to several mm and the diameter from 5 nm to 50 nm. BNCs also have a higher purity than other cellulose derived from natural fibers, because they do not contain hemicelluloses and lignin. In addition, the thermal expansion coefficient of BCs in the axial direction is small. Due to these mentioned fascinating characteristics of BC, it has been found to be effective for various applications such as films prepared from gel and reinforcing pulp papers or optically transparent plastics.

Biofibers from Other Sources

Another important biofiber source is chitin which can be found in the shells of crustacean (e.g., crab and shrimp, cuticles of insects, and cell walls of some fungi and microorganisms). Chitin has a linear structure composed of β-(1-4)-linked 2-deoxy-2-acetamido-d-glucose units (Glu-$NHCOCH_3$) [61,62,63]. After chitin is deacetylated with a degree of deacetylation over 75% the so called chitosan (Glu-NH_2) is obtained. Chitosan has excellent biodegradability and biocompatibility, hence it is being applied for biomedical uses such as kidney membrane, artificial skin, and drug delivery systems. Similar to cellulose, chemical modifications of both chitin and chitosan are of interest without changing their backbone structure and physicochemical properties. Diverse derivatives can be prepared as well [1,8,64,65,66]. Chitin nanofibers were prepared and with their modifications and applications were reported [67]. Chitin-protein composite nanofibers were recently studied by Mushi *et al.* [68].

BIOCOMPOSITES

By simply blending the above mentioned biofibers with polymeric matrix, the so-called biocomposites is produced. Generally, the polymer matrix material can be conventional thermoplastics/thermosets or bio-based polymer/resins.

Matrix Materials

Petroleum Based Polymer Matrix

Petrochemical based thermoplastics and thermoset matrices were extensively used for biocomposites production. These conventional thermoplastic matrix materials include polyethylene (PE), polypropylene (PP), polystyrene (PS), poly(ethylene terephthalate) (PET), and polyvinyl chloride (PVC), *etc.*, while some epoxy resin, phenol formaldehyde, and vinyl esters were used as thermoset matrices.

Bio-Based Polymer Matrix

Driven by efforts to develop more environmentally friendly products, there is a need to replace the conventional polymer matrix materials with bioplastics for composites. Bioplastics matrix material may be bio-based polymers readily derived from renewable feed-stocks in large quantities. They are low cost and biodegradable so that a polymer can return to nature at the end of its life. Some of the most commonly known bio-based plastics in today's market place are poly(ethylene terephthalate) (PET), polyethylene (PE), polypropylene (PP), polyamides (PA), poly(trimethylene terephthalate) (PTT), poly(lactic acid) (PLA), polyhydroxyalkanoates (PHAs), poly(propylene carbonate) (PPC), thermoplastic starch (TPS), soy protein based resins, and other biodegradable polyesters [*i.e.*, poly(caprolactone) (PCL), poly(butylene succinate) (PBS) and poly(butylene succinate-co-adipate) (PBSA)] [69,70,71,72,73,74,75,76]. Except for bio-based PE, PP, PTT, PET, and PA, all other bio-based polymers just mentioned are biodegradable at different levels. Among all of the bioplastics mentioned above, the microbial PHA, PLA, biodegradable starch-based blends, PBS/PBSA, PCL, PPC, and soy protein-based resin have gained significantly in interest since they could be completely biodegradable in the appropriate environments. In addition, apart from their good biodegradability, PHAs and PLAs can be biosynthesized from renewable resources which allows large scale production [64,69,70,72,74].

Polyhydroxyalkanoate (PHA) is a special group of polyesters produced by a wide variety of microorganisms as an internal carbon and energy reservoir as part of their survival medium. All metabolized carbon sources can be used for the production of PHA such as fatty acids and carbohydrate [77]. A large amount of literature is available to describe the biosynthesis [78,79,80,81,82,83,84], characterization of the PHA properties [72,85,86,87,88], opportunities and challenges of PHA products on the global market [89], and applications of PHA [77]. Poly(β-hydroxybutyrate) (PHB), Figure 5a, is a homopolymer of 3-hydroxybutyrate and was the first discovered member of the PHA family in 1925 by Lemoigne, while its copolymer the poly(3-hydroxybutyrate-co-3-hydroxyvalerate) (PHBV, Figure 5b) has attracted great attention with improved thermomechanical properties as compared to PHB.

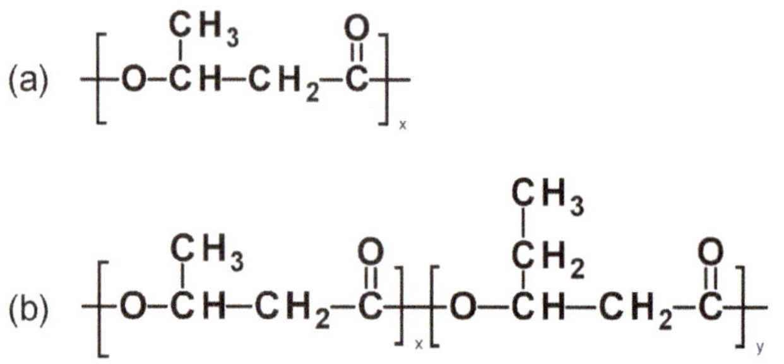

Figure 5: Chemical structures of (**a**) poly(β-hydroxybutyrate) (PHB) and (**b**) poly(3-hydroxybutyrate-co-3-hydroxyvalerate (PHBV).

Polylactic acid (PLA) is a synthetically produced biodegradable plastic from bioderived monomers from renewable resources. PLA can be synthesized via different polymerization routes, which is shown in Figure 6. Based on the stereochemistry of lactic acid (LA) and lactide monomers (Figure 6a), several distinct forms of PLA exist: poly-l-lactic acid (PLLA) is the product resulting from polymerization of l-lactide (also known as (S,S) or l-lactide), while poly-d-lactic acid (PDLA) is synthesized from d-lactide (also known as (R,R)-lactide). Commercial PLA homopolymer is generally prepared from l-lactide (LLA) which can be derived from corn starch fermentation with relatively high enantiomeric purity [90,91]. To date, the major PLA suppliers have been Cargill (in USA known as Ingreo™, under trade name Nature Works, (Minnetonka, MN, USA), Mitsui Chemicals, Inc. (Tokyo, Japan), Purac (Gorinchem, The Netherlands), and Teijin Limited (Osaka, Japan). The increased production of PLA has facilitated an increase in its research and development activities. The number of publications related to PLA increased dramatically over the past decade, which is due to the increasing demands of sustainable and eco-efficient bioplastics. A viable end-of-life option of PLA is chemically recycling back to monomers through hydrolysis into lactic acid which will enter into the polycondensation step [92], or reversible polymerization to lactide monomer using catalytic thermal depolymerization [93]. Work is in progress to upscale the depolymerizatin of PLA back to lactide, which, if successful, it is believed will reduce the market prices of PLA significantly.

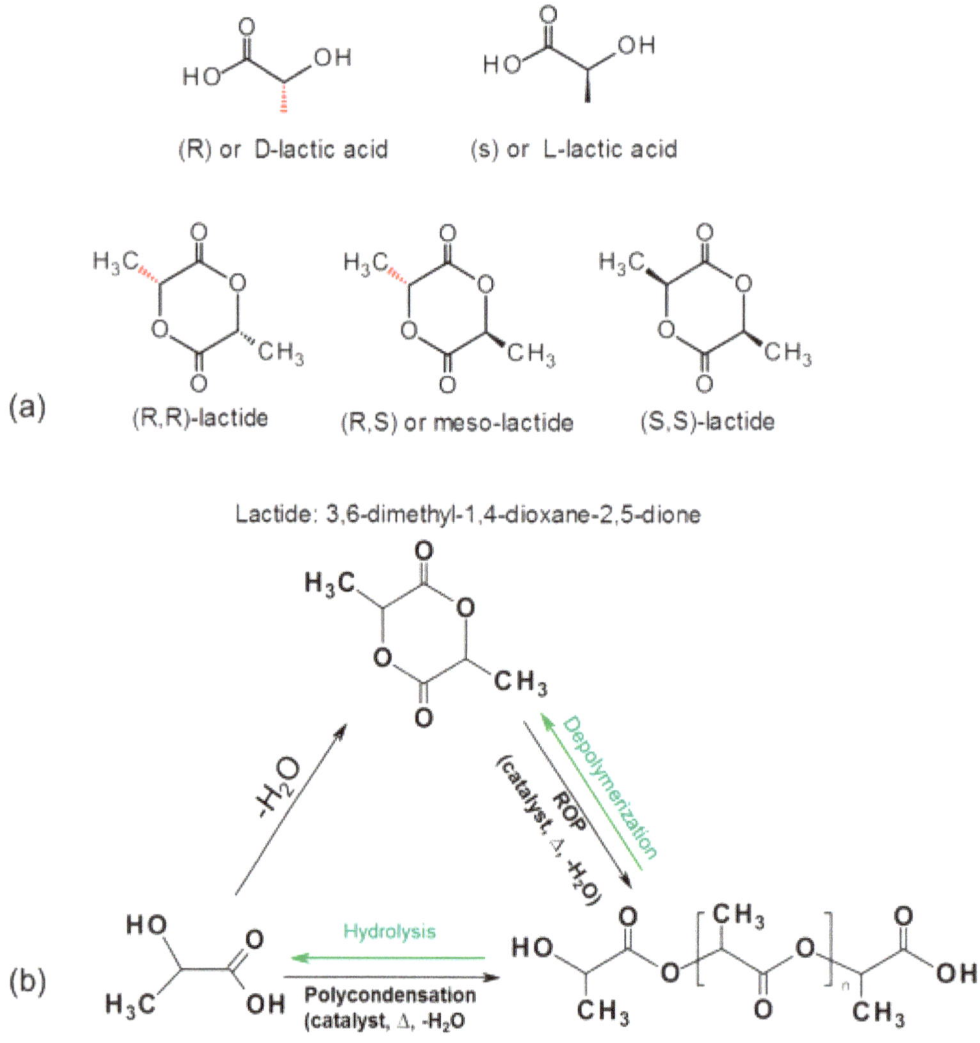

Figure 6: Stereochemistry of the lactic acid (LA) and lactide monomers (**a**) and chemistry of the interconversion between LA, lactide and poly(lactic acid) (PLA) (**b**). Note: Reversible depolymerization and hydrolysis steps are highlight in green color [94].

Green Composites

By replacing the conventional polymer matrix materials with bio-based polymers such as starch and soy protein based resins, PHAs, and PLAs, green biocomposites is achieved [17,29,95,96,97,98]. Green composites are advantageous materials as they can completely decompose after being discarded without any environmental impact at the end of life cycle. Commonly known green composites are discussed below.

PHA Based Biocomposites

Biofibers have been incorporated into the PHA polymer matrix to improve its thermomechanical properties and reduce the cost of the final products (biocomposites) [96,99,100]. PHB/PHBV based biocomposites reinforced with bast fibers [101], straw/husk [102,103], bamboo fibers [104], distiller's dried grains with solubles (DDGS) [105,106], and potato peel fermentation residue [107], have been reported. In these cases, the thermal stability was increased by addition of fibers while Shanks *et al.* [108] showed improvement in bending/storage modulus by addition of flax fibers. Barkoula and colleagues [109] studied properties of flax fibers-reinforced PHB composites and showed that PHB was toughened in these biocomposites. Furthermore, to the presence of fibers stimulated the initial biodegradation of these biocomposites by its moisture uptake. Srithep *et al.* [110] prepared nanocomposites with PHBV reinforced with nanofibrillated cellulose (NFC). Addition of NFC increased the tensile modulus significantly because it served as nucleation agent that facilitated PHBV crystallization. Compounding PHB with starch for transparent film was also studied by Godbole *et al.* [98]. The results revealed that blended films had a single glass transition temperature. The nature of all combinations was found to be crystalline. As the cost of PHB is high as compared to conventional plastics and because starch is abundantly available at a very low cost, blending of starch up to 30% in PHB would help to reduce the price of bioplastics, without sacrificing the physical properties. This biocomposite could be used as a coating material for food packaging materials. Coats *et al.* [111] prepared biocomposites with PHB-rich biomass compounded with pine wood fibers and compared properties with purified PHB based composites. The former composites exhibited comparable strength to the later type. This alternative approach to utilize PHA biomass without further purification of PHA polymers from bacterial cells could reduce the cost of the biocomposite.

PLA Based Biocomposites

Similar to PHAs based biocomposites, PLA is another attractive biodegradable polymer matrix. Various properties for different biofibers, such as hemp, flax, coir, agricultural waste, starch, recycled paper fibers and wood fibers as well as cellulose/nanocellulose and lignin have been used as reinforcement to PLA matrices [65,97,104,112,113,114,115,116,117,118,119,120,121] and reviewed [2,13,42,122]. By compounding the fibers with PLA the mechanical properties were improved as compared with neat PLA. For example, quasi-static tensile modulus of the unidirectional long artichoke fibers reinforced PLA green biocomposites was increased by 40% as compared with neat PLA [120].

Arrieta *et al.* [123] prepared the PLA/PHB/CNC blends and evaluated the mechanical, barrier and degradation properties of the biocomposites film. The ternary system showed enhanced mechanical properties, improved water resistance, and reduced oxygen and UV-light transmission. The disintegration of PLA was delayed due to addition of PHB; in contrast, CNC facilitated its degradation, which suggests potential applications for packaging materials of the biocomposites.

Other Commonly Studied Biocomposites

Another commonly studied bioplastic matrix material for biocomposites is PCL. In a recent study by Hejna *et al.* [124], agricultural waste (brewer's spent grain (BSG) and wheat bran (WB)) were compounded with PCL for biocomposites production. The influence of BSG biofiber loadings on processing, static and dynamic mechanical properties, thermal properties was investigated. All the results indicated that BSG and WB-produced biocomposites is an avenue to effectively manage agricultural wastes and develop valuable industrial and practical applications.

Wang *et al.* [125] investigated the mechanical properties and water resistance of BC reinforced TPS/PLA biocomposites and modified with sodium hexametaphosphate (SHMP). It was found this modification approach significantly improved the tensile strength and impact strength values of composites as compared to initial BC-reinforced TPS.

Approaches to Improve the Performance of Biocomposites by Modifications

Biofibers have promising properties as reinforcing materials for biocomposites production; however, the biggest obstacle for the practical application of biocomposites is the poor stress transfer between the reinforcement and polymer matrix. Generally, the stress transfer from matrix to filler/reinforcement is strongly dependent on the degree of bonding between the two phases [126,127]. The poor interfacial adhesion at the two interfaces means that the full capabilities of the composite cannot be exploited, resulting in the cellulose being just a filler and not a true reinforcement, that could weaken its mechanical properties, and thus reducing its life span [127]. Two major types of modifications have been conducted to facilitate the stress transfer at the interphase [128,129,130,131]: (i) reducing the hydrophobicity of biofibers; (ii) using coupling agent. The first step has been extensively employed and achieved either through a physical treatment (e.g., cold plasma treatment) [132] or chemical treatment (e.g., organosilanes, isocyanates, etherification and esterification). These modifications result in a decrease in moisture absorption and an increase in mechanical properties, biodurability and weatherability [127]. Chemical treatments of natural fiber for use in natural fiber-reinforced petroleum based polymer composites were reviewed elsewhere [133].

One of the most effective methods for chemical modfication of cellulose was found to attach polymer chains directly to cellulose fiber surfaces, which was named "grafting". Via grafting, biofibers are covalently bonded onto the polymer matrix using a wide variety of polymerization techniques. This can effectively improve the compatibility and stress transfer between the two phases of the created biocomposite. Hence, the performances of biocomposites can be improved significantly.

GRAFTING MODIFICATIONS OF BIOFIBERS FOR BIOCOMPOSITES

Grafting Techniques

To date, graft copolymerization of biofibers with polymer for biocomposites can be categorized into three groups: (i) grafting of fiber with a single monomer; (ii) grafting with a mixture of two or more monomers; and (iii) grafting with the polymer directly. Final properties of grafted composites can be tailored by changing the processing parameters during polymerization such as the nature of the biofiber/polymer backbone, monomer/oligomer types, and the effect of initiator. Different means have been investigated to copolymerize different monomers onto biofiber backbones, especially for cellulose and lignin. These techniques majorly include living polymerization, free radical initiation, ionic, photochemical, plasma induced initiation, plasma radiation induced grafting, and enzyme grafting. For example, maleic anhydride has been utilized as a grafting monomer to functionalize biopolymers and used as coupling agent to promote the interfacial adhesion of fiber/polymer interphases in biocomposites [96].

Grafting via Living Polymerizaiton Technique

By grafting polymers onto cellulose its stiffness can be retained and the thermoplasticity of the polymer matrix can be imparted. The applications of biocomposites are expected to be broadened accordingly. Among all the grafting techniques, ring opening polymerization (ROP), a living polymerization technique, has been mostly used to graft polymer to/from cellulose fibers, followed by radical initiation because ROP of lactide has been well established for PLA synthesis. Cellulose modification by polymer grafting via ROP has been reviewed [12,37]. This method was demonstrated to be potentially used for biocomposites applications. Apart from ROP, grafting via free radical initiation has also been applied to the grafting copolymerization process and can be used for biofibers-reinforced biocomposites.

Ring Opening Polymerization for Biocomposites

The organocatalytic ROP process leaves no metal residues in the resulting biocomposites and allows for quite good control of the final product characteristics (*i.e.*, controlled molecular weight, narrow molecular weight distribution) [134]. In Hafren and Cordova's work [135] an organic acid was effectively used to catalyze the ROP for producing PCL-graft-cellulose directly for the first time (Figure 7a). Lönnberg *et al.* [136] successfully grafted cyclic monomers, ε-caprolactone and lactide, to the cellulose surface via covalent bonding with the respective polymers by using the $Sn(Oct)_2$-catalyzed ROP process (Figure 7b). Benzyl alcohol was used as an initiator and the grafting density can be controlled by the ratio of initiator to monomers.

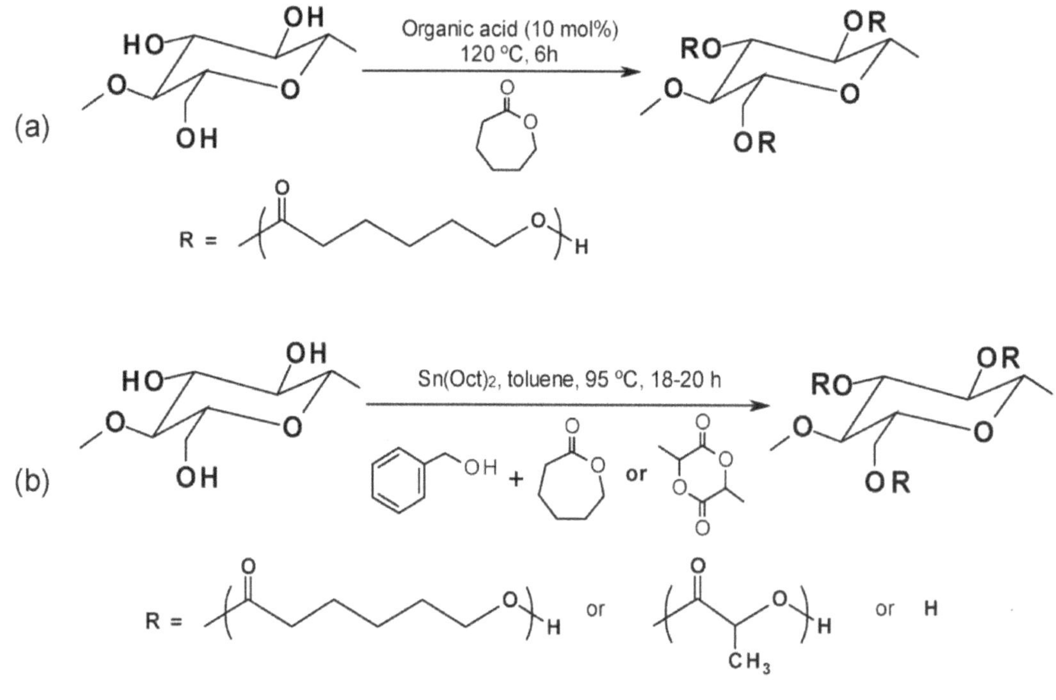

Figure 7: Organocatalytic ROP process of (**a**) cyclic ε-caprolactone and (**b**) cyclic ε-caprolactone or lactide with cellulose [132,133].

Outstanding mechanical properties were obtained by blending nanocellulose and polymer matrix at low filler loadings [137]. Diverse efforts have been undertaken to modify nanocellulose via grafting to improve the compatibility between the two phases. PCL-grafted-CNC bionanocomposite was prepared via $Sn(Oct)_2$-catalyzed ROP mechanism [138]. Fourier Transform Infrared (FT-IR) spectroscopy, Time-of-Flight Secondary Ion Mass Spectrometry (TOF-SIMS), and X-ray diffraction (XRD) analyses were used to confirm and demonstrate the efficiency of the PCLgrafting via the ROP method onto CNCs. The thermal and mechanical properties and crystalline structures of the grafted composites were studied by differential scanning calorimeter (DSC), dynamic mechanical analysis (DMA) and tensile testing. The obtained

PCL-graft-CNC showed an improved mechanical performance and the grafted composites became more ductile at room temperature. The Young's modulus was increased by 152% for bionanocomposites reinforced with 40% PCL-graft-CNC, and elongation at break was increased by 333% for composites containing 30% of the grafted copolymer as compared to the composites filled with unmodified CNCs. It was concluded that the surface-grafted CNCs could act as nucleation sites, which facilitated the crystallization of the polymer matrix. It was evidenced that the grafting modification enhanced the compatibility of CNC/polymer matrix and thus the final properties of the grafted bionanocomposites.

Ethyl cellulose-graft-poly(ε-caprolactone)-block-poly(l-lactide) (EC-g-PCL-b-PLLA) block copolymers were prepared by sequential ROP [139]. The block length was tuned by changing the molar ratios of ε-caprolactone monomer to ethyl cellulose and the l-lactide to ε-caprolactone. Thermal properties and crystalline nature of EC-g-PCL and EC-g-PCL-b-PLLA copolymers were different from those of linear PCL. The *in vitro* degradation was facilitated with the addition of PLLA blocks in the EC-g-PCL-b-PLLA copolymers as compared to linear PCL and EC-g-PCL. The biodegradation rate can be controlled by monitoring the grafting efficiency. This can enable the grafted biocomposites to be used in the controlled release applications.

In the study by Goffin *et al.* [140], CNC was grafting modified via ROP with l-lactide to obtain the PLA-graft-CNC copolymer and then filled into PLA matrix to prepare the PLA/PLA-graft-CNC green biocomposites. Above the glass transition temperature, the stiffness of PLA-graft-CNC reinforced PLA biocomposite (PLA/PLA-graft-CNC) was increased as compared to the PLA/CNC simple blended biocomposite, which was caused by the higher crystallinity due to grafting modification of CNC [140]. With the incorporation of PLA-graft-CNC into PLA/CNC composite matrix, the final product was colorless and more transparent then the unmodified CNC filled PLA biocomposite. This suggested that the grafting modification of CNC limited their thermal degradation, allowing them to be processed at higher temperature when subjected to melt processing. This broadens the PLA/CNC composite applications into new modern applications such as 3D printing materials, packaging, and electronic industries.

Nanocomposites of nanopaper fibers were obtained via surface initiated ROP of ε-caprolactone [141]. Two types of catalysts, titanium n-butoxide ($Ti(On\text{-}Bu)_4$) and tin octoate ($Sn(Oct)_2$) were used. It was found that the high surface area of nanopaper fibers gave higher grafting efficiency with $Sn(Oct)_2$. The mechanical properties of grafted copolymer were superior to those of pure PCL.

In addition to cellulose, to utilize the low-cost but renewable lignin fibers through compounding with biopolymers for biocomposite applications is gaining attention. However, it is difficult to blend lignin with other polymers due to poor biocompatibility. Lignin has an abundance of accessible OH groups on its surfaces, which can be modified as cellulose. Recently, Chung *et al.* [52] prepared the lignin-graft-PLA copolymer via ROP process. The copolymerization was catalyzed by triazabicyclodecene (TBD). The molecular structure of the PLA in the copolymer was controlled by manipulating the ratio of lignin/lactide monomers and the lignin pretreatment (acetylation). Results showed that the aliphatic OH groups were preferred over the phenolic OH groups during grafting. The obtained PLA-graft-lignin copolymer was subsequently used as a dispersion modifier for the PLA-lignin biocomposite and properties were studied and compared with the simple blended composites made from PLA and unmodified lignin. The glass transition temperature of PLA-graft-lignin was increased by 40 °C when lignin content was increased from 10% to 50%. The addition of PLA-graft-lignin into the biocomposite resulted in a respective increase in tensile strength and strain by 16% and 9% without diminishing the Young's modulus. It was found the biocomposites modified with 10% PLA-graft-lignin displayed excellent UV (UVC, UVB, and UVA) resistance.

Free Radical Grafting

Grafting of various monomers onto cellulose fibers or cellulose derivatives is an extensively used tool for cellulose modification. The grafting reaction is carried out with the aid of free radical initiators. Free radicals are created by either chemical (ceric ammonium nitrate (CAN), various persulfates, and Fenton reagent ($Fe_{(II)}$–H_2O_2)), thermal (azobisisobutyronitrile (AIBN) and organic peroxide), or irradiation (UV, gamma, and electron/plasma ion beam) initiation [11]. For example, vinyl or acrylic monomers are grafted onto cellulose backbone is an important technique to modify the cellulose and give new properties of the resulting composites. Generally, three major steps are involved in the free radical initiation polymerization: (i) initiation; (ii) propagation; and (iii) termination. In step (i) the free radicals are formed on the surface of cellulose substrate; followed by (ii) and (iii) in which the monomers react with the free radical sites to propagate and terminate with new polymer chains that are covalently bonded to the cellulose backbone. Free radicals would be generated at the newly formed polymer chains and terminated with other radical sites on polymer chains, and thus branches will be formed.

Cellulosic Grewia optiva fibers were grafted with methyl acrylate monomer to form PMA-graft-cellulose copolymers [142]. The process was carried out under microwave conditions using ferrous ammonium sulfate-potassium per sulfate as redox free radical initiator. The grafting efficiency (GE) was increased with microwave power up to 70%, while a maximum at 37% GE was reached at the optimized parameters (microwave radiation power, ratio of monomer, solvent and initiator concentrations). The grafted copolymers showed an increase in chemical resistance thermal stability. This provides a proficient method to broaden the applications of biofibers for use in biocomposites. A similar method was applied for synthesizing poly(butyl acrylate)-graft-cellulose copolymers [143]. The graft copolymers have been found to be more resistant to moisture and also showed better resistance to chemical and thermal deterioration.

Gum rosin polymer-graft-lignin composites were prepared via atom transfer radical polymerization (ATRP) [144]. The 2-bromoisobutyryl ester-modified lignin was produced and used as a macroinitiator. Three different monomers derived from dehydroabietic acid (DA) were separately attached to lignin through an esterification process. Grafting of both DA and rosin polymers significantly enhanced the hydrophobicity of lignin. Water uptake of the grafted composite was significantly reduced. The ATRP technique has also been used to synthesize CNC grafted to poly(acrylic acid) (PAA) materials with a brush type structure with controlled chain length using a Cu-mediated surface initiated controlled radical polymerization [145]. The CNCs were first modified to create an initiator moiety on the surface. This new modification route for CNCs enables further advanced uses for functionalized CNC and applications.

Zoppe *et al.* [146] prepared CNC-graft-poly(*N*-isopropylacrylamide) biocomposite brushes via surface-initiated single-electron transfer living radical polymerization (SI-SET-LRP). The CNCs were produced from sulfuric acid hydrolyzed ramie fibers (3–15 nm × 50–250 nm). The grafting was confirmed by FTIR and X-ray photoelectron spectroscopies. It was expected that the interfacial interactions between CNCs and the polymer matrix can be controlled by changing the reaction temperature and monomer loading. Their results provided insight for the development of temperature-responsive materials from CNCs.

In order to make the CNC compatible with natural rubber, Parambath Kanoth *et al.* prepared biocomposites using the free radical initiation technique [147]. Firstly, the cellulose was functionalized with mercaptoundecanoic acid to get mercaptoundecanoyl CNC bearing thiol groups, and then the photochemically initiated cross-linking reaction occurs between the double bonds of natural rubber and thiol groups of CNC. This technique improved the interfacial bonding between CNC and the rubber matrix and thus improved their thermomechanical properties. For example, the tensile strength and elongation to break were increased 2.4 and 1.6 fold, respectively, upon modification as compared to blends.

Additionally, cellulose can be treated with high energy radiation (e.g., γ-rays from radioactive isotopes or electron beams) to generate radicals together with chain scission [11,148]. Irradiation of cellulose would produce free radicals which further initiate the polymerization of vinyl and acrylic monomers, and thus the cellulose-graft-polymer biocomposites are produced. Sisal fibers were grafted with co-monomer of styrene and ethyl acrylate initiated by gamma irradiation [149]. The mechanical properties experienced deterioration upon irradiation regardless of the irradiation atmosphere, while dye uptake for basic and direct dyes were found to slightly increased at low graft efficiency and then to decrease with increasing grafting extent. For the disperse dye, the dye uptake was increased with increasing grafting levels.

Microwave radiation was used to graft *Luffa cylindrica* [150] and *Grewia optiva* fibers [151] to methyl acrylate/acrylamide. The grafted copolymer showed improved thermal, structural, chemical, and morphological properties and this technique can be potentially used as an adsorbent for water purification and other industrial applications. The main advantage of using microwave chemical grafting strategy is in the use of a liquid medium with a wide range of solvents that can be used, which allows better dispersion of biofibers in the polymer matrix. However, these manufacturing methods are not scalable and non-economic. Although radiation induced grafting is relatively easier to operate, it is always difficult to control the radical generation process.

Melt-compounding techniques, such as extrusion and injection molding, are usually used for thermoplastic processing, which are solvent free and scalable (for blown film especially). In recent studies [152,153], cellulose and PHB/PHBV composites were manufactured via *in situ* reactive extrusion with dicumyl peroxide (DCP) free radical initiation at high temperature (Figure 8). The grafted copolymers at the interfaces of cellulose and PHB/PHBV matrix performed like an interfacial coupling agent. Grafting tended to interact with both the hydrophilic fibers and the hydrophobic PHB or PHBV matrix. The grafting mechanism was confirmed by electron spin resonance (ESR) analysis and showed the presence of radicals produced by DCP radical initiation for the first time. The grafted copolymer structure was determined by nuclear magnetic resonance spectroscopy. The biocomposites were also characterized by scanning electron microscopy and dynamic mechanical analysis and indicated good interfacial bonding and compatibility between the two phases. The mechanical properties of the biocomposites were improved by grafting due to improved stress transfer between the two interphases of the cellulose fiber/polymer matrix as compared to the blend control composite [152]. The better mechanical performances of grafted biocomposites also contributed to the crosslinks formed among the PHB or PHBV polymer chains. The crystallinity of PHB, PHBV and cellulose in the grafted biocomposite were reduced as determined by FTIR spectroscopy, X-ray diffraction, and DSC analyses. The reduction of crystallinity indicated that the grafting reaction occurred not just in the amorphous region of cellulose fibers but also slightly in crystalline domains. Reduced crystal sizes suggested that the brittleness of PHB was decreased due to grafting; in other words, flexibility of PHB was introduced via grafting. Thermogravimetric analysis showed that the grafted copolymer was stabilized relative to PHB. By varying the reaction parameters, the compositions (%PHB and %cellulose) of resultant cellulose-graft-PHB copolymer are expected to be manipulated to obtain tunable properties.

This *in-situ* reactive extrusion process offers an effective approach to improve the properties of biocomposite materials from sustainable resources and has also been employed to produce thermally recyclable PLA-graft-CNC biocomposites [154]. The grafted portion was isolated and re-fed into the biocomposites thermal processing as coupling agent. The presence of chemical crosslinks between the CNC and PLA matrix gave efficient modulus transfer, improving the tensile strength and Young's modulus by 40% and 490%, respectively. This provided an alternative utilization of CNC and PLA for biodegradable and recyclable biocomposites films for packaging materials.

Kraft lignin was successfully grafted onto a high density polyethylene (HDPE) matrix by the DCP free radical initiation [155]. Maleated lignin was generated via esterification with maleic anhydride prior to the grafting reaction. Better compatibility of lignin and HDPE was observed than for the maleated lignin and HEPE composites system. Due to grafting, the thermal stability of maleated lignin and HDPE composites was improved as well. Csikós *et al.* [156] used reactive extrusion process with aid of peroxide to improve the interfacial adhesion of wood fiber and PLA biocomposites. In this study, the functionalized PLA with maleic anhydride was prepared first and then compounded and radically-initiated grafted with wood fibers for biocomposites. The functionalized PLA proved to be an efficient coupling agent in PLA/wood composites.

The mechanical strength was increased due to grafting. This opens the window of utilizing low-value lignin as renewable reinforcement for biocomposites used for structural materials for automotive or construction purposes.

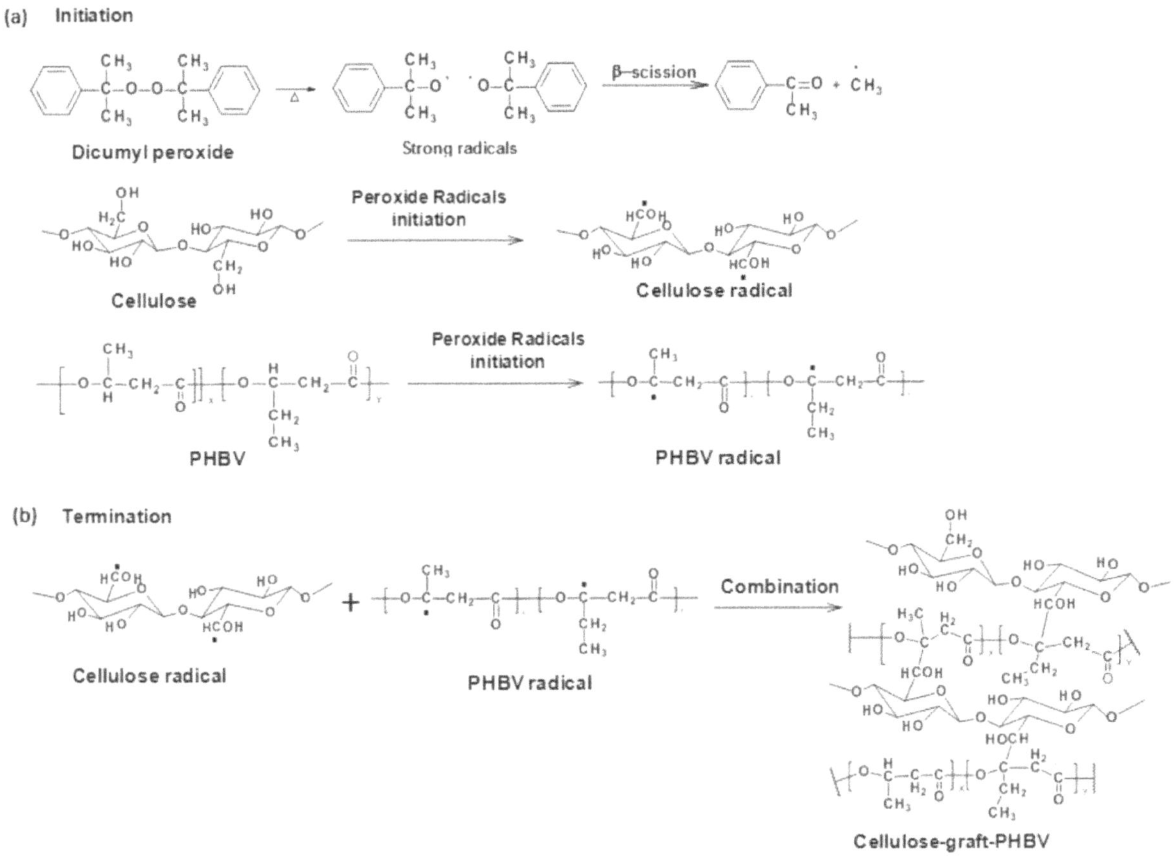

Figure 8: Reaction mechanism of peroxide radical initiated grafting of PHBV onto cellulose: (**a**) radical initiation and (**b**) reaction termination [149].

Grafting via Coupling Agent

Another approach to enhance the compatabilization between biofibers reinforcement and polymer matrix is to use a grafting or coupling agent, such as isocyanates. *N*-Octadecyl isocyanate was used as grafting agent to improve the compatibility of CNC and microfibrillated cellulose extracted from sisal fibers to PCL polymer matrix [157]. Nanocomposite films were prepared by solvent casting. The chemically grafted composites improved the compatibility between the two phases and helped dispersion of CNCs into the PCL matrix. Mechanical properties of the grafted composites were improved in terms of stiffness and ductility. Transition temperatures and crystallinity were increased with fiber content.

Methylenediphenyl diisocyanate (MDI) was used as coupling agent for biocomposites from PLA and starch [121]. The authors found the presence of MDI in the PLA/starch composite did not affect the thermal decomposition profile when the composites were stored at 25 °C and relative humidity fluctuating between 90% and 30% for up to 180 days. The mechanical properties of PLA and PLA/starch-based composites sealed in PE plastic bags were not changed during 30-days of storage in fluctuating humidity conditions (30%–90%).

Magniez *et al.* [158] found the interfacial interaction was enhanced by using poly-(ethylene glycol)-b-poly-(l-lactide) (PEG-PLLA) block copolymers as coupling agent for jute fiber and PLA biocomposites. A solvent solution of the copolymer was deposited onto the fibrous substrate. The strong interaction between the fiber-matrix interphase contributed to the intra- and inter-molecular hydrogen bonds between these two phases. By adjusting the amphiphilicity and the type of building blocks of copolymer, the surface properties of the biocomposites can be tuned.

Biocomposites from biodegradable polymers onto cellulose and cellulose derivatives have been reported [159,160]. Samain*et al.* [159] investigated the grafting of three biodegradable polymers, PLA, PCL and PHA onto cellulose paper and microcrystalline cellulose fibers. Firstly, the low molecular weight telechelic OH-terminated oligomers were prepared, followed by the graft copolymerization. Specifically, the reactive carboxylic end groups were formed for PHA by thermal treatment, while PCL oligomers were prepared via ROP. This was followed by producing oligomers with an acid chloride group at the end of PHA, PLA and PCL. These were prepared with the aid of thionyl chloride. The grafting reaction was initiated at high temperature between chloride oligomers and cellulose. In the study by Yu *et al.*, telechelic OH-terminated PHBV oligomers were prepared first by a transesterification procedure. PHBV-graft-ethyl cellulose copolymers were synthesized using 1,6-hexamethylene diisocyanate (HDI) as a grafting agent. The grafting conditions, such as the reaction time, temperature and mass ratio of PHBV oligomer to EC, were optimized to obtain the maximum grafting percentage. Compared with neat PHBV, the crystallinity of the grafted copolymer decreased and the moisture resistance was improved significantly.

Very recently, biocomposites made from PP and argan nut shell as reinforcement was studied [161]. In this work, styrene–(ethylene–butene)–styrene triblock copolymer grafted with maleic anhydride (SEBS-g-MA) was used as a coupling agent to enhance the interaction adhesion between filler and polymer matrix. SEM micrographs revealed that good filler dispersion into matrix with the aid of the coupling agent. The addition of coupling agent improved the thermal stability of PP.

In addition to these above mentioned major grafting strategies, PHBV and poly(3-hydroxyoctanoate) (PHO) have been successfully grafted onto chitosan by condensation with reaction occurred between amine groups of chitosan and carboxylic end of PHBV and PHO [61]. These grafted biocomposites could be used as new materials in medical applications, such as tissue engineering and drug delivery systems.

CONCLUSIONS

Naturally derived biofibers have been frequently used as reinforcement for polymer composite applications, especially the natural cellulose, lignin, microbial cellulose, and nanocellulose fibers, due to their low-cost and light weight. To reduce the impact on the environment, bio-based plastics are used to replace the conventional polymer matrix, giving completely green biocomposites. By incorporating various biofibers into polymer matrix, the prices of the bioplastics can be reduced and the properties can be improved. However, biofibers contain a large amount of hydroxyl groups, making them highly hydrophilic; whereas the polymer matrices materials are generally hydrophobic. The poor interfacial adhesion at the interphases limits the applications of biocomposites exposed to water, heat, and UV light. Various approaches have been employed to modify the fiber surfaces to improve the compatibility between the reinforcement and polymer matrix. On the other hand, "grafting" has been shown to be effective at imparting the desired and targeted properties of polymers onto cellulose, nanocellulose, and lignin (or their derivatives) as well as enhancing the interaction between interphases. Ring opening polymerization, coupling agent-induced grafting, and grafting via free radical initiated copolymerization techniques have been investigated. These grafting copolymerization techniques have offered an alternative modification approach to biofibers as a reinforcement of polymers for biocomposites including green composites and bionanocomposites production. This can potentially broaden the biocomposites utilized in numerous modern applications in the future, such as packaging materials, structural materials, coatings, biomedical devices, drug delivery release, electronic devices, waste-water treatment, and in the automotive industry.

ACKNOWLEDGMENTS

The authors would like to acknowledge the support from the University of Idaho and the USDA Forest Products Laboratory.

REFERENCES

1. Li, J.; Hunt, J.F.; Gong, S.; Cai, Z. High strength wood-based sandwich panels reinforced with fiberglass and foam.*BioResources* 2014, *9*, 1898–1913. [Google Scholar] [CrossRef]
2. Faruk, O.; Bledzki, A.K.; Fink, H.-P.; Sain, M. Progress report on natural fiber reinforced composites. *Macromol. Mater. Eng.* 2014, *299*, 9–26. [Google Scholar] [CrossRef]
3. Dhand, V.; Mittal, G.; Rhee, K.Y.; Park, S.-J.; Hui, D. A short review on basalt fiber reinforced polymer composites.*Compos. B Eng.* 2015, *73*, 166–180. [Google Scholar] [CrossRef]
4. Li, J.; Hunt, J.F.; Gong, S.; Cai, Z. Improved fatigue performance for wood-based structural panels using slot and tab construction. *Compos. A Appl. Sci. Manuf.* 2016, *82*, 235–242. [Google Scholar] [CrossRef]
5. Li, J.; Hunt, J.F.; Gong, S.; Cai, Z. Testing and evaluation of a slot and tab construction technique for light-weight wood-fiber-based structural panels under bending. *J. Test. Eval.* 2016, *44*, 1–10. [Google Scholar] [CrossRef]
6. Sun, L.; Wu, Q.; Xie, Y.; Song, K.; Lee, S.; Wang, Q. Thermal decomposition of fire-retarded wood flour/polypropylene composites. *J. Therm. Anal. Calorim.* 2016, *123*, 309–318. [Google Scholar] [CrossRef]
7. Li, J.; Hunt, J.F.; Gong, S.; Cai, Z. Simplified analytical model and balanced design approach for light-weight wood-based structural panel in bending. *Compos. Struct.* 2016, *136*, 16–24. [Google Scholar] [CrossRef]
8. European and global markets: 2012 and future trends. Bioplastics Magazine, May/June 2014.
9. Saha, S. Bio Composites Market: Asia Pacific Industry Analysis and Opportunity Assessment 2014–2020. Future Market Insights. Available online: http://www.Futuremarketinsights.Com/reports/asia-pacific-bio-composites-market(accessed on 12 February 2016).
10. Pilla, S. *Handbook of Bioplastics and Biocomposites Engineering Applications*; John Wiley & Sons, Inc.: Hoboken, NJ, USA, 2011; Volume 81. [Google Scholar]
11. Roy, D.; Semsarilar, M.; Guthrie, J.T.; Perrier, S. Cellulose modification by polymer grafting: A review. *Chem. Soc. Rev.*2009, *38*, 2046–2064. [Google Scholar] [CrossRef] [PubMed]
12. Carlmark, A.; Larsson, E.; Malmström, E. Grafting of cellulose by ring-opening polymerisation—A review. *Eur. Polym. J.* 2012, *48*, 1646–1659. [Google Scholar] [CrossRef]
13. Faruk, O.; Bledzki, A.K.; Fink, H.-P.; Sain, M. Biocomposites reinforced with natural fibers: 2000–2010. *Prog. Polym. Sci.*2012, *37*, 1552–1596. [Google Scholar] [CrossRef]
14. John, M.; Thomas, S. Biofibres and biocomposites. *Carbohydr. Polym.* 2008, *71*, 343–364. [Google Scholar] [CrossRef]
15. Rowell, R. *Handbook of Wood Chemistry and Wood Composites*; CRC Press: Boca Raton, FL, USA, 2005. [Google Scholar]
16. Barnett, J.; Bonham, V.A. Cellulose microfibril angle in the cell wall of wood fibres. *Biol. Rev.* 2004, *79*, 461–472. [Google Scholar] [CrossRef] [PubMed]
17. Clemons, C.M.; Caulfield, D.F. Natural fibers. In *Functional Fillers for Plastics*; Xanthos, M., Ed.;

Wiley-VCH: Weinheim, Germany, 2005; pp. 195–206. [Google Scholar]

18. Hamelinck, C.N.; Hooijdonk, G.V.; Faaij, A.P.C. Ethanol from lignocellulosic biomass: Techno-economic performance in short-, middle- and long-term. *Biomass Bioenergy* 2005, *28*, 384–410. [Google Scholar] [CrossRef]
19. Siqueira, G.; Bras, J.; Dufresne, A. Cellulosic bionanocomposites: A review of preparation, properties and applications. *Polymers* 2010, *2*, 728–765. [Google Scholar] [CrossRef]
20. Komuraiah, A.; Kumar, N.S.; Prasad, B.D. Chemical composition of natural fibers and its influence on their mechanical properties. *Mech. Compos. Mater.* 2014, *50*, 359–376. [Google Scholar] [CrossRef]
21. Mwaikambo, L.Y.; Ansell, M.P. Mechanical properties of alkali treated plant fibres and their potential as reinforcement materials. I. Hemp fibres. *J. Mater. Sci.* 2006, *41*, 2483–2496. [Google Scholar] [CrossRef]
22. Dungani, R.; Khalil, H.P.S.A.; Sumardi, I.; Suhaya, Y.; Sulistyawati, E.; Islam, M.N.; Saraya, N.L.M.; Aprilia, N.A.S. Non-wood renewable materials: Properties improvement and its application. In *Biomass and Bioenergy: Applications*; Hakeem, K.R., Jawaid, M., Rashid, U., Eds.; Springer: Basel, Switzerland, 2014. [Google Scholar]
23. Mwaikambo, L.Y. Review of the history, properties and application of plant fibres. *Afr. J. Sci. Technol.* 2006, *7*, 120–133. [Google Scholar]
24. Hon, D.N.-S.; Shiraishi, N. *Wood and Cellulosic Chemistry, Second Edition, Revised, and Expanded*; Marcel Dekker, Inc.: New York, NY, USA, 2001. [Google Scholar]
25. Kadla, J.F.; Gilbert, R.D. Cellulose structure: A review. *Cell. Chem. Technol.* 2000, *34*, 197–216. [Google Scholar]
26. Klemm, D.; Heublein, B.; Fink, H.-P.; Bohn, A. Cellulose: Fascinating biopolymer and sustainable raw material. *Angew. Chem. Int. Ed.* 2005, *44*, 3358–3393. [Google Scholar] [CrossRef] [PubMed]
27. Nishiyama, Y.; Sugiyama, J.; Chanzy, H.; Langan, P. Crystal structure and hydrogen bonding system in cellulose iα from synchrotron X-ray and neutron fiber diffraction. *J. Am. Chem. Soc.* 2003, *125*, 14300–14306. [Google Scholar] [CrossRef] [PubMed]
28. Nishiyama, Y.; Langan, P.; Chanzy, H. Crystal structure and hydrogen-bonding system in cellulose iβ from synchrotron X-ray and neutron fiber diffraction. *J. Am. Chem. Soc.* 2002, *124*, 9074–9082. [Google Scholar] [CrossRef] [PubMed]
29. Donaldson, L.A. Lignification and lignin topochemistry—An ultrastructural view. *Phytochemistry* 2001, *57*, 859–873. [Google Scholar] [CrossRef]
30. Xu, X.; Liu, F.; Jiang, L.; Zhu, J.Y.; Haagenson, D.; Wiesenborn, D.P. Cellulose nanocrystals *vs.* Cellulose nanofibrils: A comparative study on their microstructures and effects as polymer reinforcing agents. *ACS Appl. Mater. Interf.* 2013, *5*, 2999–3009. [Google Scholar] [CrossRef] [PubMed]
31. Habibi, Y.; Lucia, L.A.; Rojas, O.J. Cellulose nanocrystals: Chemistry, self-assembly, and applications. *Chem. Rev.* 2010, *110*, 3479–3500. [Google Scholar] [CrossRef] [PubMed]
32. Song, K.; Wu, Q.; Zhang, Z.; Ren, S.; Lei, T.; Dooley, K.M.; Liu, D.; Janes, M.E. Fabricating electrospun nanofibers with antimicrobial capability: A facile route to recycle biomass tar. *Fuel* 2015, *150*, 123–130. [Google Scholar] [CrossRef]
33. Šturcová, A.; Davies, G.R.; Eichhorn, S.J. Elastic modulus and stress-transfer properties of tunicate cellulose whiskers.*Biomacromolecules* 2005, *6*, 1055–1061. [Google Scholar] [CrossRef] [PubMed]
34. Tashiro, K.; Kobayashi, M. Theoretical evaluation of three-dimensional elastic constants of native and regenerated celluloses: Role of hydrogen bonds. *Polymer* 1991, *32*, 1516–1526. [Google Scholar] [CrossRef]
35. Sabo, R.C.; Elhajjar, R.F.; Clemons, C.M.; Pillai, K.M. Characterization and processing of nanocellulose thermosetting composites. In *Handbook of Polymer Nanocomposites. Processing, Performance and Application*; Pandey, J.K., Takagi, H., Nakagaito, A.N., Kim, H.-J., Eds.; Springer: Berlin, Germany; Heidelberg, Germany, 2015; pp. 265–295. [Google Scholar]
36. Li, M.-C.; Wu, Q.; Song, K.; Lee, S.; Qing, Y.; Wu, Y. Cellulose nanoparticles: Structure–morphology–rheology relationships. *ACS Sustain. Chem. Eng.* 2015, *3*, 821–832. [Google Scholar] [CrossRef]
37. Lin, N.; Huang, J.; Dufresne, A. Preparation, properties and applications of polysaccharide nanocrystals in advanced functional nanomaterials: A review. *Nanoscale* 2012, *4*, 3274–3294. [Google Scholar] [CrossRef] [PubMed]
38. Agarwal, U.P.; Sabo, R.; Reiner, R.S.; Clemons, C.M.; Rudie, A.W. Spatially resolved characterization of cellulose nanocrystal-polypropylene composite by confocal raman microscopy. *Appl. Spectrosc.* 2012, *66*, 750–756. [Google Scholar] [CrossRef] [PubMed]
39. Nishino, T.; Takano, K.; Nakamae, K. Elastic modulus of the crystalline regions of cellulose polymorphs. *J. Polym. Sci. B Polym. Phys.* 1995, *33*, 1647–1651. [Google Scholar] [CrossRef]
40. Moon, R.J.; Beck, S.; Rudie, A.W. *Cellulose Nanocrystals—A Material with Unique Properties and Many Potential Applications*; Postek, M.T., Moon, R.J., Rudie, A.W., Bilodeau, M.A., Eds.; TAPPI Press: Peachtree Corners, GA, USA, 2013; pp. 9–12. [Google Scholar]
41. Chen, L.; Wang, Q.; Hirth, K.; Baez, C.; Agarwal, U.P.; Zhu, J.Y. Tailoring the yield and characteristics of wood cellulose nanocrystals (cnc) using concentrated acid hydrolysis. *Cellulose* 2015, *22*, 1753–1762. [Google Scholar] [CrossRef]
42. Moon, R.J.; Martini, A.; Nairn, J.; Simonsen, J.; Youngblood, J. Cellulose nanomaterials review: Structure, properties and nanocomposites. *Chem. Soc. Rev.* 2011, *40*, 3941–3944. [Google Scholar] [CrossRef] [PubMed]

43. Abe, K.; Iwamoto, S.; Yano, H. Obtaining cellulose nanofibers with a uniform width of 15 nm from wood. *Biomacromolecules* 2007, *8*, 3276–3278. [Google Scholar] [CrossRef] [PubMed]

44. Saito, T.; Kimura, S.; Nishiyama, Y.; Isogai, A. Cellulose nanofibers prepared by tempo-mediated oxidation of native cellulose. *Biomacromolecules* 2007, *8*, 2485–2491. [Google Scholar] [CrossRef] [PubMed]

45. Brodin, I. Chemical Properties and Thermal Behaviour of Kraft Lignins. Master's Thesis, Department of Fibre and Polymer Technology, KTH Royal Institute of Technology, Stockholm, Sweden, 2009. [Google Scholar]

46. Thakur, V.K.; Thakur, M.K.; Gupta, R.K. Graft copolymers of natural fibers for green composites. *Carbohydr. Polym.*2014, *104*, 87–93. [Google Scholar] [CrossRef] [PubMed]

47. Liu, R.; Peng, Y.; Cao, J.; Chen, Y. Comparison on properties of lignocellulosic flour/polymer composites by using wood, cellulosc, and lignin flours as fillers. *Compos. Sci. Technol.* 2014, *103*, 1–7. [Google Scholar] [CrossRef]

48. Bahi, A.; Goudarzi, A.; Cho, M.; Lin, L.; Karaaslan, M.; Ko, F. Fabrication and application of lignin-lignin composites. In *Design, Manufacturing and Applications of Composites*, Proceedings of the Tenth Joint Canada-Japan Workshop on Composites, Vancouver, BC, Canada, 19–21 August 2014; DEStech Publications, Inc.: Lancaster, PA, USA, 2015; p. 195. [Google Scholar]

49. Mousavioun, P.; Halley, P.J.; Doherty, W.O.S. Thermophysical properties and rheology of phb/lignin blends. *Ind. Crop. Prod.* 2013, *50*, 270–275. [Google Scholar] [CrossRef]

50. Pohjanlehto, H.; Setala, H.M.; Kiely, D.E.; McDonald, A.G. Lignin xylaric acid polyurethane based polymer network systems: Preparation and characterization. *J. Appl. Polym. Sci.* 2013, *131*. [Google Scholar] [CrossRef]

51. Camargo, F.A.; Innocentini-Mei, L.H.; Lemes, A.P.; Moraes, S.G.; Duran, N. Processing and characterization of composites of poly(3-hydroxybutyrate-co-hydroxyvalerate) and lignin from sugar cane bagasse. *J. Compos. Mater.*2011, *46*, 417–425. [Google Scholar] [CrossRef]

52. Chung, Y.-L.; Olsson, J.V.; Li, R.J.; Frank, C.W.; Waymouth, R.M.; Billington, S.L.; Sattely, E.S. A renewable lignin–lactide copolymer and application in biobased composites. *ACS Sustain. Chem. Eng.* 2013, *1*, 1231–1238. [Google Scholar] [CrossRef]

53. Singha, A.; Shama, A.; Thakur, V.K. Graft copolymerization of acrylonitrile onto saccharum cilliari fiber. *e-Polymers*2009, *9*, 1258–1269. [Google Scholar] [CrossRef]

54. Thakur, V.; Thakur, M.; Singha, A. Free radical–induced graft copolymerization onto natural fibers. *Int. J. Polym. Anal. Charact.* 2013, *18*, 430–438. [Google Scholar] [CrossRef]

55. Thakur, V.; Singha, A.; Thakur, M. Pressure induced synthesis of ea grafted saccaharum cilliare fibers. *Int. J. Polym. Mater.* 2014, *63*, 17–22. [Google Scholar] [CrossRef]

56. Sen, S.; Patil, S.; Argyropoulos, D.S. Thermal properties of lignin in copolymers, blends, and composites: A review.*Green Chem.* 2015, *17*, 4862–4887. [Google Scholar] [CrossRef]

57. Thakur, V.K.; Thakur, M.K.; Raghavan, P.; Kessler, M.R. Progress in green polymer composites from lignin for multifunctional applications: A review. *ACS Sustain. Chem. Eng.* 2014, *2*, 1072–1092. [Google Scholar] [CrossRef]

58. Iguchi, M.; Yamanaka, S.; Budhiono, A. Review bacterial cellulose—A masterpiece of nature's arts. *J. Mater. Sci.* 2000,*35*, 261–270. [Google Scholar] [CrossRef]

59. Ifuku, S.; Nogi, M.; Abe, K.; Handa, K.; Nakatsubo, F.; Yano, H. Surface modification of bacterial cellulose nanofibers for property enhancement of optically transparent composites: Dependence on acetyl-group ds. *Biomacromolecules*2007, *8*, 1973–1978. [Google Scholar] [CrossRef] [PubMed]

60. Butchosa, N.; Brown, C.; Larsson, P.T.; Berglund, L.A.; Bulone, V.; Zhou, Q. Nanocomposites of bacterial cellulose nanofibers and chitin nanocrystals: Fabrication, characterization and bactericidal activity. *Green Chem.* 2013, *15*, 3404–3413. [Google Scholar] [CrossRef]

61. Arslan, H.; Hazer, B.; Yoon, S.C. Grafting of poly(3-hydroxyalkanoate) and linoleic acid onto chitosan. *J. Appl. Polym. Sci.* 2007, *103*, 81–89. [Google Scholar] [CrossRef]

62. Yu, G.-E.; Morin, F.G.; Nobes, G.A.R.; Marchessault, R.H. Degree of acetylation of chitin and extent of grafting phb on chitosan determined by solid state 15n nmr. *Macromolecules* 1999, *32*, 518–520. [Google Scholar] [CrossRef]

63. Kikkawa, Y.; Fukuda, M.; Kimura, T.; Kashiwada, A.; Matsuda, K.; Kanesato, M.; Wada, M.; Imanaka, T.; Tanaka, T. Atomic force microscopic study of chitinase binding onto chitin and cellulose surfaces. *Biomacromolecules* 2014, *15*, 1074–1077. [Google Scholar] [CrossRef] [PubMed]

64. Vroman, I.; Tighzert, L. Biodegradable polymers. *Materials* 2009, *2*, 307–344. [Google Scholar] [CrossRef]

65. Fortunati, E.; Armentano, I.; Zhou, Q.; Iannoni, A.; Saino, E.; Visai, L.; Berglund, L.A.; Kenny, J.M. Multifunctional bionanocomposite films of poly(lactic acid), cellulose nanocrystals and silver nanoparticles. *Carbohydr. Polym.* 2012, *87*, 1596–1605. [Google Scholar] [CrossRef]

66. Mushi, N.E.; Butchosa, N.; Salajkova, M.; Zhou, Q.; Berglund, L.A. Nanostructured membranes based on native chitin nanofibers prepared by mild process. *Carbohydr. Polym.* 2014, *112*, 255–263. [Google Scholar] [CrossRef] [PubMed]

67. Ifuku, S.; Saimoto, H. Chitin nanofibers: Preparations, modifications, and applications. *Nanoscale* 2012, *4*, 3308–3318. [Google Scholar] [CrossRef] [PubMed]

68. Mushi, N.E.; Butchosa, N.; Zhou, Q.; Berglund, L.A. Nanopaper membranes from chitin–protein composite nanofibers—Structure and mechanical properties. *J. Appl. Polym. Sci.* 2014, *131*. [Google Scholar] [CrossRef]
69. Mekonnen, T.; Mussone, P.; Khalil, H.; Bressler, D. Progress in bio-based plastics and plasticizing modifications. *J. Mater. Chem.* 2013, *1*, 13379–13398. [Google Scholar] [CrossRef]
70. Chen, G.-Q.; Patel, M.K. Plastics derived from biological sources: Present and future: A technical and environmental review. *Chem. Rev.* 2012, *112*, 2082–2099. [Google Scholar] [CrossRef] [PubMed]
71. Ojijo, V.; Sinha Ray, S.; Sadiku, R. Toughening of biodegradable polylactide/poly(butylene succinate-co-adipate) blends via *in situ* reactive compatibilization. *ACS Appl. Mater. Interf.* 2013, *5*, 4266–4276. [Google Scholar] [CrossRef] [PubMed]
72. Bugnicourt, E.; Cinelli, P.; Lazzeri, A.; Alvarez, V. Polyhydroxyalkanoate (pha): Review of synthesis, characteristics, processing and potential applications in packaging. *Express Polym. Lett.* 2014, *8*, 791–808. [Google Scholar] [CrossRef]
73. Lasprilla, A.J.R.; Martinez, G.A.R.; Lunelli, B.H.; Jardini, A.L.; Filho, R.M. Poly-lactic acid synthesis for application in biomedical devices—A review. *Biotechnol. Adv.* 2012, *30*, 321–328. [Google Scholar] [CrossRef] [PubMed]
74. Laycock, B.; Halley, P.; Pratt, S.; Werker, A.; Lant, P. The chemomechanical properties of microbial polyhydroxyalkanoates. *Prog. Polym. Sci.* 2013, *38*, 536–583. [Google Scholar] [CrossRef]
75. Wei, L.; McDonald, A.G. Accelerated weathering studies on the bioplastic, poly(3-hydroxybutyrate-co-3-hydroxyvalerate). *Polym. Degrad. Stab.* 2016, *126*, 93–100. [Google Scholar] [CrossRef]
76. Wei, L.; Liang, S.; Coats, E.R.; McDonald, A.G. Valorization of residual bacterial biomass waste after polyhydroxyalkanoate isolation by hydrothermal treatment. *Bioresour. Technol.* 2015, *198*, 739–745. [Google Scholar] [CrossRef] [PubMed]
77. Gumel, A.M.; Annuar, M.S.M.; Chisti, Y. Recent advances in the production, recovery and applications of polyhydroxyalkanoates. *J. Polym. Environ.* 2013, *21*, 580–605. [Google Scholar] [CrossRef]
78. Khanna, S.; Srivastava, A.K. Recent advances in microbial polyhydroxyalkanoates. *Process Biochem.* 2005, *40*, 607–619. [Google Scholar] [CrossRef]
79. Ienczak, J.L.; Schmidell, W.; de Aragão, G.M.F. High-cell-density culture strategies for polyhydroxyalkanoate production: A review. *J. Ind. Microbiol. Biotechnol.* 2013, *40*, 275–286. [Google Scholar] [CrossRef] [PubMed]
80. Reddy, C.S.K.; Ghai, R.; Kalia, V.C. Polyhydroxyalkanoates: An overview. *Bioresour. Technol.* 2003, *87*, 137–146. [Google Scholar] [CrossRef]
81. Suriyamongkol, P.; Weselake, R.; Narine, S.; Moloney, M.; Shah, S. Biotechnological approaches for the production of polyhydroxyalkanoates in microorganisms and plants—A review. *Biotechnol. Adv.* 2007, *25*, 148–175. [Google Scholar] [CrossRef] [PubMed]
82. Urtuvia, V.; Villegas, P.; González, M.; Seeger, M. Bacterial production of the biodegradable plastics polyhydroxyalkanoates. *Int. J. Biol. Macromol.* 2014, *70*, 208–213. [Google Scholar] [CrossRef] [PubMed]
83. Jendrossek, D.; Pfeiffer, D. New insights in the formation of polyhydroxyalkanoate granules (carbonosomes) and novel functions of poly(3-hydroxybutyrate). *Environ. Microbiol.* 2014, *16*, 2357–2373. [Google Scholar] [CrossRef] [PubMed]
84. Dias, J.M.L.; Lemos, P.C.; Serafim, L.S.; Oliveira, C.; Eiroa, M.; Albuquerque, M.G.E.; Ramos, A.M.; Oliveira, R.; Reis, M.A.M. Recent advances in polyhydroxyalkanoate production by mixed aerobic cultures: From the substrate to the final product. *Macromol. Biosci.* 2006, *6*, 885–906. [Google Scholar] [CrossRef] [PubMed]
85. Sudesh, K.; Abe, H.; Doi, Y. Synthesis, structure and properties of polyhydroxyalkanoates: Biological polyesters. *Prog. Polym. Sci.* 2000, *25*, 1503–1555. [Google Scholar] [CrossRef]
86. Wei, L.; Guho, N.M.; Coats, E.R.; McDonald, A.G. Characterization of poly(3-hydroxybutyrate-co-3-hydroxyvalerate) biosynthesized by mixed microbial consortia fed fermented dairy manure. *J. Appl. Polym. Sci.* 2014, *131*. [Google Scholar] [CrossRef]
87. Wei, L.; McDonald, A.G. Peroxide induced cross-linking by reactive melt processing of two biopolyesters: Poly(3-hydroxybutyrate) and poly(l-lactic acid) improve their melting processability. *J. Appl. Polym. Sci.* 2014, *132*. [Google Scholar] [CrossRef]
88. Wei, L.; McDonald, A.G. Thermophysical properties of bacterial poly(3-hydroxybutyrate): Characterized by tma, dsc, and tmdsc. *J. Appl. Polym. Sci.* 2014, *132*. [Google Scholar] [CrossRef]
89. Chanprateep, S. Current trends in biodegradable polyhydroxyalkanoates. *J. Biosci. Bioeng.* 2010, *110*, 621–632. [Google Scholar] [CrossRef] [PubMed]
90. Södergård, A.; Stolt, M. Properties of lactic acid based polymers and their correlation with composition. *Prog. Polym. Sci.* 2002, *27*, 1123–1163. [Google Scholar] [CrossRef]
91. Wool, R.; Sun, X.S. *Bio-Based Polymers and Composites*; Academic Press: New York, NY, USA, 2011. [Google Scholar]
92. Witzke, D.R.; Narayan, R.; Kolstad, J.J. Reversible kinetics and thermodynamics of the homopolymerization of l-lactide with 2-ethylhexanoic acid tin (ii) salt. *Macromolecules* 1997, *30*, 7075–7085. [Google Scholar] [CrossRef]
93. Narayan, R.; Wu, W.M.; Criddle, C.S. Lactide Production from Thermal Depolymerization of PLA with Applications to Production of PLA or Other Bioproducts. U.S. Patent 20,130,023,674, 15 March 2012. [Google Scholar]
94. Narayan, R.; Shi, X.; Graiver, D. PLA recycling via thermal depolymerization. Bioplastics Magazine, May/

June 2013. [Google Scholar]

95. Terzopoulou, Z.N.; Papageorgiou, G.Z.; Papadopouloub, E.; Athanassiadou, E.; Alexopoulou, E.; Bikiaris, D.N. Green composites prepared from aliphatic polyesters and bast fibers. *Ind. Crop. Prod.* 2015, *68*, 60–79. [Google Scholar] [CrossRef]

96. Mohanty, A.K.; Misra, M.; Drzal, L.T. Sustainable bio-composites from renewable resources: Opportunities and challenges in the green materials world. *J. Polym. Environ.* 2002, *10*, 19–26. [Google Scholar] [CrossRef]

97. Guan, J.; Hanna, M.A. Selected morphological and functional properties of extruded acetylated starch-polylactic acid foams. *Ind. Eng. Chem. Res.* 2005, *44*, 3106–3115. [Google Scholar] [CrossRef]

98. Godbole, S.; Gote, S.; Latkar, M.; Chakrabarti, T. Preparation and characterization of biodegradable poly-3-hydroxybutyrate-starch blend films. *Bioresour. Technol.* 2003, *86*, 33–37. [Google Scholar] [CrossRef]

99. Yu, L.; Dean, K.; Li, L. Polymer blends and composites from renewable resources. *Prog. Polym. Sci.* 2006, *31*, 576–602. [Google Scholar] [CrossRef]

100. Barud, H.S.; Souza, J.L.; Santos, D.B.; Crespi, M.S.; Ribeiro, C.A.; Messaddeq, Y.; Ribeiro, S.J.L. Bacterial cellulose/poly(3-hydroxybutyrate) composite membranes. *Carbohydr. Polym.* 2011, *83*, 1279–1284. [Google Scholar] [CrossRef]

101. Avella, M.; Bogoeva-Gaceva, G.; Buzõarovska, A.; Emanuela Errico, M.; Gentile, G.; Grozdanov, A. Poly (3-hydroxybutyrate-co-3-hydroxyvalerate)-based biocomposites reinforced with kenaf fibers. *J. Appl. Polym. Sci.* 2007,*104*, 3192–3200. [Google Scholar] [CrossRef]

102. Avella, M.; Martuscelli, E.; Raimo, M. Properties of blends and composites based on poly(3-hydroxy)butyrate (phb) and poly(3-hydroxybutyrate-hydroxyvalerate) (phbv) copolymers. *J. Mater. Sci.* 2000, *35*, 523–545. [Google Scholar] [CrossRef]

103. Wu, C.-S. Preparation and characterization of polyhydroxyalkanoate bioplastic-based green renewable composites from rice husk. *J. Polym. Environ.* 2014, *22*, 384–392. [Google Scholar] [CrossRef]

104. Wang, T.; Drzal, L.T. Cellulose-nanofiber-reinforced poly(lactic acid) composites prepared by a water-based approach.*ACS Appl. Mater. Interfaces* 2012, *4*, 5079–5085. [Google Scholar] [CrossRef] [PubMed]

105. Zarrinbakhsh, N.; Misra, M.; Mohanty, A.K. Biodegradable green composites from distiller's dried grains with solubles (DDGS) and a polyhydroxy(butyrate-co-valerate) (PHBV)-based bioplastic. *Macromol. Mater. Eng.* 2011, *296*, 1035–1045. [Google Scholar] [CrossRef]

106. Madbouly, S.A.; Schrader, J.A.; Srinivasan, G.; Liu, K.; McCabe, K.G.; Grewell, D.; Graves, W.R.; Kessler, M.R. Biodegradation behavior of bacterial-based polyhydroxyalkanoate (PHA) and DDGS composites. *Green Chem.* 2014,*16*, 1911–1920. [Google Scholar] [CrossRef]

107. Wei, L.; Liang, S.; McDonald, A.G. Thermophysical properties and biodegradation behavior of green composites made from polyhydroxybutyrate and potato peel waste fermentation residue. *Ind. Crop. Prod.* 2015, *69*, 91–103. [Google Scholar] [CrossRef]

108. Shanks, R.A.; Hodzic, A.; Wong, S. Thermoplastic biopolyester natural fiber composites. *J. Appl. Polym. Sci.* 2004, *91*, 2114–2121. [Google Scholar] [CrossRef]

109. Barkoula, N.M.; Garkhail, S.K.; Peijs, T. Biodegradable composites based on flax/polyhydroxybutyrate and its copolymer with hydroxyvalerate. *Ind. Crop. Prod.* 2010, *31*, 34–42. [Google Scholar] [CrossRef]

110. Srithep, Y.; Ellingham, T.; Peng, J.; Sabo, R.; Clemons, C.; Turng, L.-S.; Pilla, S. Melt compounding of poly (3-hydroxybutyrate-co-3-hydroxyvalerate)/nanofibrillated cellulose nanocomposites. *Polym. Degrad. Stab.* 2013, *98*, 1439–1449. [Google Scholar] [CrossRef]

111. Coats, E.R.; Loge, F.J.; Wolcott, M.P.; Englund, K.; McDonald, A.G. Production of natural fiber reinforced thermoplastic composites through the use of polyhydroxybutyrate-rich biomass. *Bioresour. Technol.* 2008, *99*, 2680–2686. [Google Scholar] [CrossRef] [PubMed]

112. Oksman, K.; Skrifvars, M.; Selin, J.F. Natural fibres as reinforcement in polylactic acid (PLA) composites. *Compos. Sci. Technol.* 2003, *63*, 1317–1324. [Google Scholar] [CrossRef]

113. Dong, Y.; Ghataura, A.; Takagi, H.; Haroosh, H.J.; Nakagaito, A.N.; Lau, K.-T. Polylactic acid (PLA) biocomposites reinforced with coir fibres: Evaluation of mechanical performance and multifunctional properties. *Compos. A Appl. Sci. Manuf.* 2014, *63*, 76–84. [Google Scholar] [CrossRef]

114. Masirek, R.; Kulinski, Z.; Chionna, D.; Piorkowska, E.; Pracella, M. Composites of poly (l-lactide) with hemp fibers: Morphology and thermal and mechanical properties. *J. Appl. Polym. Sci.* 2007, *105*, 255–268. [Google Scholar] [CrossRef]

115. Wong, S.; Shanks, R.; Hodzic, A. Interfacial improvements in poly(3-hydroxybutyrate)-flax fibre composites with hydrogen bonding additives. *Compos. Sci. Technol.* 2004, *64*, 1321–1330. [Google Scholar] [CrossRef]

116. Huda, M.S.; Drzal, L.T.; Misra, M.; Mohanty, A.K.; Williams, K.; Mielewski, D.F. A study on biocomposites from recycled newspaper fiber and poly(lactic acid). *Ind. Eng. Chem. Res.* 2005, *44*, 5593–5601. [Google Scholar] [CrossRef]

117. Sullivan, E.; Moon, R.; Kalaitzidou, K. Processing and characterization of cellulose nanocrystals/polylactic acid nanocomposite films. *Materials* 2015, *8*, 8106–8116. [Google Scholar] [CrossRef]

118. Li, Y.; Sun, X. Mechanical and thermal properties of biocomposites from poly(lactic acid) and ddgs. *J. Appl. Polym. Sci.*2011, *121*, 589–597. [Google Scholar] [CrossRef]

119. Tomé, L.C.; Pinto, R.J.B.; Trovatti, E.; Freire, C.S.R.; Silvestre, A.J.D.; Neto, C.P.; Gandini, A. Transparent bionanocomposites with improved properties prepared

from acetylated bacterial cellulose and poly(lactic acid) through a simple approach. *Green Chem.* 2011, *13*, 419–427. [Google Scholar] [CrossRef]

120. Botta, L.; Fiore, V.; Scalici, T.; Valenza, A.; Roberto, R. New polylactic acid composites reinforced with artichoke fibers.*Materials* 2015, *8*, 7770–7779. [Google Scholar] [CrossRef]
121. Acioli-Moura, R.; Sun, X.S. Thermal degradation and physical aging of poly(lactic acid) and its blends with starch.*Polym. Eng. Sci.* 2008, *48*, 829–836. [Google Scholar] [CrossRef]
122. Satyanarayana, K.G.; Arizaga, G.G.C.; Wypych, F. Biodegradable composites based on lignocellulosic fibers—An overview. *Prog. Polym. Sci.* 2009, *34*, 982–1021. [Google Scholar] [CrossRef]
123. Arrieta, M.; Fortunati, E.; Dominici, F.; Rayón, E.; López, J.; Kenny, J.M. PLA-PHB/cellulose based films: Mechanical, barrier and disintegration properties. *Polym. Degrad. Stab.* 2014, *107*, 139–149. [Google Scholar] [CrossRef]
124. Hejna, A.; Formela, K.; Saeb, M.R. Processing, mechanical and thermal behavior assessments of polycaprolactone/agricultural wastes biocomposites. *Ind. Crop. Prod.* 2015, *76*, 725–733. [Google Scholar] [CrossRef]
125. Wang, D.-W.; Xu, Y.-J.; Li, X.; Huang, C.-M.; Huang, K.-S.; Wang, C.-K.; Yeh, J.-T. Mechanical retention and waterproof properties of bacterial cellulose-reinforced thermoplastic starch biocomposites modified with sodium hexametaphosphate. *Materials* 2015, *8*, 3168–3194. [Google Scholar] [CrossRef]
126. Dufresne, A.; Kellerhals, M.B.; Witholt, B. Transcrystallization in MCL-PHAS/cellulose whiskers composites.*Macromolecules* 1999, *32*, 7396–7401. [Google Scholar] [CrossRef]
127. Kalia, S.; Dufresne, A.; Cherian, B.M.; Kaith, B.S.; Avérous, L.; Njuguna, J.; Nassiopoulos, E. Cellulose-based bio- and nanocomposites: A review. *Int. J. Polym. Sci.* 2011, *2011*, 1–35. [Google Scholar] [CrossRef]
128. Wei, L.; McDonald, A.G.; Freitag, C.; Morrell, J.J. Effects of wood fiber esterification on properties, weatherability and biodurability of wood plastic composites. *Polym. Degrad. Stab.* 2013, *98*, 1348–1361. [Google Scholar] [CrossRef]
129. Lu, J.Z.; Wu, Q.; McNabb, H.S. Chemical coupling in wood fiber and polymer composites: A review of coupling agents and treatments. *Wood Fiber Sci.* 2000, *32*, 88–104. [Google Scholar]
130. McDonald, A.G.; Ma, L. Plastic moldable pine fiber by benzylation. In *Wood: Types, Properties, and Uses (Environmental Science, Engineering and Technology)*; Botannini, L.F., Ed.; Nova Science Publishers, Inc.: Hauppauge, NY, USA, 2010; pp. 181–192. [Google Scholar]
131. Lu, J.Z.; Wu, Q.; Negulescu, I.I. Wood-fiber/high-density-polyethylene composites: Coupling agent performance. *J. Appl. Polym. Sci.* 2005, *96*, 93–102. [Google Scholar] [CrossRef]
132. Bledzki, A.K.; Gassan, J. Composites reinforced with cellulose based fibres. *Prog. Polym. Sci.* 1999, *24*, 221–274. [Google Scholar] [CrossRef]
133. Li, X.; Tabil, L.G.; Panigrahi, S. Chemical treatments of natural fiber for use in natural fiber-reinforced composites: A review. *J. Polym. Environ.* 2007, *15*, 25–33. [Google Scholar] [CrossRef]
134. Jerome, C.; Lecomte, P. Recent advances in the synthesis of aliphatic polyesters by ring-opening polymerization. *Adv. Drug Deliv. Rev.* 2008, *60*, 1056–1076. [Google Scholar] [CrossRef] [PubMed]
135. Hafrén, J.; Córdova, A. Direct organocatalytic polymerization from cellulose fibers. *Macromol. Rapid Commun.* 2005, *26*, 82–86. [Google Scholar] [CrossRef]
136. Lönnberg, H.; Zhou, Q.; Brumer, H.; Teeri, T.T.; Malmström, E.; Hult, A. Grafting of cellulose fibers with poly(e-caprolactone) and poly(l-lactic acid) via ring-opening polymerization. *Biomacromolecules* 2006, *7*, 2178–2185. [Google Scholar] [CrossRef] [PubMed]
137. Dufresne, A. Nanocellulose: A new ageless bionanomaterial. *Mater. Today* 2013, *16*, 220–227. [Google Scholar] [CrossRef]
138. Habibi, Y.; Goffin, A.-L.; Schiltz, N.; Duquesne, E.; Dubois, P.; Dufresne, A. Bionanocomposites based on poly(ε-caprolactone)-grafted cellulose nanocrystals by ring-opening polymerization. *J. Mater. Chem.* 2008, *18*, 5002–5010. [Google Scholar] [CrossRef]
139. Yuan, W.; Yuan, J.; Zhang, F.; Xie, X. Syntheses, characterization, and *in vitro* degradation of ethyl cellulose-graft-poly(ε-caprolactone)-block-poly(l-lactide) copolymers by sequential ring-opening polymerization. *Biomacromolecules*2007, *8*, 1101–1108. [Google Scholar] [CrossRef] [PubMed]
140. Goffin, A.-L.; Raquez, J.-M.; Duquesne, E.; Siqueira, G.; Habibi, Y.; Dufresne, A.; Dubois, P. From interfacial ring-opening polymerization to melt processing of cellulose nanowhisker-filled polylactide-based nanocomposites. *Biomacromolecules* 2011, *12*, 2456–2465. [Google Scholar] [CrossRef] [PubMed]
141. Boujemaoui, A.; Carlsson, L.; Malmström, E.; Lahcini, M.; Berglund, L.; Sehaqui, H.; Carlmark, A. Facile preparation route for nanostructured composites: Surface-initiated ring-opening polymerization of ε-caprolactone from high-surface-area nanopaper. *ACS Appl. Mater. Interfaces* 2012, *4*, 3191–3198. [Google Scholar] [CrossRef] [PubMed]
142. Thakur, V.K.; Thakur, M.K.; Gupta, R.K. Rapid synthesis of graft copolymers from natural cellulose fibers. *Carbohydr. Polym.* 2013, *98*, 820–828. [Google Scholar] [CrossRef] [PubMed]
143. Thakur, V.K.; Thakur, M.K.; Gupta, R.K. Graft copolymers from cellulose: Synthesis, characterization and evaluation.*Carbohydr. Polym.* 2013, *97*, 18–25. [Google Scholar] [CrossRef] [PubMed]
144. Wang, J.; Yao, K.; Korich, A.L.; Li, S.; Ma, S.; Ploehn, H.J.; Iovine, P.M.; Wang, C.; Chu, F.; Tang, C. Combining renewable gum rosin and lignin: Towards hydrophobic polymer composites by controlled polymerization. *J. Polym. Sci. A Polym. Chem.* 2011, *49*, 3728–3738.

[Google Scholar] [CrossRef]

145. Majoinen, J.; Walther, A.; McKee, J.R.; Kontturi, E.; Aseyev, V.; Malho, J.M.; Ruokolainen, J.; Ikkala, O. Polyelectrolyte brushes grafted from cellulose nanocrystals using cu-mediated surface-initiated controlled radical polymerization. *Biomacromolecules* 2011, *12*, 2997–3006. [Google Scholar] [CrossRef] [PubMed]
146. Zoppe, J.O.; Habibi, Y.; Rojas, O.J.; Venditti, R.A.; Johansson, L.-S.; Efimenko, K.; Osterberg, M.; Laine, J. Poly(n-isopropylacrylamide) brushes grafted from cellulose nanocrystals via surface-initiated single-electron transfer living radical polymerization. *Biomacromolecules* 2010, *11*, 2683–2691. [Google Scholar] [CrossRef] [PubMed]
147. Parambath Kanoth, B.; Claudino, M.; Johansson, M.; Berglund, L.A.; Zhou, Q. Biocomposites from natural rubber: Synergistic effects of functionalized cellulose nanocrystals as both reinforcing and cross-linking agents via free-radical thiol–ene chemistry. *ACS Appl. Mater. Interfaces* 2015, *7*, 16303–16310. [Google Scholar] [CrossRef] [PubMed]
148. Rowell, R.M.; Young, R.A.; Rowell, J. *Paper and Composites from Agro-Based Resources*; CRC Press: Boca Raton, FL, USA, 1996. [Google Scholar]
149. El-Naggar, A.M.; El-Hosamy, M.B.; Zahran, A.H.; Zondy, M.H. Surface morphology/mechanical/dyeability properties of radiation-grafted sisal fibres. *Am. Dyestuff Rep.* 1992, *81*, 40–44. [Google Scholar]
150. Gupta, V.K.; Pathania, D.; Agarwal, S.; Sharma, S. Amputation of congo red dye from waste water using microwave induced grafted luffa cylindrica cellulosic fiber. *Carbohydr. Polym.* 2014, *111*, 556–566. [Google Scholar] [CrossRef] [PubMed]
151. Gupta, V.K.; Pathania, D.; Priya, B.; Singha, A.S.; Sharma, G. Microwave induced synthesis of graft copolymer of binary vinyl monomer mixtures onto delignified grewia optiva fiber: Application in dye removal. *Front. Chem.* 2014, *2*. [Google Scholar] [CrossRef] [PubMed]
152. Wei, L.; Stark, N.M.; McDonald, A.G. Interfacial improvements in biocomposites based on poly(3-hydroxybutyrate) and poly(3-hydroxybutyrate-co-3-hydroxyvalerate) bioplastics reinforced and grafted with α-cellulose fibers. *Green Chem.* 2015, *17*, 4800–4814. [Google Scholar] [CrossRef]
153. Wei, L.; McDonald, A.G.; Stark, N.M. Grafting of bacterial polyhydroxybutyrate (PHB) onto cellulose via *in situ*reactive extrusion with dicumyl peroxide. *Biomacromolecules* 2015, *16*, 1040–1049. [Google Scholar] [CrossRef] [PubMed]
154. Dhar, P.; Tarafder, D.; Kumar, A.; Katiyar, V. Thermally recyclable polylactic acid/cellulose nanocrystal films through reactive extrusion process. *Polymer* 2016, *87*, 268–282. [Google Scholar] [CrossRef]
155. Hu, L.; Stevanovic, T.; Rodrigue, D. Unmodified and esterified kraft lignin-filled polyethylene composites: Compatibilization by free-radical grafting. *J. Appl. Polym. Sci.* 2015, *132*. [Google Scholar] [CrossRef]
156. Csikós, Á.; Faludi, G.; Domján, A.; Renner, K.; Móczó, J.; Pukánszky, B. Modification of interfacial adhesion with a functionalized polymer in pla/wood composites. *Eur. Polym. J.* 2015, *68*, 592–600. [Google Scholar] [CrossRef]
157. Siqueira, G.; Bras, J.; Dufresne, A. Cellulose whiskers *versus* microfibrils: Influence of the nature of the nanoparticle and its surface functionalization on the thermal and mechanical properties of nanocomposites. *Biomacromolecules* 2009,*10*, 425–432. [Google Scholar] [CrossRef] [PubMed]
158. Magniez, K.; Voda, A.S.; Kafi, A.A.; Fichini, A.; Guo, Q.; Fox, B.L. Overcoming interfacial affinity issues in natural fiber reinforced polylactide biocomposites by surface adsorption of amphiphilic block copolymers. *ACS Appl. Mater. Interfaces* 2013, *5*, 276–283. [Google Scholar] [CrossRef] [PubMed]
159. Samain, X.; Langlois, V.; Renard, E.; Lorang, G. Grafting biodegradable polyesters onto cellulose. *J. Appl. Polym. Sci.*2011, *121*, 1183–1192. [Google Scholar] [CrossRef]
160. Yu, H.-Y.; Qin, Z.-Y.; Wang, L.-F.; Zhou, Z. Crystallization behavior and hydrophobic properties of biodegradable ethyl cellulose-g-poly(3-hydroxybutyrate-co-3-hydroxyvalerate): The influence of the side-chain length and grafting density. *Carbohydr. Polym.* 2012, *87*, 2447–2454. [Google Scholar] [CrossRef]
161. Essabir, H.; Bensalah, M.O.; Rodrigue, D.; Bouhfid, R.; Qaiss, A.E.K. Biocomposites based on argan nut shell and a polymer matrix: Effect of filler content and coupling agent. *Carbohydr. Polym.* 2016, *143*, 70–83.

CHAPTER 2

RECENT PROGRESS AND PERSPECTIVES IN THE ELECTROKINETIC CHARACTERIZATION OF POLYELECTROLYTE FILMS

Ralf Zimmermann[1], Carsten Werner[1,2] and Jérôme F. L. Duval[3]

[1]Leibniz Institute of Polymer Research Dresden, Max Bergmann Center of Biomaterials Dresden, Hohe Strasse 6, 01069 Dresden, Germany

[2]Technische Universität Dresden, Center for Regenerative Therapies Dresden, Tatzberg 47, 01307 Dresden, Germany

[3]Laboratoire Interdisciplinaire des Environnements Continentaux (LIEC), CNRS UMR 7360, 15 avenue du Charmois, B.P. 40, F-54501 Vandoeuvre-lès-Nancy cedex, France

ABSTRACT

The analysis of the charge, structure and molecular interactions of/within polymeric substrates defines an important analytical challenge in materials science. Accordingly, advanced electrokinetic methods and theories have been developed to investigate the charging mechanisms and structure of soft material coatings. In particular, there has been significant progress in the quantitative interpretation of streaming current and surface conductivity data of polymeric films from the application of recent theories developed for the electrohydrodynamics of diffuse soft planar interfaces. Here, we review the theory and experimental strategies to analyze the interrelations of the charge and structure of polyelectrolyte layers supported by planar carriers under electrokinetic conditions. To illustrate the options arising from these developments, we discuss experimental and simulation data for plasma-immobilized poly(acrylic acid) films and for a polyelectrolyte bilayer consisting of poly(ethylene imine) and poly(acrylic acid). Finally, we briefly outline potential future developments in the field of the electrokinetics of polyelectrolyte layers.

INTRODUCTION

The immobilization of polyelectrolytes is a versatile method for tailoring the surface properties of bulk materials. The spectrum of coatings ranges from non-ordered, adsorptively-bound polyelectrolytes over tethered polyelectrolyte brushes to polyelectrolyte multilayers formed by the alternate physisorption of polycationic and polyanionic polymers onto planar or colloidal supports [1,2]. Applications of polyelectrolyte-coated surfaces include, e.g., antimicrobial materials [3], stimuli-responsive surfaces [4], liquid displays [5], drug delivery [6], biosensing [7] and tissue-engineering scaffolds [8]. Capsules fabricated from the layer-by-layer adsorption of polyelectrolytes onto a particulate template prior to subsequent removal of the core are used as multifunctional carrier systems, e.g., for drug delivery [9]. Optimizing the formation, stability and performance of polyelectrolyte coatings necessarily includes the analysis of the interrelations between their charge, structure and interactions in a broad range of environmental conditions (pH, salt concentration, nature of ions present in the solution). Electrokinetic methods are well-established and valuable techniques to achieve such a level of understanding [10,11]. The interpretation of electrokinetic data requires theoretical formalisms that go beyond the classical concept of zeta potential [12], and instead, it necessitates the account of hydrodynamics within the soft polymer coatings under lateral flow conditions. Theories for electric double

Citation: Zimmermann, R.; Werner, C.; Duval, J.F.L. Recent Progress and Perspectives in the Electrokinetic Characterization of Polyelectrolyte Films. Polymers 2016, 8, 7. doi:10.3390/polym8010007

layer (EDL) and electrokinetics at soft (i.e., ion- and water-permeable) surfaces were mainly developed in the last three decades [13,14,15,16,17,18,19,20,21,22,23,24,25,26,27,28]. Advanced theories are now available for the adequate physical interpretation of electrokinetic data collected at soft polymer/solution interfaces [21,22,23,24,25,26,27,28]. Applying such theories, it becomes possible to unambiguously unravel the electrostatic, hydrodynamic and structural properties of, e.g., micrometer-thick hydrogels [21,22,23], ionic-strength/pH-responsive thin polymer coatings [25], thermo-responsive thin films [24] or surface-supported brushes [28] carrying (or not) ionizable groups.

In this article, we review the fundamentals for the characterization of planar polyelectrolyte coatings by streaming current and surface conductivity measurements. We provide an overview of existing theory and of state-of-the-art experimental techniques used to analyze the charge and structure of polyelectrolyte systems. The impacts of the polymer charge, film permeability and segment density distribution on the measured streaming current are further illustrated and discussed on the basis of computational examples. To demonstrate the options arising from the recent theoretical developments, we discuss strategies and results reported for the investigation of plasma-immobilized poly(acrylic acid) films and for a polyelectrolyte bilayer consisting of poly(ethylene imine) and poly(acrylic acid). Finally, we briefly invoke a few potential future developments in the field.

FUNDAMENTAL PRINCIPLES AND THEORY

Streaming Current/Potential in Rectangular Microchannels

The ionization of functional groups distributed in adsorbed or covalently immobilized polyelectrolytes leads to the development of an interfacial excess charge. This charge is compensated by counter ions located in the adjacent electrolyte medium, and an electrical double layer is formed. Under the action of an applied lateral hydrodynamic flow field, the counter ions are displaced in the flow direction due to friction forces. In capillary systems, the convective ion transport originates a streaming current, Istr, that can be measured with an ampere meter via two non-polarizable (e.g., Ag/AgCl) electrodes positioned at the inlet and outlet side of the measuring cell (Figure 1). At high ohmic resistance of the external circuit, an electrical potential, the streaming potential, Ustr, may be measured, provided that an electrometer is connected to the two electrodes. The magnitude of the streaming potential is governed by the steady-state balance between the streaming current and a back conduction current determined by the specific electrical conductivity of the electrolyte embedded in the capillary system and by the surface conductivity that arises from mobile counter ions at the material surface [11].

For the analysis of the interfacial charging and structure of soft polymer films, streaming current and streaming potential are performed under conditions of laminar flow across rectangular streaming channels that are formed by two identically-coated sample carriers of length L_0 and width (Figure 1) [11]. To minimize the effects of side walls on the development of the hydrodynamic flow field within the channel, the separation distance between the two surfaces, H, is much smaller than the length and width of the surfaces ($H \ll L_0$ and $H \ll \ell$). Typical dimensions of the streaming channel are $L_0 = 20$ mm, $\ell = 10$ mm and $H = 30$ µm [29,30]. To eliminate the contributions of the electrodes to streaming current/potential, measurements should be performed as a function of the applied pressure gradient along the surfaces of interest. Proceeding so, the contributions of the electrodes (due to asymmetry) can be eliminated upon evaluation of the ratio streaming current/potential over the applied pressure gradient. It is stressed that measured electrokinetic response varies linearly with the applied pressure, except for situations where the investigated interface is electroactive and (reversible) redox active species are present in the neighboring electrolyte solution [31,32].

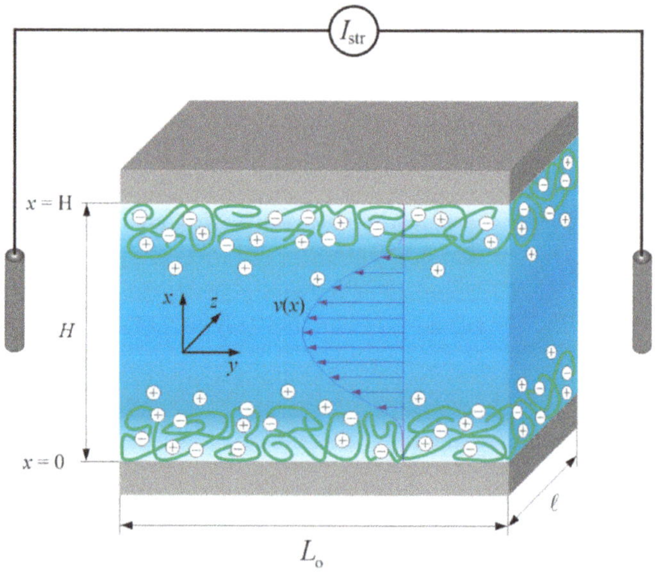

Figure 1: Schematic representation of a rectangular streaming channel for the characterization of soft polymer films supported by hard planar substrates from streaming current/potential measurements. The hydrodynamic flow field, v(x), developing in the channel under a lateral pressure gradient is indicated. The dimensions of the channel and polymer film are not shown to scale. For the sake of simplicity, charges located at the carrier surface are not represented. The scheme further details the nomenclature used in the theory reviewed in Section 2. Typical dimensions of the cell are: $L_0 = 20$ mm, $\ell = 10$ mm and $H = 5$ µm to 60 µm. Reprinted from [11].

Below, we review the fundamentals for the analysis of the streaming current and surface conductivity of soft (non-conducting) polymer films [23,24,25,26,27,28]. The theory was developed for steady-state electrostatic and hydrodynamic flow fields, and it is applicable for polymer films of sufficiently high water content [23,24,25,26,27,28]. The typical separation distance adopted between the two sample carriers excludes any overlap of the electric double layers formed between the soft films and the electrolyte solution embedded in the channel. Numerical and analytical procedures for solving the set of coupled equations determining the electrostatic potential and the hydrodynamic flow field distributions across the channel are described elsewhere [24].

Electrostatic Potential

The electrical potential across the substrate/polyelectrolyte (PE) layer/electrolyte solution interface, $\psi(x)$, is a function of the local charge density at that position x with x referring to the dimension perpendicular to the interface (Figure 1). The charges of the PE-supporting substrate and those stemming from the ionizable groups carried by the PE layer are compensated by mobile counter ions distributed inside and outside the permeable layer structure. The electrolyte is considered to comprise N ions of valence z_i with bulk concentrations c_i, with the index i running from i = 1,..., N. The position-dependent electrostatic potential across a PE layer carrying M types of ionizable groups can be evaluated from solution of the non-linear Poisson–Boltzmann equation [26]:

$$\frac{d^2y(X)}{dX^2} = -\frac{(\kappa H)^2}{\sum_{i=1}^{N} c_i z_i^2}\left\{\sum_{i=1}^{N} z_i c_i e^{-z_i y(X)} + \sum_{j=1}^{M} \frac{\rho_j/F}{1+10^{-\varepsilon_j(pK_j-pH)}e^{\varepsilon_j y(X)}} f(X)\right\}, \quad (1)$$

where y is the dimensionless electrostatic potential ($y=F\psi/RT$), κ the reciprocal Debye length, F the Faraday constant, Rthe gas constant and T the temperature, and X corresponds to the dimensionless space variable x/H. The ionization of the j-th type of functional groups (j = 1,..., M) is defined here by their dissociation pK_j value, and for the sake of conciseness, we consider that charged groups carry a unique charge ±e. The parameter εj involved in Equation (1) is −1 for the ionization of functional groups according to deprotonation reactions (i.e., of the type −RHj⇄−Rj−+H+), and εj is +1 for the ionization of groups due to protonation reactions (i.e., −RHj+⇄−Rj+H+) [26]. The function f(x) in Equation (1) represents the polymer segment density distribution across the soft interfacial layer. Analytical expressions or numerical values for f(x) may be derived from polymer theory [28] or determined experimentally [33,34,35]. Measurements performed with spatially-resolved techniques, such as ellipsometry or neutron/X-ray reflectivity [33,34,35], showed that the polymer segment density profile can be appropriately represented by the function:

$$f(x) = \frac{\omega}{2}\left(1 - \tanh\left(\frac{x-d}{\alpha}\right)\right), \quad (2)$$

where α is the characteristic decay length of the segment density across the soft layer [24] and d is the nominal surface layer thickness reached in the limit α→0. The scalar quantity ω ensures that the total amount of polymer segments within the soft film remains constant when the film structure/thickness vary as a result of shrinking/swelling processes driven by changes in solution pH, salt concentration or temperature [24] and accompanied by modulations of α. In the limiting case of a homogeneous segment distribution, i.e., α/d→0, Equation (2) corresponds to a step function-like profile [24], and the film protrudes from the supporting solid substrate up to a distance d. Under such conditions, the charge source term, |ρj=1,...,M|/F, in Equation (1) corresponds to the total concentration of ionizable groups of type j within the layer. The boundary condition at X=0 associated with the non-linear differential Equation (1) can be expressed in terms of the surface potential (or zeta potential ζ) or surface charge density of the hard (impermeable) carrier supporting the film [25]. If the charge of the substrate does not significantly contribute to the streaming current and surface conductivity, zero electric field condition may be imposed at X=0 [25]. For practically relevant systems where the separation distance between the channel walls is much larger than the extension of the PE layers and the electric Debye layer thickness, the overlap of the electrical double layers is excluded. As a result of electroneutrality far from the interface, the second boundary condition required for solving Equation (1) is simply y (X → 1/2) = 0.

Hydrodynamic Flow Field

The hydrodynamic flow profile v(x) within the thin electrokinetic cell can be evaluated on the basis of the Brinkman equation [36]. For rectangular channels with a separation distance H much smaller than the length L_o and width ℓ of the channel, the flow field is oriented parallel to the sample surfaces and symmetric with respect to the y-z-plane at x=H/2. Provided that sample carriers are coated with identical polymer films, it is therefore sufficient to consider the hydrodynamic problem from X=0 to X=1/2. Under these conditions the Brinkman equation may be written in the form [25]:

$$\frac{d^2V(X)}{dX^2} - (\lambda_o H)^2 f(X)\, V(X) = -1, \quad (3)$$

where the velocity field v is normalized according to $V(X)=v(X)/vo$ with $vo=\Delta PH^2/(\eta Lo)$, where η is the dynamic viscosity of the electrolyte. In Equation (3), λ_o corresponds to the hydrodynamic softness of the polymer film. Its reciprocal value, 1/λo, is the Brinkman length that reflects the extension of the hydrodynamic flow field into the soft film in case of homogeneous segment density distribution, i.e. α/d→0 and ω→1. Equation (3) is applicable for the analysis of soft polymer films with sufficiently high water content, a situation frequently met in practice. The friction exerted by the polymer segments located at the position x on the fluid flow is then determined by the term $(\lambda oH)^2 f(X)$. For systems with higher segment density, higher order density-dependent friction terms have to be taken into account [28]. The boundaries necessary to solve the Brinkman Equation (3) are given by the no-slip condition at the supporting hard carrier surface (X = 0) and by the symmetry of the flow field at the position X = 1/2 [25].

Streaming Current and Streaming Potential

The streaming current I_{str} can be calculated from the spatial integral of the product between the charge density distribution of counter ions (first term on the right-hand side of Equation (1)) and the hydrodynamic flow field, V(X) (Equation (3)), according to [24,25]:

$$\frac{I_{str}}{\Delta P} = \frac{2\ell F H^3}{\eta L_o}\int_0^{1/2}\sum_{i=1}^{N} z_i c_i e^{-z_i y(X)} V(X)\, dX. \quad (4)$$

Analytical expressions for $I_{str}/\Delta P$ may be derived under conditions in line with the application of the Debye–Hückel approximation (valid for sufficiently low potentials y) and for films with a homogeneous segment density distribution [24]. For hard surfaces, the solution of Equation (4) leads to the famous Smoluchowski equation that relates the streaming current to the electrokinetic or zeta potential [11]. Equation (4) can be converted into an expression for the streaming potential Ustrvia Ohm's law, taking into account the terms for the specific electrical conductivity of the electrolyte and for the surface conductivity [11].

Surface Conductivity

The accumulation of mobile counter ions inside and outside a charged soft surface layer leads to elevated electrical conductivity in the substrate/PE/electrolyte interfacial region that is designated as (excess) surface conductivity, K^{σ}. This excess surface conductivity originates from the (lateral) migrative transport of ions located in the interfacial electric double layer due to the tangential field U_{str}/L (term K_m^{σ}):

$$K_{\mathrm{m}}^{\sigma} = \frac{HF^2}{RT}\int_0^{1/2}\sum_{i=1}^{N}|z_i|\,D_i c_i \beta_i(X)\left[e^{-z_i y(X)}-1\right]dX, \tag{5}$$

and from the electro-osmotic charge transport due to the action of the field Ustr/Lo on mobile ions within and outside the soft surface layer (term E_{eo}^{σ}):

$$K_{\mathrm{eo}}^{\sigma} = \frac{H\varepsilon_0\varepsilon_r RT(\kappa H)^2}{\eta}\int_0^{1/2} V_{\mathrm{eo}}(X)\sum_{i=1}^{N} z_i c_i e^{-z_i y(X)}dX, \tag{6}$$

where D_i is the diffusion coefficient of ion i in the bulk electrolyte solution, $\beta_i(X)$ the ratio between the mobilities of ion i at position X and in the bulk electrolyte solution, and $\varepsilon_0\varepsilon_r$ is the dielectric permittivity of the medium [25]. For soft films with sufficiently low polymer density it is legitimate to assume $\beta_i(X)\sim 1$. The electro-osmotic flow field, $V_{eo}(X)$, can be calculated on the basis of the Brinkman equation (3) after substituting the right hand side by the relevant charge source term and V(X) by $V_{eo}(X)$ [25,26]. The overall surface conductivity, K^{σ}, is then given by the sum $K_{\mathrm{m}}^{\sigma}+K_{\mathrm{eo}}^{\sigma}$.

In case of rectangular streaming channels (Figure 1), the surface conductivity can be determined from streaming current and streaming potential measurements at a single or several channel heights according to the following equation:

$$\frac{I_{\mathrm{str}}/\Delta P}{U_{\mathrm{str}}/\Delta P}\frac{L_{\mathrm{o}}}{2\ell} = \frac{H}{2}K^{\mathrm{B}}+^{\sigma}, \tag{7}$$

where K^B is the specific electrical conductivity of the bulk electrolyte [29]. The ratio of the pressure-normalized streaming current ($I_{str}/\Delta P$) and streaming potential ($U_{str}/\Delta P$) represents the channel conductance. This quantity can be alternatively determined under conditions of zero flow from the measurement of the electrical current when applying externally an electrical potential difference along the microchannel [23]. If the channel conductance is evaluated at several channel heights, K^{σ} then simply corresponds to the intercept in the (linear) plot channel conductance vs. channel height [29] and KBdirectly follows from the slope. The comparison between so-determined K^B and values estimated from the concentrations and known molar conductivities of the ions present in solution can be helpful for verification purposes.

Surface Conductivity and Donnan Potential

Polyelectrolyte layers can be treated as a 3D meshwork where ionizable groups are immobilized. The ionization of these groups is related to an enhanced concentration of (counter) ions within the layer, which in turn causes an electrical potential difference between the PE layer phase and the electrolyte. For sufficiently thick PE layers, i.e., whose thickness is much larger than the screening Debye length 1/κ, with a homogeneous segment distribution (i.e., α/d→0), the potential difference can be viewed as the analogue of the Donnan potential difference across a semipermeable membrane [37]. Dukhinet al. provided analytical expressions that relate a given concentration of ionizable groups in the PE layer (denoted here as |ρ0|/F) to the dimensionless Donnan potential yD, across the PE layer/electrolyte solution interface [38]. For PE layers with anionic groups (negative yD), the corresponding equations can be written in the form:

$$e^{-y_{\mathrm{D}}} = \sqrt{\frac{|\rho_0|}{cF}10^{pH-pK}+1} \quad \text{for} \quad |y_{\mathrm{D}}|<2 \tag{8}$$

$$e^{-y_{\mathrm{D}}} = \frac{10^{pH-pK}}{2}\left(\sqrt{1+4\frac{|\rho_0|}{cF}10^{pK-pH}}-1\right) \quad \text{for} \quad |y_{\mathrm{D}}|>2 \tag{9}$$

where c is the total ion concentration in the bulk electrolyte (ions of the neutral background electrolyte and ions resulting from the variation of the solution pH have to be taken into account) [38]. Under the conditions specified above, the concentration of ionizable groups, $|\rho_0|/F$, can be quantified by measuring the surface conductivity and layer thickness under conditions of complete ionization. Equations (8) and (9) then allow for the evaluation of pH and salt concentration effects on the magnitude of the Donnan potential.

ELECTROKINETIC ANALYSIS OF POLYELECTROLYTE LAYERS

Illustrative Computational Examples

In this section, we briefly discuss the basic processes governing the dependence of the pressure-normalized streaming current, $I_{str}/\Delta P$, on electrolyte concentration, flow penetration length scale $1/\lambda_o$ (Figure 2A,B), charge density $|\rho_0|$ in (anionic) polyelectrolyte film (Figure 2C) and segment density distribution therein (Figure 2D). For the sake of demonstration, we consider a monovalent electrolyte of bulk concentration c and films with a given (negative) density of charges (i.e., no variation of the film charge with the pH).

In order to first highlight the hydrodynamic contribution of the polyelectrolyte film (thickness d) to $I_{str}/\Delta P$, Figure 2A refers to the situation of an uncharged film ($|\rho_0|=0$) supported by a hard carrier with (negative) surface potential, the latter identifying with the carrier zeta-potential ζ that can be measured in the absence of the film [24]. The ratio $I_{str}/\Delta P$ for this interfacial system is reported as a function of c expressed here in terms of the dimensionless quantity κd that compares the Debye layer thickness $\kappa^{-1} = \left[2F^2c/(\varepsilon_o\varepsilon_r RT)\right]^{-1/2}$ with the film thickness d. At fixed κd, Figure 2A illustrates that $I_{str}/\Delta P$increases in magnitude with increasing $1/\lambda_o$. Upon increase of the penetration of the pressure-driven flow in the film, which is related to a decrease of the film hydrodynamic drag, the counter ions located in the vicinity of the charged supporting substrate become accessible to the flow and $I_{str}/\Delta P$ increases accordingly. The amount of electrokinetically active ions thus increases with increasing $1/\lambda_o$, and so does $I_{str}/\Delta P$. As expected, at fixed $1/\lambda_o$ $|I_{str}/\Delta P|$ decreases with increasing κdbecause of the screening of the carrier surface charge by ions present in the medium. Doing so, $I_{str}/\Delta P$reaches at sufficiently large κd an asymptotic plateau, the value of which is determined by $1/\lambda_o$. In this limit where κd>>1, the electric double layer significantly recedes within the film coating and $I_{str}/\Delta P$ becomes essentially determined by details of flow velocity and electrostatic potential distributions in the direct neighborhood of the carrier surface. Under such conditions, it can be shown that the streaming current reduces to [24]:

$$I_{str} = \frac{\beta\tilde{\zeta}}{\left[2(\kappa H)^2\right]}\left\{\frac{1}{\cosh(\lambda_o d)} + \frac{2}{\lambda_o H}\tanh(\lambda_o d)\right\}, \tag{10}$$

Where $\beta = 4\ell Fc\Delta PH^3/(\eta L_o)$ and $\tilde{\zeta} = \zeta F/RT$ is the dimensionless zeta-potential of the substrate. In the extreme 1/λo→∞, the situation is that of a free-draining film fully permeable to flow. From a hydrodynamic point of view, the (uncharged) film is thus immaterial and its presence does not affect the electrokinetic signature of the rigid surface. The streaming current then simplifies into the well-known Smoluchowski formulation $I_{str} = \beta\tilde{\zeta}/\left[2(\kappa H)^2\right]$. In the other limit where $1/\lambda_o \to 0$, the film is stricto sensu impermeable. The streaming current can then be written as a function of the electrostatic potential at the outer film surface (x=d in Figure 1) via the modified Smoluchowski equation $I_{str} = \beta\tilde{\zeta}e^{-\kappa d}/\left[2(\kappa H)^2\right]$ that holds at sufficiently low $\tilde{\zeta}$ where linearization of the Poisson-Boltzmann equation is legitimate. Figure 2A makes it clear that the screening of the carrier surface charge by the film is defined by the interplay between $1/\lambda_o$ and d (hydrodynamic screening), and by the respective magnitudes of 1/κ and d (electrostatic screening).

Another way to formulate these two types of processes is to define the so-called electrokinetic thickness of the film coating, denoted as δ_E and expressed via the following relationship

$$\tanh(F\bar{\zeta}/4RT) = \tanh(F\zeta/4RT)\exp(-\kappa\delta_E), \tag{11}$$

The effective electrokinetic potential $\tilde{\zeta}$ involved in Equation (11) is obtained from the Smoluchowski-based conversion of the streaming current rigorously evaluated in the presence of the film layer, i.e. $\bar{\zeta} = 2RT(\kappa H)^2 I_{str}/(F\beta)$. The presence of the uncharged film leads to a partial or complete suppression of the flow in the vicinity of the charged rigid surface so that ||ζ¯||<|ζ|. In this picture, ζ¯ is thus basically determined from the outer shift of a virtual "plane of shear" for which the decrease in zeta potential is a measure, and this shift is identified to δE. Examination of measured electrokinetic data on the basis of Equation (11) only, should be considered as phenomenological rather than based on physical arguments. Instead, the formulation given by Equation (11) necessarily implies that δ_E reflects the details of the electrostatic and hydrodynamic flow field profiles within the soft surface layer. Figure 2B collects the dependence of δ_E on κd under the conditions of Figure 2A. In line with the preceding arguments, at fixed 1/λo δE decreases with increasing κd because the electrokinetically active ions (i.e. probed by the tangential flow) becomes gradually confined to the close vicinity of the carrier surface, and the more so the larger $1/\lambda_o$ is. It is observed that at large κd, δ_E decreases linearly with κ (at fixed d) in double logarithmic representation. This trend is confirmed by the analytical expression that may be developed for δ_E at κd>>1 and low ζ¯[24]:

$$\delta_E = -\frac{1}{\kappa}\ln\left\{\frac{1}{\cosh(\lambda_o d)} + \frac{2}{\lambda_o H}\tanh(\lambda_o d)\right\}, \tag{12}$$

In the limit κd<<1 where film thickness is small compared to the Debye layer extension from the rigid carrier surface, δ_E reaches a plateau whose value is solely determined by the hydrodynamic film permeability (Figure 2B). In this limit, δ_E identifies with the so-called hydrodynamic thickness δ_H of the film that further reduces to the physical film thickness d for situations where film resistance to flow is infinite (i.e. $1/\lambda_o \to 0$).

Figure 2C shows the evolution of $I_{str}/\Delta P$ on κd for cases where both supporting carrier and polyelectrolyte film are charged. Data are reported for various values of 1/λo with $|\rho_0|/F$=0.1 mM (poorly charged film, black lines) and |ρ0|/F=10mM (red lines). At fixed |ρ0|/F, the dependence of $I_{str}/\Delta P$ on κd and $1/\lambda_o$ is qualitatively similar to that discussed in Figure 2A. Quantitatively, the introduction of structural charges in the polyelectrolyte films goes in pair with an enhancement of the local electrostatic potential under given salt concentration conditions, and thus with an increase in the amount of counterions accumulated therein. As a result, the streaming current increases in magnitude and the larger 1/λois, the most pronounced is this increase in |Istr/ΔP|. The dependence of Istr/ΔP on κd becomes weaker as the density of charges carried by the film increases. This is due to a reduction of the electrostatic screening of the carrier surface charge as the electric double layer composition becomes increasingly governed by the electrostatic properties of the polyelectrolyte layer. Similarly to the situation analyzed in Figure 2A, $I_{str}/\Delta P$ reaches an asymptotic plateau at κd>>1 and the value of this plateau is not only a function of ζ˜ and $1/\lambda_o$ (see Equation (10)) but also involves ρ_0 as detailed in [24]. Finally, the limits $1/\lambda_o \to 0$ and $1/\lambda_o \to \infty$ corresponding to impermeable and free-draining films, respectively, can be rationalized via a Smoluchowski-like equation of the form $I_{str} = \beta\tilde{\psi}_o/\left[2(\kappa H)^2\right]$ where the dimensionless potential $\tilde{\psi}_o = F\psi_o/RT$ is evaluated at x=d (case $1/\lambda_o \to 0$) or x=0 (case $1/\lambda_o \to \infty$). In both limits, it is stressed that ψ˜o is a function of both the surface carrier potential ζ and the film charge densityρ0. As an illustration, in the limit of low potentials, $\tilde{\psi}_o$is defined by

$\tilde{\psi}_o = \left[\tilde{\zeta} + \frac{\rho_0}{2cF}(\cosh(\kappa d) - 1)\right]e^{-\kappa d}$ at $1/\lambda_o \to 0$ and $\tilde{\psi}_o = \tilde{\zeta} + F\rho_0 d^2/(2\varepsilon_0\varepsilon_r RT)$ at $1/\lambda_o \to \infty$,expressions that are derived from the relevant analytical formulations reported in [24]. In addition the former expression further reduces to $\tilde{\psi}_o = y_D/2$ for $\kappa d >> 1$ where yD/2 corresponds to the (dimensionless) potential at the film surface (x=d) with yD=ρ_0/2cF the dimensionless Donnan potential in the Debye-Hückel approximation.

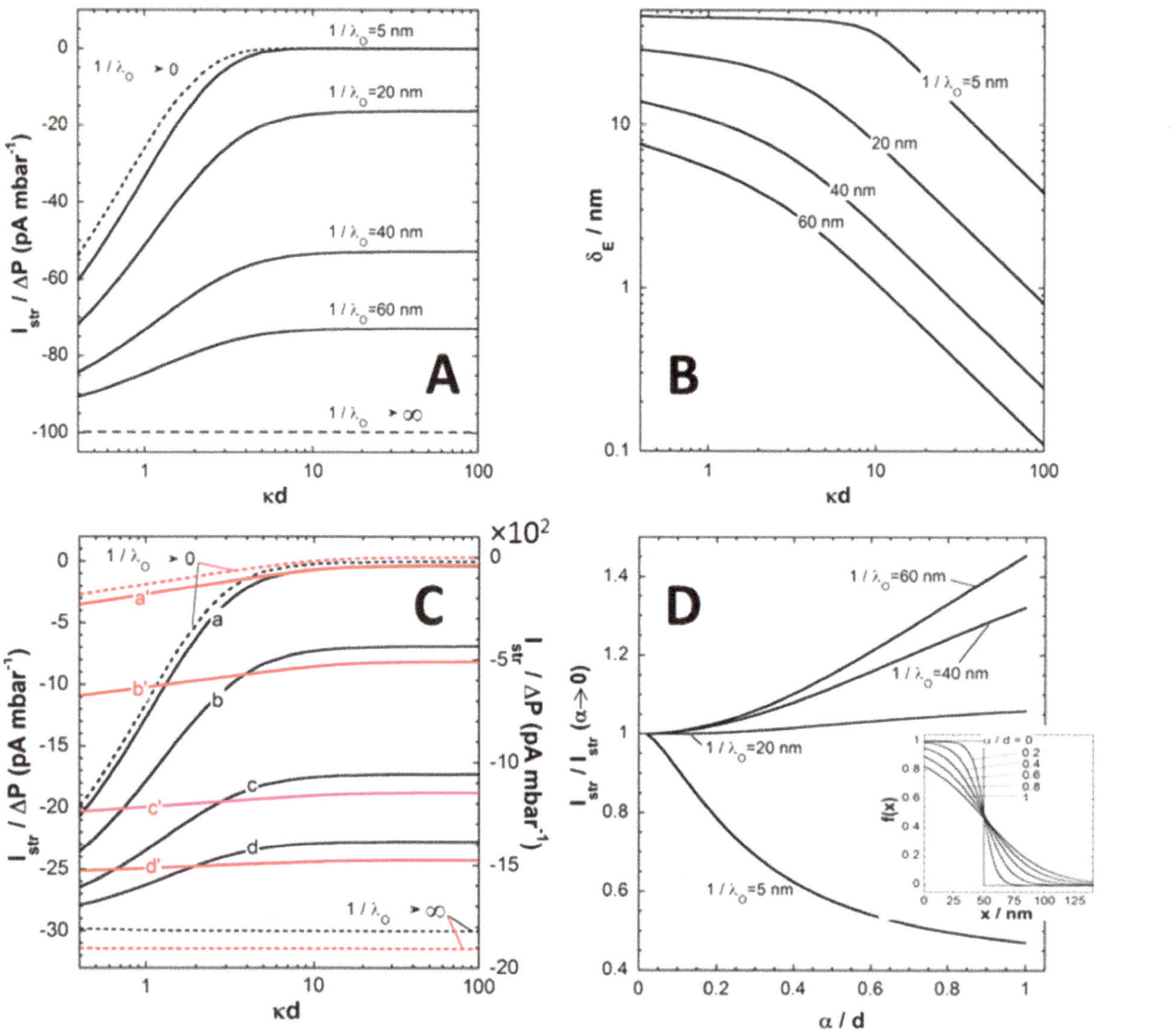

Figure 2: (**A**) Ratio between streaming current (I_{str}) and applied pressure (ΔP) versusionic strength evaluated from theory for different values of the hydrodynamic penetration length $1/\lambda_O$ (indicated). (**B**) Electrokinetic layer thickness δ_E estimated from theory under the conditions of panel **A** at various $1/\lambda_o$ (indicated). Model parameters in (**A**) and (**B**): ζ=−90 mV, z=1 (monovalent electrolyte), T = 22 °C, d=50 nm, H=30 μm, ℓ=10 mm, LO=20 mm, εr=79.5, η=0.954 mPa s^{-1}, α→0(homogeneous segment density distribution in the film), ρO=0 (uncharged film). (**C**) I_{str}/ΔP versus ionic strength evaluated from theory for |ρO/F|=0.1 mM (black lines, left y-axis) and |ρO/F|=10 mM (red lines, right y-axis) at (a, a') $1/\lambda_O$=5 nm, (b, b') $1/\lambda_O$=20 nm, (c, c') $1/\lambda_O$=40 nm and (d, d') $1/\lambda_O$=60 nm. The limits $1/\lambda_O$→0 and 1/λO→∞ are explicitly indicated. Other model parameters: as in (**A**)-(**B**) except ζ=−10 mV (**D**) Ratio I_{str}/I_{str}(α→0) as a function of α/d at various $1/\lambda_O$(indicated). Other model parameters: as in (**A**) and (**B**) except that ζ=−10 mV, |ρ_O/F|=10 mM and solution ionic strength = 1 mM. In the inset of panel (**D**), representation of the segment density distribution f(x) for various values of α/d(indicated) and d=50 nm. Adapted from [24].

Finally, we report in Figure 2D the impact of the interfacial film diffuseness α (see Equation (2), and insert in Figure 2D) on streaming current Istr for various $1/\lambda_o$ values at fixed (monovalent) electrolyte concentration (c=1 mM). It is stressed that variations in α lead to modifications of the segment density profile f(x) as given in Figure 2D with the condition of constant total amount of polymer segments in the film (condition ensured via the scalar ω involved in Equation (2)). Results are given in the form $I_{str}(\alpha)/I_{str}(\alpha\to 0)$ versus α/d in order to reveal how I_{str} evaluated under conditions where segment density distribution is diffuse, compares with that obtained for homogeneous polyelectrolyte film. At low to moderate $1/\lambda_o$, the ratio $I_{str}(\alpha)/I_{str}(\alpha\to 0)$ decreases with increasing α/d. This effect results from the friction exerted by the segments protruding from the film bulk phase on the flow in the region where flow velocity field is most significant. The reduction in streaming current thus results from an increased hydrodynamic drag of the outer film region and therewith from a marked immobilization of otherwise electrokinetically active ions.

On the opposite, at large $1/\lambda_o$ (typically $1/\lambda_o$~d or $1/\lambda_o$>>d), the ratio $I_{str}(\alpha)/I_{str}(\alpha\to 0)$ increases with increasing α/d.

Indeed, extending the thickness of the diffuse interface also leads to an increase of the magnitude of the local electrostatic potential in the region x>d (see insert). For large $1/\lambda o$, this electrostatic feature overwhelms the hydrodynamic modulations of the flow velocity associated to changes in segment density distribution and, consequently, it results in an increase of $I_{str}(\alpha)/I_{str}(\alpha\rightarrow 0)$ with increasing α/d. For values of $1/\lambda o$ intermediate between the two aforementioned extremes, the balance between effects of α/d on interfacial hydrodynamics and electrostatics determines the magnitude of the ratio $I_{str}(\alpha)/I_{str}(\alpha\rightarrow 0)$ [24]. In situations of practical interest where $1/\lambda_o<d$, heterogeneity in the outer distribution of polymer segments density primarily causes a reduction in $|I_{str}|$ due to hydrodynamic screening, as further discussed below for poly(acrylic acid) thin films.

Charging and Structure of Poly(Acrylic Acid) Thin Films

The interfacial structure of PE films is governed by the level of ionization of their functional groups and the related increase in osmotic pressure. In this section, we review a strategy for the analysis of the charging and structure of surface-immobilized PE layers on the basis of streaming current, surface conductivity and swelling data measured at covalently-immobilized poly(acrylic acid) (PAA) films [26]. The PAA was attached to hard surfaces using a low-pressure plasma treatment [26].

Figure 3 shows the pH-dependence of streaming current at KCl solution concentrations in the range 0.1 to 10 mM. According to the polymer structure, the charging of PAA films is dominated by the ionization of carboxyl groups in the repeat unit (please note that unsymmetrical water ion adsorption [39] can contribute to the overall film charge at low pH). In line with the highly anionic nature of PAA, the isoelectric point (IEP) of the films was found below pH 2. In 0.1 mM KCl solution, the streaming current exhibits a pronounced non-monotonous dependence on pH with the presence of a maximum at pH ~6 (Figure 3), while data collected at lower salt concentrations conform to commonly-reported trends (see below). The maximum in streaming current gradually vanishes with increasing KCl solution concentration. To decipher the reasons for this unconventional electrokinetic fingerprint, surface conductivity and film swelling were systematically measured under the conditions adopted for the electrokinetic experiments [26].

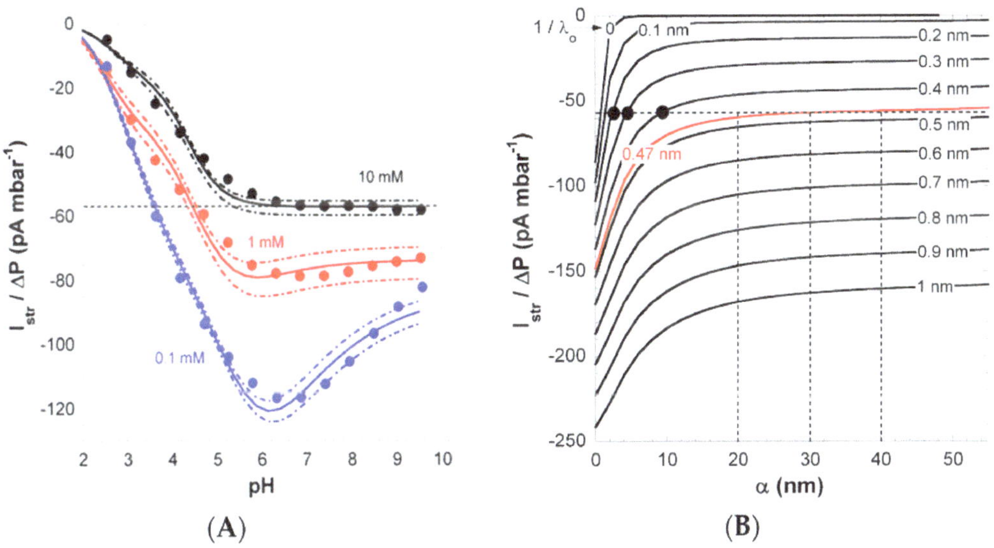

Figure 3: (**A**) Pressure-normalized streaming current, Istr/ΔP, for PAA films as a function of pH in 0.1, 1 and 10 mM KCl solutions (indicated). Experimental data are represented by symbols. Solid and dotted lines were calculated on the basis of the theory by adjustment of the pH- and ionic-strength dependence of the interphasial diffuseness α that best brackets experimental values (dotted lines). (**B**) Dependence of $I_{str}/\Delta P$ on α and $1/\lambda_o$ at 10 mM KCl concentration under condition where full dissociation of the film charges is reached (pH > 6). For each $1/\lambda_o$ value tested, the value adopted for the charge density $\rho o/F$ in the film is that which fits the surface conductivity K^{σ} at large pH (see Figure 2). The unambiguous evaluation of $1/\lambda_o$ is done upon analysis of the intersection points between $I_{str}/\Delta P$ vs. α curves with the $I_{str}/\Delta P$ value measured at large pH in 10 mM KCl (dotted horizontal line in panel **A**and **B**), with as result $1/\lambda_o$ = 0.47 nm. The corresponding range of α consistent with $I_{str}/\Delta P$ at large pH in 10 mM KCl is then straightforwardly determined (see vertical dotted lines). The full black circles correspond to conditions that do not allow recovery of the $I_{str}/\Delta P$ vs. pH curve displayed in panel **A** (which was verified from theoretical simulations). Reprinted from [26].

Figure 4A shows the dependence of the measured surface conductivity K^{σ} on pH in 1 mM KCl solution concentration. $K\sigma$increases significantly with increasing pH above 4 and reaches a ~330 nS plateau value at pH ≥ 8. The surface conductivity and the layer thickness (not shown here) exhibit very similar dependence on solution pH, underlining that both quantities are inherently functions of the ionization level of carboxyl groups within the PAA film [26]. The increase in K^{σ} with pH results from the accumulation of counterions within the film and at the interfacial EDL region outside the film, thereby compensating the structural film charge stemming from ionized carboxyl groups. Under action of an electric field, these ions migrate along the PE surface layer and thus contribute to $K\sigma$. At sufficiently large pH (pH ≥ 8), dissociation of carboxyl groups is complete, which explains the presence of the plateau in $K\sigma$ (Figure 2A).

A mean film thickness of ~18 nm was estimated from ellipsometry measurements at pH 2 [26]. For a given salt concentration, the film thickness increased with increasing pH and reached 60 to 70 nm at pH 10 [26]. In line with the stronger ionization of densely-packed carboxyl groups at higher salt concentrations (see the charge source term in Equation (1)), film swelling occurred in a lower pH range upon the increase of the KCl solution concentration [26]. As stated above, the increase in film thickness closely followed the increase in surface conductivity, meaning that surface conductivity mirrors the increasing counter ion concentration in the PAA film, a process that leads to increased osmotic pressure and enhanced film swelling.

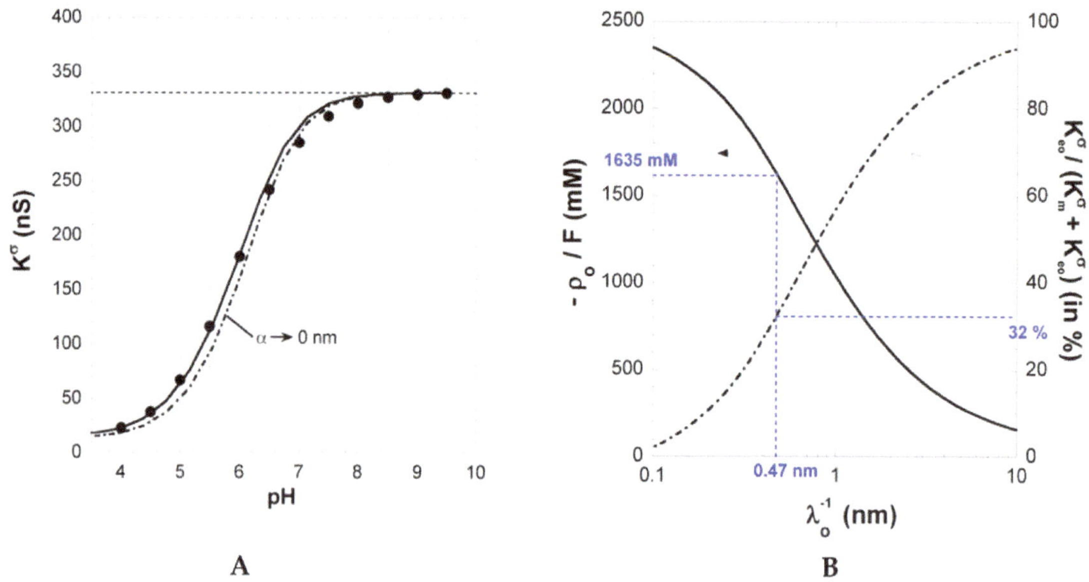

Figure 4: (**A**) Surface conductivity K^σ vs. pH for PAA film in 1 mM KCl solution (symbols). The solid line represents the reconstruction of the data by theory. Parameters: ρ_o/F= -1635 mM, $1/\lambda_o$= 0.47 nm, pK_1= 3.3, and α = 27 nm (solid line) as well as $\alpha \rightarrow 0$ (dotted line). The plateau value reached by K^σ at large pH (horizontal dashed line) was used as a basis for the determination of (λ_o;ρ_o). (**B**) Set of (λ_o;ρ_o) couples that best reproduce the experimentally determined surface conductivity K^σ at large pH. The contribution of electroosmosis to the total surface conductivity is given in terms of the ratio $K^\sigma_{eo}/(K^\sigma_m + K^\sigma_{eo})$ on the right axis. The (λ_o;ρ_o) couple consistent with K^σ and $I_{str}/\Delta P$ data is highlighted in blue. See the text and [26] for further details. Reprinted from [26].

With increasing charge density, electroosmosis may significantly contribute to the surface conductivity of the PE layer. Kσis then determined by the electrostatic properties of the film (contribution Kσm) and by its permeability to flow via the electroosmotic term Kσeo. Indeed, similar to the pressure-driven flow field V(X), the electroosmotic flow field Veo(X)depends on the quantity $1/\lambda_o$ (see Equation (3) and Section 2.5). Duval et al. developed a procedure (detailed below) to evaluate the migration (Kσm) and electroosmotic (Kσeo) components of the surface conductivity Kσwith consistent account of the measured variations of Istr/ΔP with changing pH and salt concentration [26].

The pronounced maximum observed for the streaming current at 0.1 mM KCl (Figure 3) was shown to originate from heterogeneous film swelling and the corresponding increase in the interphase diffuseness α, the latter length scale quantifying how far PAA polymer segments protrude from bulk film toward the outer electrolyte solution [26]. Indeed, as discussed in Section 3.1, an increase in α leads to enhanced friction exerted by the soft polymer interphase on flow, and thus, this results in decreased streaming current above a critical pH value, where swelling is most significant. Below that critical pH, the streaming current classically increases with pH as the result of the gradual dissociation of carboxylic groups. The maximum in the streaming current thus stems from interphase film swelling and the accompanied effects on developed hydrodynamic flow. Due to significant screening of film electrostatics, at fixed pH, the streaming current in a 10 mM KCl solution concentration is lower (in magnitude) compared to values measured in 1 and 0.1 mM KCl. More interestingly, the maximum disappears due to significant screening of film charges and to reduced Debye length compared to α, thus rendering impossible the electrokinetic probing of the segment density distribution at high salt concentrations. At pH ≥ 6, $I_{str}/\Delta P$ reaches a constant value, which suggests that ionization of the carboxyl groups is complete in this pH range.

Figure 4 shows that the increase in K^σ with pH basically compounds that measured for Istr/ΔP (in magnitude) in 10 mM KCl solution. The latter result was exploited in Ref. [26] for deriving in a consistent way the relevant PAA electrohydrodynamic parameters $1/\lambda_o$ and ρ_o/F. Indeed, at pH ≥ 6, the magnitude of K^σ is solely defined by these two quantities and hardly depends on α, which was verified a posteriori. The set of (λ_o;ρ_o) couples that best recover the plateau value reached by K^σ at large pH is given in Figure 2B together with the corresponding ratio $K^\sigma_{eo}/^\sigma$. The latter increases with increasing hydrodynamic penetration, expressed in terms of the Brinkman length $1/\lambda_o$ and the magnitude of the charge density required to reproduce K^σ at large pH decreases with increasing $1/\lambda_o$. Considering the surface conductivity alone is therefore not sufficient for an unambiguous determination of the parameters $1/\lambda_o$ and ρ_o/F. Duval et al. [26] have shown that a consistent solution may be obtained by the additional evaluation of the dependence of $I_{str}/\Delta P$ on α in 10 mM

KCl under complete ionization conditions (pH $\geq$ 6). This analysis was performed for the different couples (λ_o;ρ_o) that are in agreement with Kσ data measured at large pH (as shown in Figure 2B). In line with theory on electrokinetics of diffuse soft thin films [25,26], Istr/ΔP decreases with increasing α and increases with increasing 1/λo, as a result of the associated changes in the hydrodynamic flow-field (case discussed in Figure 2D at low to moderate $1/\lambda_o$). A comparison of the simulation results obtained for $I_{str}/\Delta P$ vs. α (Figure 1B) with the streaming current plateau value measured at large pH in 10 mM KCl solution (Figure 1A) finally provided the searched ρ_o/F and $1/\lambda_o$ values, i.e., $1/\lambda_o$ = 0.47 nm and ρ_o/F= 1635 mM [26]. The corresponding ratio $K^{\sigma}_{eo}/(K^{\sigma}_{m}+K^{\sigma}_{eo})$ is 0.32, confirming that electroosmotic charge transport contributes significantly to the overall surface conductivity K^{σ} for polymer films with moderate to high charge density. Using $1/\lambda_o$ = 0.47 nm and ρ_o/F = 1635 mM, the softness parameter α (and its modulation with pH and salt concentration) and the dissociation pK value pertaining to PAA carboxyl groups were subsequently obtained from reconstruction of the full dependence of K^{σ} and $I_{str}/\Delta P$ on pH and salt concentration [26].

In summary, the developed strategy for the coupled analysis of the pH- and salt concentration-dependent streaming current, surface conductivity and swelling data allows for a comprehensive characterization of the charge and structure of polyelectrolyte layers. Strategies for polymer coatings without ionizable groups were further developed; see e.g., [24,28]. The gained information helps to better understand the fundamental properties and interactions of polymer systems, but it is also of interest for designing application-oriented materials, e.g., responsive or anti-fouling coatings.

Electrohydrodynamics of Polyelectrolyte Multilayers

Electrokinetic measurements are frequently applied to characterize the formation, charge and stability of polyelectrolyte multilayers (PEMs) [39,40,41,42]. Because of the inherent coupling between electrostatics and hydrodynamics in the development of soft surface electrokinetics, the unambiguous interpretation of data acquired on PEM materials requires concepts that extend the basic theory outlined in Section 2. Here, we review recent theoretical developments on electrokinetics of planar PEMs, and these developments are shown to reproduce, at least qualitatively, the peculiar electrokinetic properties of PEM that exhibit some degree of charge stratification absent from the theory outlined in Section 2.

The theory for the streaming current of PEMs was developed by Duval et al. for systems consisting of N alternately deposited cationic and anionic layers of thickness denoted as δj, each being assigned an index j, with j=1,...,N, where j=1 stands for the most internal layer directly supported by the substrate surface and j=N for the outermost layer at the PEM/electrolyte solution interface [27]. Each layer j is considered to carry a single type of ionisable groups with a given pKand ε value (denoted in [27] as pK_j and ε_j, see analogy to Equation (1)). Furthermore, the theory is applicable for any spatial distribution of polymer segments and structural charge density from the hard supporting surface to the outer solution medium without limitations on thickness and charge magnitude pertaining to each layer constituting the PEM. The theoretical framework [27] further ensures a smooth transition of the segment density between adjacent layers, each being attributed a given hydrodynamic softness λ_j and charge density ρj/F, the analogue of the quantities λ_o and ρ_o/Fintroduced in Section 2 for monolayered films.

The electrostatic potential distribution across PEMs may be obtained by solving the non-linear Poisson–Boltzmann equation that includes terms for the structural charges resulting from the ionizable groups in each sub-layer and for mobile counter ions distributed inside and outside the PEM [27]. The hydrodynamic flow field can be calculated on the basis of the Brinkman Equation (3) taking into account the impact of the local density of polymer segments on the friction they exert on flow under electrokinetic conditions. The expressions given in Section 2 for the calculation of the streaming current and surface conductivity remain valid for PEMs.

In analogy to monolayered systems, the magnitude and sign of the streaming current of PEMs is determined by the spatial extension of the electrokinetically-active film region, i.e., the zone where film electrostatic properties are probed by the pressure-driven flow penetrating the film at a finite thickness. The extension of that electrokinetically-active outer film region depends not only on the hydrodynamic film permeability determined by the polymer segment density distribution, but also on the Debye layer thickness fixed by solution ionic strength. Streaming current measurements performed at PEMs may therefore reflect electrohydrodynamic contributions from internal layers embedded within PEMs and not only those from layers that directly face the outer electrolyte solution. As a consequence, the comprehensive electrokinetic characterization of PEMs requires measurements at varying pH values and different salt concentrations of a neutral background electrolyte in order to tune the Debye layer thickness and therewith the outer film region that is effectively probed by electrokinetic flow. Applying this strategy, the contributions of the different layers constituting the PEM may be evaluated, and the differentiated impacts of, e.g., outer polyanionic and inner polycationic layers are reflected by a salt concentration-dependent shift in the point of zero streaming current (PZSC), defined by the pH value where $I_{str}/\Delta P$=0. This remarkable PEM electrokinetic feature is now discussed in some more detail.

Illustrative simulation results for the dependence of $I_{str}/\Delta P$ on pH and electrolyte concentration are shown in Figure 5A for a polyelectrolyte bilayer film consisting of a polycationic layer (thickness δ_1=30 nm, pK_1=9, ε_1=1) immobilized on a hard and uncharged carrier, and a polyanionic layer (thickness δ_2=5 nm, pK_2=4, ε_2=−1) located at the solution side of the bilayer construction [27]. For the sake of comparison, the variation of Istr/ΔP with pH and electrolyte concentration is shown for single polycationic and polyanionic layers. For the monolayered polycationic and polyanionic systems, the position of the PZSC and the dependence of $I_{str}/\Delta P$ on solution pH are conform to the intrinsic ionization characteristics of the polyelectrolytes and to the compression of the electric double layer with increasing solution ionic strength that leads to reduction of $|I_{str}/\Delta P|$ at given pH [26,27].

The electrokinetic fingerprint of the polyelectrolyte bilayer shows two remarkable features: (i) For a given electrolyte concentration, $I_{str}/\Delta P$ changes its sign from negative values at high pH to positive values at low pH. The PZSC (i.e., the pH where sign reversal occurs) is shifted to higher values with decreasing salt concentration. These results thus reveal an increasing contribution of the inner cationic layer to the streaming current at sufficiently low ionic strengths. In addition, Figure 5A highlights the necessity to vary pH and salt concentration over broad ranges of values in order to achieve an unambiguous electrokinetic characterization of PEMs. (ii) The curves $I_{str}/\Delta P$ vs. pH obtained at different salt concentrations exhibit a common intersection point at pH ~8.27. The occurrence of this feature is expected for multilayered systems provided that the thickness of the constituting layers is independent of the concentration of the background electrolyte and of pH. Changes in the layer structure would lead to varying contributions of the various layers to $I_{str}/\Delta P$and, as a consequence, to the absence of common intersection point. Figure 5B illustrates the dependence of the PZSC on salt concentration for different densities of charges $|\rho_2|$ in the topmost (anionic) layer, while all other model parameter values are identical to those adopted in Figure 5A. At a fixed salt concentration, the origin of the change in PZSC towards acidic pH values with increasing the charge density (in magnitude) of the outermost anionic layer is the decrease of the electrical potential in the electrokinetically-active film region. Under such conditions, the potential in that zone becomes significantly affected by the electrostatic features of the external layer, which leads to a decrease in PZSC with increasing $|\rho_2|$. At a fixed charge density of the anionic layer, increasing salt concentration leads to an electric double layer that becomes mostly confined within the outer layer. The latter then predominantly governs the electrokinetic properties of the PEM as a whole, which results in the observed decrease of the PZSC with increasing salt concentration. This decrease is most pronounced for cases where the magnitudes of charge densities in cationic and anionic layers are significantly different from each other, which exacerbates their differentiated respective contributions upon increasing solution ionic strength. A detailed analysis given in [27] further demonstrates that the PZSC of PEMs is strongly determined by the dissymmetry in p_K values of the ionogenic groups in each constituting layer, by the respective thickness of the layers and by their segment density distribution. Furthermore, the relevance of the theoretical framework outlined in this section for PEMs (in particular, the unconventional dependence of their PZSC on salt concentration) was confirmed by experimental results obtained on a polyelectrolyte bilayer consisting of PEI (poly(ethylene imine)) and PAA and on an anionic lipid bilayer supported by a polycationic PEI cushion [27].

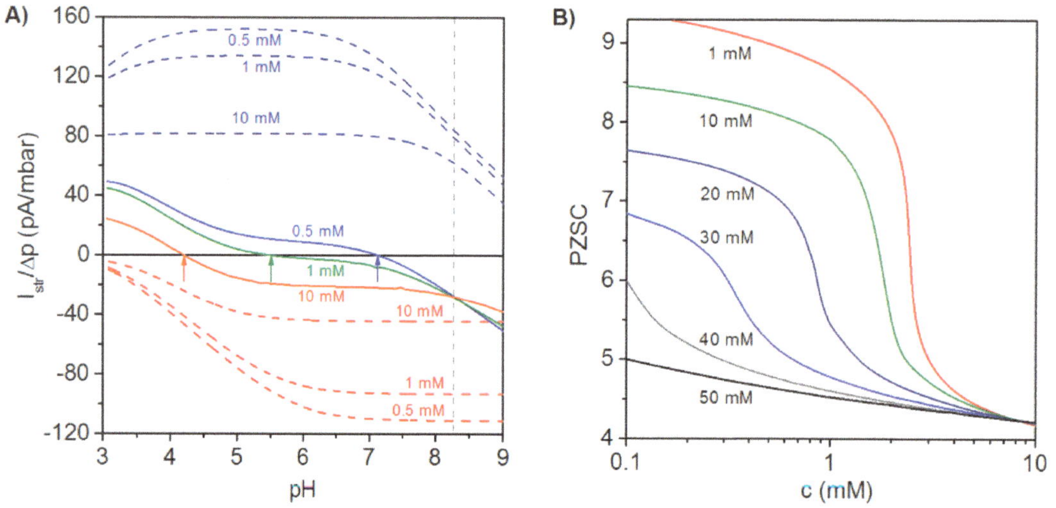

Figure 5: (**A**) Ratio of streaming current over applied pressure ($I_{str}/\Delta P$) vs. pH evaluated at three electrolyte concentrations (indicated) for a polycationic layer (blue dashed lines), a polyanionic layer (red dashed lines) and a polycation (j=1)/polyanion (j=2) bilayer film (colored solid lines). Model parameters for the bilayer system: $pK_1=9$, $\varepsilon_1=+1$, $\delta_1=30$ nm, $pK_2=4$, $\varepsilon_2=-1$, $\delta_2=5$ nm, $|\rho_{1,2}/F|=20$mM, $1/\lambda_{1,2}=4.5$ nm; in each layer, a homogeneous distribution of polymer segments is considered. Model parameters for the polycation layer and polyanion layer are similar to those adopted for the polycationic and polyanionic layers forming the bilayer film. Arrows highlight the decrease of point of zero streaming current (PZSC) for the bilayer film with increasing electrolyte concentration. (**B**) PZSC vs. electrolyte concentration for a bilayer system consisting of a polycationic layer with fixed charge density and a polyanionic layer with varying charge density (indicated in terms of equivalent concentration of charges). Other parameters as in the left panel. Adapted from [27].

In summary, the PZSC of PEMs is determined by the complex interplay between pH- and ionic strength-dependent electrohydrodynamic processes that govern the charging, charge compensation and structural properties of the various layers constituting the PEM. For this reason, the PZSC measured for soft multilayered films does not correspond to the pH value where the overall film charge is zero. As a result, the comprehensive electrokinetic characterization of PEMs requires systematic measurements performed as a function of pH and salt concentration. Finally, it is stressed that the electrohydrodynamics and the point of zero mobility of charge-stratified particles exhibit similar characteristics as those discussed here for planar systems. A comprehensive overview about theory, experiments and simulations on charge-stratified particulate systems can be found in [43,44,45,46]. As an illustrative example, the point of zero electrophoretic mobility for poly(amidoamine) starburst dendrimers with a positive internal core and a negatively-charged outer shell layer strongly decreases with increasing salt concentration for the same reasons invoked here for planar charge-stratified polyelectrolyte systems.

CONCLUSIONS AND PERSPECTIVES

The electrohydrodynamics of soft surfaces is strongly determined by the distribution of charged polymer segments across the film/electrolyte interface. Electrokinetics therefore provides valuable options for analyzing the interfacial charge and structure of soft surfaces in contact with aqueous solutions. Here, we report and discuss the theory and strategies for the analysis of the interfacial charge and structure of planar polyelectrolyte layers from the measurement of streaming current, surface conductivity and film swelling properties. General trends are illustrated and discussed on the basis of simulation results. As specific examples, we report data of the analysis of charge and structure of PAA thin films and discuss remarkable electrokinetic features of charge-stratified interphases on the basis of simulation results. Overall, electrokinetics is shown to be a spatially-resolved technique suitable for deciphering the interfacial charge and structure of soft polymeric coatings within a so-called electrokinetically-active outer film region whose extension is fixed by the electric double layer thickness and by the film permeability. Future developments will have to aim at a better understanding of the interactions between multivalent ions or charged molecules and polyelectrolyte layers. In combination with other methods, surface conductivity measurements can be expected to provide additional information about the amount of ionized groups involved in the formation of salt bridges between oppositely-charged layers and to allow the analysis of changes in layer structure upon variations of adjacent electrolyte composition. The authors also anticipate that electrokinetic analysis of electroactive polymer films in the presence of redox active analytes in solutions could be decisive to investigate electron-transfer reactions at modified soft electrodes under lateral flow conditions. Depending on (bipolar) electrode kinetic properties, such systems are indeed expected to give rise to the non-linear dependence of streaming current on applied pressure. While analyses of this type have been conducted on bare (uncoated) conducting materials with resulting applications on the detection of redox active analytes, basic investigations of electron-conducting soft polymer electrokinetics are to the best of our knowledge still lacking in the literature. Finally, tackling the electrokinetics of soft planar polyelectrolytes at a molecular level using theories that go beyond the mean-field Poisson–Boltzmann level could pave the way for apprehending ion specificity effects, as well as the impacts of film structure on ions/particles selectivity or flow permeability.

REFERENCES

1. Zhao, B.; Brittain, W.J. Polymer brushes: Surface-immobilized macromolecules. Prog. Polym. Sci. 2000, 25, 677–710.
2. Decher, G. Fuzzy Nanoassemblies: Toward layered polymeric multicomposites. Science 1997, 277, 1232–1237.
3. Munoz-Bonilla, A.; Fernandez-Garcia, M. Polymeric materials with antimicrobial activity. Prog. Polym. Sci. 2012, 37, 281–339.
4. Molina, M.; Asadian-Birjand, M.; Balach, J.; Bergueiro, J.; Miceli, E.; Calderón, M. Stimuli-responsive nanogel composites and their application in nanomedicine. Chem. Soc. Rev. 2015, 44, 6161–6186.
5. Machida, S.; Urano, T.I.; Sano, K.; Kawata, Y.; Sunohara, K.; Sasaki, H.; Yoshiki, M.; Mori, Y. A chiral director field in the nematic liquid crystal phase induced by a poly(.gamma.-benzyl glutamate) chemical reaction alignment film.Langmuir 1995, 11, 4838–4843.
6. Li, X.; Du, P.; Peng Liu, P. Layer-by-layer polyelectrolyte complex coated poly(methacrylic acid) nanogels as a drug delivery system for controlled release: Structural effects. RSC Adv. 2014, 4, 56323–56331.
7. Shao, Y.; Xu, S.; Zheng, X.; Wang, Y.; Xu, W. Optical fiber LSPR biosensor prepared by gold nanoparticle assembly on polyelectrolyte multilayer. Sensors 2010, 10, 3585–3596.
8. Croisier, F.; Jérôme, C. Chitosan-based biomaterials for tissue engineering. Eur. Polym. J. 2013, 49, 780–792.
9. Delcea, M.; Möhwald, H.; Skirtach, A.G. Stimuli-responsive LbL capsules and nanoshells for drug delivery. Adv. Drug Delivery Rev. 2011, 63, 730–747.
10. Barbati, A.C.; Kirby, B.J. Soft diffuse interfaces in electrokinetics—Theory and experiment for transport in charged diffuse layers. Soft Matter 2012, 8, 10598–10613.
11. Zimmermann, R.; Dukhin, S.S.; Werner, C.; Duval, J.F.L. On the use of electrokinetics for unraveling charging and structure of soft planar polymer films. Curr. Opin. Colloid Interface Sci. 2013, 18, 83–92.
12. Lyklema, J. Fundamentals of Interface and Colloid Science, Vol. II: Solid-Liquid Interfaces; Academic Press: London, UK, 1995.
13. Cohen Stuart, M.A.; Waajen, F.H.W.H.; Dukhin, S.S. Electrokinetic effects of adsorbed neutral polymers. Colloid Polym. Sci. 1984, 262, 423–426.
14. Donath, E.; Voigt, A. Streaming current and streaming potential on structured surfaces. J. Colloid Interface Sci. 1986,109, 122–139.
15. Ohshima, H.; Kondo, T. Electrokinetic flow between two parallel plates with surface charge layers: Electro-osmosis and streaming potential. J. Colloid Interface Sci. 1990, 135, 443–448.
16. Starov, V.M.; Solomentsev, Y.E. Influence of gel layers on electrokinetic phenomena: 1. Streaming potential. J. Colloid Interface Sci. 1993, 158, 159–165.
17. Starov, V.M.; Solomentsev, Y.E. Influence of gel layers on electrokinetic phenomena: 2. Effect of ions interaction with the gel layer. J. Colloid Interface Sci. 1990, 158, 166–170.
18. Ohshima, H. Electrophoretic mobility of soft particles. Colloids Surf. A 1995, 103, 249–255.
19. Cohen, J.A.; Khorosheva, V.A. Electrokinetic measurement of hydrodynamic properties of grafted polymer layers on liposome surfaces. Colloids Surfaces A 2001, 195, 113–127.
20. Hill, J.R.; Saville, D.A.; Russel, W.B. Electrophoresis of spherical polymer-coated colloidal particles. J. Colloid

Interface Sci. 2003, 258, 56–74.

21. Duval, J.F.L.; van Leeuwen, H.P. Electrokinetics of diffuse soft interfaces. 1. Limit of low donnan potentials. Langmuir2004, 20, 10324–10336.
22. Duval, J.F.L. Electrokinetics of diffuse soft interfaces. 2. Analysis based on the nonlinear poisson−boltzmann equation.Langmuir 2005, 21, 3247–3258.
23. Yezek,L.;Duval,J.F.L.;vanLeeuwen,H.P.Electrokinetics of diffuse soft interfaces. III. Interpretation of data on the polyacrylamide/water interface. Langmuir 2005, 21, 6220–6227.
24. Duval, J.F.L.; Zimmermann, R.; Cordeiro, A.L.; Rein, N.; Werner, C. Electrokinetics of diffuse soft interfaces. IV. Analysis of streaming current measurements at thermoresponsive thin films. Langmuir 2009, 25, 10691–10703.
25. Zimmermann, R.; Kuckling, D.; Kaufmann, M.; Werner, C.; Duval, J.F.L. Electrokinetics of a poly(N-isopropylacrylamid-co-carboxyacrylamid) soft thin film: Evidence of diffuse segment distribution in the swollen state.Langmuir 2010, 26, 18169–18181.
26. Duval, J.F.L.; Küttner, D.; Nitschke, M.; Werner, C.; Zimmermann, R. Interrelations between charging, structure and electrokinetics of nanometric polyelectrolyte films. J. Colloid Interf. Sci. 2011, 362, 439–449.
27. Duval, J.F.L.; Küttner, D.; Werner, C.; Zimmermann, R. Electrohydrodynamicsofsoftpolyelectrolytemultilayers: Point of zero-streaming current. Langmuir 2011, 27, 10739–10752.
28. Zimmermann, R.; Romeis, D.; Bihannic, I.; Cohen-Stuart, M.; Sommer, J.U.; Werner, C.; Duval, J.F.L. Electrokinetics as an alternative to neutron reflectivity for evaluation of segment density distribution in PEO brushes. Soft Matter 2014,10, 7804–7809.
29. Werner, C.; Körber, H.; Zimmermann, R.; Dukhin, S.S.; Jacobasch, H.-J. Extended electrokinetic characterization of flat solid surfaces. J. Colloid Interf. Sci. 1998, 208, 329–346.
30. Zimmermann, R.; Osaki, T.; Schweiss, R.; Werner, C. Electrokinetic microslit experiments to analyse the charge formation at solid/liquid interfaces. Microfluid. Nanofluid. 2006, 2, 367–379.
31. Duval, J.; van Leeuwen, H.P.; Cecilia, J.; Galceran, J. Rigorous analysis of reversible faradaic depolarization processes in the electrokinetics of the metal / electrolyte solution interface. J. Phys. Chem. B 2003, 107, 6782–6800.
32. Duval, J.F.L.; Buffle, J.; van Leeuwen, H.P. Quasi-reversible faradaic depolarisation in the electrokinetics of the metal/solution interface. J. Phys. Chem. B 2006, 110, 6081–6094.
33. Toomey, R.; Freidank, D.; Rühe, J. Swelling behavior of thin, surface-attached polymer networks. Macromolecules 2004,37, 882–887.
34. Junk, M.J.N.; Anac, I.; Menges, B.; Jonas, U. Analysis of optical gradient profiles during temperature- and salt-dependent swelling of thin responsive hydrogel films. Langmuir 2010, 26, 12253–12259.
35. Sanjuan, S.; Perrin, P.; Pantoustier, N.; Tran, Y. Synthesis and swelling behavior of pH-responsive polybase brushes.Langmuir 2007, 23, 5769–5778.
36. Brinkman, H.C. A calculation of the viscous force exerted by a flowing fluid swarm of particles. Appl. Sci. Res. 1947,A1, 27–34.
37. Donnan, F.G. Theorie der membrangleichgewichte und membranpotentiale bei vorhandensein von nicht dialysierenden elektrolyten. Ein beitrag zur physikalisch-chemischen physiologie (Theory of the balances of membranes and potential of membranes at the existence of non dialysing electrolytes—A contribution to physical chemical physiology). Z. Elektroch, Angew. Phys. Chem. 1911, 17, 572–581.
38. Dukhin, S.S.; Zimmermann, R.; Werner, C. Intrinsic charge and Donnan potentials of grafted polyelectrolyte layers determined by surface conductivity data. J. Colloid Interface Sci. 2004, 274, 309–318.
39. Zimmermann, R.; Freudenberg, U.; Schweiß, R.; Küttner, D.; Werner, C. Hydroxide and hydronium ion adsorption—A survey. Curr. Opin. Colloid Interface Sci. 2010, 15, 196–202.
40. Schönhoff, M. Self-assembled polyelectrolyte multilayers. Curr. Opin. Colloid Interface Sci. 2003, 8, 86–95.
41. Díez-Pascual, A.M.; Shuttleworth, P.S. Layer-by-layer assembly of biopolyelectrolytes onto thermo/pH-responsive micro/nano-gels. Materials 2014, 7, 7472–7512.
42. Ferriz-Manas, M.; Schlenoff, J.B. Zeta potential of polyelectrolyte multilayers using the spinning disk method.Langmuir 2014, 30, 8776–7783.
43. Langlet, J.; Gaboriaud, F.; Gantzer, C.; Duval, J.F.L. Impact of chemical and structural anisotropy on the electrophoretic mobility of spherical soft multilayer particles: The case of bacteriophage MS2. Biophys. J. 2008, 94, 3293–3312.
44. Rochette, C.N.; Crassous, J.J.; Drechsler, M.; Gaboriaud, F.; Eloy, M.; de Gaudemaris, B.; Duval, J.F.L. Shell structure of natural rubber particles: Evidence of chemical stratification by electrokinetics and cryo-TEM. Langmuir 2013, 29, 14655–14665.
45. Moussa, M.; Caillet, C.; Town, R.M.; Duval, J.F.L. Remarkable electrokinetic features of charge-stratified soft nanoparticles: Mobility reversal in monovalent aqueous electrolyte. Langmuir 2015, 31, 5656–5666.
46. Raafatnia, S.; Hickey, O.A.; Holm, C. Mobility reversal of polyelectrolyte-grafted colloids in monovalent salt solutions.Phys. Rev. Lett. 2014, 113, 238301.

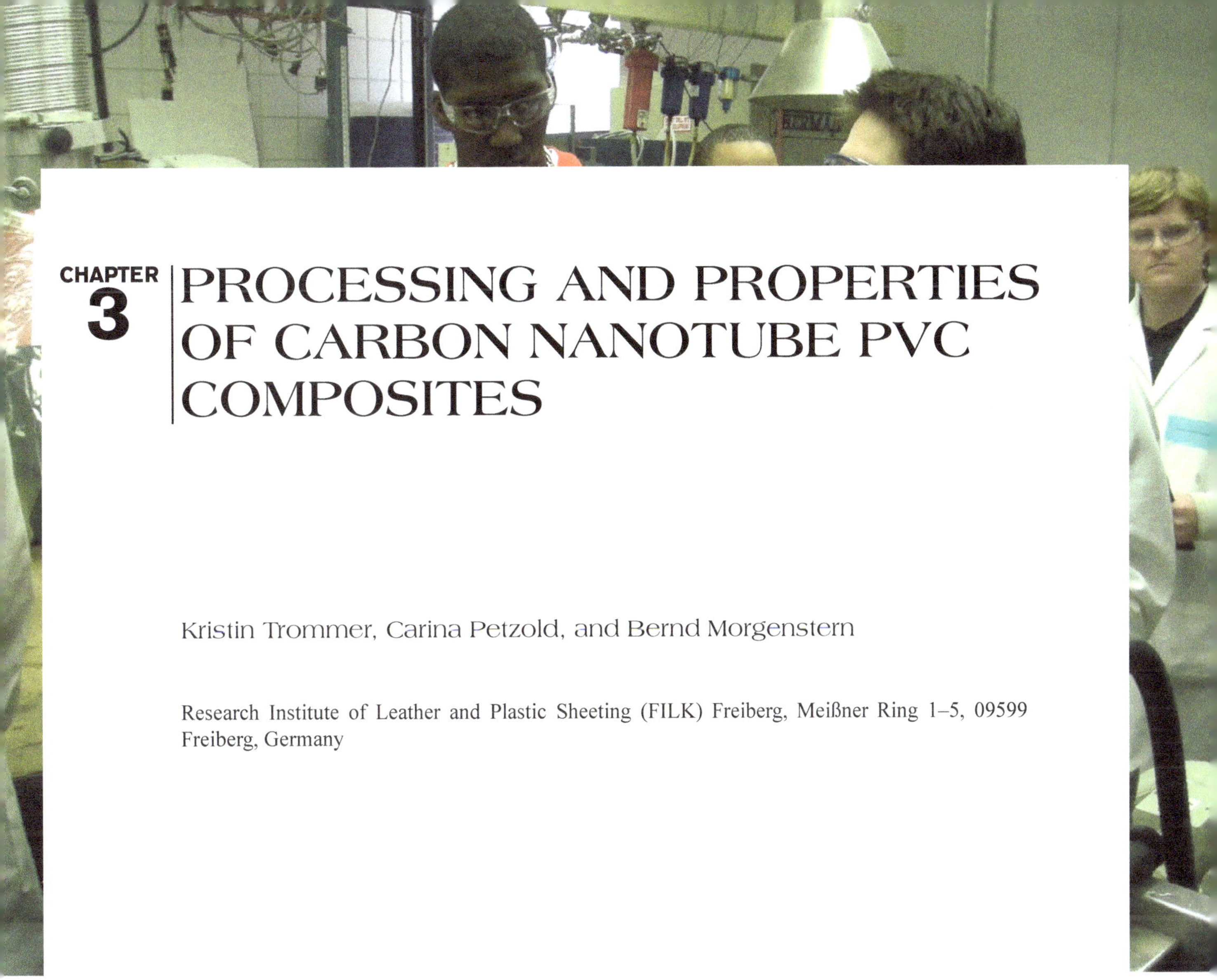

CHAPTER 3 PROCESSING AND PROPERTIES OF CARBON NANOTUBE PVC COMPOSITES

Kristin Trommer, Carina Petzold, and Bernd Morgenstern

Research Institute of Leather and Plastic Sheeting (FILK) Freiberg, Meißner Ring 1–5, 09599 Freiberg, Germany

ABSTRACT

Commercially available multiwalled carbon nanotubes (MWCNT) were incorporated in coating masses based on PVC by means of three roll mill. The best results could be obtained using the 5 µm gap. Thin PVC sheets were formed via knife coating having an electrical conductivity up to 1,500 S/m that are applicable as electric heating elements. For the use in the antistatic range, CNT contents ≤0.5% are sufficient. Rheological measurements indicate the quality of particle processing. AFM investigations are suitable to investigate the alignment of the nanoparticles in the bulk polymer. Using this method, the decrease of agglomerates as well as the splitting of CNT bundles within further mass processing could be visualized.

INTRODUCTION

Sheets and coatings made of plasticized polyvinyl chloride (PVC-P) are currently applied to a variety of products for various utilizations. In 2010, PVC-P was set to the top of the total production of coated fabrics with a share rate of 48%. This corresponds to approximately 200–230 million m^2. Among them there are more than 25% canvas materials; 10% to 15% are products of textile architecture (roof sheeting and waterproofs), advertising goods (banners and billboards), and also tents as well as marquees and sun screens. Further important applications include conveyor belts, floor coverings, inflatable items, and artificial leather for the automotive industry or upholstery in public spaces. Because of the various purposes, the requirements concerning the materials' properties are very different. Depending on the exact field of application, the coated materials are required to have a variety of special properties. These include, for instance, cut and abrasion resistance, flame retardant, heat protection, antistatic or electrical conductivity, and resistances to soiling, dust, gas, UV-radiation, or chemicals. In order to generate such properties, the polymer matrix is provided with additive compounds in liquid or particle form. Prior art is the use of filler materials with particle sizes in the micrometer range. Such particles enable the setting of the desired function. At the same time, they also change polymeric properties such as high mechanical performance (tensile strength, elongation at break, and cold flexibility), ageing, or migration. The use of additive compounds with a particle size distribution in the micrometer range of approximately 20 µm especially restricts a thin layer development. Because of the particle sizes, defect-free layers with thicknesses below 60 µm are hardly feasible. However, often the desired material properties can be set in a

Citation: Kristin Trommer, Carina Petzold, and Bernd Morgenstern, "Processing and Properties of Carbon Nanotube PVC Composites," Journal of Applied Chemistry, vol. 2014, Article ID 307274, 10 pages, 2014. doi:10.1155/2014/307274

targeted way by using small amounts of nanoparticles. Carbon-based nanoparticles with a high aspect ratio are often applied to reach properties like antistatic or electrical conductivity. Carbon nanotubes (CNT) or graphene is especially suitable. Several publications already described analyses concerning the percolation ranges of these particles in polymers. The review articles [1, 2] provide a comprehensive overview. So far, the focus for integration into paste-like polymeric masses was on the combination of nanoparticles with epoxy resins. PVC layers are hardly investigated. In [3, 4] the electrical and the thermal behavior of PVC-MWCNT nanocomposites are analyzed; in [5, 6] the influence of the glass transition temperature and the mechanical properties are described.

The present paper shows the development of electrically conductive plasticized PVC layers by incorporation of multiwalled carbon nanotubes (MWCNT) in PVC solutions. The preparation of the nanoparticles is the most challenging part in producing the MWCNT polymeric composites. Technical MWCNT show, with having a diameter of 5–30 nm and a length of >1 μm, a very high aspect ratio and they form agglomerates. In order to generate electrically conductive paths in the polymeric material, these agglomerates have to be split and spread homogeneously in the polymeric matrix. In contrast to processing in low viscous media, such as water, the remaining agglomerates cannot be separated from the polymeric mass. Usual processes for the splitting of the agglomerates are sonification as well as calendering [6–13]. The maintenance of the high aspect ratio of the MWCNT is the most important criterion for the selection of the agglomerates. If the shearing forces on the filamentary particles are too high, this may result in a break. In this paper the dependencies between the process conditions during particle preparation and the properties of the pastes as well as the electrical conductivity of the polymeric composite are analyzed. The nanoparticles in the processed mass were characterized by using atomic force microscopy (AFM). This method is increasingly used for the characterization of CNT nanoparticles in polymers because it enables visualizing structures up to the lower nanometer range [14–17].

The preparation of nanoparticle-functionalized electrically conductive plasticized PVC layers is continuously possible from roll to roll, whereas the PVC layer can be formed as flexible textile coating or as separate film.

MATERIALS AND METHODS

Materials

The PVC sheets were prepared of E-PVC EP 7060 (Vinnolit GmbH & Co. KG) and Hexamoll DINCH (BASF) as plasticizers. KRTTS (titanium triisostearoylisopropoxide, Simagchem Corp.) (additive 1) and Byk 2155 (BYK Chemistry GmbH) (additive 2) were used as additives. The properties of the used conductive particles are summarized in Table 1. The particles were used without any further modification.

Table 1: Properties of the different particles (manufacturer data)

Product	Nanocyl NC 7000	Baytubes C150P	WM-CNT CP-0017-SG	Acetylene black P50 uv
Manufacturer	Nanocyl SA	Bayer material science AG	IoLiTec ionic liquids technologies GmbH	SKW Stickstoffwerke Piesteritz GmbH
Indication	CNT-A	CNT-B	CNT-C	D
Carbon purity	90%	95%	95%	99.5%
Diameter	9.5 nm	4–13 nm	11–20 nm	Not specified
Length	1.5 μm	1 μm	5–15 μm	Not specified
Surface area	250–300 m^2/g	Not specified	40–300 m^2/g	70 m^2/g
Bulk density	66 kg/m^3	120–150 kg/m^3	Not specified	1,900 kg/m^3

Characterization Methods

Measurement of the Surface Resistance

According to DIN 54345 part 1, the surface resistance was measured as four-terminal method (Kelvin method) using the measuring device Milli-To2 of Fischer Elektronik GmbH & Co. Four-terminal method is an electrical impedance measuring technique that uses separate pairs of current-carrying and voltage-sensing electrodes. The key advantage of four-terminal sensing is that the separation of current and voltage electrodes eliminates the impedance contribution of the wiring and contact resistances. The equipment allows resistivity measurements up to 2×10^6 Ω. After climate conditioning (24 h, 23°C, and 50% rel. humidity), the samples were applied with a voltage of 750 mV for 60 s. The thickness of the samples was 50 μm or 100 μm. Due to the small layer thickness and the low ratio between thickness and electrode distance of less than one of hundreds, the resistivity as well as the electrical conductivity can be calculated from the measuring value of the square resistance by multiplying the thickness.

AFM Measurements

An atomic force microscope of the type Nanowizard 3 of the company JPK Instruments was used for the structure measurements in the nanometer range. The scanning unit of the apparatus is able to detect a grid of up to 100×100×15 microns (*XYZ* direction). The measurements were performed in tapping mode. For the evaluation, the images of height, amplitude, and phase were used.

Analysis of the Viscosity

The rheological analyses were carried out by using a Modular Compact Rheometer MCR 300 of Anton Paar GmbH with comprehensive evaluation software. The measuring range of the rheometer comprises a torque range from 0.5 μNm to 120 mNm. For measurements, the cone/plate measuring system CP50-1 was used. The flow curves were recorded at 23°C within a shear range of 0–1,000 s^{-1}.

Preparation of PVC Nanocomposites

Material Preparation

In order to obtain an optimal dispersion of the CNT, the particles were incorporated into a PVC solution. For this purpose, a 10% PVC solution was prepared of 180 g EP 7060 and 1,620 g cyclohexanone during the first step. The solvent was put into a temperature-controlled vessel and under stirring conditions the PVC powder was added. The mixture was stirred for 3 hours at 30°C and 1,000 r/min. Afterwards the PVC was completely dissolved. This solution was used as stock solution from which the respective quantities were taken for further processing to CNT-containing masses. Generally 240 g of the PVC solution (24 g PVC = 100 phr) was intensely mixed with 50 phr of DINCH (plasticizer), 1.5 phr of process additive Byk 1770 and disperser additive. The ratios refer, respectively, to 100 phr of PVC. The quantity of disperser additives depends on the quantity of CNT. Disperser additives and CNT particles were always used at a proportion of 1 : 1. The CNT particles were added at the end to the homogeneous polymeric solution with ratios of 0.5% up to 6% based on the total solid content of the mixture. Furthermore the homogenous dispersion was processed at a three roll mill (120 EH-250 of EXAKT) in several passages. The roll mill consists of two roller gaps which can be operated with a defined roller gap or a defined force that presses the rolls together. Several operation modi were set to optimize the process parameters: gap mode of 10 μm (process parameter O) and 5 μm (process parameter A) as well as force mode of 5 N (process parameter B) and 12 N (process parameter C). For each set of parameters, a part of the mass was separated for characterization and the further processing into coating sheets.

Preparation of PVC Sheets

From the prepared PVC masses, CNT films were prepared with a thickness of 50 μm and 100 μm by blade coating. For that the compounds were spread out on release paper at a Mathis Labcoater in order to form thin layers which were dried for 3 minutes at 100°C and cured for 5 minutes at 160°C. Afterwards the cured layers were delaminated from the release paper to become a thin film of 50 μm or 100 μm. These sheets were characterized by using the described methods.

Rheology

The rheological behavior of dispersions with nanoparticles was analyzed within a shear range of 10^{-3}–10^{3} s^{-1}. All samples were equivalent in composition of the polymeric solution in which the nanoparticles were dispersed. Only the type of disperser additive was varied. Samples with a particle content of 1–6 ma% CNT based on the total solid content were characterized. As expected, the viscosity rises with increasing particle content. The mass shows a structural viscous behavior, if the particle agglomerates were processed into smaller fibrous structures. If this process is not successful and if there are still spherical particles, the dispersions show a nearly Newtonian behavior. In these cases, the flow curves of the PVC solution modified with CNT run parallel to the curve of the pure PVC solution. Figure 1 shows this fact using the example of the CNT-B particles. On Figure 1(a), the flow curves of differently processed CNT dispersions containing 3% (Newtonian behavior) as well as 4% (structural viscous behavior) of CNT-B filler material are shown. The light microscopic images (Figure 1(b)) show that the masses with Newtonian behavior contain spherical particle agglomerates, whereas the intrinsically viscous masses contain fibrous particles with a high aspect ratio. Obviously the processing conditions were only suitable for the masses with 4% CNT-B in order to achieve the desired structures. So the flow curve delivers the first information whether the particle processing into smaller structures with high aspect ratio was successful.

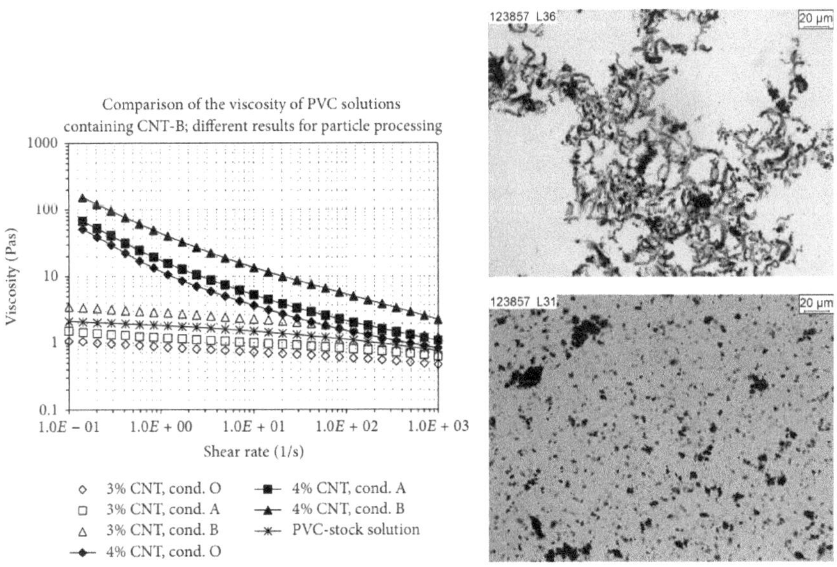

(a) (b)

Figure 1: (a) Rheological behavior of CNT-containing PVC solutions which contain particles with high (4% CNT) and low (3% CNT) aspect ratio. (b) LIMI images of the CNT-containing dispersions; top with 4% CNT-B; bottom with 3% CNT-B.

RESULTS AND DISCUSSION

Characterization of the Paste-Like Coating Compound

The splitting of the CNT-C particles into fibrous structures cannot be achieved in any test with the applied process method on the three roll mill. All compounds filled with CNT-C particles showed Newtonian flow behavior and the light microscopic images were similar to those shown in Figure 1(b) bottom. Contrary to the CNT-C particles, the CNT-A particles could always be split into fibrous structures by treating on the three roll mill. According to the processing steps O, A (gap mode) and B, C (force mode), each of these compounds was comprehensively characterized in rheological terms. For all flow curves, the particle contents of the compounds prepared in gap mode differ only slightly. The same applies for a CNT content of up to 4% also for processing in force mode. During transition from gap mode to force mode, a significant increase in viscosity is observed. For a particle content of more than 5%, viscosity of the compound processed at a lower force (B) differs from that of the compound processed at high force (C). The flow curve (B) runs between the curves resulting from gap processing (O, A) and force processing C. For compounds which were processed in force mode C, viscosity in the shear range between 1 s^{-1} and 1000 s^{-1} increases by about one power of ten (Figure 2). With a shear rate of 1 s^{-1}, the viscosity of the compounds processed in gap mode under the condition O also differs from that processed in force mode, condition C, up to one power of ten. In order to achieve such an effect, much more particles have to be created in force mode than in gap mode. This can be caused either by improving the separation or by breaking the CNT.

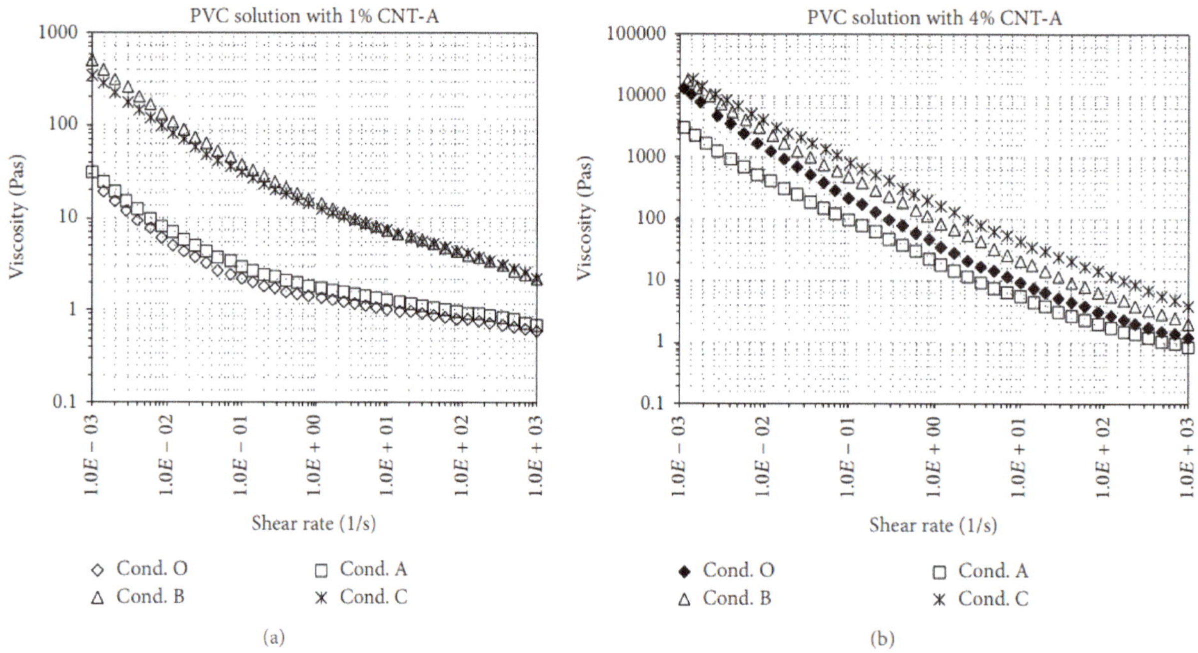

Figure 2: Rheological behavior of PVC solutions with different CNT contents according to different manufacturing processes.

The higher the particle quantity is, the steeper the flow curve is. For processing films and coatings, a shear range between 50 s^{-1} and 1000 s^{-1} is important. The measured viscosity below 10 Pas enables an application by using the blade coating process.

The further characterization of the compounds as well as the sheets produced from them concerning electrical conductivity and particle distribution showed that the processing method A is the most suitable. Rheological analyses over 8 weeks showed that the long-term stability for all investigated concentrations is given. For this purpose, the flow curves of the compounds were analyzed after one, seven, and fourteen days, as well as after four and eight weeks. Over this period of time the flow curves are nearly identical. Segregation or sedimentation of the particles could not be observed.

Particle Size and Alignment

In order to get an overview of the distribution behavior of the particles within the matrix and of their composition with regard to their size and form, the polymeric dispersions were examined by light microscopy (LIMI). Figure 3 shows the LIMI images for a PVC compound containing 3% CNT-A after the processing stages O, A, B, and C.

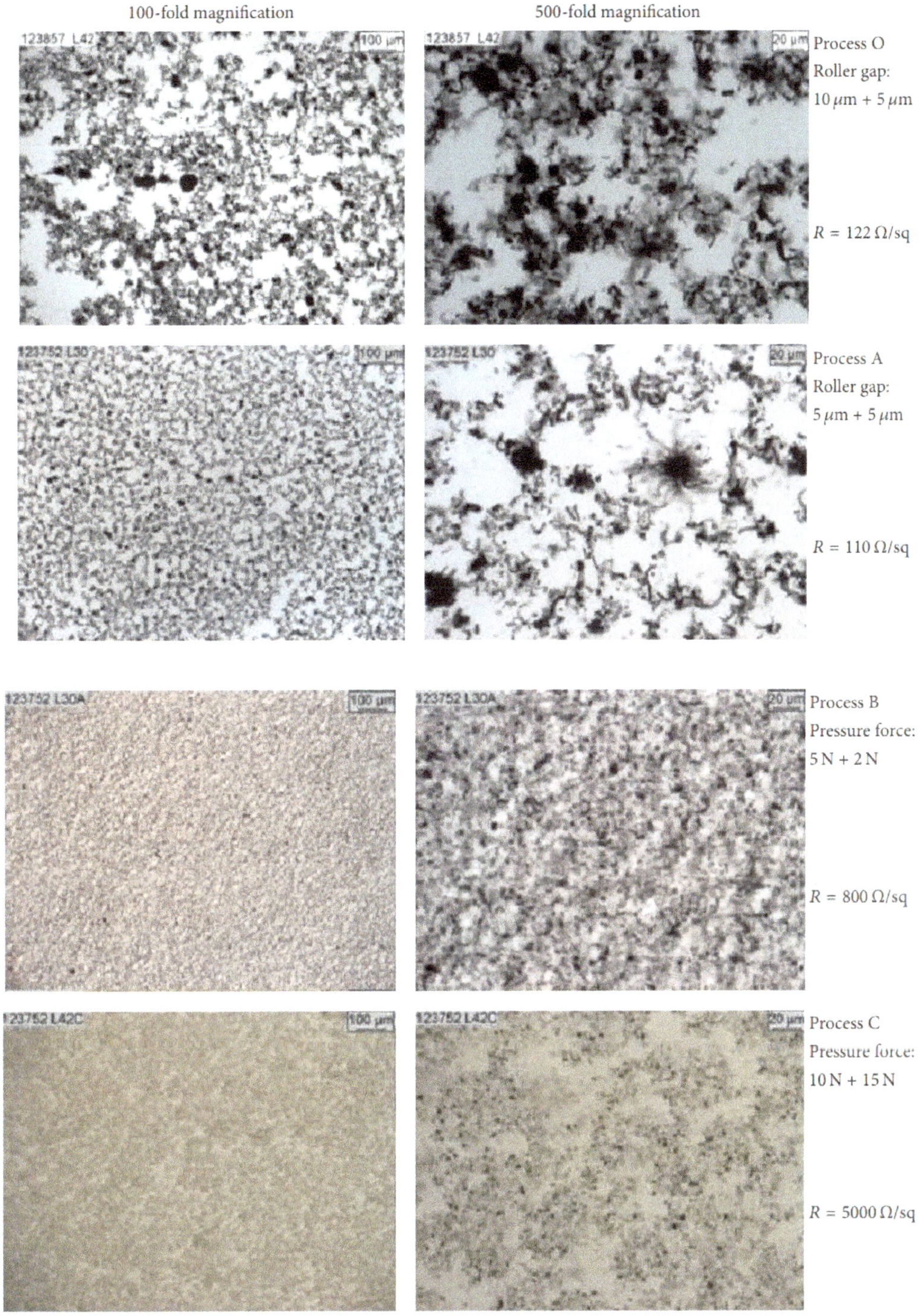

Figure 3: LIMI images of differently processed PVC compounds containing 3% CNT-A (R= surface resistance).

As expected, the particle agglomerates get smaller with diminished roller gap through which the samples are calendered. In the gap mode, this can be set to a minimum of 5 μm. In force mode, the gap can be smaller. A network of conductive structures is formed after processing the compounds in gap mode O and A. However, the conductive structures are still rough. In addition to the fibrous structures, there are spherical agglomerates as well. Furthermore, sections with a size range between 10 μm and 20 μm can be seen where no conductive particles exist. Transition from gap to force mode entails a clear change in the compound. The structures are much smaller and seem to be more evenly distributed in the polymer matrix. If the process takes place in force mode, the significant change in the polymeric compound correlates to the effects found in the rheological behavior. For these samples, the resolution achievable with LIMI is not sufficient anymore to analyze the structures. There is no information whether the CNT are better separated or split up already. For clarifying the CNT structures in the polymeric compounds, AFM analyses were consulted with evaluation of the three analysis modes: height, amplitude, and phase. By analogy with the LIMI images, the polymeric compounds were brought to a glass slide and the CNT structures were analyzed.

Figure 4 shows the height, amplitude, and phase images of a PVC compound with CNT-A each after the processes A, B, and C. The height image clearly shows a decrease of the particulate structures from A via B to C. The dimensions of the particles can be determined from the shown section of 40 × 40 μm. They measure 7.6 μm (STw 0.89) for process step A (gap mode), 5.7 μm (STw 0.52) for process step B, and only 2.65 μm (STw 0.28) for process step C. The amplitude and phase images measure 10 × 10 μm. The amplitude image shows elongated structures. In combination with the phase image, these structures can be clearly assigned to CNT bundles or matrix effects. After process step A, there are CNT bundles spreading to all sides of the polymeric matrix emanating from the CNT agglomerate. It must be taken into consideration that the agglomerates of the polymeric compounds are isolated due to samples preparation for the AFM images. The CNT structures in the gelled, compact PVC layers are closer to each other. Therefore an intensive interaction of the CNT bundles can be assumed. The diameter of the CNT structures starting from the agglomerates and aligning into the polymeric matrix measures 180 nm on average. The structures show a length of up to 4 μm. These values suggest that CNT bundles were separated from the agglomerates, which consist of 10 to 15 nanotubes. The fiber bundles were measured on the basis of a 5 × 5μm section in the height image (Figure 5). The structures do not present surface effects of the matrix, but they consist of different materials, as shown in the phase image, so that these fibrous structures can be assigned to the CNT.

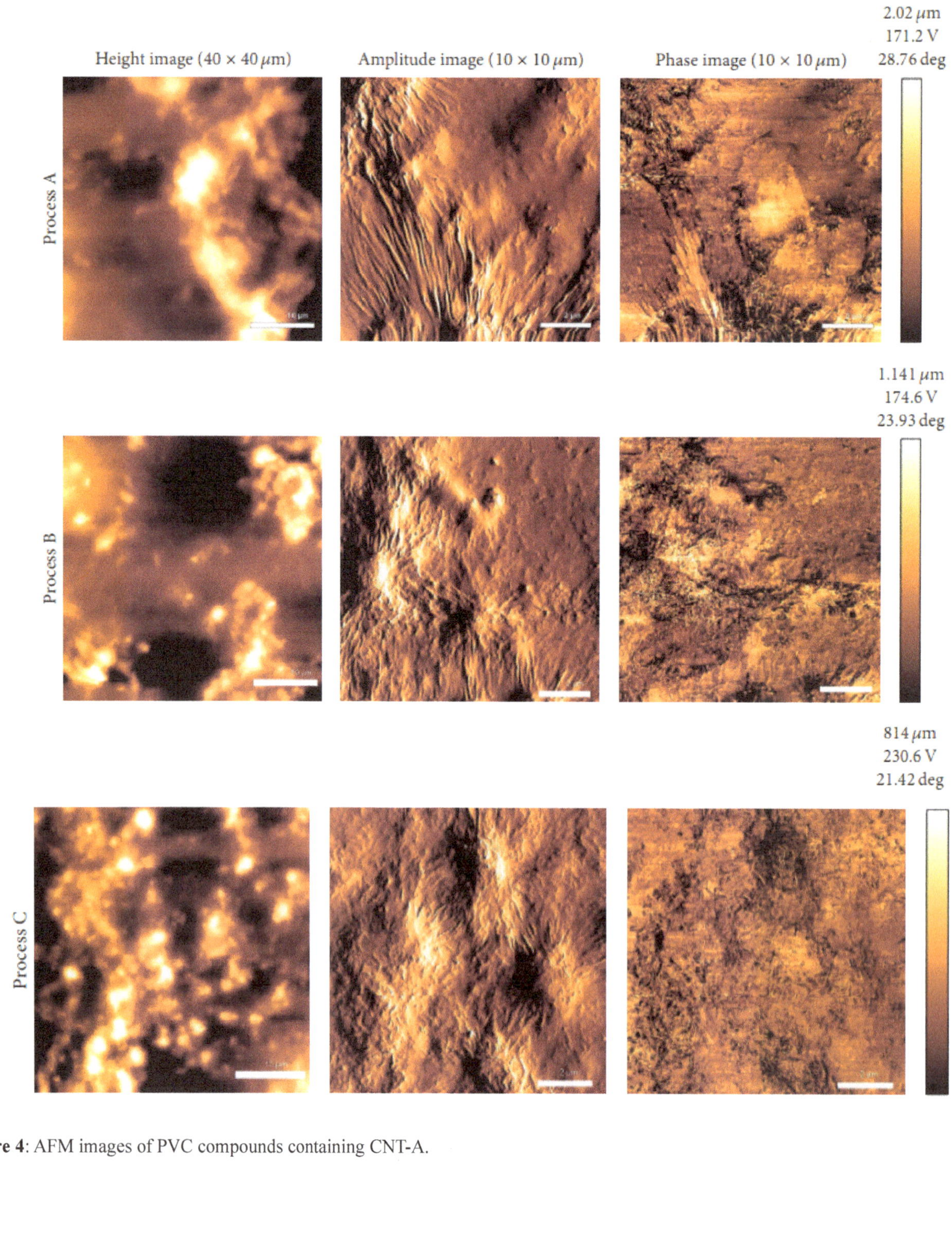

Figure 4: AFM images of PVC compounds containing CNT-A.

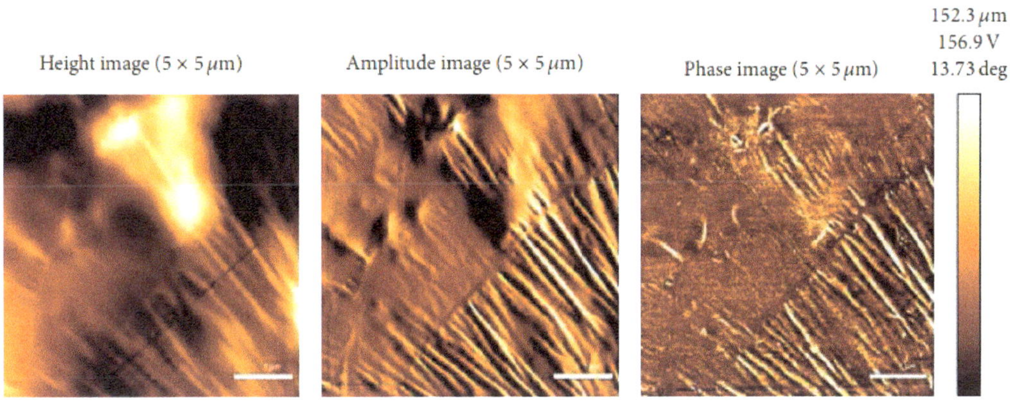

Figure 5: AFM images of PVC compound containing CNT-A; the cross section of the fibrous structures was determined along the black line in the amplitude image.

The AFM images demonstrate that by applying a processing method with a gap of 5 µm (process A) CNT bundles can be separated from the agglomerates and aligned. A splitting into primary CNT is not achieved by this way. Also, still larger agglomerates occur in the matrix.

The same applies to the force mode (processes B and C). The agglomerates are further separated, even though the application of force at the roller gap causes a break of the CNT into small fibrous segments. The AFM images in Figure 4 for process B and C show entangled CNT in the area of small agglomerates as well as between the CNT fragments. Longer CNT bundles were not detected anymore. This results in the conclusion that particle processing should ideally be conducted with a small gap and minimum force input. If the force is limited to <0.4 N, no damage of the CNT was observed.

Electrical Conductivity of PVC Nanocomposites Films

From the PVC compound containing nanoparticles, films of two different thicknesses (50 µm and 100 µm) were produced using the blade coating process. The surface resistance of these layers was determined according to DIN 54345 and used as value for the quality of incorporating the nanoparticles. The lower the surface resistance, the more favorable the distribution of the particles in the matrix. The surface resistance of those samples with the same layer thickness can be compared directly. For comparing samples with a different thickness the surface resistance was converted into the resistivity taking into account the thickness (1). From this again the electrical conductivity can be calculated (2)

$$\rho = R \times d, \quad \rho \text{ resistivity}, d \text{ thickness} \tag{1}$$

$$\sigma = \frac{1}{\rho}, \quad \sigma \text{ electrical conductivity.} \tag{2}$$

In order to investigate the influence of process parameters in material preparation on the film properties, films were prepared from the coating composition of each passage on the three roll mill and determined the surface resistance. The process parameters are given in Table 2.

Table 2: Process parameters for compound preparation at the three-roll mill

Designation	Gap 1	Gap 2
G15 + G10	Gap: 15 μm	Gap: 10 μm
G10 + G5	Gap: 10 μm	Gap: 5 μm
G5 + G5	Gap: 5 μm	Gap: 5 μm
G5 + F1	Gap: 5 μm	Force: 1 N
F3 + F5	Force: 3 N	Force: 5 N
F10 + F15	Force: 10 N	Force: 15 N

The results for the resistivity show already low values for the first processing step. As expected, the resistivity declines with decreasing gap setting. In principle, the curve progression is similar for all analyzed films. The minimum adjustable gap mode is 5 µm. As long as the roller gab is defined, the resistivity is reduced. Shifting to force mode processing, an increase of the resistivity is caused. This behavior is demonstrated by Figure 6showing the measuring values for PVC masses containing 3% and 4% CNT-A. In both cases, the resistivity increases during transition to force mode. Processing with a gap of 5 µm was determined as optimum. Therefore all coating compositions that differ in formulation but should be compared were prepared by this way. With a filler content of 3% CNT-A using this processing method a resistivity of 10^{-2} Ωm and with 4% CNT-A a resistivity of 5×10^{3} Ωm could be achieved. The diagrams in Figure 6 show that the values of the resistivity calculated from the surface resistance of films with a thickness of 50 µm (light column) and 100 µm (dark column) correlate well.

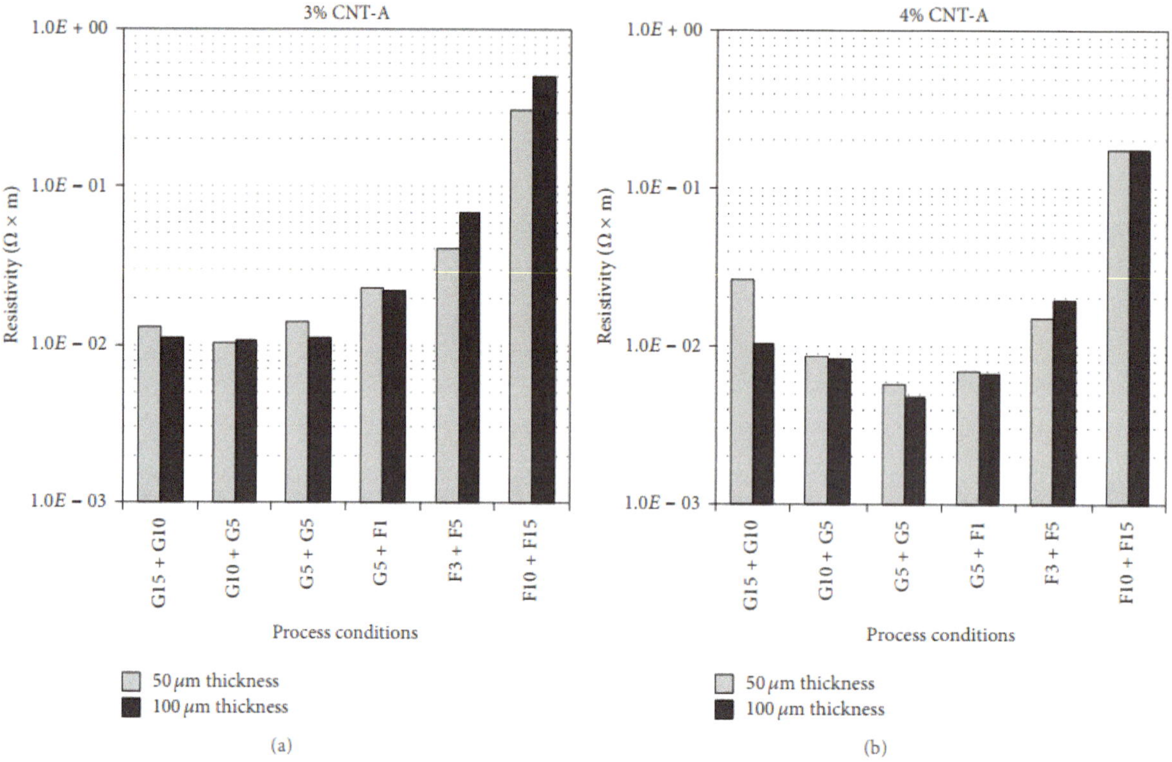

Figure 6: Resistivity of PVC films containing 3% (a) and 4% (b) CNT-A processed under various conditions; G—gap mode; F—force mode (see Table 2).

Figure 7(a) shows the percolation graph for the particles CNT-A between 0% and 6% in the PVC film. The diagram shows that the resistivity is reduced already with a particle content of 0.5% from 10^8 Ωm (unfilled PVC sheet) down to 10 Ωm. For a content of 1% CNT-A, the resistivity is decreased further to 1 Ωm. For up to 3% particle content, further adding of one percent CNT-A, respectively, causes a further reduction of the resistivity by about one power of ten. Films containing 3% of CNT-A show a resistivity of 10^{-2} Ωm. If the CNT content is increased to 6%, the resistivity reduces to 2×10^{-3} Ωm. This corresponds to an electrical conductivity of 480 S/m. The compounds were prepared with titanate-based (additive 1) and silane-based additive (additive 2) and without dispersing additive. Differences in the resulting electrical properties were only determined for particle contents ≤2% (Figure 7(b)). The films containing the particles CNT-A without additive have the lowest values for the resistivity although the differences are rather small.

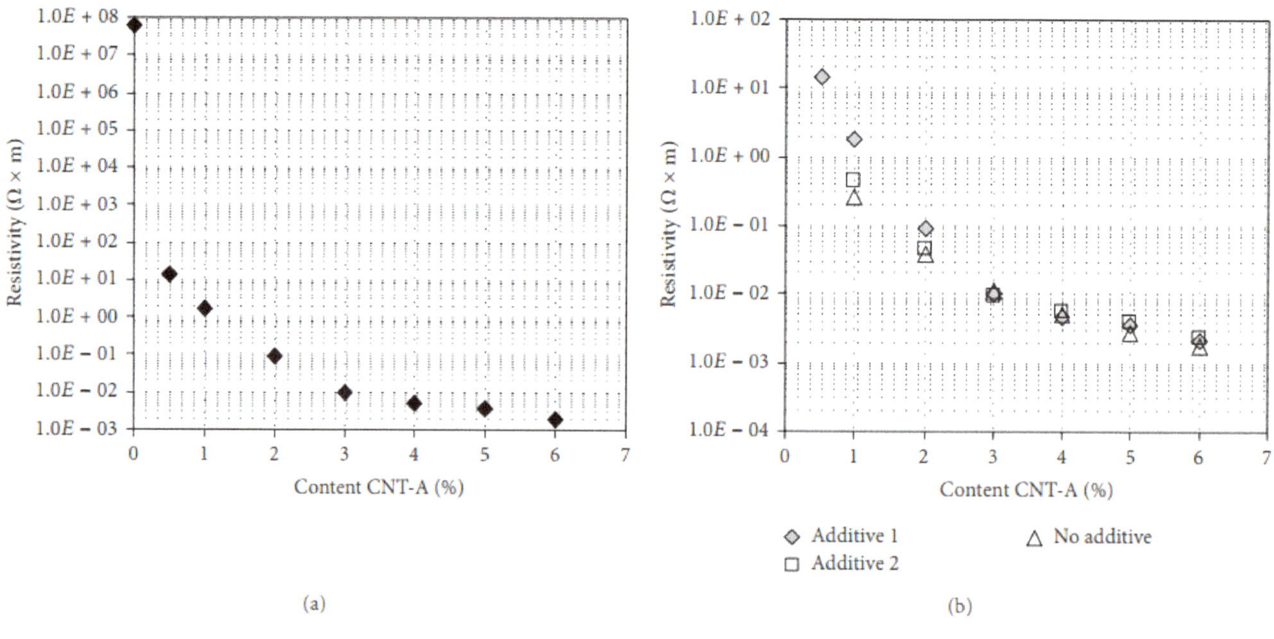

Figure 7: Resistivity of PVC sheets containing up to 6% CNT-A; (a) CNT-A in PVC containing additive 1 and (b) CNT-A in PVC various additives.

By combining CNT-A and particles with particle sizes in the micrometer range (particle D), a synergistic effect was achieved. The interaction of both particle types in the matrix causes a clear reduction of the resistivity. Different mass contents were set, and the effect on the resistance was analyzed. The result is shown in Figure 8. 2% and 5% carbon black (D) were added to a coating composition of 3% CNT-A. The resistivity of the sheets entailed a reduction from about 11×10^{-3} Ωm (3% CNT-A) to 6.4×10^{-3} Ωm (3% CNT-A + 2% D) or to 3.3×10^{-3} Ωm (3% CNT-A + 5% D). The latter roughly corresponds to the value which is also achieved with 5% CNT-A. In this case, 2% CNT can be replaced by 5% conductive carbon black. Adding 5% of conductive carbon black to a polymeric matrix containing 6% CNT-A reduced the resistivity from about 2×10^{-3} Ωm (6% CNT-A) to 6.6×10^{-4} Ωm (6% CNT-A + 5% D). This corresponds to an electrical conductivity of about 1,500 S/m. Compared to this, a film containing only 5% of conductive carbon black and no CNT particles has a high resistivity of about 0.445 Ωm (white column in Figure 8). The described effect could only be achieved with microparticles. Adding flaky nanoparticles only caused a reduction of the resistivity at a similar extent as by adding CNT-A. A synergistic effect can be achieved with the conductive carbon black; however, it was not found. The synergistic effect of CNT and microparticles was also observed in other polymers and is described as, for example, epoxy, high density polyethylene, and polyamide 12 [18–20].

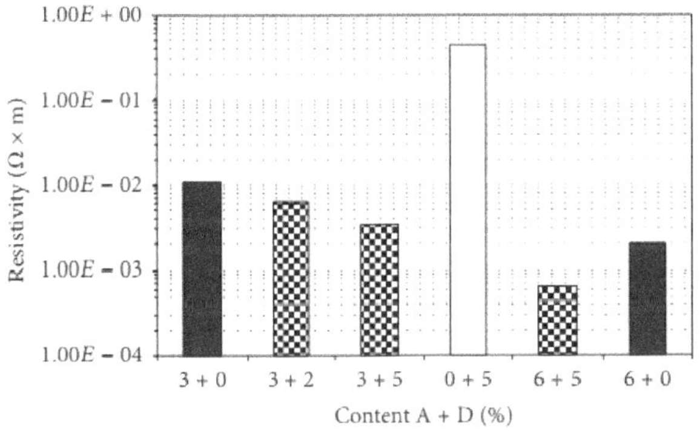

Figure 8: Resistivity of PVC films containing CNT (black), acetylene black (white), and both CNT and acetylene

CONCLUSIONS

Electrically conductive PVC films could be prepared by incorporating carbon nanotubes into the PVC solutions. The investigation showed that the preparation at the three-roll mill is suitable to split the CNT agglomerates into fibrous structures with a high aspect ratio.

For a first assessment, whether elongated structures were achieved, assessment of the flow curve is suitable. A structural viscous behavior is an indicator for the existence of such structures. Preparing of the compounds in gap mode with a minimum gap of ≤5 μm is considered to be the optimum. It could be shown that the long, fibrous CNT structures break into short pieces during processing in force mode. As a consequence the viscosity of the masses drastically increases and the films produced of that show a high resistivity. The atomic force microscopy (AFM) is suitable to visualize the arrangement and the size of the particles in the processed compounds.

Optimally treated compounds show a storage stability of at least two months which makes production and storage independently from usage.

Compound processing in gap mode using the particles CNT-A thin PVC sheets could be prepared, achieving an electrical conductivity of up to 480 S/m. Synergistic effects are obtained by adding conductive carbon black with particle sizes in the micrometer range. An electrical conductivity of 1,500 S/m was measured for the combination of 6% CNT-A and 5% conductive carbon black. This conductivity is sufficient to use these films as electric heating element after suitable contacting, for example, electrically heatable layer in PVC artificial leather or in textile laminates. The conductive PVC layer can be applied as a coat in a multilayer structure as well as separate film useable for laminating processes.

By adding conductive carbon black, the proportion of CNT particles can be optimized and thus the costs can be reduced. For the use in the antistatic range, CNT contents ≤0.5% are sufficient.

ACKNOWLEDGMENT

The research project "Modification of plasticized PVC with functional nanoparticles," reg. no. VF110026 was partially promoted from budget funds of the Bundesministerium für Wirtschaft und Technologie (BMWi—Federal Ministry of Economics and Technology). The authors would like to thank the support awarded.

REFERENCES

1. Z. Spitalsky, D. Tasis, K. Papagelis, and C. Galiotis, "Carbon nanotube-polymer composites: chemistry, processing, mechanical and electrical properties," Progress in Polymer Science, vol. 35, no. 3, pp. 357–401, 2010.
2. W. Bauhofer and J. Z. Kovacs, "A review and analysis of electrical percolation in carbon nanotube polymer composites," Composites Science and Technology, vol. 69, no. 10, pp. 1486–1498, 2009.
3. Y. Mamunya, A. Boudenne, N. Lebovka, L. Ibos, Y. Candau, and M. Lisunova, "Electrical and thermophysical behaviour of PVC-MWCNT nanocomposites," Composites Science and Technology, vol. 68, no. 9, pp. 1981–1988, 2008.
4. G. Broza, K. Piszczek, K. Schulte, and T. Sterzynski, "Nanocomposites of poly(vinyl chloride) with carbon nanotubes (CNT)," Composites Science and Technology, vol. 67, no. 5, pp. 890–894, 2007.
5. T. Sterzyński, J. Tomaszewska, K. Piszczek, and K. Skórczewska, "The influence of carbon nanotubes on the PVC glass transition temperature," Composites Science and Technology, vol. 70, no. 6, pp. 966–969, 2010.
6. J.-H. Shi, B. Yang, K. P. Pramoda, and S. H. Goh, "Enhancement of the mechanical performance of poly(vinyl chloride) using poly(n-butyl methacrylate)-grafted multi-walled carbon nanotubes," Nanotechnology, vol. 18, no. 37, Article ID 375704, 2007.
7. B. P. Möller, Herstellung, Charakterisierung und Weiterverarbeitung von Carbon Nanotube Dispersionen

[Dissertation], Würzburg, Germany, 2013.

8. J. Z. Kovacs, K. Andresen, J. R. Pauls et al., "Analyzing the quality of carbon nanotube dispersions in polymers using scanning electron microscopy," Carbon, vol. 45, no. 6, pp. 1279–1288, 2007.
9. J. Z. Kovacs, B. S. Velagala, K. Schulte, and W. Bauhofer, "Two percolation thresholds in carbon nanotube epoxy composites," Composites Science and Technology, vol. 67, no. 5, pp. 922–928, 2007.
10. F. H. Gojny, M. H. G. Wichmann, U. Köpke, B. Fiedler, and K. Schulte, "Carbon nanotube-reinforced epoxy-composites: enhanced stiffness and fracture toughness at low nanotube content," Composites Science and Technology, vol. 64, no. 15, pp. 2363–2371, 2004.
11. N. Starostina, M. Brodsky, S. Prikhodko, C. M. Hoo, M. L. Mecartney, and P. West, "AFM capabilities in characterization of particles and surfaces: from angstroms to microns," Pacific Nanotechnology, vol. 59, no. 3, pp. 225–232, 2008.
12. J. O. Aguilar, J. R. Bautista-Quijano, and F. Avilés, "Influence of carbon nanotube clustering on the electrical conductivity of polymer composite films," Express Polymer Letters, vol. 4, no. 5, pp. 292–299, 2010.
13. P. Pötschke, A. R. Bhattacharyya, A. Janke et al., "Melt mixing as method to disperse carbon nanotubes into thermoplastic polymers," Fullerenes Nanotubes and Carbon Nanostructures, vol. 13, no. supplement 1, pp. 211–224, 2005.
14. P. Pötschke, A. R. Bhattacharyya, and A. Janke, "Advanced Multifunctional Nanocarbon Materials and Nanosystems Carbon nanotube-filled polycarbonate composites produced by melt mixing and their use in blends with polyethylene," European Polymer Journal, vol. 40, pp. 137–148, 2004.
15. J. S. Bunch, T. N. Rhodin, and P. L. McEuen, "Noncontact-AFM imaging of molecular surfaces using single-wall carbon nanotube technology," Nanotechnology, vol. 15, no. 2, pp. S76–S78, 2004.
16. A. U. Yalovenko, "AFM investigation of carbon nanotubes," in Proceedings of the SPM 2003, 02.-05.03, pp. 171–173, Nizhny Novgorod, Russia, 2003.
17. K. Safarova, A. Dvorak, R. Kubinek, M. Vujtek, and A. Rek, "Usage of AFM, SEM and TEM for the research of carbon nanotubes," in Modern Research and Educational Topics in Microscopy, A. Mendéz-Vilas and J. Díaz, Eds., vol. 2, pp. 513–519, FORMATEX Microscopy Books, 2007.
18. R. Socher, B. Krause, S. Hermasch, R. Wursche, and P. Pötschke, "Electrical and thermal properties of polyamide 12 composites with hybrid fillers systems of multiwalled carbon nanotubes and carbon black,"Composites Science and Technology, vol. 71, no. 8, pp. 1053–1059, 2011.
19. J.-H. Lee, S. K. Kim, and N. H. Kim, "Effects of the addition of multi-walled carbon nanotubes on the positive temperature coefficient characteristics of carbon-black-filled high-density polyethylene nanocomposites," Scripta Materialia, vol. 55, no. 12, pp. 1119–1122, 2006.
20. J. Sumfleth, X. C. Adroher, and K. Schulte, "Synergistic effects in network formation and electrical properties of hybrid epoxy nanocomposites containing multi-wall carbon nanotubes and carbon black,"Journal of Materials Science, vol. 44, no. 12, pp. 3241–3247, 2009.

CHAPTER 4

PROPERTIES OF PTFE TAPE AS A SEMIPERMEABLE MEMBRANE IN FLUOROUS REACTIONS

Brendon A. Parsons[1,2], Olivia Lin Smith[1], Myeong Chae[1] and Veljko Dragojlovic[1]

[1]Wilkes Honors College of Florida Atlantic University, 5353 Parkside Drive, Jupiter, FL 33458, USA and [2]Department of Chemistry, University of Washington, Box 351700, Seattle, WA 98195-1700, USA

ABSTRACT

In a PTFE tape phase-vanishing reaction (PV-PTFE), a delivery tube sealed with PTFE tape is inserted into a vessel which contains the substrate. The reagent diffuses across the PTFE tape barrier into the reaction vessel. PTFE co-polymer films have been found to exhibit selective permeability towards organic compounds, which was affected by the presence of solvents. In this study, we attempted to establish general trends of permeability of PTFE tape to different compounds and to better describe the process of solvent transport in PV-PTFE bromination reactions. Though PTFE tape has been reported as impermeable to some compounds, such as dimethyl phthalate, solvent adsorption to the tape altered its permeability and allowed diffusion through channels of solvent within the PTFE tape. In this case, the solvent-filled pores of the PTFE tape are chemically more akin to the adsorbed solvent rather than to the PTFE fluorous structure. The solvent uptake effect, which was frequently observed in the course of PV-PTFE reactions, can be related to the surface tension of the solvent and the polarity of the solvent relative to the reagent. The lack of pores in bulk PTFE prevents solvents from altering its permeability and, therefore, bulk PTFE is impermeable to most solvents and reagents. However, bromine, which is soluble in liquid fluorous media, diffused through the bulk PTFE. A better understanding of the PTFE phase barrier will make it possible to further optimize the PV-PTFE reaction design.

INTRODUCTION

Fluorous reactions are a relatively new addition to the portfolio of synthetic methods [1–5]. In a fluorous triphasic reaction, introduced by Curran in 2001, a fluorous phase screen was used to simultaneously carry out the reaction and separate the products [6–7]. Soon after, Ryu and Curran reported the fluorous phase-vanishing (PV) method, which unlike the triphasic fluorous reaction, does not require the use of fluorous reactants [8]. In the course of the reaction, one phase vanishes, hence the name. An important requirement of PV reactions is that, in order to ensure an effective reaction, the rate of passive diffusion of the reagent through the fluorous phase to the substrate layer must be significantly greater than the diffusion of the substrate to the reagent layer [9]. Ryu and Curran's work was followed by numerous improved and modified designs including a quadraphasic/tetraphasic PV reaction [10], a photochemical PV reaction [11–12], a stacked phase-vanishing reaction [13], a PV reaction on neat substrates [11,14–16] and, most recently, a Grignard reaction [17]. Jana and Verkade reported a variation of the PV method that allows the use of

Citation: Parsons BA, Smith OL, Chae M, Dragojlovic V. Properties of PTFE tape as a semipermeable membrane in fluorous reactions. Gouverneur V, ed. Beilstein Journal of Organic Chemistry. 2015;11:980-993. doi:10.3762/bjoc.11.110.

lighter reagents without the need for a U-tube by dissolving one of the reagents in a heavy solvent [18]. Curran and Werner established that such a reaction operates by an extractive mechanism, rather than the passive diffusion-controlled reagent transport observed in typical phase-vanishing techniques [19]. Curran followed up with a study of miscibilities and solubilities of fluorous/organic phases [20].

Following the improvements that resulted from use of heavier polymeric liquid perfluoro compounds reported by Ryu et al. [10], we introduced PTFE (polytetrafluoroethylene, Teflon) tape as a phase screen [21–23]. In a PTFE tape phase-vanishing reaction (PV-PTFE), a delivery tube containing the reagent is sealed on both ends with PTFE tape. The PTFE tape covering the top of the delivery tube prevents the loss of the reagent over the course of the reaction. The delivery tube is then inserted into a vessel which contains substrate, either neat or dissolved in a solvent. The reagent diffuses across the PTFE tape barrier into the outer vessel (Fig. 1). The reaction generally takes place in the vessel, although in some instances due to the diffusion of the substrate through the PTFE tape the product also formed in the delivery tube. The PV-PTFE method is inexpensive, simple to use and more environmentally friendly. Furthermore, the delivery of the reagent can easily be stopped and the reaction products are easy to recover. PV-PTFE reactions have no limitations due to the densities of the reagents or substrates and they can be used in sequential or tandem reactions, or reactions under reflux [21–23]. We also utilized alternative PTFE-covered vials to simplify reagent handling [24].

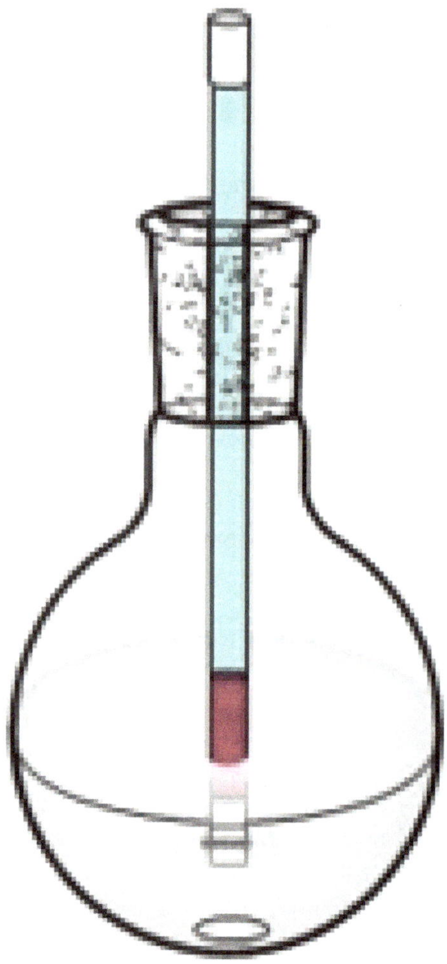

Figure 1: PV-PTFE reaction design.

The PV-PTFE method can be made more efficient by better understanding the phase barrier and its role in phase-vanishing reactions. Other PTFE films have been found to exhibit selective permeability towards organic compounds, and this permeability is affected by the presence of solvents. Weber and coworkers have reported such an investigation into the properties of Teflon AF (poly(2,2-bis(trifluoromethyl)-4,5-difluoro-1,3-dioxide-*co*-tetrafluoroethylene), a co-polymer with 13% PTFE content [25–26]. A Teflon AF 2400 tubing was also used in a reaction to deliver a gaseous reactant by Ley [27]. Weber carried out a quantitative study on Teflon AF films that were carefully cast and well-characterized [25–26]. As we used commercially available PTFE tape for which tolerances are considerably higher, drawing general reproducible conclusions may be more difficult.

PTFE has found use as both a catalyst carrier and a means of catalyst recovery by Gladysz [28] and as a support when taking IR spectra [29]. The most common everyday laboratory use of PTFE is as a protective coating of magnetic stirring bars. Bulk PTFE is chemically identical to PTFE tape. However, Dihn and Gladysz reported that it behaves very differently in reactions involving fluorous catalyst recovery, and that study into the morphology of PTFE was needed [30]. To better understand how bulk and tape PTFE perform, we investigated their properties with respect to some common reagents and reaction conditions. In this paper, which was designed as a preliminary study, we attempted to establish general trends of the permeability of PTFE to different compounds, to investigate whether organic solvents alter it, and to better describe the process of solvent transport in PV-PTFE bromination reactions. We also compared PTFE tape to bulk PTFE, which is typically encountered as an inert protective coating, such as on laboratory stir bars.

RESULTS AND DISCUSSION

Solvent Transport in the Course of a Simple PV-PTFE Diffusion of Bromine into Solvent

To better understand the solvent uptake phenomenon observed in PV-PTFE reactions with dichloromethane as a solvent [21] we investigated the addition of bromine through a PTFE-sealed delivery tube to a solvent that did not contain any substrate. In the delivery of bromine into dichloromethane, one may expect that the denser reagent in the delivery tube would simply diffuse downward into the less dense solvent. Indeed, when a delivery tube of bromine was placed in an empty flask, bromine vapors were observed diffusing through the PTFE tape and descending to the bottom of the flask [22]. However, when the tip of the delivery tube was immersed in dichloromethane, initially the solvent in the flask was quickly drawn up into the delivery tube. At this point, relatively little bromine had diffused into the solvent in the flask (Fig. 2). Next, after the column height of the bromine-solvent mixture reached a maximum, it slowly decreased. During this phase, most of the bromine was delivered to the flask.

(a) (b) (c) (d) (e)

Figure 2: Solvent uptake in the delivery of bromine into dichloromethane (a) 0 min, (b) 0.50 min, (c) 0.83 min, (d) 1.50 min, (e) 3.50 min, (f) 4.17 min, (g) 7.18 min, (h) 10.18 min, (i) 17.18 min, (j) 23.57 min. Note that a part of the delivery tube was obscured.

The solvent uptake also occurs with other organic solvents. We repeated the experiment with ethyl acetate and hexanes. Ethyl acetate exhibited solvent transport similar to that observed with dichloromethane. However, the column height reached a lower maximum and both the uptake and drawdown processes were slower (Fig. 3). Unfortunately, we were unable to obtain good results for hexanes, as the uptake of the solvent was rapid and the column of solvent overflowed the delivery tube.

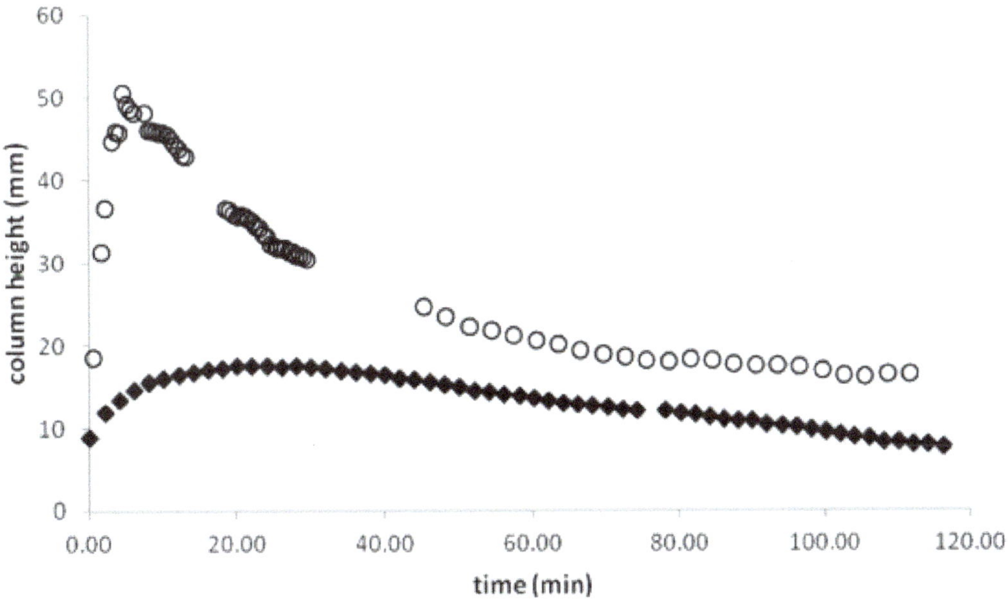

Figure 3: Solvent column heights of bromine delivery into dichloromethane (○) and ethyl acetate. (♦).

PTFE tape is easily stretched, which might cause a change in the size of pores. This may explain why in different trials we obtained slightly different results (Fig. 4). Variations were considerably greater in trials involving dichloromethane.

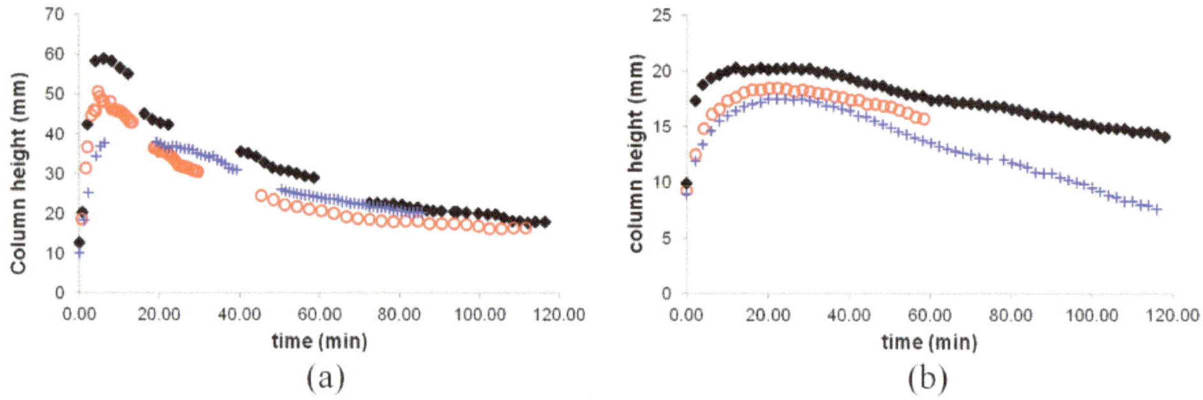

Figure 4: Reproducibility of bromine delivery into a) dichloromethane and b) ethyl acetate. In each case three different trials are shown.

Solvent transport in the course of PV-PTFE Bromination of Cyclohexene

To examine the solvent transport effect in the course of a reaction, we performed the delivery of bromine to an equimolar amount of cyclohexene dissolved in a solvent. Due to the previously described problem with hexanes overflowing the delivery tube, we investigated only reactions in dichloromethane and ethyl acetate. Fig. 5 represents the heights of the liquid columns in the delivery tubes over the course of the reaction. Again, there are gaps present due to the ground glass joints preventing the observation of the level of the bromine–solvent mixture in the delivery tube. As with the previous experiments, the trials with dichloromethane were less reproducible in terms of the maximum column height reached and the amount of time it took for the solvent drawdown to complete. Trials involving ethyl acetate were more reproducible. Most of the variation was in the time it took for the solvent column to reach its maximum. The amount of solvent that entered the delivery tube and the time until the delivery tube was drawn down were relatively consistent.

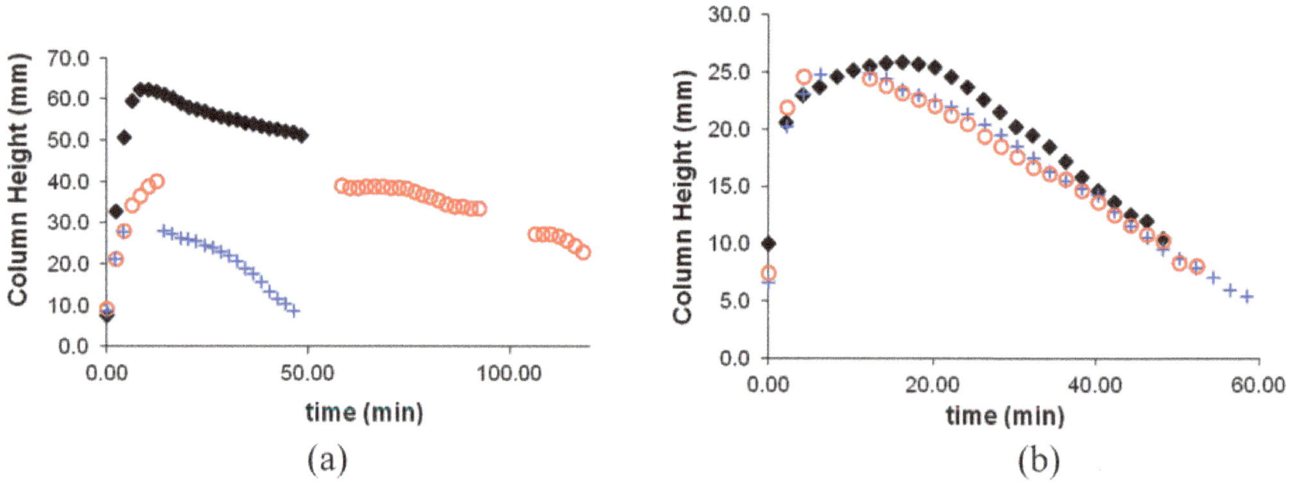

Figure 5: Height of the solvent column in the course of the bromination of cyclohexene in (a) dichloromethane and (b) ethyl acetate. In each case three different trials are shown.

Compared to the simple delivery of bromine into ethyl acetate, the bromination of cyclohexene in ethyl acetate resulted in a bromine–solvent column which rose faster and reached a higher maximum (Fig. 6). Additionally, in the presence of a substrate, the drawdown process was faster.

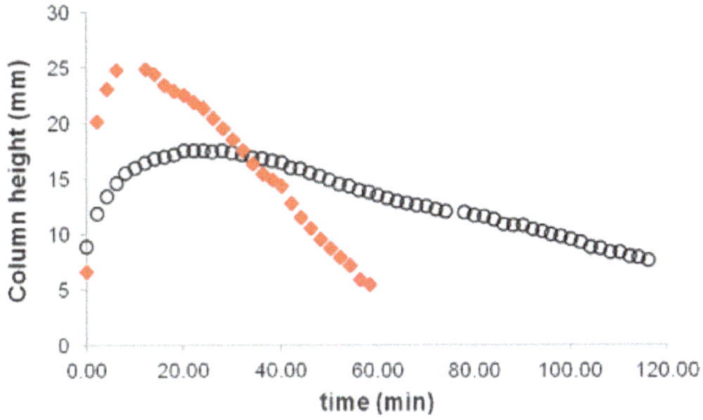

Figure 6: Height of the solvent column in the course of the bromination of cyclohexene in ethyl acetate (♦) and the delivery of bromine to neat ethyl acetate (○).

Therefore, we have found that the solvent uptake effect reported earlier [21] also occurs without a substrate. We have also found that the process of solvent transport varies with the solvent used. The implication for the PV-PTFE method is that even once a solvent is chosen, the rate at which the reagent is delivered may still vary between individual trials. A possible solution to this problem is to place bromine, or the reagent in general, dissolved in a suitable amount of solvent, into the delivery tube rather than using a neat reagent.

The solvent uptake process can be controlled to an extent by varying the depth to which the delivery tube is immersed in the solution. If the tube is immersed to a greater depth, the solvent can be drawn into the tube until equilibrium is reached. However, if the delivery tube is immersed into the solution to a shallow depth, the uptake into the delivery tube draws down the level of solvent in the reaction vessel, and the uptake stops once the level of the solution in the reaction vessel is below the bottom of the delivery tube. Then, the solution flows from the delivery tube back into the reaction vessel until the liquid level reaches the delivery tube again, at which point the uptake of the solvent resumes (Fig. 7).

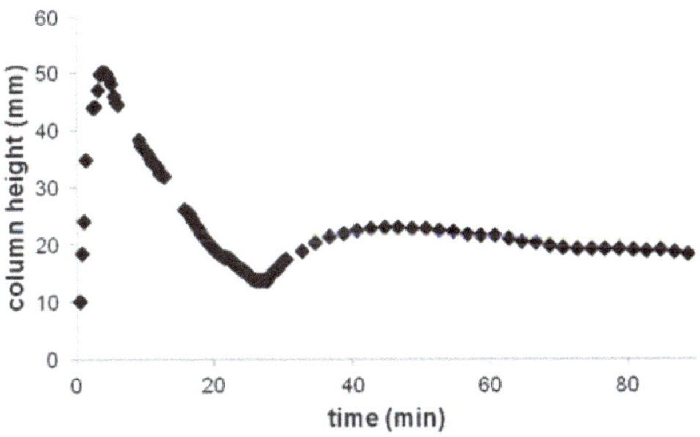

Figure 7: Solvent uptake when the delivery tube is inserted to a shallow depth. The solvent uptake stopped once the level of the solution reached the bottom of the delivery tube. After some solution flowed from the tube back into the reaction vessel and the liquid level reached the tube, the uptake of the solvent resumed.

By having the delivery tube just touch the surface of the solution in the reaction vessel, one can almost completely eliminate the solvent uptake as demonstrated by the PV-PTFE iodolactonization of the unsaturated diester **1** with iodine monochloride in dichlormethane as a solvent (Scheme 1 and Fig. 8).

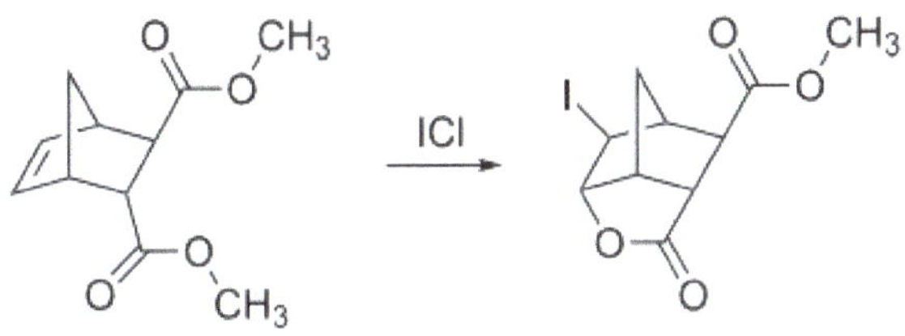

1 **2**

Scheme 1: Iodolactonization of unsaturated diester 1 with iodine monochloride in dichlormethane.

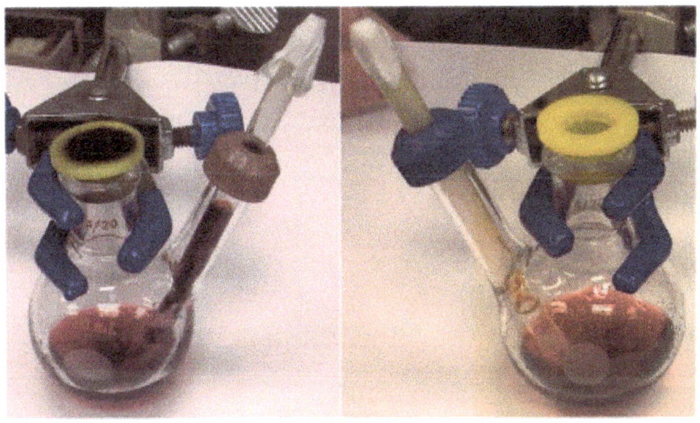

(a) (b)

Figure 8: (a) The delivery tube is immersed into the solution and there is a considerable solvent uptake. (b) The delivery tube is just touching the surface of the solution and there was almost no solvent uptake.

Solvent-assisted transport through PTFE Tape

In the earlier work we reported that PTFE tape is impermeable to dimethyl phthalate [21]. However, we suspected that solvent effects may alter the permeability of PTFE tape even when it is otherwise impermeable to a compound. In order to visualize the flow of solvents, a small amount of methyl red was dissolved in dimethyl phthalate. When a delivery tube containing dyed dimethyl phthalate was suspended in an empty beaker, the column height did not change at all, even after three days. However, when the same delivery tube was immersed in dichloromethane, the dyed dimethyl phthalate was able to flow out of the delivery tube almost immediately (Fig. 9–d). When a delivery tube of dyed dimethyl phthalate was similarly immersed in hexanes, the phthalate did not cross the PTFE tape barrier (Fig. 9–h).

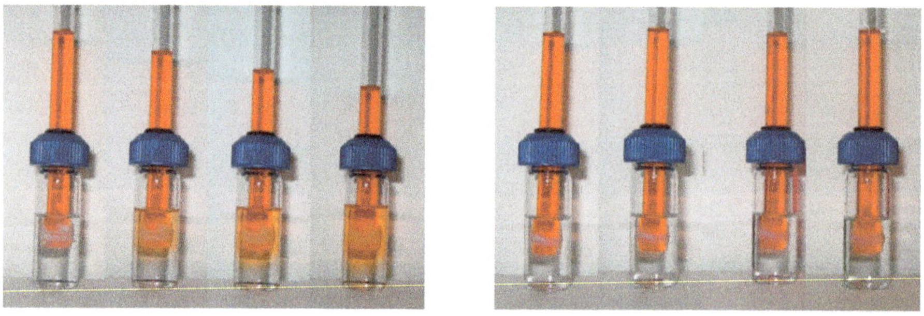

(a) (b) (c) (d) (e) (f) (g) (h)

Figure 9: Transport of dyed dimethyl phthalate in dichloromethane after (a) 0 h, (b) 1 h, (c) 2 h, (d) 3 h and dyed dimethyl phthalate in hexanes after (e) 0 h, (f) 1 h, (g) 2 h, (h) 3 h.

The delivery tube immersed in ethyl acetate exhibited a solvent-assisted phthalate delivery that was somewhat different compared to one observed with dichloromethane. The column height of the delivery tube increased indicating that, while dyed phthalate was exiting the delivery tube, ethyl acetate was also drawn into it. At some point (≈1–2 h into the process), the column height reached a maximum, which was followed by a relatively slow transport of the contents of the delivery tube into the outer vial (Fig. 10).

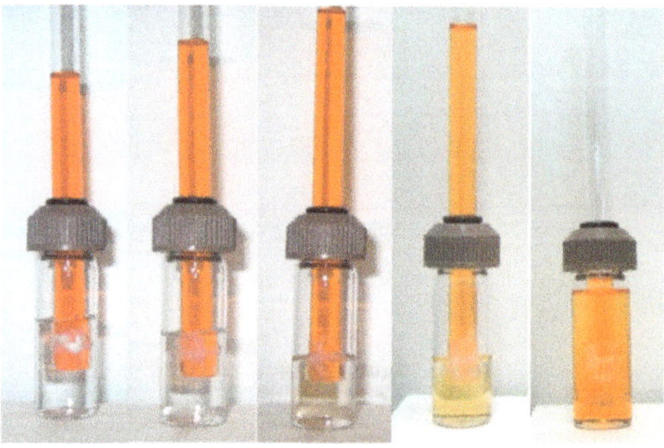

(a) (b) (c) (d) (e)

Figure 10: Transport of dyed dimethyl phthalate in ethyl acetate after (a) 0 h, (b) 0.17 h, (c) 1 h, (d) 3 h, (e) 30 h.

As with the experiments on the transport of bromine, it is apparent that the nature of the solvent plays a major role in the transport of dimethyl phthalate across the PTFE tape barrier. PTFE tape is characterized at the microscopic level as a highly-porous solid. These pores are estimated from scanning electron micrographs to be about 0.5 μm in size, though the pores may change in size and number if the tape is stretched [29,31]. Weber and coworkers found that solvent-saturated Teflon AF films are not truly fluorous and that the solvent creates a supported-liquid membrane which drastically changes the diffusion properties of solutes through the film [25–26]. Based on our experimental observations, the structure of PTFE tape, and results of investigation of Teflon AF reported by Weber, we propose that compatible solvents adsorb to the PTFE tape and fill in the voids in the tape. The solvent-filled pores of the PTFE tape then are more chemically akin to the solvent rather than the PTFE tape. Therefore, compounds that do not diffuse through PTFE tape, such as dimethyl phthalate, will be able to diffuse through in the presence of a suitable solvent. To be suitable as a transport medium, a solvent must meet two criteria.

First, the solvent must be capable of adsorbing to the PTFE tape. This characteristic is visibly displayed by "wetting" of the PTFE tape by the solvent and is related to the contact angle that results between the solvent and the solid. PTFE tape is wetted spontaneously by any solvent which has a critical surface tension (γ_C) of less than 27×10^{-3} N m^{-1} [32]. In the course of our experiments, we observed that dichloromethane (26.50×10^{-3} N m^{-1}) [33], hexanes (18.43×10^{-3} N m^{-1}) [33], and ethyl acetate (23.9×10^{-3} N m^{-1}) [34] all wet PTFE tape. Acetonitrile (29.29×10^{-3} N m^{-1}) [35], water (72.80×10^{-3} N m^{-1}) [33], and dimethyl phthalate (41.85×10^{-3} N m^{-1}) [36] did not wet PTFE tape.

Second, the compound of interest must be fairly soluble in the solvent which has filled the void space in the PTFE tape. Dimethyl phthalate is insoluble in solvents which are non-polar (aliphatic hydrocarbons such as hexanes) or highly-polar (water), but dissolves well in medium-polarity solvents such as dichloromethane and ethyl acetate [37]. Even though all the solvents used in this experiment adsorbed on the PTFE tape, dimethyl phthalate crossed the PTFE tape barrier only when the medium-polarity solvents (dichloromethane and ethyl acetate) were used. When either the non-polar hexanes or polar water (vide infra: chemiluminescence experiment) were employed, dimethyl phthalate did not diffuse through the tape-adsorbed solvent at a measurable rate.

Selective reagent transport in PV-PTFE Chemiluminescence

To visualize the transport across a PTFE barrier, we designed a PV-PTFE chemiluminescence experiment as a qualitative test. The known chemiluminescence reaction of the diaryl oxalate esters oxidized by hydrogen peroxide in the presence of rubrene was investigated (Scheme 2) [38–40]. Two solutions were prepared. One solution contained a mixture of diaryl oxalate and rubrene in dimethyl phthalate. The second solution was hydrogen peroxide dissolved in either water, or a water–organic solvent mixture.

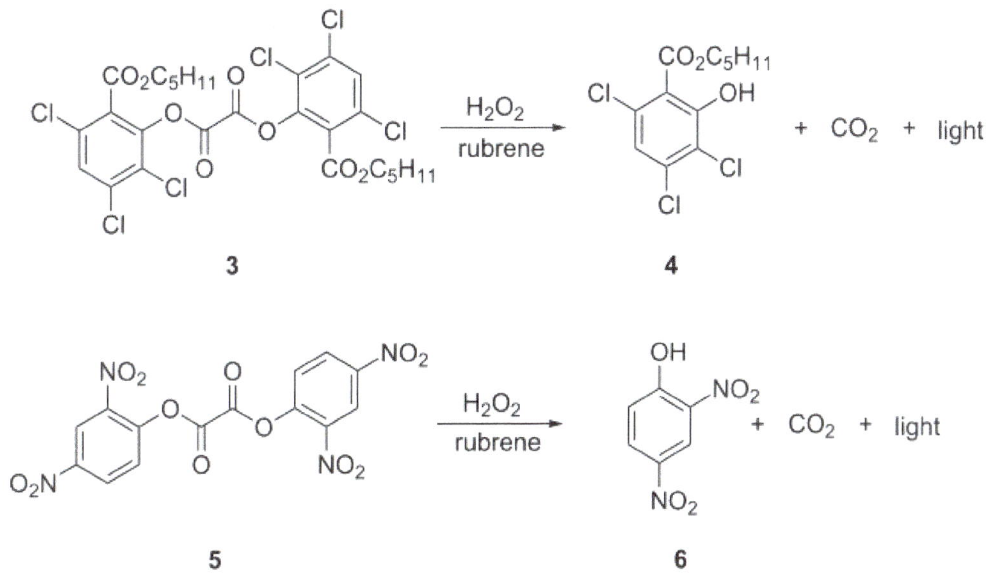

Scheme 2: Chemiluminescence reaction of diaryl oxalate esters oxidized by hydrogen peroxide in the presence of rubrene.

The solutions in the vial and delivery tube were alternated to ensure that the direction of diffusion was not gravity or density-dependent. We observed that the chemiluminescence reaction was directional. Only the oxalate–rubrene solution exhibited luminescence during the reaction, regardless of whether it was in the delivery tube or the vial (Fig. 11). The peroxide solution remained dark.

Figure 11: When the diaryl oxalate was oxidized by aqueous peroxide solution, chemiluminescence was observed only in the oxalate–rubrene solution, which was (a) in the delivery tube in the experiment on the left and (b) in the vial in the experiment on the right. (c) The experimental set up with the lights on and (d) lights off. The orange solution contains an oxalate/rubrene mixture in dimethyl phthalate and the clear solution is 10% aqueous H_2O_2.

The reactions with the aqueous peroxide solution proceeded with no change in the column height of the oxalate–rubrene solution in the delivery tube (Fig. 12), even over an extended period of time. This indicates that solvent transport was not necessarily a requirement for the delivery of the H_2O_2 reagent. From the directionality of the PV-PTFE chemiluminescence reaction and lack of change in solvent volume in the delivery tube, we conclude that PTFE tape is permeable to small H_2O_2 molecules but not to the considerably larger oxalate or rubrene molecules.

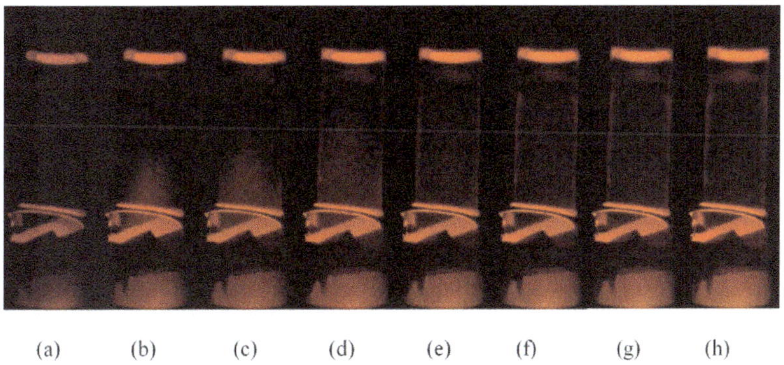

Figure 12: Progression of PV-PTFE chemiluminescence with aqueous peroxide solution in the vial after (a) 10 min, (b) 30 min, (c) 50 min, (d) 70 min, (e) 90 min, (f) 110 min, (g) 130 min, (h) 180 min.

In the reactions where hydrogen peroxide was dissolved in acetonitrile/water (2:1) or *tert*-butanol/water (2:1) mixtures, the column height changed during the reaction. This effect was similar to the phenomenon observed in the course of PV-PTFE brominations and solvent-assisted transport of dimethyl phthalate. When the peroxide–solvent solution was in the delivery tube, the column height decreased over the course of the reaction; when the peroxide–solvent solution was in the vial, the column height of the oxalate–rubrene solution in the delivery tube increased (Fig. 13). Thus, some acetonitrile and *tert*-butanol diffused through the PTFE barrier. Therefore, we observed that some solvents such as acetonitrile are able to diffuse through PTFE tape, even if though they do not wet the tape by adsorption [32].

(a) (b) (c) (d) (e) (f) (g) (h)

Figure 13: Progression of PV-PTFE chemiluminescence with acetonitrile–aqueous peroxide solution in the vial after (a) 10 min, (b) 15 min, (c) 20 min, (d) 25 min and *tert*-butanol–aqueous peroxide solution in the vial after (e) 5 min, (f) 10 min, (g) 15 min, (h) 20 min.

Finally, when peroxide in *tert*-butanol/water (2:1) was the oxidant and 9,10-bis(phenylethynyl)anthracene the acceptor, solvent-assisted diffusion of dimethyl phthalate was visualized (Fig. 14). One can observe bright green streams of dense dimethyl phthalate solution as it diffuses through PTFE tape and pools at the bottom of the vial.

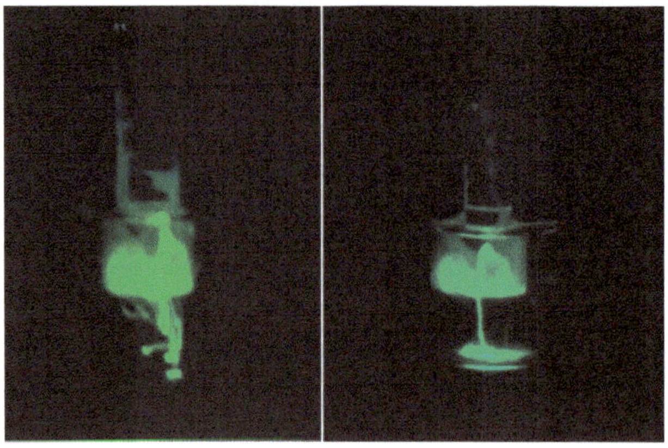

Figure 14: Diffusion of dimethyl phthalate assisted by *tert*-butanol through PTFE was visualized in a chemiluminescence reaction in which 9,10-bis(phenylethynyl)anthracene and an oxalate ester in dimethylphthalate diffused into a solution of hydrogen peroxide (10%) in a mixture of tert-butanol and water (2:1).

Solvent-free Diffusion of Bromine through PTFE Tape

Solvent-free diffusion of bromine through PTFE tape was observed when a PV-PTFE bromination was carried out in the absence of a solvent on a neat substrate [22]. Furthermore, when a delivery tube of bromine was placed in an empty container, it was relatively quickly filled with bromine vapors. As an additional test of bromine diffusion through PTFE tape in the absence of a solvent, we utilized the reaction of bromine with aluminum, which discolors the metal. A single drop of bromine was placed on two sections of an aluminum bar; one section was exposed aluminum (Fig. 15), the other section was covered with a single layer of PTFE tape (Fig. 16). The bromine was allowed to completely evaporate (Fig. 15 and Fig. 16), at which point the PTFE tape was removed. We found that the discoloration of the aluminum that was covered with PTFE tape was consistent with the discoloration of the bare aluminum (Fig. 15 and Fig. 16). Therefore, bromine readily and at a high rate diffused through PTFE tape in the absence of a solvent.

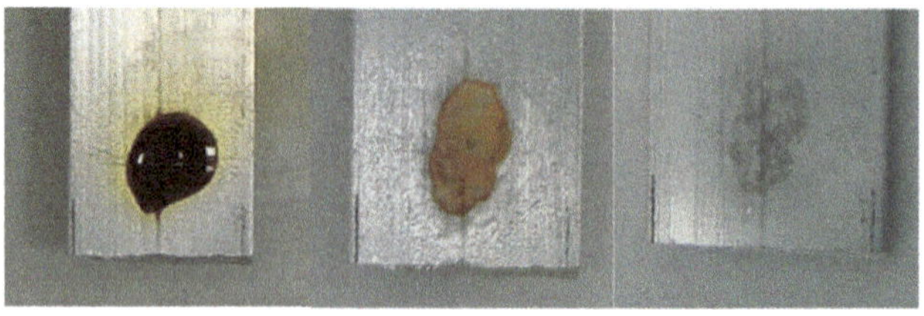

(a) (b) (c)

Figure 15: Corrosion of aluminum resulting from bromine applied directly to metal.

(a) (b) (c)

Figure 16: Discoloration of aluminum from bromine applied to PTFE tape on metal.

Comparison of PTFE tape to Bulk PTFE

Dihn and Gladysz reported that in the course of a recovery of fluorous catalysts, the catalysts attached to the PTFE tape, but not the PTFE-coated stir bar that was used in the reaction [30]. This difference in properties may be due to the microscopic structural differences between PTFE tape and the bulk PTFE used in inert coating applications. Unlike the porous structure of PTFE tape [29], bulk PTFE is a solid matrix with little in the way of surface irregularities [41]. Conversely, PTFE tape is characterized at the microscopic level as a highly-porous solid. As mentioned earlier, the pores are estimated from scanning electron micrographs to be about 0.5 μm in size and they may change in size and number if the tape is stretched [29,31]. The non-porous nature of bulk PTFE precludes the adsorption of other materials. This may be the reason why solvents do not spontaneously wet bulk PTFE as they do with PTFE tape. However, we predicted that since bromine is somewhat soluble in liquid fluorous media [19], it may exhibit similar solubility in solid PTFE and may be capable of diffusing through a bulk PTFE barrier.

We prepared a bulk PTFE vessel by cutting apart a laboratory stirring bar and removing the iron magnet. After cutting stirring bars open, we found that some of the iron bars were corroded (Fig. 17). While we cannot exclude a possibility that the manufacturer used corroded bars, a more likely explanation is that they underwent corrosion with use. PTFE shell was cleaned, filled with bromine, closed with a septum and immersed in a vial containing dichloromethane (Fig. 18 and b). To ensure that any bromine vapors that may leak through the stopper or a connection of it to PTFE vessel did not pass into the dichloromethane-containing vial, we used a Wheaton connector as a vial stopper and secured the PTFE container in it. Thus, the stopper on the PTFE container was external to the vial and any leak at it would be released outside the vial. We found that bromine did diffuse through the shell and that there was considerable variation between different shells of the same size (Fig. 18 and b). To confirm that this was not a solvent-induced effect, we also repeated the experiment in an empty sealed vial. Bromine vapors filled the vial (Fig. 18) and eventually liquid bromine was observed (Fig. 18). Thus, though PTFE tape is chemically identical to bulk PTFE, we see that its microscopic structure markedly alters its behavior.

Figure 17: After stirring bars were cut open, some iron bars were found to be corroded.

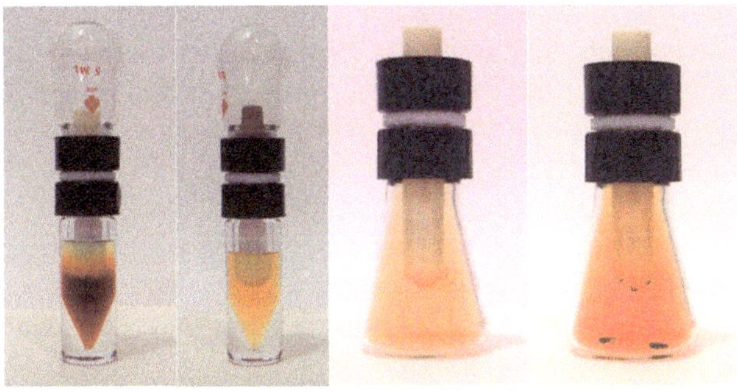

(a) (b) (c) (d)

Figure 18: (a) Diffusion of bromine through a bulk PTFE from stirring bar into dichloromethane after 2 h. (b) Diffusion of bromine through a bulk PTFE from a different stirring bar of the same size into dichloromethane after 14 h. (c) Diffusion of bromine through a bulk PTFE into an empty flask after 2 h. (d) Diffusion of bromine through a bulk PTFE into an empty flask after 12 h.

We also tested the permeability of PTFE tubing in a set up that consisted of a two-neck flask and a PTFE tube. The PTFE tube entered the flask through two septa while both of its openings were sealed and were external to the flask (Fig. 19). In this experiment, we again observed diffusion of the bromine across a bulk PTFE barrier.

Figure 19: Diffusion of bromine through a PTFE tube.

While bromine was observed to diffuse through the bulk PTFE, we did not observe any diffusion of solvents, including dichloromethane and ethyl acetate, when they were similarly contained in the PTFE container made from a stirring bar.

Even though bromine diffuses through the bulk PTFE, as most solvents and substrates cannot diffuse through it, a reaction of the reagent with the stirring bar is not expected to have a direct effect on the reaction outcome. To test this hypothesis, we carried out a reaction of bromine with benzene. In the absence of a catalyst there is only a very slow reaction. However, in the presence of a Lewis acid catalyst there is a fast bromination reaction. Since iron reacts with bromine to produce iron(III) bromide, a stirring bar may conceivably catalyze the reaction provided that both bromine and benzene can diffuse through bulk PTFE. However, we found that with a stirring bar present, there was no acceleration of the reaction compared to the control (Fig. 20). There was a small increase in the conversion of benzene to bromobenzene starting after 8 days. For a comparison, a reaction was carried out with a cut stirring bar, so that its iron core was exposed, resulted in a very fast reaction and the formation of bromobenzene in a high yield (Fig. 20).

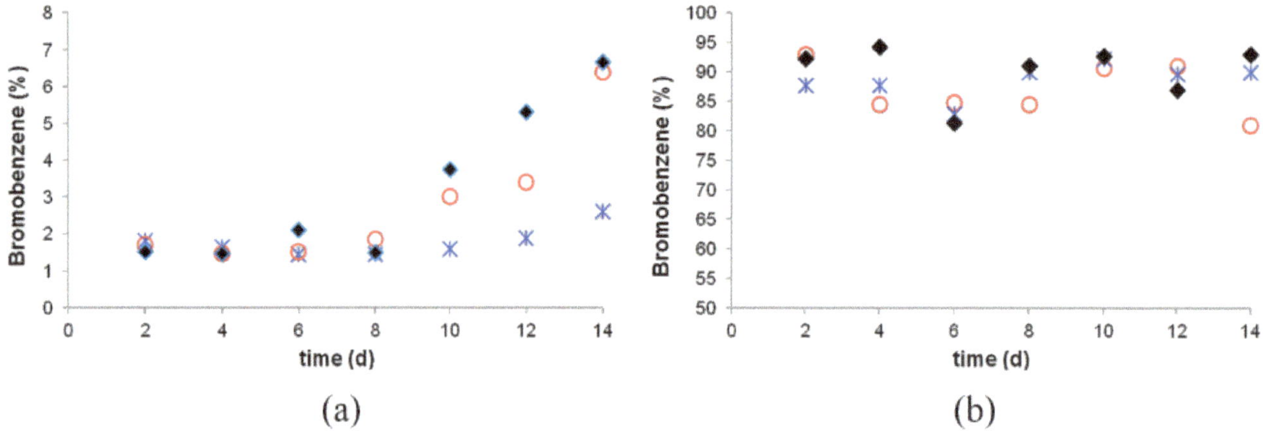

Figure 20: (a) The reaction of benzene and bromine in the absence of a stirring bar (✕), in the presence of a new stirring bar (○) and a used stirring bar (♦). (b) The reaction of benzene and bromine stirred by a cut stirring bar with the iron core exposed (three trials shown).

Magnetic stirring bars have been in use for a rather long time and, had they had any systematic effect on the reaction outcome, it likely would have been noticed a long ago. Nonetheless, one cannot exclude the possibility that there may be an effect due to diffusion through the PTFE coating in rare cases. Moreover, the diffusion of bromine and similar reagents through bulk PTFE may still affect the reaction outcome, as the stirring bar may consume some of the reagent. Finally, we would like to point out that we experienced some reproducibility problems (see Supporting Information File 1) and additional investigation of possible effects of PTFE-coated stirring bars on the reaction outcome may be warranted.

CONCLUSION

To better understand how PTFE tape functions as a semipermeable membrane, we examined its behavior and compared it to bulk PTFE. We included both simple diffusion processes and chemical reactions to examine the effects of solvents and reagents on the permeability of the tape. We confirmed that when PV-PTFE reactions were carried out in a solvent, there was a solvent uptake effect. We found that the solvent uptake phenomenon varied according to the solvent, and that the effect is somewhat variable, especially with dichloromethane. By altering the solvent used in a PV-PTFE reaction, one may be able to control the rate of delivery of a reagent.

Though PTFE tape has been reported as impermeable to some compounds, such as dimethyl phthalate, we found that solvent adsorption to the tape alters its permeability. We believe that the porous microscopic structure of PTFE tape adsorbs solvents which fill in the void spaces. When the pores are filled with a suitable solvent, the compounds that cannot diffuse through PTFE tape instead dissolve in the adsorbed solvent, and then diffuse through channels of solvent within the PTFE tape. In this case, the PTFE tape is chemically more akin to the adsorbed solvent than its own fluorous structure. From the limited data available, it appears that this solvent effect can be related to the surface tension of the solvent and the polarity of the solvent relative to the reagent. If the surface tension is within the range necessary to adsorb to the pores of the PTFE tape, and the polarity is such that the reagent is soluble in the solvent, the otherwise impermeable reagent will be able to diffuse across the PTFE tape barrier.

Finally, the microscopic structural differences between PTFE tape and bulk PTFE are responsible for the observed differences in functionality. The lack of pores in bulk PTFE prevents solvents from altering its permeability. However, bromine diffused through the bulk PTFE. While our investigation was limited to bromine, other reagents soluble in liquid fluorous media should be tested with bulk PTFE.

EXPERIMENTAL

Solvent transport experiments

The 10 mL round bottom flask and delivery tube set-up (Fig. 2) was placed next to a ruler within the camera frame. A Canon PowerShot S500 Digital Elph was set to take pictures on a time interval using Canon RemoteCapture software. The images were recorded in macro mode with auto-focus and flash enabled and ISO set to automatic. The maximum column height reached by the solution was quantified by measuring the number of vertical pixels in the column and correlating this number to pixel length of the ruler. The round bottom flask was filled with 5.0 mL of organic solvent. The delivery tube was inserted into the flask so that the lower tip of the delivery tube was immersed in the solvent. This process was repeated in triplicate with each solvent.

Solvent-free Diffusion of Bromine through PTFE Tape

A single drop of bromine was placed on two sections of an aluminum bar; one section was exposed aluminum, the other section was wrapped with a single layer of PTFE tape. Bromine was allowed to completely evaporate, at which point the PTFE tape was removed. This section of the aluminum was then visually inspected for corrosion consistent with the section of bare metal exposed to bromine.

Comparison of PTFE tape to bulk PTFE

A PTFE-coated stir bar was cut apart, the inside magnet was removed, and any contaminants were removed by immersion in concentrated hydrochloric acid. Bromine (0.30 mL) was added to the hollow PTFE shell and its top closed with a rubber septum. The bromine-filled shell was then immersed in dichloromethane. In an alternative experiment, PTFE tubing was filled with bromine and run through a two-necked flask containing dichloromethane, such that the PTFE portion of the tubing was immersed in the solvent and both ends of the tubing were outside of the sealed flask.

Effect of a PTFE Stirring Bar on Bromination of Benzene

A stock solution of 15.0 mL of benzene, 0.50 mL of bromine, and 0.50 mL of *n*-pentadecane (internal standard) was prepared in the dark. Seven 4 mL vials were prepared and 2.1 mL of the stock solution was added to each vial under a red light. Vial 1 was clear and the remaining were amber. Vials 1 and 2 were control experiments while stirring bars were added to the vials 3–7. The stirring bars were: 3) cut so that the iron bar was exposed, 4) unused small (0.2 cm), 5) unused large (0.4 cm), 6) used large, and 7) used small. All reactions were carried out in the dark. Periodically, samples were taken from each vial under a red light and analyzed by means of GC–MS.

SUPPORTING INFORMATION

Experimental Procedures and Additional Experimental Details

Chemicals and supplies All reagents (bromine, cyclohexene, 30% aqueous hydrogen peroxide, rubrene, 9,10-bis(phenylethynyl)anthracene, bis(2,4-dinitrophenyl)oxalate and bis(2-carbopentyloxy-3,5,6- trichloropheny)oxalate) and solvents (benzene, dichloromethane, ethyl acetate, hexane, dimethyl phthalate, acetonitrile, and tert-butanol) were obtained from Sigma-Aldrich or Fisher Scientific and were used as purchased. Dimethyl ester 1 was prepared according to a previously published procedure [1]. Taega Technologies High Density PTFE tape (1-inch width) was purchased from Fisher Scientific (Cat. No. 14610121). The PTFE syringe tube was purchased from Aldrich (gauge 16, 12 in long, cat. no. Z126772).

Construction of Delivery Tubes

A delivery tube was made of a glass tube which was sealed on each end with a PTFE tape. In the preparation of a delivery tube, one end of the tube was first covered with the PTFE tape; this was the end of the tube which was inserted into the substrate-containing vessel. The reagent was then added to the tube, and the other end was sealed with PTFE tape to reduce any evaporative loss. Care must be taken to stretch the PTFE over the end of the tube in only the lengthwise direction of the tape. PTFE tape is highly stretchable in the widthwise direction and such stretching reduces the thickness of the tape, which may alter the rate at which reagent diffuse across it. A variety of glass tubes are suitable for construction of delivery tubes. It was most economical to construct the delivery tubes from disposable glass Pasteur pipettes. Van Zee and Dragojlovic provide an illustrated guide to the construction of delivery tubes in their Supporting Information section [2].

Solvent Transport in The Course of a Simple PV-PTFE Diffusion of Bromine into Solvent

Experimental: The 10 mL round bottom flask and delivery tube set-up (Figure S1) was placed next to a ruler within the camera frame. A Canon PowerShot S500 Digital Elph was set to take pictures on a time interval using Canon RemoteCapture software. The images were recorded in macro mode with auto-focus and flash enabled and ISO set to automatic. The maximum column height reached by the bromine solution was quantified by measuring the number of vertical pixels in the column and correlating this number to pixel length of the ruler. Additionally, a stopwatch was also placed in-frame so that the time to reach the maximum height was known. In each experiment, 0.10 mL (0.20 mmol) of bromine was placed into the delivery tube, which was sealed on both ends with PTFE tape. The round bottom flask was filled with 5.0 mL of organic solvent. The delivery tube was inserted into the flask so that the lower tip of the delivery tube was immersed in the solvent. This process was repeated in triplicate with each of three solvents: dichloromethane, ethyl acetate, and hexanes.

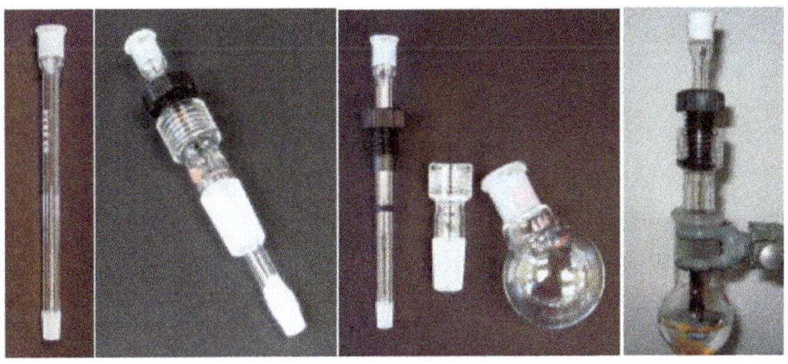

Figure S1: Set-up for PV-PTFE delivery of bromine into solvent.

Solvent Transport in the Course of PV-PTFE Bromination of Cyclohexene

Experimental: Cyclohexene (0.20 mmol) was dissolved in 5.0 mL of solvent (dichloromethane, ethyl acetate, and hexanes) in a 10 mL round bottom flask. Using the same set-up as in the previous experiment, 0.20 mmol of bromine was delivered from a PTFE-tipped delivery tube to react with the cyclohexene dissolved in the solvent. This reaction was repeated in triplicate with dichloromethane and ethyl acetate.

Additional information about Figure 8 in the manuscript. Figure 8 in the manuscript shows the iodolactonization of the unsaturated diester 1 shown below.

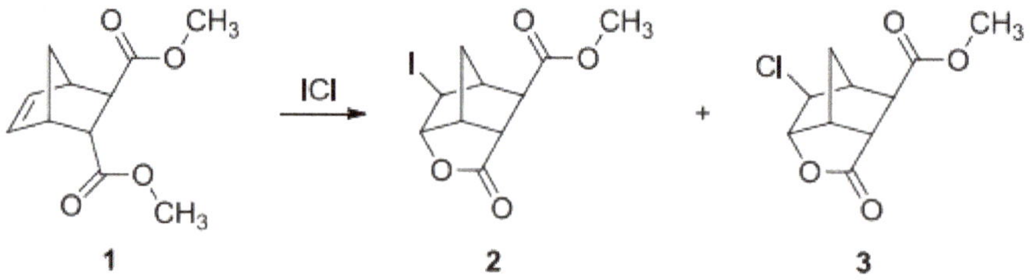

Iodolactone 2 was obtained in 93–99% yield in various trials. The conversion of the starting materials was complete and the product was pure. No other byproducts, including chlorolactone 3, were observed. Work up consisted of treatment with aqueous thiosulfate and filtration of the crude product through a short plug of silica gel. Iodolactone 2 is a known compound and we reported its analytical data in an earlier publication [3].

Solvent-assisted Transport through PTFE Tape

Experimental: A small amount of methyl red was dissolved in dimethyl phthalate (10.0 mL). Several delivery tubes were prepared by cutting off the tips of glass pipettes, capping the flat tops with PTFE tape, then inverting the pipettes so that they could be filled with the dyed phthalate. The prepared delivery tube were secured into 4 mL vial caps, then each was immersed in 2.0 mL of an organic solvent (dichloromethane, ethyl acetate, or hexanes) (Figure S2).

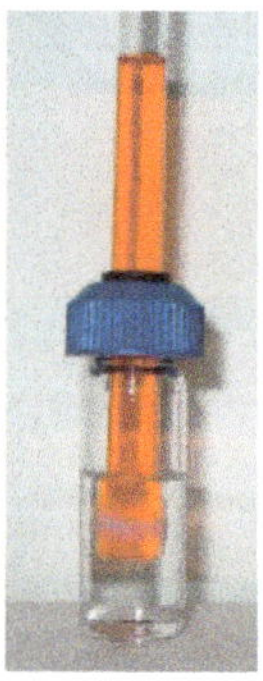

Figure S2: Set-up for dimethyl phthalate transport experiments.

Selective reagent transport in PV-PTFE Chemiluminescence

Experimental: The set-up of this experiment was identical to that of the previous section testing the permeability to dimethyl phthalate. Two solutions were prepared for each reaction. An oxidizer solution of 10% hydrogen peroxide (H_2O_2) was prepared by diluting by volume 30% aqueous H_2O_2 with each of three solvents: water, acetonitrile, or tert-butanol. The second solution was 0.030 M solution of either bis(2-carbopentyloxy-3,5,6-trichlorophenyl)oxalate or bis(2,4-dinitrophenyl)oxalate and 0.0025 M rubrene in dimethyl phthalate. The reactions were performed in two different experimental set ups. The first reaction was set up with the peroxide solution in the delivery tube and the oxalate–rubrene solution in the vial (Figure S3a). In the second set up, the oxalate–rubrene solution was placed in the delivery tube with the peroxide solution in the vial (Figure S3b). Observations were made by means of time interval photography, using the same equipment as in previous experiments. The images were recorded in macro mode with auto-focus enabled, flash disabled, and ISO set to 400.

Figure S3: Set-up for chemiluminescence experiments.

Additional information about Figures 11–14 from the manuscript. The light generated by the reactions in Figures 11a, 11b and 12 is orange in color due to the dyeing effect of rubrene in relatively high (0.0025 M) concentration. The concentration of rubrene was reduced to 0.00041 M for the Figures 11c, 11d, 13 and 14. Following this change, the chemiluminescence reactions gave off light of the expected yellow color. In Figure 14, 0.0020 M solution of 9,10-bis(phenylethynyl)anthracene was used. A photograph alternative to Figure 14, in which rubrene was used, was published in our earlier review article [4].

Comparison of PTFE tape to bulk PTFE

Effect of a PTFE stirring bar on bromination of benzene: Only bromobenzene was observed as a reaction product. Use of a large excess of benzene prevented possible problems with di- and tribromination. Bromination of pentadecane was not observed. The experiment suffered from numerous reproducibility problems and was rather difficult to remove all the external factors. All the reactions were carried out in the dark. Initially, rubber septa were used to seal the reaction vials. However, septa first swell, then hardened and finally cracked releasing some fumes out of the reaction mixtures. We replaced septa with plastic (Bakelite) caps. That in turn meant that we had to open each vial to take a sample for analysis. As we did so, some hydrogen bromide, generated in the course of cut stirring bar reactions, entered other vials and catalyzed bromination of benzene. At the end, only the vials in which the same experiment was carried out were kept in the same dark cupboard. That ensured reasonably reproducible reaction conditions. Still, about 25–30% of the experiments were outliers and bromobenzene was produced in higher amounts. Results presented in the manuscript are representative examples, excluding the outliers.

REFERENCES

1. Gladysz J A, Curran D P. Tetrahedron. 2002;58:3823–3825. doi: 10.1016/S0040-4020(02)00222-3.
2. Gladysz J A, Curran D P, Horváth I T, editors. The Handbook of Fluorous Chemistry. Weinheim: Wiley-VCH; 2004.
3. Horváth I T, editor. Fluorous Chemistry. Vol. 308. Berlin, Germany: Springer; 2012. ((Topics in Current Chemistry)).
4. Iskra J. Lett Org Chem. 2006;3:170–175. doi: 10.2174/157017806775789976.
5. Ryu I, Matsubara H, Nakamura H, Curran D P. Chem Rec. 2008;8:351–363. doi: 10.1002/tcr.20161.
6. Nakamura H, Linclau B, Curran D P. J Am Chem Soc. 2001;123:10119–10120. doi: 10.1021/ja011716c.
7. Luo Z, Swaleh S M, Theil F, Curran D P. Org Lett. 2002;4:2585–2587. doi: 10.1021/ol026232f.
8. Ryu I, Matsubara H, Yasuda S, Nakamura H, Curran D P. J Am Chem Soc. 2002;124:12946–12947. doi: 10.1021/ja027965y.
9. Nakamura H, Usui T, Kuroda H, Ryu I, Matsubara H, Yasuda S, Curran D P. Org Lett. 2003;5:1167–1169. doi: 10.1021/ol034060w.
10. Rahman M T, Kamata N, Matsubara H, Ryu I. Synlett. 2005:2664–2666. doi: 10.1055/s-2005-917103.
11. Podgoršek A, Stavber S, Zupan M, Iskra J. Eur J Org Chem. 2006:483–488. doi: 10.1002/ejoc.200500355.
12. Matsubara H, Tsukida M, Ishihara D, Kuniyoshi K, Ryu I. Synlett. 2010:2014–2018. doi: 10.1055/s-0030-1258482.
13. Ma K, Li S, Weiss R G. Org Lett. 2008;10:4155–4158. doi: 10.1021/ol801327n. Correction: Ma, K.; Li, S.; Weiss, R. G. *Org. Lett.*2009,*11,* 1461. doi:10.1021/ol900201t.
14. Iskra J, Stavber S, Zupan M. Chem Commun. 2003:2496–2497. doi: 10.1039/b305629e.
15. Windmon N, Dragojlovic V. Tetrahedron Lett. 2008;49:6543–6546. doi: 10.1016/j.tetlet.2008.09.007.
16. Windmon N, Dragojlovic V. Beilstein J Org Chem. 2008;4:No. 29. doi: 10.3762/bjoc.4.29.
17. Matsubara H, Niwa Y, Matake R. Synlett. 2015;26:1276–1280. doi: 10.1055/s-0034-1380381.
18. Jana N K, Verkade J G. Org Lett. 2003;5:3787–3790. doi: 10.1021/ol035391b.
19. Curran D P, Werner S. Org Lett. 2004;6:1021–1024. doi: 10.1021/ol0499248.
20. Chu Q, Yu M S, Curran D P. Tetrahedron. 2007;63:9890–9895. doi: 10.1016/j.tet.2007.06.092.
21. Van Zee N J, Dragojlovic V. Org Lett. 2009;11:3190–3193. doi: 10.1021/ol901450h.
22. Pels K, Dragojlovic V. Beilstein J Org Chem. 2009;5:No. 75. doi: 10.3762/bjoc.5.75.
23. Van Zee N J, Dragojlovic V. Chem – Eur J. 2010;16:7950–7958. doi: 10.1002/chem.201000269.
24. Abney A J, Dragojlovic V. J Chem Educ. 2012;89:293–295. doi: 10.1021/ed200080m.
25. Zhao H, Ismail K, Weber S G. J Am Chem Soc. 2004;126:13184–13185. doi: 10.1021/ja047453e.
26. Zhao H, Zhang J, Wu N, Zhang X, Crowley K, Weber S G. J Am Chem Soc. 2005;127:15112–15119. doi: 10.1021/ja052875p.
27. O'Brien M, Baxendale I R, Ley S V. Org Lett. 2010;12:1596–1598. doi: 10.1021/ol100322t.
28. Seidel F O, Gladysz J A. Adv Synth Catal. 2008;350:2443–2449. doi: 10.1002/adsc.200800403.
29. Oberg K A, Palleros D R. J Chem Educ. 1995;72:857–859. doi: 10.1021/ed072p857.
30. Dinh L V, Gladysz J A. Angew Chem, Int Ed. 2005;44:4095–4097. doi: 10.1002/anie.200500237.
31. Van Der Linden W E. Anal Chim Acta. 1983;151:359–369. doi: 10.1016/S0003-2670(00)80097-5.
32. Yang Y, Liang Y C. J Power Sources. 2007;165:185–195. doi: 10.1016/j.jpowsour.2006.11.059.
33. [Jan 4;2015];Surface tension values of some common test liquids for surface energy analysis. Available from: http://www.surface-tension.de.
34. [Jan 4;2015];Surface tension values of liquids. Available from:http://user.engineering.uiowa.edu/~cfd/pdfs/tables/1-39b.pdf.
35. [Jan 4;2015];Surface Tension of Acetonitrile from Dortmund Data Bank. Available from:http://www.ddbst.com/en/EED/PCP/SFT_C3.php.
36. [Jan 4;2015];Dimethyl phthalate. Available from: http://www.microkat.gr/msdspd90-99/Dimethyl%20phthalate.htm.
37. [Jan 4;2015];Pubchem Open Chemistry Database, Dimethyl phthalate. Available from:http://pubchem.ncbi.nlm.nih.gov/compound/Dimethyl_phthalate#section=Flash-Point.

38. Orosz G. Tetrahedron. 1989;45:3493–3506. doi: 10.1016/S0040-4020(01)81028-0.

39. Mohan A G, Turro N J. J Chem Educ. 1974;51:528–529. doi: 10.1021/ed051p528.

40. Duarte R, Nielsen J T, Dragojlovic V. J Chem Educ. 2004;81:1010–1015. doi: 10.1021/ed081p1010.

41. Morra M, Occhiello E, Garbassi F. Langmuir. 1989;5:872–876. doi: 10.1021/la00087a050.

CHAPTER 5

PREPARATION AND CHARACTERIZATION OF NANOCOMPOSITE POLYMER MEMBRANES CONTAINING FUNCTIONALIZED SNO2 ADDITIVES

Roberto Scipioni[1], Delia Gazzoli[1], Francesca Teocoli[2], Oriele Palumbo[3], Annalisa Paolone[3], Neluta Ibris[4], Sergio Brutti[3,4] and Maria Assunta Navarra[1]

[1]Dipartimento di Chimica, Sapienza Università di Roma, Piazzale Aldo Moro 5, 00185 Roma, Italy

[2]Centro di Ricerca Hydro-Eco, Sapienza Università di Roma, Piazzale Aldo Moro 5, 00185 Roma, Italy

[3]Istituto dei Sistemi Complessi-Consiglio Nazionale delle Ricerche–ISC-CNR U.O.S. Sapienza, Piazzale A. Moro 5, 00185 Roma, Italy

[4]Dipartimento di Scienze, Università della Basilicata, Viale dell'Ateneo Lucano 10, 85100 Potenza, Italy

ABSTRACT

In the research of new nanocomposite proton-conducting membranes, SnO_2 ceramic powders with surface functionalization have been synthesized and adopted as additives in Nafion-based polymer systems. Different synthetic routes have been explored to obtain suitable, nanometer-sized sulphated tin oxide particles. Structural and morphological characteristics, as well as surface and bulk properties of the obtained oxide powders, have been determined by means of X-ray diffraction (XRD), scanning electron microscopy (SEM), Fourier Transform Infrared (FTIR) and Raman spectroscopies, N_2 adsorption, and thermal gravimetric analysis (TGA). In addition, dynamic mechanical analysis (DMA), atomic force microscopy (AFM), thermal investigations, water uptake (WU) measurements, and ionic exchange capacity (IEC) tests have been used as characterization tools for the nanocomposite membranes. The nature of the tin oxide precursor, as well as the synthesis procedure, were found to play an important role in determining the morphology and the particle size distribution of the ceramic powder, this affecting the effective functionalization of the oxides. The incorporation of such particles, having sulphate groups on their surface, altered some peculiar properties of the resulting composite membrane, such as water content, thermo-mechanical, and morphological characteristics.

INTRODUCTION

Perfluorosulphonated polymer systems, *i.e.*, Nafion membranes developed by Du Pont, find one of their most challenging application in polymer electrolyte membrane (PEM) fuel cells (FCs), that are a promising technology for both mobile and stationary clean and efficient energy production [1,2,3,4,5,6,7]. Despite undiscussed benefits, such as high proton conductivity, mechanical, and chemical stability, conventional Nafion membranes show properties and performances in FC that limit its mass commercialization. In particular, they show limited duration under cycling voltage, low humidity, and freezing-thawing conditions. Furthermore efficient protons conduction

Citation: Scipioni, R.; Gazzoli, D.; Teocoli, F.; Palumbo, O.; Paolone, A.; Ibris, N.; Brutti, S.; Navarra, M.A. Preparation and Characterization of Nanocomposite Polymer Membranes Containing Functionalized SnO_2 Additives. Membranes 2014, 4, 123-142. doi:10.3390/membranes4010123

in Nafion membranes is obtained only at high hydration, thus, limiting operating temperatures of PEMFCs to around 80 °C at relative humidity (RH) approaching 100%.

One promising strategy to improve the performance of PEMs is based on the incorporation of nanometer-sized particles of hygroscopic inorganic acids or oxides, which act as a water reservoir into the polymer matrix [8,9]. According to the cluster network model for ionomers, the Nafion microstructure can be visualized as formed by hydrophobic domains (*i.e.*, the fluorinated backbone) and hydrophilic domains, the latter consisting of ionic clusters comprising the sulfonic acid moiety and the absorbed water [10,11]. These clusters are interconnected by a network of short and narrow channels and the size of the clusters is strongly dependent on the water content of the system [12]. Considering the mechanisms of proton transport [13], one may assume that the presence of the inorganic additive, when uniformly dispersed as nano-sized particles in the Nafion matrix, may favorably affect the ionic conductivity. Indeed, the hydration water of the additive can form bridges between shrunken clusters [14] and, in virtue of its acidity, the filler can provide additional pathways for proton hopping from one cluster to another.

Malhotra and Datta first proposed the use of inorganic solid acids (*i.e.*, heteropolyacids, HPAs) in conventional Nafion, with the objective of serving the dual function of improving water retention, as well as providing additional acidic sites for proton conduction [15].

Similarly to HPAs, sulfated metal oxides have become subjects of intensive studies, being them even more stable than other inorganic solid acid. In virtue of its extraordinary high acidity, sulphated zirconium oxide, $SZrO_2$, has attracted interest as fuel cell membrane additive [16,17,18,19,20,21,22,23]. Deeply studied as catalyst for the conversion of organic compounds [24], $SZrO_2$ has been rarely investigated as proton conductor [25]. Hara *et al.* found that surface SO_4 species existed as bidentate complexes strongly bonded with ZrO_2 in a $SZrO_2$ powder calcined at 620 °C. SO_4-species are electron-withdrawing and, thus, the Lewis acidity of zirconium is strengthened. The Lewis points can easily convert to Brønsted acid points in the presence of water, finally acting as proton donors [26]. Currently, sulphated zirconia is recognized as one of the strongest superacid among all known solids with its Hammett acidity, $H_o < -16$. An extraordinary strong surface acidity, even higher than that of $SZrO_2$, has been attributed to sulphated tin oxide, $SSnO_2$ [27]. Nevertheless, papers concerning the use of sulphated tin oxide as catalyst have been quite few [27] being it more difficult to obtain especially when large surface area is required [28]. As reported by Arata and co-workers, the main difficulty consists of the preparation of the oxide gels from tin salts, needed for the subsequent treatment with sulfate ions [29].

To our knowledge, there is only a recent example on the use of $SSnO_2$ in Nafion membranes for direct methanol fuel cells, demonstrating a reduced methanol crossover, improved proton transport, and enhanced membrane/electrode interface stability [30].

In this work highly hydrophilic nanosized tin oxide particles without and with sulphated surfaces were obtained by properly adapting well-known synthesis procedures where two different tin oxide precursors were considered [31,32]. The incorporation of the obtained powders into Nafion-based membranes is here described. The goal is to develop a comprehensive description of the properties of these composite materials in order to drive the selection of advanced organic-inorganic hybrid electrolytes for PEM fuel cell applications.

RESULTS AND DISCUSSION

In the following presentation and discussion of results, all samples, both powders and membranes, are labeled as specified in Section 3.1 and Section 3.2 of the experimental section.

Structure and Morphology of the Bare and Sulphated Tin Oxide Powders

The X-ray diffraction (XRD) patterns of the synthesized ceramic samples, both bare and sulphated, are presented in Figure 1.

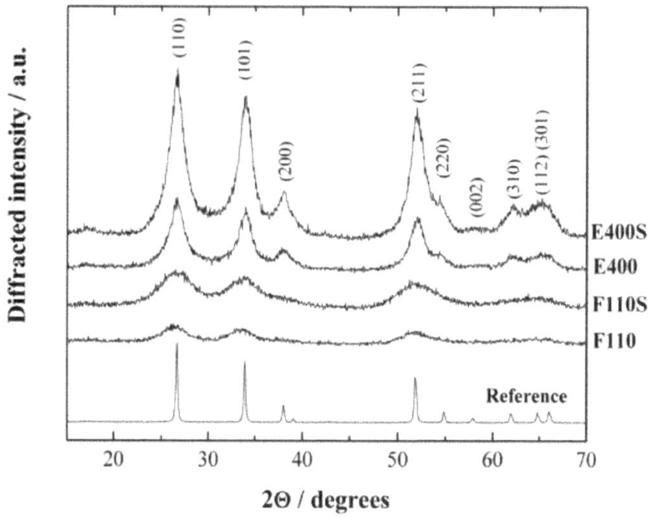

Figure 1: X-ray diffraction (XRD) patterns of the tin oxide samples, bare and sulphated. Reference: SnO_2 bulk oxide.

The XRD patterns of all materials have been indexed in agreement with the tetragonal rutile structure of the bulk SnO_2(reference spectra, Sigma Aldrich 99% purity). All the synthesized samples show broad peaks revealing the presence of particles in the nanosized range. The mean crystallite sizes (<d>) have been calculated by the standard Scherrer equation [33] starting from the profile broadening (Δ, full width at half maximum) of the (110), (101), and (211) peaks using the usual formula: $\langle d \rangle = \lambda/(\Delta\cos\theta)$, being $\lambda = 1.5418$ Å the wavelength of the CuKα radiation. The resulting values are 1.7 and 2.8 nm for F110 and F110S samples, and 5.0 and 6.1 nm for E400 and E400S samples, respectively. The various degree of crystallinity reflects the thermal history of the samples. In particular it is to be noted that the sulphated oxides show in both cases (E400S and F110S) larger crystallite sizes compared to the bare pristine oxide (E400 and F110, respectively), this being due to the further annealing step at 500 °C performed during the sulphation procedure. Additionally, the nanosized structure is retained by both E400S and F110S samples.

The amount of sulphate groups chemically bonded to the surfaces has been established by thermogravimetric measurements, Figure 2.

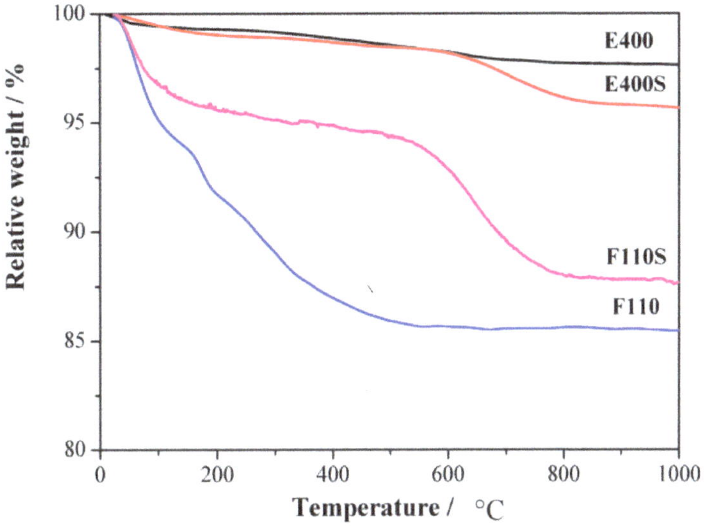

Figure 2: Thermal gravimetric (TG) analysis responses for the four synthesized oxides.

Mass losses are observed in different thermal ranges. The first mass loss below 250 °C is observed in all the four samples and it is related to the removal of physisorbed and chemisorbed water as well as OH groups: the largest mass loss is observed, as expected, for the F110 sample, being it the one treated at relatively low temperature (*i.e.*, 110 °C, see experimental section). A second mass loss is observed between 400 and 500 °C only for the E400 sample and it is likely related to residual organic traces from the organometallic precursor. A similar mass loss in the 400–500 °C range is not evident for the sulphated parent sample E400S, most likely due to the removal of the contaminations during the high temperature treatment involved in the sulphation process. Another mass loss is observed above 600 °C only for the two E400S and F110S sulphated samples. This mass loss is related to the removal of the sulphate groups bonded to the surface of the tin oxide particles [30]. The two samples show a loss of about 3% and 7% in mass for E400S and F110S, respectively. The larger concentration of chemisorbed sulphate groups in the F110S sample compared to the E400S goes in parallel with the smaller crystallite size of the pristine F110 bare oxide before the sulphation process in respect to the E400 one (see XRD results above). Such evidence suggests that the amount of sulphate groups chemisorbed on the SnO_2 surface increases with the nanometrization of the oxide particles.

The SEM micrographs of the four oxides at 20 k magnifications are reported in Figure 3.

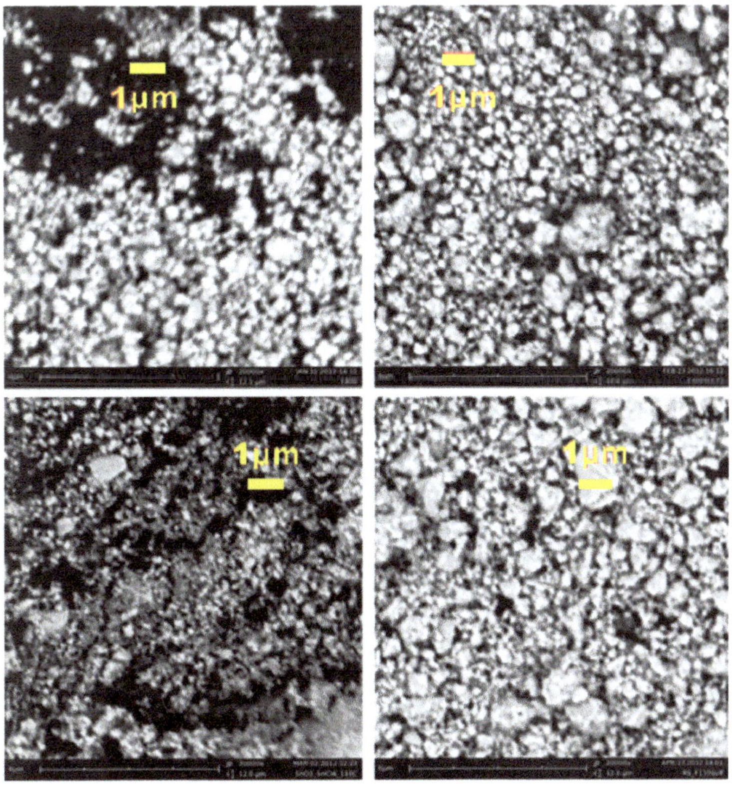

Figure 3: Scanning electron microscopy (SEM) micrographs of the four synthesized oxides. From left to right and from top to bottom: E400, E400S, F110, and F110S, respectively.

The morphology in the microscopic range of all the powders is similar: all the samples consist of sub-micrometric particles (secondary particles) with a remarkable dispersion in size. Such large particles are likely formed by the aggregation of the nanometric crystallites estimated by the XRD data. The sulphation of the surfaces does not lead to drastic alterations: the surface-modified materials show only a slight increase in the size of the secondary particles. This picture is also confirmed by the trend of the surface areas, the measured values being 65 ± 2 m^2 g^{-1} (E400), 52 ± 3 m^2 g^{-1} (E400S), 160 ± 5 m^2 g^{-1} (F110) and 100 ± 1 m^2 g^{-1} (F110S). These values correspond to estimated spherical particle diameters of 13, 17, 5, and 9 nm for the E400, E400S, F110, and F110S samples, respectively. It is interesting to observe that the surface area of the F110S samples is approximately the double of the other sulphated E400S sample. This difference accounts for the approximate double amount of sulphate groups bonded to the F110S sample in comparison to the E400S one.

Spectroscopic Characterization of the Bare and Sulphated Tin Oxide Powders

All the SnO_2 samples, both bare and sulphated, have been investigated by vibrational spectroscopies (FTIR and Raman) in order to investigate the modes of the surface groups and the nature of the chemical bond between the sulphates and the SnO_2 surfaces.

The FTIR spectra of the four samples are reported in the Figure 4 where three main spectral features are observed. All samples show a broad composite band due to the vibrational modes of the Sn–O–Sn oxide lattice below 800 cm^{-1} and a band at about 1620 cm^{-1} attributed to the bending mode of surface water molecules and –OH groups [34].

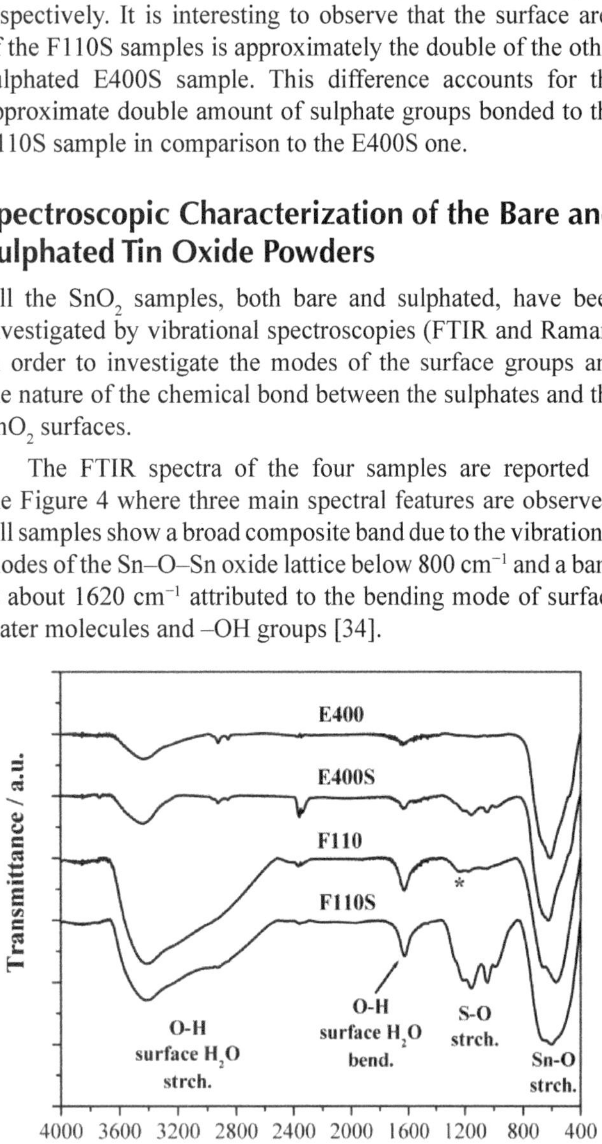

Figure 4: Fourier Transform Infrared (FTIR) spectra of the bare and sulphated oxides. The (*) symbol corresponds to the Sn–OH bending mode.

The presence of physisorbed water and OH terminal groups on the surface of all the samples, even after calcination at 400 °C and/or 500 °C, is confirmed by the large broad band due to the stretching of the OH bond between 2400 and 3600 cm^{-1}. The F110 sample shows in addition a composite response between 1250 and 800 cm^{-1} that is likely due to the bending of Sn–OH bonds as discussed in [35].

For the two E400S and F110S sulphated samples a series of FTIR bands between 1400 and 800 cm^{-1} due to the sulphate stretching modes are evident. As expected, their intensity is stronger for the F110S sample that has a double amount of sulphate groups bonded to the surface compared to the E400S sample. It is to be noted that the position and relative intensity of the bunch of narrow bands due to sulphate groups are very similar for the two samples: this evidence suggests that the nature of the chemical bond between the SnO_2 oxide and the sulphate groups bonded is almost the same in the two samples.

The evolution of surface vibration modes, revealed by Raman spectroscopy, are presented in Figure 5 for the F110 and F110S samples, as a representative case. The spectrum of bulk SnO_2 is also shown for comparison.

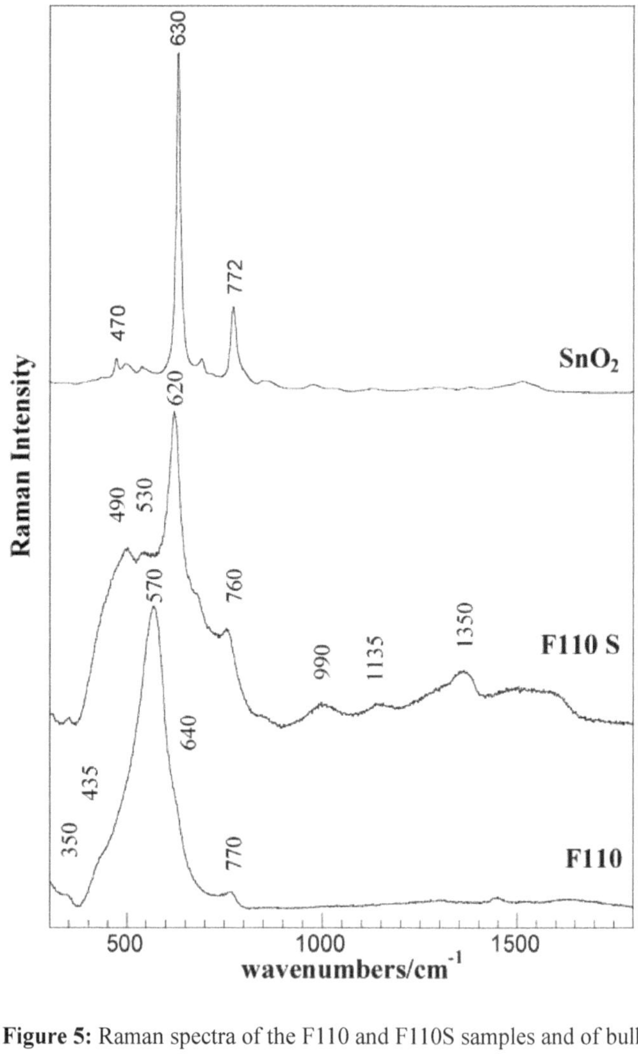

Figure 5: Raman spectra of the F110 and F110S samples and of bulk SnO_2 (Aldrich).

The Raman spectrum of the starting F110 sample mainly consists of a broadened strong band centered at about 570 $cm^{-1}$and of low intensity bands at about 350 cm^{-1}, 435 cm^{-1}, 640 cm^{-1}, and 770 cm^{-1}. According to factor group analysis, four first-order Raman active modes (A_{1g}, B_{1g}, B_{2g}, E_g) are predicted for crystalline SnO_2 (tetragonal rutile structure) at 100 cm^{-1}(B_{1g}), 479 cm^{-1} (E_g), 638 cm^{-1} (A_{1g}), and 779 cm^{-1} (B_{2g}) [36]. The bands at about 570 and 350 cm^{-1} are attributed to interface or surface phonon modes of nanostructured SnO_2 material [37,38], being this in agreement with the results of XRD and morphological analyses. As, at decreasing particle size, a shift to lower wavenumbers is expected, the strong band at about 570 cm^{-1} can be related to the A_{1g} mode of the SnO_2 structure, whereas the barely detectable bands at about 435 cm^{-1}, 640 cm^{-1}, and at about 770 cm^{-1} are attributed to normal interior phonon modes of the bulk SnO_2 structure.

The sulphation process yielded a strong modification in the Raman spectra of the F110 sample. In addition to the changes in the SnO_2 structure, revealed in the 300–800 cm^{-1} spectral region, the presence of sulphate-species were detected in the range 930–1600 cm^{-1} of the F110S sample. The sharp and intense bands at about 490, 620, and 760 cm^{-1} are ascribed to SnO_2in crystalline form (E_g, A_{1g}, and B_{2g} modes) [36], the amorphous nanostructured starting material is, however, still present as evidenced by the large band underlying the whole region representative of surface phonon modes.

The bands in the 900–1500 cm^{-1} range, at about 1000, 1150, and 1360 cm^{-1}, disclose the presence of various surface sulfate species. Although bands assignment is still conflicting, the bands at about 1000 and 1350 cm^{-1} can be attributed to S–O and S=O modes of the surface sulfate groups [39]. The band at about 1135 cm^{-1} can also be attributed to S–O stretching modes, but in polynuclear sulfate groups and/or stretching modes in bidentate sulfate species [40]. Furthermore, the broad band centered at about 1600 cm^{-1} can be assigned to the OH vibration of adsorbed molecular water.

Physico–Chemical Characterization of the Nafion-based Membranes

Five membranes have been prepared without and with the four filler powders. The filler loading has been chosen as 5 wt % in agreement with previous results published by Chen *et al.*, where Nafion membranes containing 5 wt % of sulphated tin oxide were shown to have optimized proton conductivity with respect to undoped and 10 wt %-added membranes [30]. Filler amount not higher than 4–5 wt % has been demonstrated to be the most effective also for other metal oxide-doped Nafion membranes, both in terms of homogeneous dispersion of the additive within the polymer matrix and of improved stiffness of the resulting composite system [21,41].

A summary of the membrane compositions and the corresponding water uptake (WU) and ionic exchange capacity (IEC) values are reported in the Table 1 together with data derived from the thermal analyses.

Table 1: Water uptake (WU), ionic exchange capacity (IEC), and parameters derived from the thermal analysis for the five prepared membranes.

Code	Filler (5 wt %)	IEC mequiv g^{-1}	WU	ΔH/J g^{-1} (DSC)	T_{tr}/°C (DSC)
N	none	0.880	37%	76.9	185
N-E400	E400	0.834	38%	105	163
N-E400S	E400S	0.812	37%	93.3	153
N-F110	F110	0.827	40%	114	146
N-F110S	F110S	0.820	40%	181	153

The ionic exchange capacity decreases slightly for all the composite membranes compared with the benchmark recast Nafion (N). This is likely ascribed to the fact that the ceramic powder addition is expected to increase the density of the membrane thus reducing the amount of acid sites for mass units. The small differences in IEC values among membranes containing sulphated or unsulphated filler are not considered of relevance. Indeed, due to the nature of the experiment adopted for the IEC determination (see Experimental section), incomplete ion exchange can even be considered, mainly when compounds with very high proton affinity are being used.

The effect of the fillers on the hydration/dehydration of the membranes is proved by water uptake and thermal analysis results. The WU is apparently slightly higher in the case of the composite membranes containing the F110 additive, both sulphated and non. However these data do not provide information on the nature and the strength of the bonds between water, filler, and Nafion ionic domains. The thermal profiles of the membranes from TG and DSC experiments are shown in Figure 6.

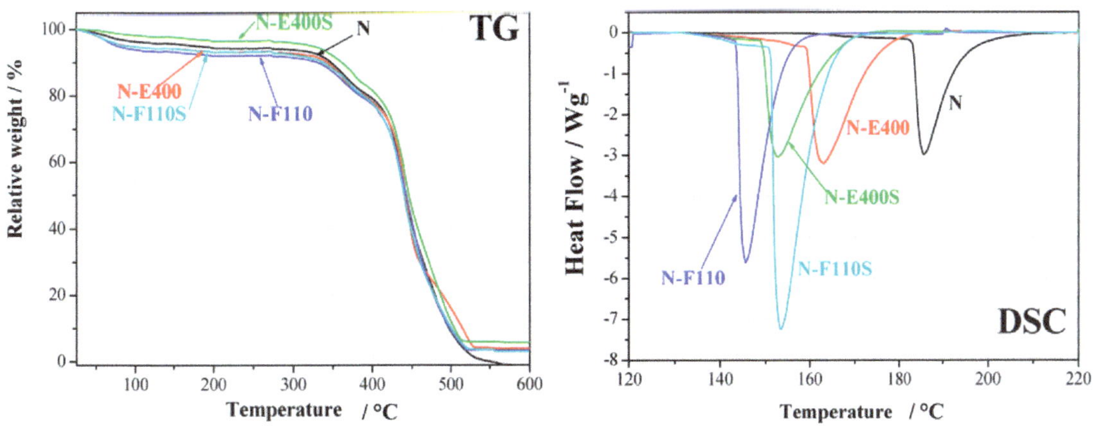

Figure 6: Thermal analysis results (TG and DSC) for the five membranes.

The TG analyses show, as expected, two main mass loss processes: below 200 °C and above 300 °C, respectively. The first process is related to the removal of water from the polymer matrix, whereas the processes above 300 °C to the degradation and combustion of the polymer matrix. The residual mass above 500 °C is due to the ceramic filler; values, slightly differing from the theoretical amount of 5 wt %, are ascribed to small non-homogeneity within the samples. The composite membranes added with the sulphated oxide fillers, *i.e.*, N-E400S and N-F110S, show a mass loss around 120 °C slightly smaller than those observed for membranes added with the bare oxides, N-E400 and N-F110, respectively. The membrane hydration above 100 °C is a key technological property as such a temperature range is a practical target for PEM fuel cell operation. This experimental evidence may suggest a content of water smaller for the N-E400S and N-F110S samples than for the N-E400 and N-F110 parent membranes. On the other hand, one cannot exclude the presence of highly coordinated or bonded water, strongly interacting with the fillers, not easy to be removed at the quoted temperature. Owing to this and to the nature of TGA experiments (see Experimental section), where membrane samples are not equilibrated under stable temperature and/or relative humidity states, the mass losses at 120 °C cannot be taken as a metric for evaluating water content in the membranes.

DSC experiments show for all samples the expected broad endothermic peaks between 140 and 190 °C. This thermal effect is due to an order-disorder transition in the ionic domains of the Nafion polymer membrane [42]. It is known to be strongly affected by the amount of water within the membrane both in terms of transition temperature (T_{tr}, smaller for larger water contents) and enthalpy variation (ΔH, larger for higher hydration degrees) [21]. All the membranes added with the inorganic fillers show smaller transition temperatures and larger enthalpy variations in comparison with the Nafion benchmark membrane. Thus, the addition of SnO_2-based inorganic fillers univocally leads to an increase of the amount of water within the composite membranes under the experimental conditions here quoted (*i.e.*, without any control of external relative humidity). In particular the comparison among the DSC data for the N-E400, the N-F110 and the benchmark N membrane suggests that the N-F110 sample keeps the larger amount of water. This is in agreement with the TG results (see Figure 2) and with the WU measurements (see data in Table 1).

Focusing the attention to the sulphated membranes, the DSC data show contradictory trends. In particular, the N-E400S sample shows a decrease of the transition temperature and of the enthalpy variation compared to the bare N-E400 membrane, whereas the N-F110S sample shows an increase of the transition temperature and of the enthalpy variation compared to the bare N-F110 membrane (see data in Table 1).

In the attempt to understand this behavior, relative interactions among components on a molecular level and the water affinity of the four oxide fillers have to be considered. As shown by the TG data reported in Figure 2, the mass loss due to water desorption from the oxides is different in the four cases. In particular, the E400 and the E400S oxides show a similar mass loss <1% up to 160 °C, whereas the F110 and the F110S samples suffer larger losses, *i.e.*, 6% and 4%, respectively. Owing to this, it is likely that also the amount of water in the N-E400 and N-E400S is quite similar. In this view it is possible to speculate that the H_2O molecules are more strongly bonded in the N-E400S membrane compared to the N-E400 sample and that the addition of the sulphated E400S filler enhances the retention of water above room temperature. This is in agreement with the decrease of the order-disorder transition temperature observed in the DSC data between N-E400 and N-E400S. The parallel contradictory smaller enthalpy variation may be related to the larger strength of the chemical bond between water molecules and the oxides particles in the case of the N-E400S membrane, this reducing the expected effect of water towards the polymer ionic clusters in terms of cohesive interactions. In fact, the sulphate groups on SnO_2 may strongly coordinate the residual water around the inorganic fillers particles, thus, altering the mean amount of free water dispersed throughout the Nafion matrix.

Turning back to the F110 and F110S samples, the TG data reported in Figure 2 suggest larger mass losses for the F110 oxide in comparison to F110S, excluding in the estimate the loss due to sulphate group removal. In this view it is likely that the amount of water in the composite matrix is much larger for the N-F110 membrane compared to the N-F110S. This is in agreement with the lower order-disorder transition temperature observed in the DSC for the N-F110 sample compared to the N-F110S membrane. Here, the higher enthalpy variation observed for the N-F110S sample in comparison with the N-F110 one may be related to enhanced interactions between water molecules and ionic clusters, this suggesting the existence of differently coordinated and interacting water in the various membranes. In summary the addition of the SnO_2-based inorganic filler promotes hydration into the Nafion-based membranes, as shown by both WU measurements and thermal analyses. Furthermore the surface modification of the filler, due to the addition of sulphate groups, apparently modifies the strength and the nature of the water-filler and water-ionic cluster bonds.

Atomic Force Microscopy (AFM) and Dynamic Mechanical Characterization of the Composite Membranes

The morphologies of the composite membranes, as well as of the benchmark Nafion, all in their dried state (*i.e.*, 1 h treatment at 70 °C in air), were investigated by AFM in tapping mode.

In Figure 7 the tapping mode AFM topographies of the five membranes in their dried state are compared together with the size distribution of the round-shaped pseudo-particles (see below) derived from the image analysis by the software ImageJ [43] of five different AFM micrographies at different magnifications for each sample. AFM experiments have been carried out on both sides of the membranes without observing remarkable differences.

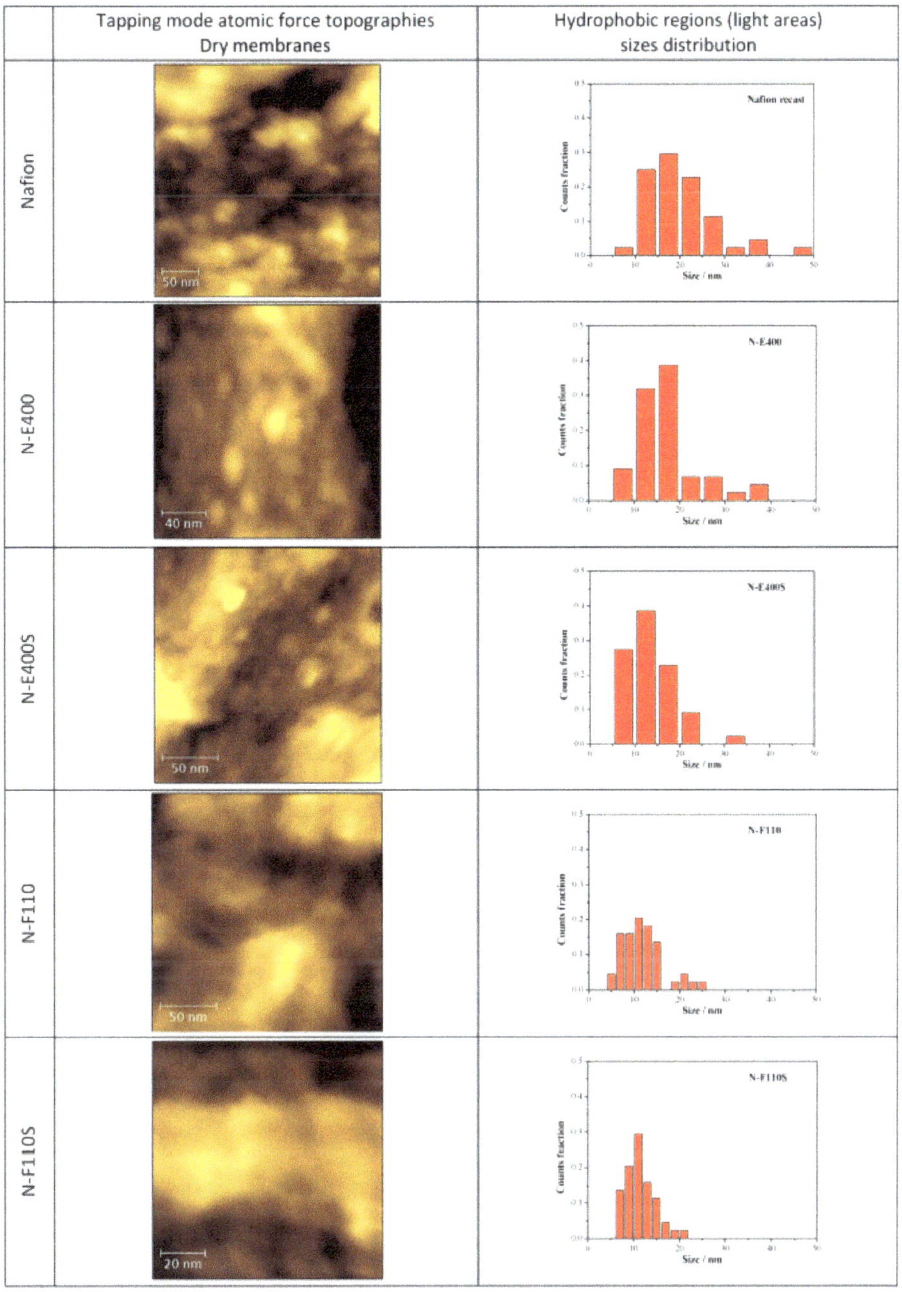

Figure 7: Atomic force microscopy (AFM) topographies measured in tapping mode of the five dried membranes and size distribution of the hydrophobic regions as derived from the image analysis.

The N, N-E400, and N-E400S dried membranes show quite similar surfaces with round shaped aggregated pseudo-particles of different sizes (light regions). This morphology (*i.e.*, agglomerated round shaped pseudo-particles separated by darker areas) can be ascribed to the alternation of the hydrophobic (light regions)/hydrophilic (dark regions) domains across the polymer matrix due to the interaction of the sulphonic groups dispersed within the hydrophobic network of the Nafion polymeric backbone [44,45,46,47]. The dimension (<d>, mean diameter of the hydrophobic regions (light round-shaped pseudo-particles) is much smaller and more homogenously dispersed in the case of the two membranes added with E400 (<d> = 16.6 ± 6.7 nm) and E400S oxides (<d> = 13.9 ± 5.4 nm) in comparison with the benchmark N (<d> = 20.0 ± 8.9 nm). Furthermore, the hydrophilic domains (dark areas) surrounding the pseudo-particles increase their size and morphological regularity passing from the recast Nafion membrane to the N-E400 and the N-E400S membranes. Apparently, the addition of the E400 and E400S oxides leads to a more homogeneous dispersion of the hydrophilic/hydrophobic domains and to a reduction of the size of the latter. The size of the light areas matches those observed for similar samples in the literature (e.g., [47]) where cluster-like structures with diameter of 5–30 nm are typically observed. Furthermore, superstructure of spherical domains with an average diameter of tenths of nanometers containing 10–12 nm grains are usually detailed in AFM images [47]. Among these grains, dark interstitial regions of few nanometers are observed being the latter an estimate of the ionic cluster size. Apparently, our observation matches this picture.

The alterations of the morphology of the three samples are induced by the oxide particles. Although these changes may be easily put into correlation with the possible larger hydration of the hydrophilic domains, they may result from two different and interacting phenomena difficult to decouple, *i.e.*, (a) the alteration of the re-casting mechanism during membrane preparation (see experimental part) that reflects on the local organization of the ionic/hydrophobic domains; (b) the possible larger water retention in the membrane after drying due to the stronger bonding of the water molecules mediated by the oxide surfaces.

On passing one may observe that the size of the oxide particles (both E400 and E400S, see XRD and morphological analyses results) matches that of the round shaped ionic domains. Unfortunately our AFM investigation did not give clear evidence of filler particles: in our experience, it is difficult to distinguish between the ionic domain and the similar-in-size filler particles without referring to the AFM phase-images, where the harder oxide particles may give evidences. As a consequence, the here reported AFM study did not provide any undoubted evidence of the presence of filler particles embedded in the polymer matrix. In this view a further interpretation of the possible origin of the larger hydration of the ionic channels based on our AFM data would be too speculative.

A larger surface morphology alteration is observed for both the N-F110 and N-F110S. In particular, although as expected the overall morphologies of the N-F110 and the N-F110S samples still resemble those of the benchmark Nafion and the N-E400 and N-E400S membranes, the surfaces are apparently smoother with less resolved alternation of the hydrophilic/hydrophobic domains. The size-distribution plots suggest a mean size of the hydrophobic pseudo-particles of about 12.0 ± 4.4 and 11.2 ± 3.2 nm for the N-F110 and the N-F110S samples, respectively. Moreover also the hydrophilic regions surrounding the light hydrophobic domains show round-shaped morphologies and sizes matching those of the light areas. In addition, in the cases of the N-F110 and the N-F110S membranes the modification of the morphologies may be ascribed to the increase of the hydration of the ionic channels resulting from the alteration of the re-cast process and/or to the larger water retention after drying. Moreover it is interesting to underline that, differently from the cases of the N-E400 and N-E400S membranes, in both the N-F110 and N-F110S cases the particle size of the inorganic fillers estimated from surface area measurements is much smaller in comparison with the size of the round-shaped pseudo-particles and the surrounding darker areas. One may speculate that, in these two cases (N-F110 and N-F110S), the oxide particles may float well within the hydrophilic channels thus improving the water retention without inducing a large mechanical stress on the surrounding polymer matrix.

Turning to the mechanical properties, the storage modulus (E') and elastic energy dissipation (tanδ) of the nanocomposite membranes (N-E400, N-E400S, N-F110, and N-F110S), measured starting from a "wet membrane" state on heating and subsequent cooling, are reported in Figure 8.

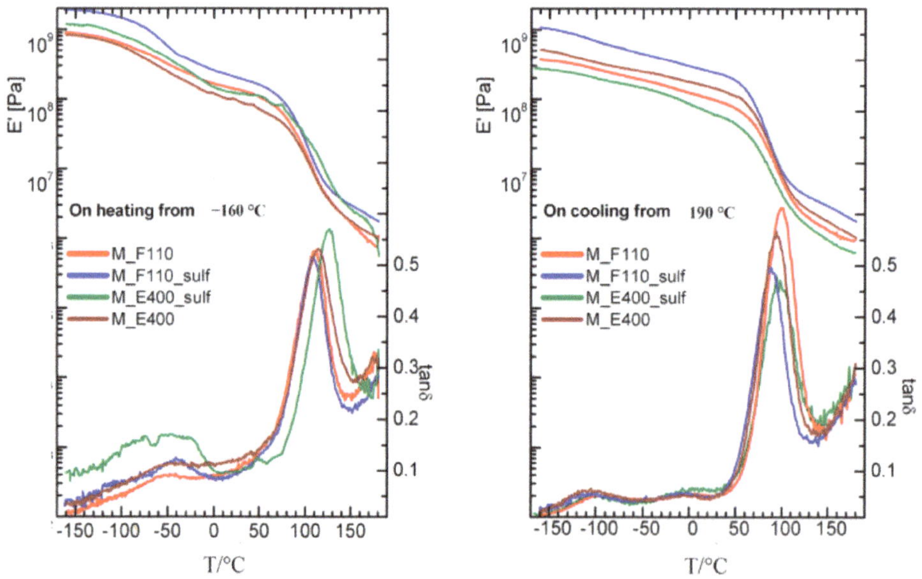

Figure 8: Storage modulus and elastic energy dissipation of nanocomposite membranes measured for f = 1 Hz on heating (left panel) and subsequent cooling (right panel).

On heating starting from −160 °C, the α relaxation (the intense peak in tanδ and the two order of magnitude drop in the modulus) is detected around 110 °C in the F110-added (either sulphated or not) and in the E400-added nanocomposite membranes, while it results slightly shifted to higher temperature (120 °C) in the sulphated E400-added sample. This relaxation corresponds to the glass transition of the hydrophilic domains (polar regions) of the Nafion polymer matrix. On cooling, the peak associated to the alpha relaxation is measured at about 80 °C (see right panel of the figure) in all membranes, showing thermal hysteresis. Moreover, on heating between −90 and 30 °C, all the nanocomposite membranes show a broad peak, accompanied by an increase in the modulus. This feature has been already observed in a wet undoped Nafion membrane [48] and attributed to the friction between the liquid water bound to the walls of the hydrophilic domains and the solid ice inside the channels. The intensity of this peak is higher in the membrane with the sulphated oxide as filler, and is maximum in the sulphated samples prepared by the E400 procedure (N-E400S). It can be speculated that the higher intensity of the broad peak below 30 °C in N-E400S is related to the distribution and to the particular status of water molecules within the composite matrix. Even though a sort of preferential interactions between water and this particular filler (E400S) in Nafion membrane has been also revealed by DSC measurements (see discussion above), further insights are needed to better explain this evidence, thus complementary investigations are currently in progress.

The effect of hydration on the properties of ionomers and polyelectrolytes is, in general, a very complex, still open question. It is known that many variables, such as external relative humidity, temperature, and, as recently demonstrated by Kreuer, internal pressure contribute to define the hydration thermodynamics and the transport properties across the membrane [49]. When inorganic nanoparticles are introduced into a polymer electrolyte, the picture becomes even more complex and hydration is altered [50].

Our findings show that the storage modulus and tanδ measured for all the nanocomposite membranes on cooling immediately after heating at 180 °C are close and similar to those already reported for a dry, undoped Nafion membrane [48] as they do not show anymore neither the broad peak nor the corresponding increase in the modulus due to the reinforcing action of solid water. Moreover, when measured on cooling, the gamma and beta relaxations are more evident around −70 and 0 °C, respectively, as the corresponding peaks are more separated. The beta relaxation has been ascribed in the literature to the glass transition of the Nafion non-ionic matrix, while the gamma relaxation was assigned to short-range molecular motions of the $-CF_2-$ backbone.

There are some interesting examples in literature where the viscoelastic properties of Nafion have been studied as a function of the hydration level and of temperature and correlated to the transport properties across the polymer membrane [49,51]. In the present work, the effect of the particles addiction on the mechanical properties of the membranes can be discussed by the measurements of the storage modulus in the cooling run (right panel of Figure 8), where most of the water may have evaporated. Most of the composite samples present values higher than that of pure recast Nafion (results not shown here) and, in particular the N-F110S sample displays the highest modulus, suggesting a reinforcing action due to the incorporation of the F110S sulphated tin oxide particles. An increase of the Young modulus in a Nafion membrane with the addition of SnO_2 nanoparticles was already observed by Nørgaard *et al.* [41] and related to a stiffening effect due to interfacial interactions between the inorganic phase and sulfonic acid groups in the polymer.

The applicability of the proposed composite membranes as effective electrolytes needs, of course, to be demonstrated. To this purpose, a detailed characterization, including electrochemical measurements of proton conductivity and fuel cell operation, will be reported in another publication aimed at elucidating the practical impact of the selected electrolytes under critical conditions [52]. It's here worth mentioning that interesting conductivity values were obtained for the two sulfated tin oxide-added membranes, ranging from 7×10^{-3} to 9×10^{-3} S cm^{-1} at room temperature under 100% RH.

EXPERIMENTAL SECTION

Preparation of SnO_2-Based Inorganic Powders

Highly hydrated nanosized SnO_2 powders were obtained by two different methods:

- Sol-gel synthesis from Sn(II)-2-ethyl-hexanoate [31]. 5 mL of the organic Sn-precursor (Sigma Aldrich, 97%) were added dropwise to 50 mL of n-propanol. A volume of 1 mL of concentrated NH_3 (30% weight) was added dropwise to the Sn(II) alcoholic solution under vigorous stirring. The yellowish precipitate was separated by centrifugation and then calcined at 400 °C in air. The obtained material was denoted as E400.
- Aqueous hydrolysis of $SnCl_4$ [32]. $SnCl_4$ (Fluka, 99%) was dissolved in water (2 M solution) and then $Sn(OH)_4$ was precipitated by adding dropwise to 20 mL of the Sn(IV) solution 10 mL of concentrated (30% weight) water solution of NH_3. The resulting precipitate was carefully washed with de-ionized water to remove the NH_4^+ and Cl^- residual contaminations. The final white gel was dried under dynamic vacuum overnight at 110 °C and then hand grinded. The sample was designated as F110.

The sulphated tin-oxides (named E400S and F110S) were obtained by chemical treatment followed by calcinations of the E400 and F110 samples. In both cases the starting oxide was suspended in de-ionized water and stirred continuously in order to enhance the water adsorption. After 24 h of vigorous stirring the water suspension was treated with H_2SO_4 (final concentration 0.5 M) for 30 min under stirring. After decantation, the material was filtered and then calcined in air for 3 h at 500 °C.

In summary four inorganic fillers were prepared:

- E400 and E400S from alcoholic hydrolysis of Sn(II)-2-ethyl-hexanoate;
- F110 and F110S from the water hydrolysis of $SnCl_4$.

Preparation of the Composite Membranes

Nafion membranes, both without and with the inorganic filler, were prepared following a solution casting procedure. The commercial Nafion dispersion (5 wt % in water/alcohol, E.W. 1100, Ion Power, GmbH, Munchen, Germany) was treated with*N*,*N*-dimethylacetamide at 80 °C in order to replace solvents. When required, inorganic filler powders were added to the final Nafion solution and stirred to homogenize the suspension. An optimal additive loading of 5% with respect to the Nafion polymer weight was chosen for our investigations [30]. The mixture was casted on a Petri dish and dry membranes were obtained by solvent evaporation at 90 °C. The casted polymer membranes, having a geometrical surface area of *ca.* 28 cm^2, were hot-pressed at 175 °C for 15 min at 10 tons to improve robustness and finally activated and purified by subsequent immersions in boiling hydrogen peroxide (3 vol %), sulfuric acid (0.5 M), and water. All the prepared membranes showed very similar thicknesses of about 90 μm.

Membranes containing the inorganic fillers are labeled in the text as N-E400, N-E400S, N-F110, and N-F110S, respectively. Undoped membranes, used as internal benchmark, will be referred to as recast Nafion and labeled as N.

Experimental Methods

The XRD patterns on the synthesized inorganic ceramic powders were recorded with a step size of 0.02° and a time per step of 3 s, with a Rigaku Ultima+ diffractometer equipped with a CuKα radiation source and a graphite monochromator in the 2θ range of 15°–85°.

The morphology of powders and membranes was studied by the SEM Phenom FEI with a LaB_6 source and by the Park AFM XE120 instrument in tapping intermittent contact mode (TM), respectively. It is important to underline that phase-distance curves were carefully monitored upon imaging in order to avoid the occurring of unstable regimes [45] and the rise of height artifacts in AFM topographies.

Single point N_2-adsorption experiments for surface area determination were carried out by the Quantachrome Monosorb instrument. All measurements were performed in triplicate on two different sample batches for each material (F110, F110S, E400, and E400S). Each sample was carefully pre-treated before N_2 adsorption at 120 °C for 9 h in order to remove physisorbed moisture.

FTIR spectroscopy studies were performed by using a Bruker Alpha FTIR apparatus in transmission mode.

Raman spectra were collected at room temperature in back-scattering geometry with an inVia Renishaw micro-Raman spectrometer equipped with an air-cooled CCD detector and super-Notch filters. A 488.3 nm emission line from an Ar^+ ion laser was focused on the sample by a Leica DLML microscope, using 5× or 20× objectives. The power of the incident beam is about 5 mW. Five 10 s accumulations were generally acquired for each sample. The resolution was 2 cm^{-1} and spectra were calibrated using the 520.5 cm^{-1} line of a silicon wafer.

Thermal analyses were carried out by differential scanning calorimetry (DSC) using a DSC 821 Mettler-Toledo instrument, and by thermal gravimetric (TG) analysis performed with a TGA/SDTA851 Mettler-Toledo instrument. DSC measurements were realized on membrane samples at a scan rate of 20 °C min^{-1} in a nitrogen flux. TG analysis was executed in air flux on both tin oxide particles and membranes, at a heating rate of 5 °C min^{-1} and 20 °C min^{-1}, respectively. Both, TG and DSC analyses of polymer membranes, composite and bare, were performed on membrane samples stored in water, by simply removing surface water drops with the help of technical paper wipes prior to measurements.

Dynamic mechanical analysis was realized by a Perkin-Elmer DMA8000 apparatus in the so-called "tension configuration". Small pieces 4–7 mm wide, 10–12 mm long, 0.10–0.15 mm thick were cut from the various membranes stored in water. The storage modulus (E') and the elastic energy loss (tanδ) were measured, with a vibration frequency f = 1 Hz, as a function of temperature in the range of -160–190 °C with a scan rate of 4 °C/min.

Membrane total water uptake (WU) was measured at room temperature. Samples were dried at 70 °C under vacuum overnight and then weighted (w_{dry}). The weight of fully hydrated membranes ($w_{swollen}$) was obtained immediately after the treatment in boiling water at the end of the activation procedure. The WU value was evaluated by the formula

$$WU = (w_{swollen} - w_{dry})/(w_{dry}) \times 100\%$$

The ion-exchange capacity (IEC), which provides the number of milli-equivalents of ions in 1 g of dry membrane, was evaluated by a classical titration method: dry membrane samples were immersed in NaCl aqueous solution and the exchanged protons were neutralized with OH^-.

CONCLUSIONS

The properties of nanocomposite ceramic-added polymer membranes were investigated by means of complementary physical chemical characterization tools. The synthesis and the investigation of highly hydrophilic nanosized tin oxide fillers, without and with sulphated surfaces, and their incorporation into a selected polymer matrix have been described. Two nanometric SnO_2 materials, differing in crystallite size and surface area, have been successfully prepared starting from different precursors. Both materials have been sulphated by a tailored procedure in order to obtain superacidic SnO_2nanoparticles. All the inorganic fillers have been incorporated into Nafion matrices to obtain the desired nanocomposite, hybrid membranes. It was demonstrated that both the particle dimension and the sulphation extent of the additive determined a different hydration degree, this in turn affecting the final properties of the doped membranes. Even though this work was not specifically aimed at elucidating the hydration concern within composite Nafion systems some interesting evidences were found when comparing samples of slightly different nature under the same experimental conditions here quoted. Indeed, calorimetric, dynamic-mechanical and atomic force investigations suggested the occurrence of distinctive interactions among the various components (*i.e.*, ceramic, polymer, and absorbed water), being these relative interactions strongly dependent on the type of SnO_2 considered. In particular, smaller crystallite size and higher surface sulphation of the ceramic particles seemed to promote a larger water affinity, a more regular organization of the polymer ionic channels and an increase in the storage modulus of the composite membrane.

All these findings are highly encouraging for possible applications of the systems here proposed as advanced electrolytes in proton exchange membrane fuel cells, mainly in view of the targeted requirement concerning applications under low relative humidity.

ACKNOWLEDGMENTS

This work has been performed in the framework of a project titled "Synthesis and physical chemical characterizations of proton conducting, nano-composite polymer electrolytes for fuel cell application" funded by Sapienza University of Rome (Progetto Ateneo 2010, prot. C26A108XME). M.A.N. also acknowledges the financial support of Sapienza University of Rome for the project "BORS: Bioinspired Organic Redox Shuttles as additives to increase lithium ion batteries efficiency and lifecycle" (Progetto Ateneo 2011, prot. C26A11FZA7).

REFERENCES

1. Kreuer, K.D. Hydrocarbon membranes. In *Handbook of Fuel Cell Fundamentals, Technology, and Applications*; Vielstich, W., Lamm, A., Gasteiger, H.A., Eds.; John Wiley: Chichester, UK, 2003; Volume 3, Part 3; Chapter 33; p. 420.
2. Doyle, M.; Rajendran, G. Perfluorinated membranes. In *Handbook of Fuel Cell Fundamentals, Technology, and Applications*; Vielstich, W., Lamm, A., Gasteiger, H.A., Eds.; John Wiley: Chichester, UK, 2003; Volume 3, Part 3; Chapter 30; p. 351.

3. Babir, F.; Gomez, T. Efficiency and economics of proton exchange membrane (PEM) fuel cells. *Int. J. Hydrog. Energy*1996, *21*, 891–901.
4. Rikukawa, M.; Sanui, K. Proton-conducting polymer electrolyte membranes based on hydrocarbon polymers. *Progr. Polym. Sci.* 2000, *25*, 1463–1501.
5. Springer, T.E.; Zawodzinski, T.A.; Gottesfeld, S. Polymer Electrolyte Fuel Cell Model. *J. Electrochem. Soc.* 1991, *138*, 2334–2342.
6. Gottesfeld, S.; Zawodzinski, T. Polymer Electrolyte Fuel Cells. In *Advances in Electrochemical Science and Engineering*; Alkire, R.C., Gerischer, H., Kolb, D.M., Tobias, C.W., Eds.; WILEY-VCH Verlag GmbH: Weinheim, Germany, 1997; Volume 5, Chapter 4; p. 195.
7. Mathias, M.F.; Makharia, R.; Gasteiger, H.A.; Conley, J.J.; Fuller, T.J.; Gittleman, C.J.; Kocha, S.S.; Miller, D.P.; Mittelsteadt, C.K.; Xie, T.; *et al.* Two Fuel Cell Cars in Every Garage? *Interface* 2005, *14*, 24–35.
8. Brown, A.S.C.; Hargreaves, J.S.J. Sulfated metal oxide catalysts. Superactivity through superacidity? *Green Chem.* 1999,*1*, 17–20.
9. Lu, Y.; Yang, Y.; Sellinger, A.; Lu, M.; Huang, J.; Fan, H.; Haddad, R.; Lopez, G.; Burns, A.R.; Sasaki, D.Y.; *et al.* Self-assembly of mesoscopically ordered chromatic polydiacetylene/silica nanocomposites. *Nature* 2001, *410*, 913–917.
10. Gierke, T.D.; Munn, G.E.; Wilson, F.C.J. The morphology in nafion perfluorinated membrane products, as determined by wide- and small-angle X-ray studies. *J. Polym. Sci. Polym. Phys. Ed.* 1981, *19*, 1687–1704.
11. Kreuer, K.D. On the development of proton conducting polymer membranes for hydrogen and methanol fuel cells. *J. Membr. Sci.* 2001, *185*, 29–39.
12. Hsu, W.Y.; Gierke, T.D. Elastic theory for ionic clustering in perfluorinated ionomers. *Macromolecules* 1982, *15*, 101–105.
13. Kreuer, K.D. Proton Conductivity: Materials and Applications. *Chem. Mater.* 1996, *8*, 610–641.
14. Ramani, V.; Kunz, H.R.; Fenton, J.M. Investigation of Nafion®/HPA composite membranes for high temperature/low relative humidity PEMFC operation. *J. Membr. Sci.* 2004, *232*, 31–44.
15. Malhotra, S.; Datta, R. Membrane-Supported Nonvolatile Acidic Electrolytes Allow Higher Temperature Operation of Proton-Exchange Membrane Fuel Cells. *J. Electrochem. Soc.* 1997, *144*, L23–L26.
16. Navarra, M.A.; Abbati, C.; Scrosati, B. Properties and fuel cell performance of a Nafion-based, sulfated zirconia-added, composite membrane. *J. Power Sources* 2008, *183*, 109–113.
17. Thampan, T.M.; Jalani, N.H.; Choi, P.; Datta, R. Systematic Approach to Design Higher Temperature Composite PEMs.*J. Electrochem. Soc.* 2005, *152*, A316–A325.
18. Zhai, Y.; Zhang, H.; Hu, J.; Yi, B. Preparation and characterization of sulfated zirconia (SO_4^{2-}/ZrO_2)/Nafion composite membranes for PEMFC operation at high temperature/low humidity. *J. Membr. Sci.* 2006, *280*, 148–155.
19. Ren, S.; Sun, G.; Li, C.; Song, S.; Xin, Q.; Yang, X. Sulfated zirconia–Nafion composite membranes for higher temperature direct methanol fuel cells. *J. Power Sources* 2006, *157*, 724–726.
20. D'Epifanio, A.; Navarra, M.A.; Weise, F.C.; Mecheri, B.; Farrington, J.; Licoccia, S.; Greenbaum, S. Composite Nafion/Sulfated Zirconia Membranes: Effect of the Filler Surface Properties on Proton Transport Characteristics. *Chem. Mater.* 2010, *22*, 813–821.
21. Navarra, M.A.; Croce, F.; Scrosati, B. New, high temperature superacid zirconia-doped Nafion™ composite membranes. *J. Mater. Chem.* 2007, *17*, 3210–3215.
22. Navarra, M.A.; Abbati, C.; Croce, F.; Scrosati, B. Temperature-dependent Performances of a Fuel Cell Using a Superacid Zirconia-doped Nafion Polymer Electrolyte. *Fuel Cells* 2009, *9*, 222–225.
23. Giffin, G.A.; Piga, M.; Lavina, S.; Navarra, M.A.; D'Epifanio, A.; Scrosati, B.; Noto, V. Characterization of sulfated-zirconia/Nafion® composite membranes for proton exchange membrane fuel cells. *J. Power Sources* 2012, *198*, 66–75.
24. Bianchi, C.L.; Ardizzone, S.; Cappelletti, G. Surface state of sulfated zirconia: the role of the sol–gel reaction parameters. *Surf. Interface Anal.* 2004, *36*, 745–748.
25. Tominaka, S.; Momma, T.; Scrosati, B.; Osaka, T. Sulfated zirconia as a proton conductor for fuel cells: Stability to hydrolysis and influence on catalysts. *J. Power Sources* 2010, *195*, 4065–4071.
26. Hara, S.; Miyayama, M. Proton conductivity of superacidic sulfated zirconia. *Solid State Ion.* 2004, *168*, 111–116.
27. Matsuhashi, H.; Hino, M.; Arata, K. Solid catalyst treated with anion. XIX. Synthesis of the solid superacid catalyst of tin oxide treated with sulfate ion. *Appl. Catal.* 1990, *59*, 205–216.
28. Du, Y.; Liu, S.; Zhang, Y.; Yin, C.; Di, Y.; Xiao, F.-S. Mesostructured sulfated tin oxide and its high catalytic activity in esterification and Friedel-Crafts acylation. *Catal. Lett.* 2006, *108*, 155–158.
29. Matsuhashi, H.; Miyazaki, H.; Kawamura, Y.; Nakamura, H.; Arata, K. Preparation of a solid superacid of sulfated tin oxide with acidity higher than that of sulfated zirconia and its applications to aldol condensation and benzoylation.*Chem. Mater.* 2001, *13*, 3038–3042.
30. Chen, F.; Mecheri, B.; D'Epifanio, A.; Traversa, E.; Licoccia, S. Development of Nafion/Tin oxide composite MEA for DMFC applications. *Fuel Cells* 2010, *10*, 790–797.
31. Yoshimizu, H.; Kodaira, K.; Shimada, S.; Matsushita, T. Preparation and properties of antimony-doped SnO2 films by thermal decomposition of tin 2-ethylhexanoate. *J. Mater. Sci.* 1986, *21*, 2731–2734.
32. Giesekke, W.; Gutowsky, H.S.; Kirkov, P.; Laitinen, H.A. A proton magnetic resonance and electron diffraction study of the thermal decomposition of tin(IV) hydroxides. *Inorg. Chem.* 1967, *6*, 2731–1297.

33. Scherrer, P. Bestimmung der Grösse und der inneren Struktur von Kolloidteilchen mittels Röntgensrahlen. *Nachr. Ges. Wiss. Göttingen* 1918, *26*, 98–100.

34. Morterra, C. An infrared spectroscopic study of anatase properties. Part 6. Surface hydration and strong Lewis acidity of pure and sulphate-doped preparations. *J. Chem. Soc. Faraday Trans. I* 1988, *84*, 1617–1637.

35. Pusawale, S.N.; Deshmukh, P.R.; Lokhande, C.D. Chemical synthesis and characterization of hydrous tin oxide (SnO_2:H_2O) thin films. *Bull. Mater. Sci.* 2011, *34*, 1179–1183.

36. Peercy, P.S.; Morosin, B. Pressure and temperature dependences of the Raman-active phonons in SnO_2. *Phys. Rev. B*1973, *7*, 2779–2786.

37. Diéguez, A.; Romano-Rodríguez, A.; Vilà, A.; Morante, J.R. The complete Raman spectrum of nanometric SnO_2particles. *J. Appl. Phys.* 2001, *90*, 1550–1557.

38. Pagnier, T.; Boulova, M.; Sergent, N.; Bouvier, P.; Lucazeau, G. Nanopowdwes and nanostructured oxides: phase transition and surface reactiviy. *J. Raman Spectrosc.* 2007, *38*, 756–761.

39. Pârvulescu, V.; Coman, S.; Grange, P.; Pârvulescu, V.I. Preparation and characterization of sulfated zirconia catalysts obtained via various procedures. *Appl. Catal. A Gener.* 1999, *176*, 27–43.

40. Radwan, D.; Saad, L.; Mikhail, S.; Selim, S.A. Catalytic evolution of sulfated zirconia pillared clay in n-hexane transformation. *J. Appl. Sci. Res.* 2009, *5*, 2332–2342.

41. Nørgaard, C.F.; Nielsen, U.G.; Skou, E.M. Preparation of Nafion 117™-SnO_2 Composite Membranes using an Ion-Exchange Method. *Solid State Ion.* 2012, *213*, 76–82.

42. De Almeida, S.H.; Kawano, Y. Thermal behavior of Nafion membranes. *J. Therm. Anal. Calorim.* 1999, *58*, 569–577.

43. Abramoff, M.D.; Magalhães, P.J.; Ram, S.J. Image processing with ImageJ. *Biophoton. Int.* 2004, *11*, 36–42.

44. Bussian, D.A.; O'Dea, J.R.; Metiu, H.; Buratto, S.K. Nanoscale current imaging of the conducting channels in proton exchange membrane fuel cells. *Nano Lett.* 2007, *7*, 227–232.

45. He, Q.; Kusoglu, A.; Lucas, I.T.; Clark, K.; Weber, A.Z.; Kostecki, R. Correlating humidity-dependent ionically conductive surface area with transport phenomena in proton-exchange membranes. *J. Phys. Chem. B* 2011, *115*, 11650–11657.

46. James, P.J.; Antognozzi, M.; Tamayo, J.; McMaster, T.J.; Newton, J.M.; Miles, M.J. Interpretation of Contrast in Tapping Mode AFM and Shear Force Microscopy. A Study of Nafion. *Langmuir* 2001, *17*, 349–360.

47. James, P.J.; Elliot, J.A.; McMaster, T.J.; Newton, J.M.; Elliott, A.M.S.; Hanna, S.; Miles, M.J. Hydration of Nafion studied by AFM and X-ray scattering. *J. Mater. Sci.* 2000, *35*, 5111–5119.

48. Teocoli, F.; Paolone, A.; Palumbo, O.; Navarra, M.A.; Casciola, M.; Donnadio, A. Effects of water freezing on the mechanical properties of Nafion membranes. *J. Polym. Sci. B Polym. Phys.* 2012, *50*, 1421–1425.

49. Kreuer, K.-D. The role of internal pressure for the hydration and transport properties of ionomers and polyelectrolytes. *Solid State Ion.* 2013, *252*, 93–101.

50. Yang, C.; Srinivasan, S.; Bocarsly, A.B.; Tulyani, S.; Benziger, J.B. A comparison of physical properties and fuel cell performance of Nafion and zirconium phosphate/Nafion composite membranes. *J. Membr. Sci.* 2004, *237*, 145–161.

51. Kreuer, K.-D.; Portale, G. A Critical Revision of the Nano-Morphology of Proton Conducting Ionomers and Polyelectrolytes for Fuel Cell Applications. *Adv. Funct. Mater.* 2013, *23*, 5390–5397.

52. Brutti, S.; Scipioni, R.; Navarra, M.A.; Panero, S.; Allodi, V.; Giarola, M.; Mariotto, G. SnO_2-Nafion nanocomposite polymer electrolytes for fuel cell applications. *Int. J. Nanotechnol.* 2014. in press

CHAPTER 6

POLY[(3-HEXYLTHIOPHENE)-BLOCK-(3-SEMIFLUOROALKYL-THIOPHENE)] FOR POLYMER SOLAR CELLS

Ichiko Yamada[1], Koji Takagi[2], Yasuhiko Hayashi[1], Tetsuo Soga[1], Norio Shibata[1] and Takeshi Toru[3]

[1]Department of Frontier Materials, Graduate School of Engineering, Nagoya Institute of Technology, Gokiso, Showa-ku, Nagoya 466-8555, Japan

[2]Department of Materials Science and Engineering, Graduate School of Engineering, Nagoya Institute of Technology, Gokiso, Showa-ku, Nagoya 466-8555, Japan

[3]Department of Research, Nagoya Industrial Science Research Institute, 1-13 Yotsuyadori, Chikusa-ku, Nagoya 464-0819, Japan

ABSTRACT

We report the synthesis of poly[(3-hexylthiophene)-*block*-(3-(4,4,5,5,6,6,7,7,7-nonafluoroheptyl)thiophene)], P(3HT-*b*-3SFT), carried out by the Grignard Metathesis Method (GRIM). The copolymers composition was determined by ^{1}H and ^{19}F NMR spectroscopies, and gel permeation chromatography (GPC). The thin films of P(3HT-*b*-3SFT) were investigated by ultraviolet-visible absorption spectroscopy and atomic force microscopy (AFM). We also fabricated bulk-hetero junction (BHJ) solar cells based on blends of P(3HT-*b*-3SFT) and [6,6]-phenyl-C_{61}-butyric acid methyl ester (PCBM). Although the composition ratio of P3SFT in P(3HT-*b*-3SFT) was low, the influence of P3SFT on the morphology and properties of solar cells was significant. The annealing process for the BHJ solar cells induced the formation of large domains and led to poor solar cell performance. The BHJ solar cells, based on PCBM and P(3HT-*b*-3SFT), prepared by the non-annealing process, had a maximum power conversion efficiency of 0.84% under 100 mW/cm^2 (AM 1.5 solar illumination) in air.

INTRODUCTION

The bulk hetero junction (BHJ) solar cells, consisting of π-conjugated polymers as donor and fullerene derivatives as acceptor, have the advantage of low cost fabrication and easy processing due to solution based low temperature processes. The polymer-based solar cells have been attracting much attention and have been investigated by many researchers [1–7].

The mechanism by which organic thin-film solar cells convert sun-light into electricity is as follows. Absorption of photon leads to the formation of the electron-hole pair (exciton). Exciton diffuses to the interface of the donor and acceptor domains, where exciton is dissociated, that is, the charge separation occurs. The separated charges (free holes and electrons) are transported to the charge extracting electrodes

Citation: Yamada, I.; Takagi, K.; Hayashi, Y.; Soga, T.; Shibata, N.; Toru, T. Poly[(3-hexylthiophene)-block-(3-semifluoroalkylthiophene)] for Polymer Solar Cells. Int. J. Mol. Sci. 2010, 11, 5027-5039, doi:10.3390/ijms11125027

through their respective transport materials (polymer and fullerene), to supply an external direct current [8].

The nanoscale morphology using the organic film as an active layer is an important factor in influencing the performance of the BHJ solar cells. The smaller phase domain of donor and acceptor materials has a larger interfacial area in the active layer, which leads to an efficient exciton separation. In addition, an interpenetrating and bicontinuous network of donor and acceptor materials allow the pathway to improve charge carrier conduction through the active layer to the respective electrodes. Various studies have been attempted to control the nanoscale-interpenetrating network of donor and acceptor materials in order to improve device performance. The character of spin-casting solvents and the additive in the solvents, affects phase segregation of two components [6,9,10]. The thermal annealing also controls the nanoscale organization and induces segregation of polymer and fullerene derivative [7,9]. The devices using donor-acceptor hybrid diblock copolymer have been studied [11–13]. In this study, we attempted to construct an interpenetrating networks structure of the donor and acceptor phase in the photoactive layer using, not only self-assembly of diblock copolymer, but also fluorine property.

It is well known that several diblock copolymers containing various conjugated segments self-assemble and form a nano-scale separated structure. This behavior improves the crystallinity of polymer film for organic electronic devices, so some diblock copolymers based on alkyl-thiophene were synthesized [14–17], and the organic devices have been fabricated [18,19].

On the other hand, the fluorine atom has the unique property of extremely low polarizability, which adds a lipophobic and hydrophobic nature to the fluorinated compounds, called "fluorophilic". Due to the repelling nature of the fluorine to the non-fluorine compounds, the fluorine and non-fluorine phases are separated in the liquid state. This phenomenon is also observed in the hybrids of fluorinated and non-fluorinated elements in the solid state [20,21]. We report on diblock copolymers that consist of thiophenes with hydrocarbon and fluorinated carbon chains to construct the interpenetrating network structures of the donor and acceptor phase in the photoactive layer. We envisaged that in the active layer of solar cells composed of the diblock copolymer and fullerene derivative, the semifluoroalkyl segments would come close to each other by the intimate interaction among the alkylthiophene parts, and the repulsive effect between the semifluoroalkyl and alkyl groups. The fullerene derivatives would reside near the P3HT segments. A schematic image is shown in Figure 1.

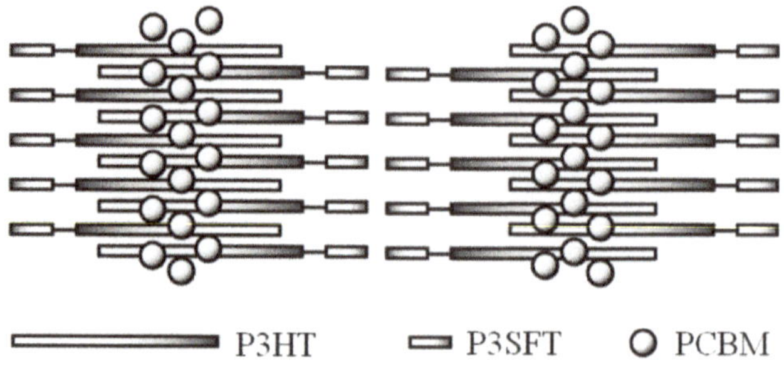

Figure 1: The image of self-assembly of diblock copolymer with semifluorinated carbon and hydrocarbon segments and PCBM on crystallization.

The diblock copolymer, poly[(3-hexylthiophene)-*block*-(3-(4,4,5,5,6,6,7,7,7-nonafluoroheptyl) thiophene)] (P(3HT-*b*-3SFT)) we synthesized, consists of poly(3-hexylthiophene) (P3HT) and semifluoroalkyl chain-substituted polythiophene, poly(3-(4,4,5,5,6,6,7,7,7-nonafluoroheptyl)thiophene) (P3SFT). P3SFT has a spacer between the thiophene ring and the fluoroalkyl chain, which was expected to relieve the electron-withdrawing effect of the fluorine atoms and, hence, lower the oxidation potential of the thiophene ring, so that P3SFT has electronic properties similar to 3-alkylthiophenes [22]. The composition and characteristics of the diblock copolymer and the performance of the solar cells fabricated with P(3HT-*b*-3SFT):PCBM blend were studied.

RESULTS AND DISCUSSION

Synthesis of Poly[(3-hexylthiophene)-block-(3-(4,4,5,5,6,6,7,7,7-nonafluoroheptyl)thiophene)], P(3HT-b-3SFT)

Diblock copolymer of P(3HT-*b*-3SFT) was synthesized from two monomers, 2,5-dibromo-3-(4,4,5,5,6,6,7,7,7-nonafluoroheptyl)thiophene (D3SFT) (5) and 2,5-dibromo-3-hexylthiophene (D3HT) (6), by the Grignard metathesis method (GRIM) [14,23,24]. D3SFT (5) were prepared by the methods reported previously [25] (Scheme 1). D3HT (6) was also prepared by bromination of 3-hexylthiophene [22]. The synthetic route for the P(3HT-*b*-3SFT) (10) is shown in Scheme 2.

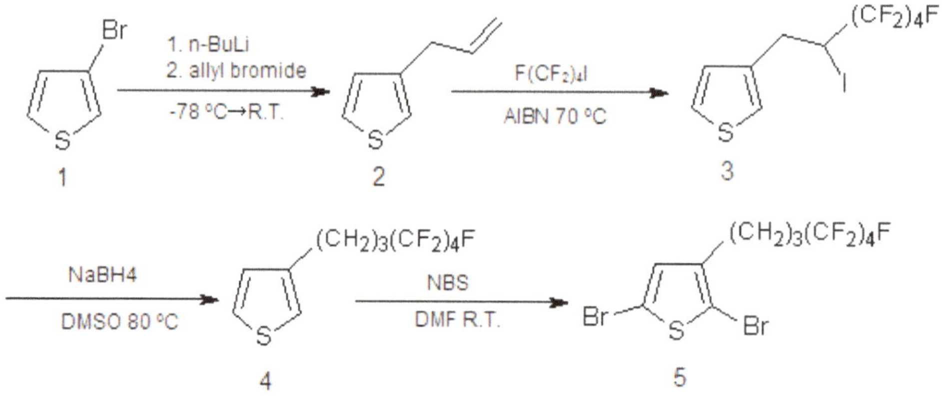

Scheme 1: Synthetic route for 2,5-dibromo-3-(4,4,5,5,6,6,7,7,7-nonafluoroheptyl)thiophene.

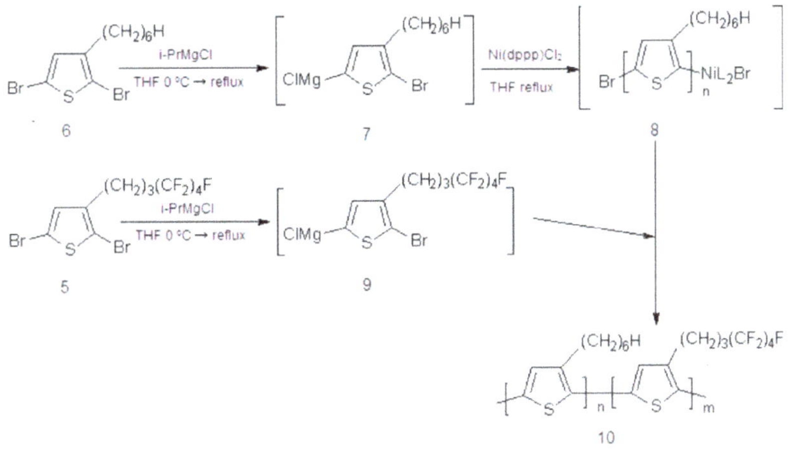

Scheme 2: Synthetic route for poly[(3-hexylthiophene)-*block*-(3-(4,4,5,5,6,6,7,7,7-nonafluoroheptyl)thiophene)].

In the first step, the P3HT homopolymer (8) was synthesized, because the P3SFT homopolymer was immediately deposited from the polymerization system. In the second step, the GRIM product of D3SFT (9) was added to the reaction solution to give the diblock copolymer, P(3HT-*b*-3SFT) (10), where the feed molar ratio of D3SFT to D3HT was 50:100. P(3HT-*b*-3SFT), prepared in an increased ratio of the D3SFT, was deposited during the polymerization reaction, and the resulting copolymer was not dissolved in any of the solvents (e.g., THF, $CHCl_3$ and ODCB).

We succeeded in the synthesis of two P(3HT-*b*-3SFT)s having different molecular weights by using a catalyst of 1.0 and 2.0 mol% Ni(dppp)Cl_2. The P3HT homopolymer was also prepared as a reference by the same method described above.

Polymer Characterization

P3HT and P(3HT-*b*-3SFT)s were investigated by ^{1}H NMR and ^{19}F NMR spectroscopies. In the ^{19}F NMR spectrum of P(3HT-*b*-3SFT), the fluorine signals indicate the semifluoroalkyl chain in diblock copolymer (Supplementary Figure S1; for convenience, only the diblock copolymer prepared by 2 mol% catalyst is shown). The compositions of both copolymers P(3HT-*b*-3SFT)s were determined from the integral ratio of the signals of P3HT and P3SFT in the ^{1}H NMR spectrum. Supplementary Figures S2 and S3 show the ^{1}H NMR spectra of P3HT and P(3HT-*b*-3SFT). The signal at 1.9–2.3 ppm appears only in the ^{1}H NMR spectrum of P(3HT-*b*-3SFT). According to the ^{1}H NMR spectrum of 3-(4,4,5,5,6,6,7,7,7-nonafluoroheptyl) thiophene [22], the signals are estimated corresponding to the β,γ-methylene protons of P3SFT. Thus, the ratios of the P3HT and P3SFT segments in P(3HT-*b*-3SFT)s, synthesized by using 1 mol% and 2 mol% Ni(dppp)Cl_2 catalysts, were determined to be 100:3 and 100:12, respectively, on the basis of the signals corresponding to P3ST and P3SFT. The contents of P3SFT, however, do not correspond to the feed molar ratio of D3HTT:D3SFT = 100:50, showing that D3SFT is rather inactive toward the polymerization reaction and was not very involved in the polymer chain. In fact, the yield of P(3HT-*b*-3SFT)s was low (approximately 30% yield). It should be noted that the solubility of P(3HT-*b*-3SFT) in THF and $CHCl_3$ was observed to be lower than P3HT itself, in spite of the low content of the 3-semifluorohepthlthiophene segment in P(3HT-*b*-3SFT)s.

The number-average (Mn) and the weight-average (Mw) molecular weight of diblock copolymers were measured by gel permeation chromatography (GPC) with THF as an eluent. Figure 2 shows the GPC profiles of P(3HT-*b*-3SFT)s, and the obtained molecular weights are integrated in Table 1. P(3HT-*b*-3SFT) synthesized by using 1 mol% Ni(dppp) Cl_2 catalyst, has Mn = 25,600, Mw = 36,200, which are roughly 2.5 times the Mn and Mw of P(3HT-*b*-3SFT) synthesized by using 2 mol% Ni(dppp)Cl_2 catalyst. This result suggests that the molecular weight of copolymers could be controlled by the amount of Ni(dppp)Cl_2.

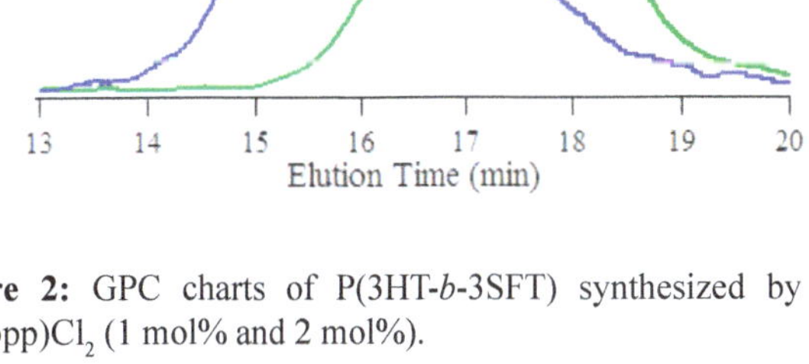

Figure 2: GPC charts of P(3HT-*b*-3SFT) synthesized by using Ni(dppp)Cl_2 (1 mol% and 2 mol%).

Table 1: Summary of GPC measurements of P3HT prepolymer and diblock copolymer, P(3HT-*b*-3SFT).

Polymer	Mn	Mw	Mw/Mn
P(3HT-*b*-3SFT) (1 mol% Ni(dppp)Cl_2)	25,600	36,200	1.4
P(3HT-*b*-3SFT) (2 mol% Ni(dppp)Cl_2)	10,500	13,500	1.2

Film Characterization of Polymer

Thin films of P(3HT-*b*-3SFT) block copolymers and the P3HT homopolymer were fabricated to investigate their optical properties and surface morphologies. Thin films of polymers were prepared by spin coating from *o*-dichlorobenzene (ODCB) on quarts substrates. In the UV-Vis spectra, P(3HT-*b*-3SFT)s, prepared by the 1 mol% and 2 mol% Ni(dppp)Cl_2catalysts, and P3HT show the maximum absorption (λ_{max}) at 526, 524 and 556 nm, respectively, with a shoulder peak at approximately 600 nm as indicated in Figure 3. The P3SFT homopolymer film has been reported to show λ_{max} of 482 nm [22]. Even though the content of P3SFT in P(3HT-*b*-3SFT) is small, the influence of P3SFT on optical property of P(3HT-*b*-3SFT) is so significant, that the λ_{max} value of P(3HT-*b*-3SFT) shifts toward the λ_{max} of P3SFT homopolymer.

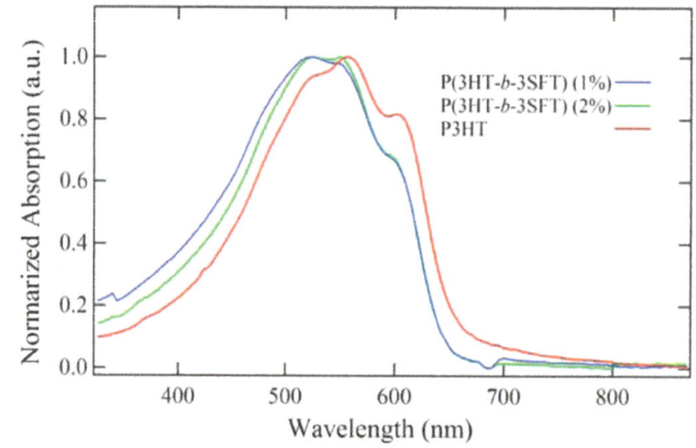

Figure 3: UV-Vis spectra of thin films of P(3HT-*b*-3SFT)s synthesized by using Ni(dppp)Cl_2 (1 mol% and 2 mol%) and P3HT.

The surface morphology of the thin film of P3HT and P(3HT-*b*-3SFT)s were observed by the atomic force microscopy (AFM). Figure 4 shows AFM phase images of thin films of P3HT and P(3HT-*b*-3SFT). Roughness values for all polymers obtained from AFM images are summarized in Table 2. Similar roughness values for P3HT and P(3HT-*b*-3SFT)s were obtained, whereas the phase image of P(3HT-*b*-3SFT) was found to be quite different from that of P3HT. The phase image of the P3HT film shows a homogeneous surface, while the P(3HT-*b*-3SFT) film shows round-shape domains on the surface. These domains are apparently obtained due to the phase separation of the hydrocarbon and fluorinated carbon segments by the repulsive nature of the fluorine atom (Figure 5); and the difference of the domain size is caused by the molecular weight of the copolymer as well as the content of P3SFT in P(3HT-*b*-3SFT). These results suggest that even the small P3SFT segment of P(3HT-*b*-3SFT) is enough to influence the morphology of the polymer thin film.

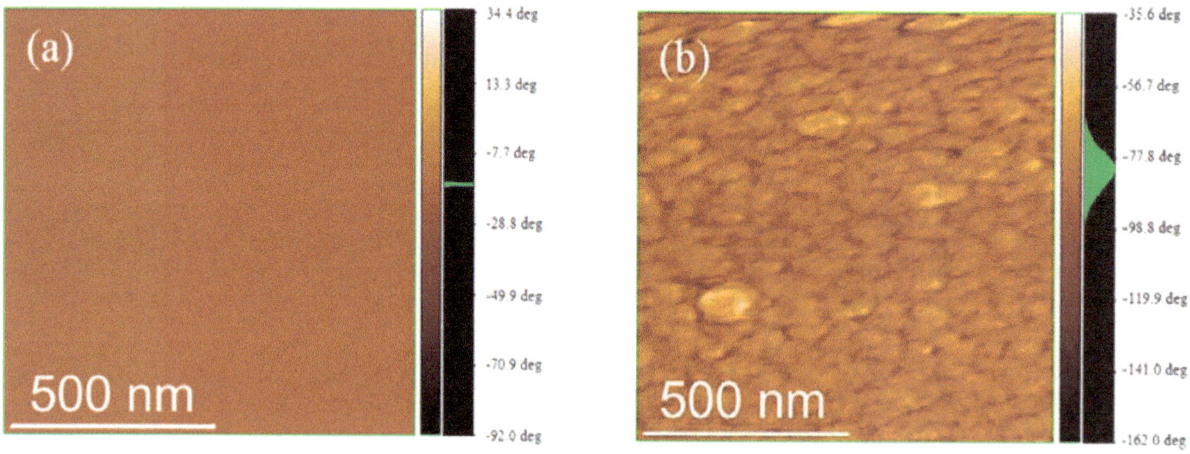

Figure 4: AFM phase images. (**a**) P3HT film; (**b**) P(3HT-*b*-3SFT) film (image size: 1 μm × 1 μm).

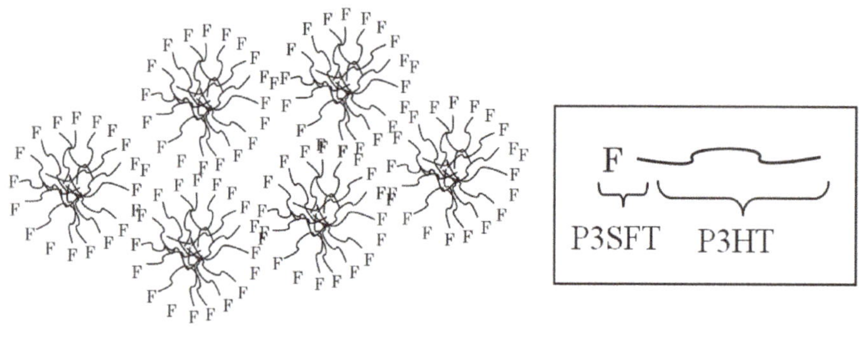

Figure 5: The model image of phase separation of P(3HT-*b*-3SFT).

Table 2: The roughness of films from atomic force microscopy (AFM) analysis

Polymer	Ra (nm)	RMS (nm)	Average Domain Area in Phase Image (μm^2)
P3HT	1.22	1.51	-
P(3HT-*b*-3SFT) (1 mol% $Ni(dppp)Cl_2$)	2.03	2.80	0.050
P(3HT-*b*-3SFT) (2 mol% $Ni(dppp)Cl_2$)	1.98	2.58	0.019

Properties of Photovoltaic Cells

The BHJ solar cells based on the blends of P(3HT-*b*-3SFT) and [6,6]-phenyl-C_{61}-butyric acid methyl ester (PCBM) were fabricated and the solar cell properties were characterized. The device structure was glass/ITO/PEDOT:PSS/blend of P(3HT-*b*-3SFT) and PCBM/Al. The property of solar cells was measured by using a solar simulator (100 mW/cm^2, AM 1.5). We fabricated the solar cells with P(3HT-*b*-3SFT) (1 mol% and 2 mol% $Ni(dppp)Cl_2$):PCBM (1:1 by weight) which were annealed at 100 °C for 20 min. Figure 6 shows the current density-voltage characteristics of the BHJ solar cells. The corresponding photovoltaic parameters (short circuit current density J_{sc}, open circuit voltage V_{oc}, fill factor *FF* and solar cell efficiency E_{ff}) are summarized in Table 3. The device with P(3HT-*b*-3SFT) (1 mol% $Ni(dppp)Cl_2$) exhibits higher J_{sc} and *FF* values than that with P(3HT-*b*-3SFT) (2 mol% $Ni(dppp)Cl_2$). On the other hand, the V_{oc} value of the solar cell with P(3HT-*b*-3SFT) (2 mol% $Ni(dppp)Cl_2$) was higher than that with P(3HT-*b*-3SFT) (1 mol% $Ni(dppp)Cl_2$). The surface morphology of the solar cells by AFM was then studied. The phase image of P(3HT-*b*-3SFT) (1 mol% $Ni(dppp)Cl_2$) has smaller domains (average domain area:0.006 μm) than the P(3HT-*b*-3SFT) (2 mol% $Ni(dppp)Cl_2$) (average domain area:0.012 μm). This result is contrary to the AFM phase image of only P(3HT-*b*-3SFT) shown in Figure 5, and the solar cell which has smaller domain shows better solar cell performance. It seems that the blend of PCBM interrupts aggregation of diblock copolymers, as a result the domain size of copolymers for solar cells is smaller than those of copolymer thin films. However we cannot ascertain why the domain size of P(3HT-*b*-3SFT) (1 mol% $Ni(dppp)Cl_2$) is smaller than that of P(3HT-*b*-3SFT) (2 mol% $Ni(dppp)Cl_2$).

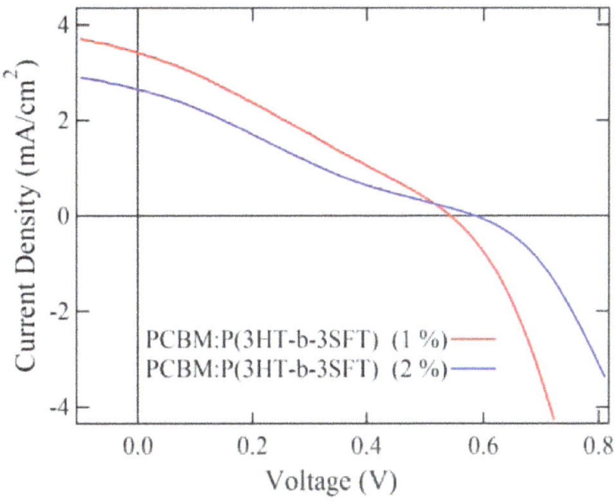

Figure 6: *J-V* characteristics of polymer solar cells with PCBM:P(3HT-*b*-3SFT)s.

Table 3: The characteristics of polymer solar cells with P(3HT-*b*-3SFT)

Materials	V_{oc} (V)	J_{sc} (mA/cm^2)	*FF*	E_{ff} (%)
PCBM:P(3HT-*b*-3SFT) (1 mol% $Ni(dppp)Cl_2$)	0.54	3.42	0.28	0.52
PCBM:P(3HT-*b*-3SFT) (2 mol% $Ni(dppp)Cl_2$)	0.59	2.65	0.23	0.35

We also fabricated the solar cells with P(3HT-*b*-3SFT) (1 mol% $Ni(dppp)Cl_2$), annealed at 140 °C, and data were compared with the non-annealed version. The current density-voltage characteristics of these BHJ solar cells are shown in Figure 7. The corresponding photovoltaic parameters are summarized in Table 4. In our previous study [26], the best annealing temperature for P3HT/PCBM solar cells was 100 °C. However, the non-annealed solar cell with P(3HT-*b*-3SFT) showed best performance with J_{sc} = 4.10 mA/cm^2, V_{oc} = 0.58 V, *FF* = 0.35, E_{ff} = 0.84%.

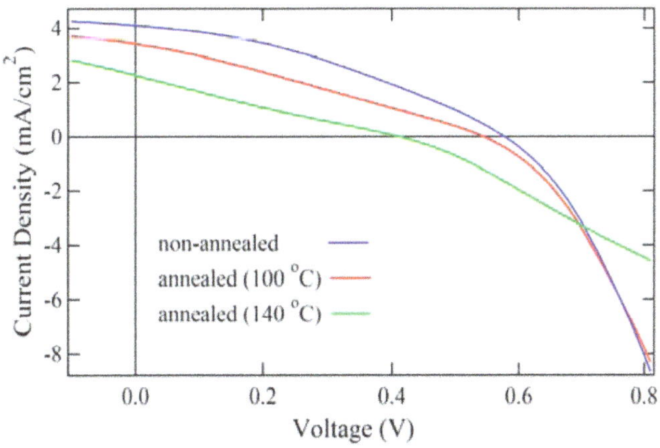

Figure 7: *J–V* characteristics and spectral responsivity of polymer solar cells with PCBM:P(3HT-*b*-3SFT) synthesized by using 1 mol% $Ni(dppp)Cl_2$.

Table 4: The characteristics of polymer solar cells with P(3HT-*b*-3SFT)

Annealing Temp	V_{oc} (V)	J_{sc} (mA/cm^2)	*FF*	E_{ff} (%)
R.T.	0.58	4.10	0.35	0.84
100 °C	0.54	3.42	0.28	0.52
140 °C	0.41	2.27	0.23	0.21

Figure 8 shows the AFM images of morphology for solar cells with P(3HT-*b*-3SFT)/PCBM prepared under various annealing conditions. The AFM images of the solar cells annealed at 100 °C and 140 °C showed round shape domains, and domains of the device annealed at 140 °C are larger than at 100 °C. Since these domains disturb formation of the bicontinuous network in the BHJ solar cells, the self-assembly of P(3HT-*b*-3SFT) decrease solar cells properties by annealing.

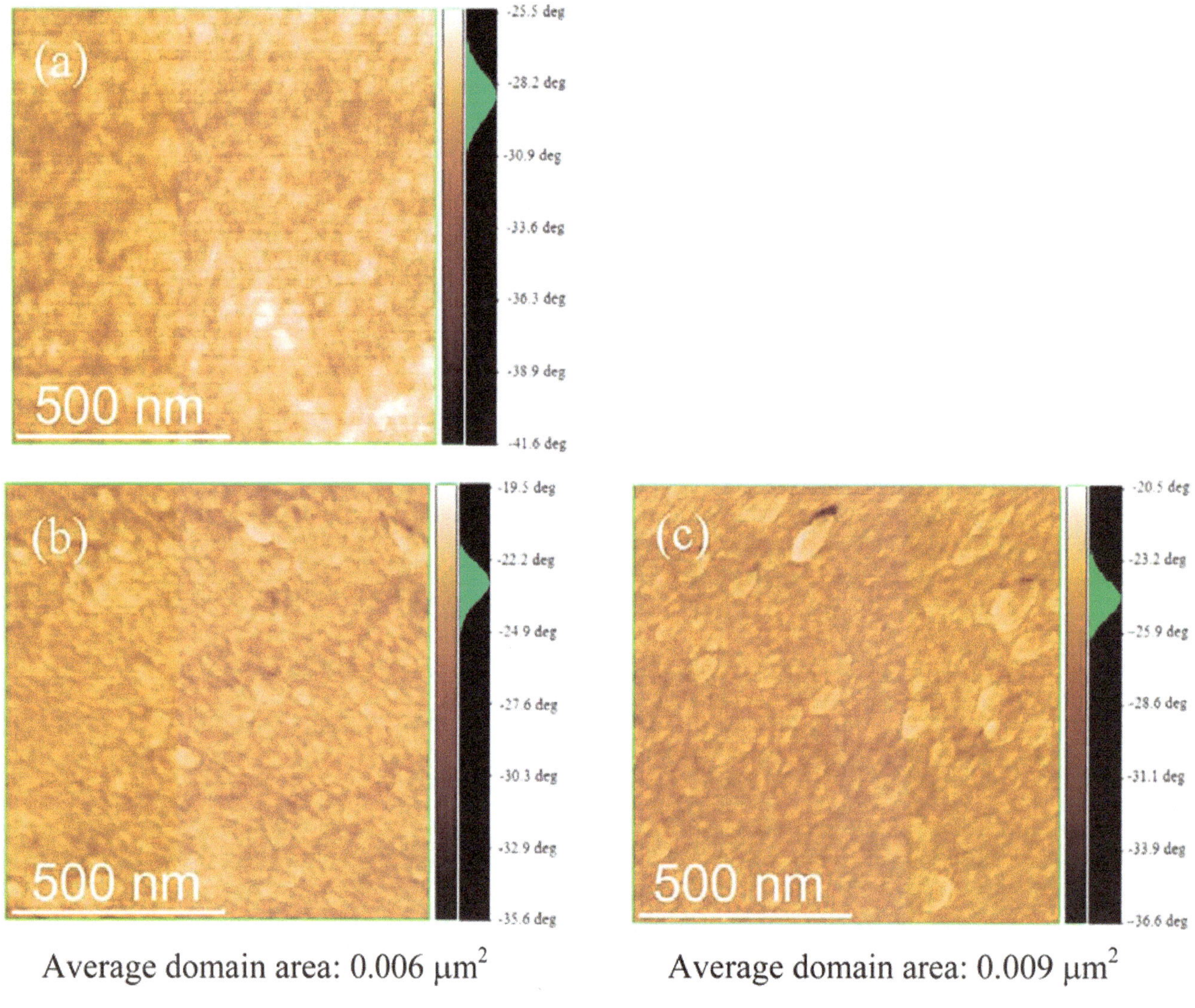

Figure 8: AFM phase images of the solar cells with PCBM:P(3HT-*b*-3SFT). (**a**) non-annealed; (**b**) 100 °C; (**c**) 140 °C (image size: 1 μm × 1 μm).

EXPERIMENTAL SECTION

Material and Instrumentation

Isopropylmagnesium Chloride in Tetrahydrofuran, (1 mol/L) were purchased from Tokyo Chemical Industry Co., Ltd. Ni(dppp)Cl_2 was purchased from Sigma-Aldrich Co. 2,5-dibromo-3-hexylthiophene and 2,5-dibromo-3-(4,4,5,5,6,6,7,7,7-nonafluoroheptyl)thiophene were synthesized by previously reported methods [22,25].

^{1}H and ^{19}F NMR spectra in $CHCl_3$ were measured using a Varian Gemini-200, 300 and ^{1}H and ^{13}C NMR spectra in $CHCl_3$for polymers were measured using Bruker AV-500. Molecular weight (Mn, Mw, Mw/Mn) were determined by Tosoh Gel Permeation Chromatography (GPC) with a UV detector using THF as the eluent at 40 °C. UV-Vis spectra of thin films were recorded using a JASCO V-570. Atomic Force Microscopy (AFM) images were done with a JEOL JSM-5600 scanning electron microscope operating in tapping mode. The current-voltage (*J-V*) measurements of the polymer photovoltaic cell were carried out using a solar simulator under the illumination of AM 1.5, 100 mW/cm^2.

Synthesis of Poly(3-hexylthiophene)-b-poly(3-(4,4,5,5,6,6,7,7,7-nonafluoroheptyl)thiophene) (10)

All glass apparatuses were dried prior to use. Two round bottomed flasks equipped with a three-way stopcock were dried by heating under reduced pressure and cooled to room temperature under Ar. 2,5-dibromo-3-hexylthiophene (D3HT) (6) (400 mg, 1.23 mmol) was placed in one of the flasks, and 2,5-dibromo-3-(4,4,5,5,6,6,7,7,7-nonafluoroheptyl)thiophene (D3SFT) (308 mg, 0.62 mmol) (5) was placed in the other. They were then evacuated under reduced pressure to remove any water and oxygen inside. Dry THF (7.0 mL) was added into the D3HT flask (1), and the solution was stirred at 0 °C. Isopropyl magnesium chloride (1.0 M solution in THF, 1.23 mL, 1.23 mmol) was added to the solution, and the mixture was heated to reflux under Ar (7). After 45 min, a suspension of Ni(dppp)Cl_2 (13.3 mg, 0.025 mmol, 2.0 mol%) in dry THF (8.0 mL) was added, and the reaction mixture was heated to reflux for 1 h.

Dry THF (3.0 mL) was added to the other flask for D3SFT (5), and the solution was stirred at 0 °C. Isopropylmagnesium chloride (1.0 M solution in THF, 0.62 mL, 0.62 mmol) was added to the solution, and the mixture was stirred at reflux under argon for 45 min. This mixture (9) was added to the polymerization mixture of D3HT (8) via a syringe. The mixture was stirred at reflux under Ar for 1 h. The reaction mixture was cooled to room temperature, and then poured into methanol (100 mL) and residue was filtered and purified via Soxhlet extraction successively with methanol, acetone, and chloroform. The chloroform fraction was concentrated under vacuum to give P(3HT-*b*-3SFT) as a purple solid (138.6 mg, 33.3%) (10). P(3HT-*b*-3SFT) synthesized by using 1.0 mol% Ni(dppp)Cl_2, afforded a 30.0% yield.

Film Preparation for Measurement of UV-Spectrum and AFM

To investigate morphology and the optical properties of P(3HT-*b*-3SFT)s and P3HT as a reference, thin films of the polymers were prepared by spin coating from *o*-dichlorobenzene on quarts substrates at room temperature, then annealed at 140 °C for 10 min inside a N_2-filled glove box. Measurement of AFM and UV-Vis spectroscopy was carried out on the same sample.

Fabrication and Measurement of Solar Cells

The indium tin oxide (ITO)-coated glass substrate was cleaned successively with acetone and methanol under ultrasonication for 5 min, and cleaned by nitrogen gas. Then the substrate was cleaned by ultraviolet ozone treatment for 1 h. Poly(3,4-eth ylenedioxythiophene):poly(styrenesulfonate) (PEDOT:PSS) (Baytron P AL 4083, HC Starck) was spin-coated on ITO at 2,000 rpm in air. Then the device was transferred to the N_2-filled glove box, and dried at 100 °C for 5 min. The polymer/PCBM mixture was prepared by dissolving together in ODCB (30 mg/mL, 1:1 wt:wt) and stirring overnight. The active layer of the device was formed by spin casting at 2,000 rpm from the mixture on the PEDOT/PSS layer. Then the films were annealed at 100 °C or 140 °C, each for 20 min. The top electrode of Al was deposited on the active layer by a vacuum deposition. Then the device was annealed at 140 °C for 10 min inside a N_2-filled glove box.

CONCLUSIONS

We have synthesized poly[(3-hexylthiophene)-*block*-(3-(4,4,5,5,6,6,7,7,7-nonafluoroheptyl) thiophene)] (P(3HT-*b*-3SFT)) by the GRIM, coupled with nickel-catalyzed coupling polymerization of D3HT followed by D3SFT, which were composed of P3HT and marginal P3SFT. Even though the amount of P3SFT is very small, it had a significant influence on the property of P(3HT-*b*-3SFT).

We also fabricated solar cells based on the blends of P(3HT-*b*-3SFT)s and PCBM, which showed phase separation as expected. The solar cells based on P(3HT-*b*-3SFT)s, synthesized by using 1 mol% Ni(dppp)Cl_2, indicated better photovoltaic properties than those based on P(3HT-*b*-3SFT)s (2 mol% Ni(dppp)Cl_2). This result may be due to the difference of domain size in the active layer, and the solar cell based on P(3HT-*b*-3SFT) (1 mol% Ni(dppp)Cl_2) is smaller. In addition, we investigated the influence of annealing temperatures on the performance of solar cells and found that the annealing process induced formation of large domains, and the properties of the annealed device were decreased.

Based on this study, we will study the solar cell composed of P(3HT-*b*-3SFT) and fullerene derivatives with perfluoroalkyl substituents, which is expected to have a different morphology due to the interactive effects of both perfluoro chains.

ACKNOWLEDGMENTS

The authors would like to thank N. Kishi at NIT for useful discussions.

REFERENCES

1. Sariciftci, NS; Smilowitz, L; Heeger, AJ; Wudl, F. Photoinduced Electron Transfer from a Conducting Polymer to Buckminsterfullerene. *Science* 1992, *258*, 1474–1476.
2. Yu, G; Gao, J; Hummelen, JC; Wudl, F; Heeger, AJ. Polymer Photovoltaic Cells: Enhanced Efficiencies via a Network of Internal Donor-Acceptor Heterojunctions. *Science* 1995, *270*, 1789–1791.
3. Shaheen, SE; Brabec, CJ; Sariciftci, NS; Padinger, F; Fromherz, T; Hummelen, JC. 2.5% efficient organic plastic solar cells. *Appl. Phys. Lett* 2001, *78*, 841.
4. Hayashi, Y; Yamada, I; Takagi, S; Takasu, A; Soga, T; Jimbo, T. Influence of Structure and C_{60} Composition on Properties of Blends and Bilayers of Organic Donor-Acceptor Polymer/C_{60} Photovoltaic Devices. *Jpn. J. Appl. Phys*2005, *44*, 1296–1300.
5. Thompson, BC; Fréchet, J. Polymer-Fullerene Composite Solar Cells. *Angew. Chem. Int. Ed* 2008, *47*, 58–77.
6. Liang, Y; Xu, Z; Xia, J; Tsai, S; Wu, Y; Li, G; Ray, C; Yu, L. For the Bright Future-Bulk Heterojunction Polymer Solar Cells with Power Conversion Efficiency of 7.4%. *Adv. Mater* 2010, *22*, E135–E138.
7. Chen, L; Hong, Z; Li, G; Yang, Y. Recent Progress in Polymer Solar Cells: Manipulation of Polymer:Fullerene Morphology and the Formation of Efficient Inverted

Polymer Solar Cells. *Adv. Mater* 2009, *21*, 1434–1449.

8. Hoppe, H; Sariciftci, NS. Organic solar cells: An overview. *J. Mater. Res* 2004, *19*, 1924–1945.
9. Yang, X; Loos, J; Veenstra, SC; Verhees, WJH; Wienk, MM; Kroon, JM; Michels, MAJ; Janssen, RAJ. Nanoscale Morphology of High-Performance Polymer Solar Cells. *Nano Lett* 2005, *5*, 579–583.
10. Lee, JK; Ma, WL; Brabec, CJ; Yuen, J; Moon, JS; Kim, JY; Lee, K; Bazan, GC; Heeger, AJ. Processing Additives for Improved Efficiency from Bulk Heterojunction Solar Cells. *J. Am. Chem. Soc* 2008, *130*, 3619–3623.
11. Sivula, K; Ball, Z; Watanabe, N; Fréchet, J. Amphiphilic Diblock Copolymer Compatibilizers and Their Effect on the Morphology and Performance of Polythiophene:Fullerene Solar Cells. *Adv. Mater* 2006, *18*, 206–210.
12. Lee, JU; Cirpan, A; Emrick, T; Russell, TP; Jo, WH. Synthesis and photophysical property of well-defined donor-acceptor diblock copolymer based on regioregular poly(3-hexylthiophene) and fullerene. *J. Mater. Chem* 2009, *19*, 1483–1489.
13. Zhang, Q; Cirpan, A; Russell, TP; Emrick, T. Donor-Acceptor Poly(thiophene-*block*-perylene diimide) Copolymers: Synthesis and Solar Cell Fabrication. *Macromolecules* 2009, *42*, 1079–1082.
14. Ohshimizu, K; Ueda, M. Well-Controlled Synthesis of Block Copolythiophenes. *Macromolecules* 2008, *41*, 5289–5294.
15. Zhang, Y; Tajima, K; Hirota, K; Hashimoto, K. Synthesis of All-Conjugated Diblock Copolymers by Quasi-Living Polymerization and Observation of Their Microphase Separation. *J. Am. Chem. Soc* 2008, *130*, 7812–7813.
16. Zhang, Y; Tajima, K; Hashimoto, K. Nanostructure Formation in Poly(3-hexylthiophene-*block*-3-(2-ethylhexyl)thiophene)s. *Macromolecules* 2009, *42*, 7008–7015.
17. Wu, P; Ren, G; Li, C; Mezzenga, R; Jenekhe, SA. Crystalline Diblock Conjugated Copolymers:Synthesis, Self-Assembly, and Microphase Separation of Poly(3-butylthiophene)-*b*-poly(3-octylthiophene). *Macromolecules* 2009, *42*, 2317–2320.
18. Chueh, C; Higashihara, T; Tsai, J; Ueda, M; Chen, W. All-conjugateddiblockcopolymerofpoly(3-hexylthiophene)-*block*-poly(3-phenoxymethylthiophene) for field-effect transistor and photovoltaic applications. *Org. Electron* 2009, *10*, 1541–1548.
19. Wu, P; Ren, G; Kim, FS; Li, C; Mezzenga, R; Jenekhe, SA. Poly(3-hexylthiophene)-*b*-poly(3-cyclohexylthiophene):Synthesis, microphase separation, thin-film transistors, and photovoltaic applications. *J. Polym. Sci. Part A* 2010, *48*, 614–626.
20. Ono, T; Hayakawa, Y. Fluorine-Based Crystal Engineering. *Fluor.-Contain. Synthons* 2005, *29*, 498–513.
21. Ono, T; Hayakawa, Y; Yasuda, N; Uekusa, H; Ohashi, Y. A Hydro-Fluoro Hybrid Compound, Hermaphrodite, Is Fluorophilic or Fluorophobic? A New Concept for the Fluorine-Based Crystal Engineering. *Curr. Fluoroorg. Chem* 2007,*10*, 170–189.
22. Hong,X;Tyson,JC;Middlecoff,JS;Collard,DM.Synthesis and Oxidative Polymerization of Semifluoroalkyl-Substituted Thiophenes. *Macromolecules* 1999, *32*, 4232–4239.
23. Loewe, RS; Ewbank, PC; Liu, J; Zhai, L; McCullough, RD. Regioregular, Head-to-Tail Coupled Poly(3-alkylthiophenes) Made Easy by the GRIM Method: Investigation of the Reaction and the Origin of Regioselectivity. *Macromolecules*2001, *34*, 4324–4333.
24. Miyakoshi, R; Yokoyama, A; Yokozawa, T. Catalyst-Transfer Polycondensation. Mechanism of Ni-Catalyzed Chain-Growth Polymerization Leading to Well-Defined Poly(3-hexylthiophene). *J. Am. Chem. Soc* 2005, *127*, 17542–17547.
25. Wang, B; Watt, S; Hong, M; Domercq, B; Sun, R; Kippelen, B; Collard, DM. Synthesis, Properties, and Tunable Supramolecular Architecture of Regioregular Poly(3-alkylthiophene)s with Alternating Alkyl and Semifluoroalkyl Substituents. *Macromolecules* 2008, *41*, 5156–5165.
26. Sukeguchi, D; Singh, SP; Reddy, MR; Yoshiyama, H; Afre, RA; Hayashi, Y; Inukai, H; Soga, T; Nakamura, S; Shibata, N; Toru, T. New diarylmethanofullerene derivatives and their properties for organic thin-film solar cells. *Beilstein J. Org. Chem* 2009, *5*, 7.

CHAPTER 7

GROWTH KINETICS OF MONODISPERSE POLYSTYRENE MICROSPHERES PREPARED BY DISPERSION POLYMERIZATION

Fan Li,[1,2] Chong Geng,[1] and Qingfeng Yan[1]

[1]Department of Chemistry, State Key Laboratory of New Ceramics and Fine Processing, Tsinghua University, Beijing 100084, China

[2]Department of Material Science and Engineering, Tsinghua University, Beijing 100084, China

ABSTRACT

Dispersion polymerization has been widely applied to the synthesis of monodisperse micron-sized polymer colloidal spheres. Many efforts have been devoted to studying the influence of initial conditions on the size and uniformity of the resultant microspheres, aiming to synthesize micron-size monodisperse colloidal spheres. However, the inner contradiction between the size and the size distribution of colloidal spheres hinders the realization of this goal. In this work, we drew our attention from the initial conditions to the growth stage of dispersion polymerization. We tracked the size evolution of colloidal sphere during the dispersion polymerization, through which we established a kinetic model that described the relationship between the monomer concentration and the reaction time. The model may provide a guideline to prepare large polymer colloidal spheres with good monodispersity by continuous monomer feeding during the growth stage to maintain the concentration of monomer at a constant value in a dispersion polymerization process.

INTRODUCTION

As a new functional macromolecule material, monodisperse polymer microspheres have many applications in environmental conservation, biomedicine, colloid science, electronic information material, and many other areas [1–3] due to their superiority of good sphericity, size tunability, large specific surface area, and excellent absorbability [4–6]. Particularly, monodisperse micron-sized spheres are ideal materials as advanced coatings [7], fillers of chromatographic column [8, 9], standard particles of electron microscope and Coulter particle size testers, and spacers in LCD [10], which has drawn increasing interest to synthesize monodisperse micron-sized spheres [11, 12].

The first report on monodisperse polymer colloidal spheres was the polystyrene spheres prepared by Vanderhoff and Brandford [13]. So far, many synthesis strategies have

Citation: Fan Li, Chong Geng, and Qingfeng Yan, "Growth Kinetics of Monodisperse Polystyrene Microspheres Prepared by Dispersion Polymerization," Journal of Polymers, vol. 2013, Article ID 754687, 7 pages, 2013. doi:10.1155/2013/754687

been developed such as emulsion polymerization, suspension polymerization, dispersion polymerization, and seed swelling polymerization. In view of the demand of micron size and monodispersity, the former two methods are excluded because they cannot meet the two requirements at the same time. Emulsion polymerization can only gain monodisperse colloidal spheres with the size below 1 μm. Suspension polymerization may achieve large microspheres with size ranging from 10 μm to 100 μm while the microspheres possess a bad monodispersity [14]. The latter two methods are suitable for the synthesis of uniform colloidal spheres with micron size. Seed swelling polymerization, including the dynamics swelling developed by Okubo et al. [15] and two-step swelling initiated by Ugelstad et al. [16], can attain colloidal spheres with larger sphere diameter. However, in terms of the whole process, seed swelling polymerization is complex. In addition, the monodispersity is determined not only by the swelling process itself, but also by the initial monodispersity of the seeds prepared by dispersion polymerization in advance. Thus, the key point is how to gain larger microspheres with uniform size by dispersion polymerization.

To attain larger microspheres with excellent monodispersity, many efforts have been made to explore the influence of the initial condition of the reaction system on the resultant microspheres, such as the dosage of monomer, initiator, and dispersing agent. Generally, with the increase in the monomer's initial concentration, the sphere diameter increases and the distribution of sphere diameter broadens. Thus, there is a suitable range for monomer's initial concentration to achieve uniform and large microspheres [17, 18]. Similarly, increasing the concentration of initiator leads to higher reaction rate and larger sphere diameter while it may also broaden the distribution of the sphere diameter simultaneously [17, 19, 20]. In addition, the dispersing agent which prevents microspheres from gelatinization and stabilizes the reaction system requires a sufficient supply to guarantee its functions. However, the increase in the amount of the dispersing agent results in a smaller size despite a better monodispersity [21–24]. In this view, the inner contradiction between the size and the size distribution of microspheres hinders the further increase of the size of monodisperse microspheres. It is thus believed that the inner contradiction cannot be solved solely by adjusting the initial conditions.

Recently, Song et al. [25] employed two-stage dispersion polymerization to synthesize large and uniform dye-contained polymer spheres. In this method, the dye-comonomer, which significantly destroyed the final monodispersity if fed into the reaction at the beginning, was added to the reaction system after the nucleation stage. Through this strategy, the influence of the dye-comonomer on the nucleation was avoided and a narrow size distribution of the final colloidal spheres was guaranteed. It was expected that two-stage or multistage dispersion polymerization strategy paved a way to the synthesis of micron-sized large and monodisperse polymer colloidal spheres. However, when and how to feed the extra monomer into the reaction system to achieve this goal remains debatable since most previous works were mainly based on trial and error.

In this work, we focused on the whole process in one-stage dispersion polymerization and tracked the size evolution of colloidal sphere at different reaction time. Based on the experimental results, we established a theoretical model to describe the relationship between the monomer concentration and the reaction time during dispersion polymerization. The model was expected to provide guidance to the two-stage or the multi-stage dispersion polymerization, aiming to synthesize micron-sized large and monodisperse polymer colloidal spheres.

MATERIALS AND METHODS

Materials

All chemicals, including styrene (St, 99%, Aldrich), anhydrous ethanol (99.95%, Aldrich), polyvinylpyrrolidone (PVP, MW 40000, TCI), and azobis(isobutyronitrile) (AIBN, Sinopharm Chemical Reagent Co., Ltd), were used as received without further purification. Deionized (DI) water (resistivity greater than 18.2 MΩ·cm, Ultra Pure UV, China) was used in all experiments.

Synthesis of PS Colloidal Microspheres by Using Dispersion Polymerization

Briefly, the synthesis was conducted in a three-neck round-bottom flask equipped with a condenser and immersed in a water bath with a preset temperature. Firstly, the monomer of 17 mL styrene and the stabilizer of 1.5 g PVP were dissolved in 98 mL anhydrous ethanol and then added to the flask. The solution was stirred with a rate of 200 rpm and the temperature was kept at 70°C while nitrogen gas was continuously led into the flask to avoid the oxidation of the initiator AIBN in the whole process. After 30 minutes, an initiator solution of 0.15 g AIBN dissolved in 28 mL ethanol was introduced into the system. The reaction continued for 24 h and then was stopped by cooling at ambient conditions. To track the evolution of sphere size during dispersion polymerization, the mixture in the reaction flask was sampled at specific intervals and dried on a glass substrate. The sphere size was measured using an optical microscope.

Characterization

The optical microphotographs of the PS colloidal microspheres were taken using an optical microscope (Olympus, BX51TRF), which was connected to a CCD camera (Pixelink-B742) and a computer for real-time image recording.

RESULTS AND DISCUSSION

The Morphology of PS Colloidal Spheres Prepared by Dispersion Polymerization

As can be seen from the optical microscope photograph shown in Figure 1, micron-size PS colloidal spheres were obtained by using the one-stage dispersion polymerization method. The PS colloidal spheres were spontaneously organized into a closely packed hexagonal lattice on the glass substrate upon evaporation of solvent, indicating the high monodispersity of the PS spheres. To reduce the error in the measurement of the size of a single PS sphere, we chose the closely packed area and measured the direct distance between the several tangent microspheres in line and calculated the average distance between neighboring two spheres. According to the measured data in Figure 1, the average diameter of the resultant PS colloidal spheres was calculated to be 2.72 μm.

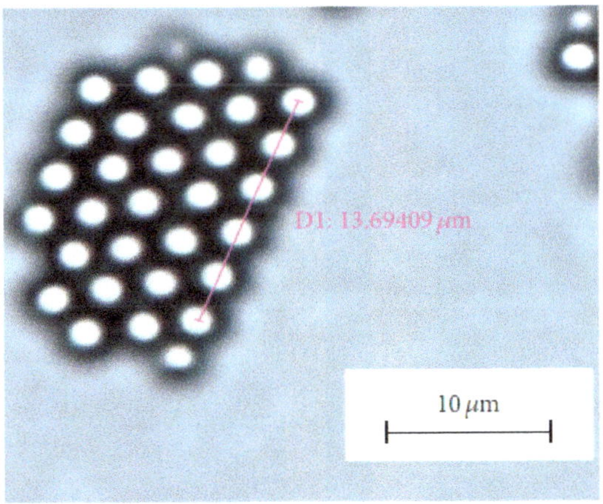

Figure 1: Optical microphotograph of PS colloidal spheres prepared by dispersion polymerization.

Growth Kinetics of PS Spheres during Dispersion Polymerization

The process of dispersion polymerization can be basically divided into several stages. At the beginning of the reaction, monomer, dispersing agent, and initiator are all dissolved in the solvent. When the temperature reaches the decomposition temperature of the initiator, the initiator decomposes and produces radicals. The radicals trigger the polymerization process, which results in formation of the oligomers. When the chain length of the oligomers exceeds a critical value, nucleation occurs and the oligomers precipitate from the solvent. At the same time, dispersing agents are absorbed by the nuclei and form graft copolymers with the polymer chains, which makes microspheres stable in the solvent. Then the synthesis enters the stage of growth. Monomer and copolymer in the solvent are persistently captured by polymer nuclei to provide the motive power of the growth, which keeps on until termination of the reaction. In the process of dispersion polymerization, the nucleation stage is a transition period for the system from homogenous phase to heterogeneous phase. This period determines the number of microspheres in the system which should be constant during the polymerization to guarantee the uniformity of final microspheres [21, 22, 26–28].

Knowing the evolution of monomer concentration during dispersion polymerization is of importance especially for synthesis of large and uniform colloidal spheres. Considering the reaction mechanism involved in dispersion polymerization, however, a large proportion of the monomers exist in the solvent in the form of oligomers whose chain length does not exceed the critical value for nucleation. Thus, direct measurement of the monomer concentration is not feasible. Herein, we establish a theoretical model to describe monomer concentration at different reaction time based on the diameter of microspheres measured at different time intervals.

The following are the major assumptions used in formulating the model. Firstly, in view of the good monodispersity of the final microspheres, the reaction is considered as an ideal dispersion polymerization, in which nucleation number is just determined by the nucleation stage, and secondary nucleation does not occur in the later growth stage. It is also assumed that the number of microspheres remains during growth stage equal to the nucleation number at the beginning of reaction. Secondly, although the microspheres in the growth stage may mostly seize the oligomer chains rather than the monomers, it is considered that the increasing rate of the microspheres' weight or volume is proportional to the equivalent concentration of the monomers, some of which are in the form of free oligomers. Thirdly, considering the actual slight volume change during the mixture process of styrene and ethanol, which can hardly be observed in the experiment, the deviation of volume is neglected in our model to reduce the complexity of the calculation. Thus, a simple approximation was adopted by the replacement of the total volume of the mixture of several liquids with the sum of their own volumes.

According to the conservation of mass, the relationship between the product and the reactant at any time t is given by

$$N\rho_{\text{sphere}} V_{\text{sphere}} = m_0 - CV, \tag{1}$$

where N is the number of microspheres in the system at time t, ρ_{sphere} is the density of polystyrene microspheres, m_0 is the initial mass of styrene monomer added into the system, C is the present concentration of surplus styrene in the system, and V is the total volume of the system at time t. V_{sphere}, the average volume of one sphere, can be described by the formula of spheroidal volume

$$V_{\text{sphere}} = \frac{\pi}{6} d^3, \tag{2}$$

where d is the average diameter of microspheres at time t. The concentration of styrene in the system at time t can be expressed using

$$C = \frac{\rho_{\text{styrene}} V_{\text{styrene}}}{V} = \frac{\rho_{\text{styrene}} \left(V - V_{\text{ethanol}}\right)}{V}, \tag{3}$$

where V_{styrene} stands for the surplus volume of styrene in the system, while V_{ethanol} stands for the surplus volume of ethanol, which is equal to the initial volume with the loss in the reaction ignored. By putting together the three equations above, out comes the relationship between the concentration of monomer and the average diameter of microspheres at time t, which can be given by

$$C = \frac{\rho_{\text{styrene}} \left(m_0 - N\rho_{\text{sphere}} \cdot (\pi/6)\, d^3\right)}{m_0 - N\rho_{\text{sphere}} \cdot (\pi/6)\, d^3 + \rho_{\text{styrene}} V_{\text{ethanol}}}. \tag{4}$$

Based on our experimental data, we input relevant parameters. The initial parameters were the same as the data described in the experimental section, while the density of microspheres was taken as that of polystyrene. According to the former assumption, the number of microspheres was a constant value which is equal to the nucleation number. Thus, we set the time t at the end of the reaction with the microspheres' diameter d taken as the final size which was 2.72 μm in our work, while the final concentration C_{final} was assumed to be one percent of the initial value. In this way, the number of microspheres was approximately calculated as follows:

$$N = \frac{\rho_{\text{styrene}} V_0 - C_{\text{final}} V_{\text{ethanol}}}{\rho_{\text{sphere}} \cdot (\pi/6)\, d_{\text{final}}^3} \approx 1.347 \times 10^{12}. \quad (5)$$

Thus, with the input of the initial parameters and the estimated number of microspheres N, the relationship between the monomer concentration and the microspheres diameter at any time was attained. To introduce the time dimension, we measured the diameter of the PS spheres sampled from the reaction system at different time intervals. According to the relationship expressed in (4), we gained the corresponding monomer concentration, which is shown in Table 1.

Table 1: The sphere diameters and monomer concentration at different time

Reaction time (h)	Sphere diameter (μm)	Monomer concentration (g/mL)
0	0	0.105
1	1.170	0.098
1.5	1.364	0.094
2	1.360	0.094
3	1.760	0.080
4	1.960	0.069
5	2.114	0.060
7	2.376	0.039
9	2.410	0.036
11	2.652	0.010
23	2.720	0.001

In order to further deduce the relationship between the monomer concentration and the reaction time, we calculated different function forms of monomer concentration versus reaction time and processed the data with linear fitting. The results are shown in Figure 2. As can be seen in Figure 2, the linear fitting of ln C and t has the largest correlation coefficient of 0.965. As a consequence, we consider that the natural log of the monomer concentration is linearly associated with the reaction time. In other words, there exists an exponential relationship between the monomer concentration and the reaction time, which can be expressed using

$$C(t) = C_0 e^{-kt}, \quad (6)$$

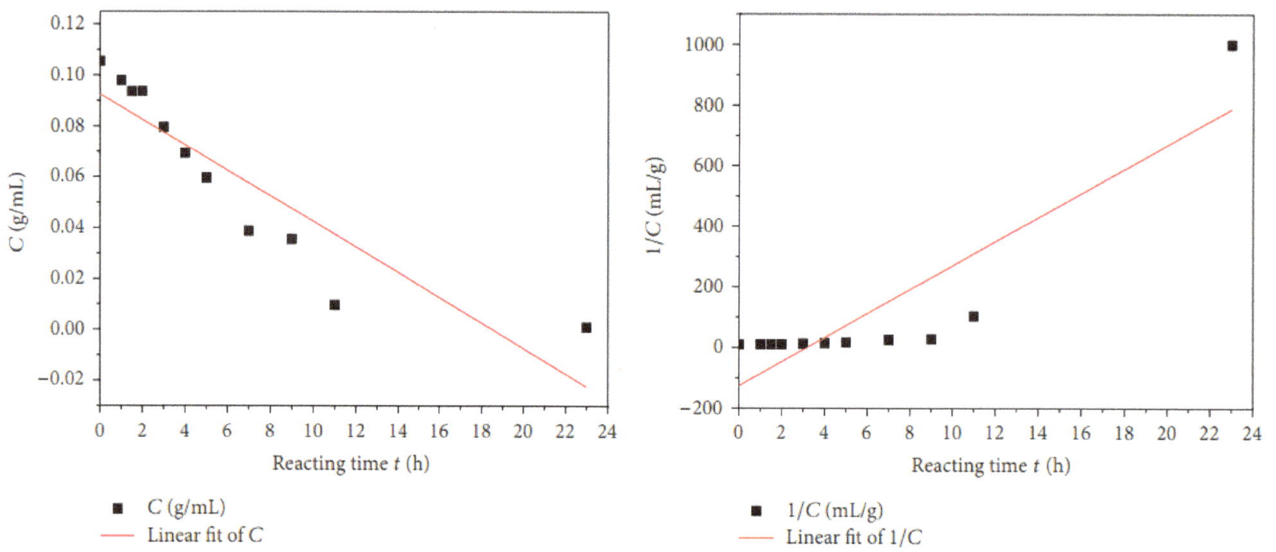

(a) (b)

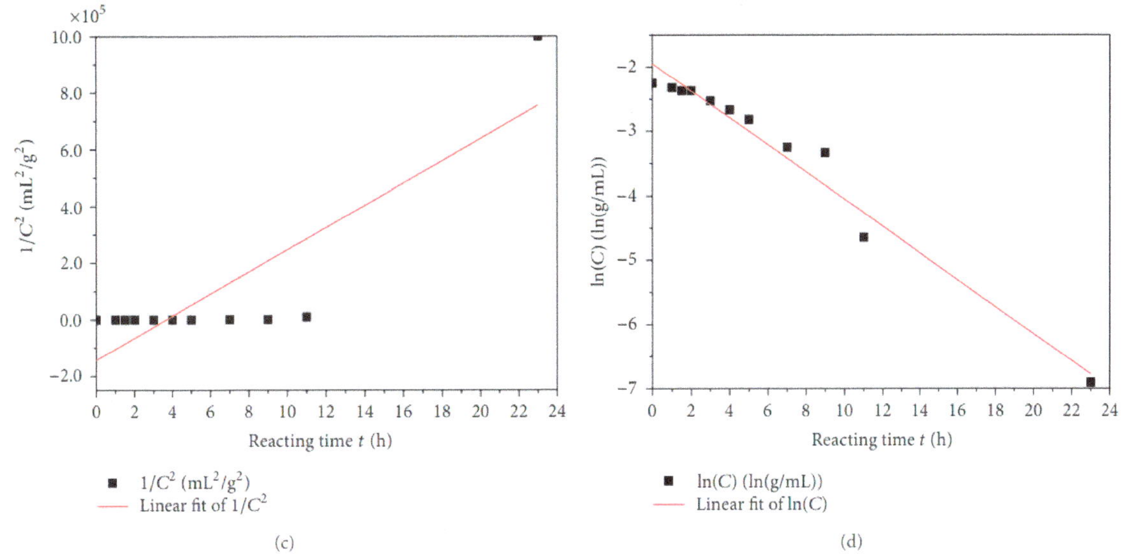

Figure 2: Linear fitting results of different function forms of concentration C with reaction time t. (a) linear fitting of C and t, $r = 0.800$ (b) linear fitting of $1/C$ and t, $r = 0.763$ (c) linear fitting of $1/C2$ and t, $r = 0.703$ (d) linear fitting of ln C and t, $r = 0.965$.

where the initial monomer concentration C_0 can be attained by the initial parameters in our experiment. Then, a program was written to find optimal value of k which minimized the sum of squared relative errors between the theoretical curve and the experimental points (t_i, C_i). That was one value that made the value of $\sum_i [(C_i - (t_i))/C(t_i)]^2$ minimum. The final result was obtained as follows:

$$C = 0.105 \times e^{-0.10t}. \tag{7}$$

Combining (4) with (7), the relationship between the colloidal sphere diameter and the reaction time could be derived. The resultant curve is shown in Figure 3 in comparison to the experimental data. As can be seen from Figure 3, the theoretical curve matches the experimental data well.

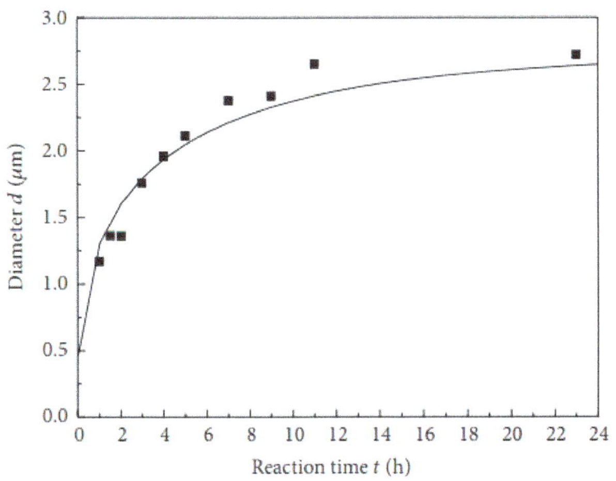

Figure 3: The evolution of spheres diameter with extended reaction time. The squares stand for the experimental points.

Application of the Kinetic Model on Synthesis of Large and Monodisperse Colloidal Spheres via Dispersion Polymerization

Through the deformation of (3), the total volume of the reaction system can be obtained as follows:

$$V = \frac{\rho_{\text{styrene}} V_{\text{ethanol}}}{\rho_{\text{styrene}} - C}. \tag{8}$$

Table 2: Different monomer feeding rates required at various beginning time of continuous monomer feeding.

t (h)	C (g/mL)	dC/dt (g/mL/h)	V (mL)	ΔV (mL/h)
1	0.0950	−0.00950	140.9	1.48
2	0.0860	−0.00860	139.4	1.32
3	0.0778	−0.00778	138.0	1.18
4	0.0704	−0.00704	136.8	1.06
5	0.0637	−0.00637	135.7	0.95
6	0.0576	−0.00576	134.7	0.86
7	0.0521	−0.00521	133.9	0.77
8	0.0472	−0.00472	133.1	0.69
9	0.0427	−0.00427	132.4	0.62
10	0.0386	−0.00386	131.8	0.56

ΔV means the monomer feeding rate needed if the beginning of the continuous monomer feeding is set at time t.

With the relationship between the monomer concentration and the reaction time described in (7), the monomer consumption rate at any time of the reaction can be easily deduced by taking the derivative of the monomer concentration with respect to time:

$$\Delta V = \frac{1}{\rho_{\text{styrene}}} \cdot \frac{dC}{dt} \cdot V. \tag{9}$$

Thus, assuming that the initiator does not lose efficacy during the reaction, the monomer concentration will be held at a constant value if one continuously feeds extra monomer into the system according to the consumption rate described in (9). This will be helpful to keep the system at a high reaction rate, leading to the enlargement of the diameter of the final PS colloidal microspheres. In fact, the initiator efficiency will inevitably descend during the process of dispersion polymerization; so in actual process, we can meet the requirement of constant efficiency by replenishing initiator during the reaction. According to the analysis above, we calculated the relationship between the beginning time of the monomer feeding and the monomer feeding rate within the present system of our experiment, which is shown in Table 2. In this way, even if an extra monomer is added into the reaction system, the monomer concentration will still be low enough to avoid new nucleation as long as one begins the continuous feeding of extra monomer in the growth stage rather in the nucleation state. As a consequence, the monodispersity of the colloidal spheres is ensured. At the same time, continuous feeding of extra monomer inevitably leads to colloidal spheres with large size.

CONCLUSION

In summary, a kinetic model has been established to describe the exponential relationship between the monomer concentration and the reaction time during synthesis of colloidal spheres by using dispersion polymerization. Accordingly, one may gain the quantitative relationship between the monomer consumption rate and the reaction time.

Taking advantage of this model, it is possible to attain larger colloidal spheres by continuously feeding monomer into the reaction system during the growth stage according to the monomer consumption rate. In this way, the reaction rate is kept at a high constant all the time until the reaction is artificially ended. Furthermore, secondary nucleation can be effectively avoided because both the average concentration and the local concentration of monomer in the reaction system are less than the critical value. Ultimately, large and uniform colloidal spheres can be prepared using the dispersion polymerization technique.

ACKNOWLEDGMENTS

This work was supported by the Natural Science Foundation of China (Project nos. 50903046 and 51173097) and the National Key Basic Research Program of China (no. 2013CB632902). The Tsinghua University Initiative Scientific Research Program and the Tsinghua National Laboratory for Information Science and Technology (TNList) Crossdiscipline Foundation are also acknowledged for partial financial support.

REFERENCES

1. S. K. Mi, H. L. Gyu, H. Jae-Min, and L. Hyunjung, "Synthesis of monodisperse PS-co-PDMS microspheres by dispersion polymerization," Materials Science and Engineering, vol. 27, no. 5–8, pp. 1247–1251, 2007.
2. S. O. Kyoung, B. won, and K. Hwayoung, "Dispersion polymerization of N-vinylcarbazole using siloxane-based and fluorine-based surfactants in compressed liquid dimethyl ether," Polymer, vol. 48, no. 6, pp. 1450–1454, 2007.
3. I. Miraballes-Martínez and J. Forcada, "Synthesis of latex particles with surface amino groups," Journal of Polymer Science A, vol. 38, pp. 4230–4237, 2000.
4. R. P. A. Dullens, E. M. Claesson, and W. K. Kegel, "Preparation and properties of cross-linked fluorescent poly(methyl methacrylate) latex colloids," Langmuir, vol. 20, no. 3, pp. 658–664, 2004.
5. H. Kawaguchi, "Functional polymer microspheres," Progress in Polymer Science, vol. 25, no. 8, pp. 1171–1210, 2000.
6. R. M. Santos and J. Forcada, "Acetal-functionalized polymer particles useful for immunoassays," Journal of Polymer Science A, vol. 35, no. 9, pp. 1605–1610, 1997.
7. K. Ishii, "Synthesis of microgels and their application to coatings," Colloids and Surfaces A, vol. 153, no. 1–3, pp. 591–595, 1999.
8. E. Caro, R. M. Marce, P. A. G. Cormack, D. C. Sherrington, and F. Borrull, "Molecularly imprinted solid-phase extraction of naphthalene sulfonates from water," Journal of Chromatography A, vol. 1047, no. 2, pp. 175–180, 2004.
9. Y. Watebe, K. Hosoya, N. Tanaka, and T. Kubo, "Shielded molecularly imprinted polymers prepared with a selective surface modification," Journal of Polymer Science A, vol. 43, no. 10, pp. 2048–2060, 2005.
10. S. Omi, M. Saito, T. Hashimoto, M. Nagai, and G.-H. Ma, "Preparation of monodisperse polystyrene spheres incorporating polyimide prepolymer by dispersion polymerization in the presence of L-ascorbic acid," Journal of Applied Polymer Science, vol. 68, no. 6, pp. 897–907, 1998.
11. K. Ishizu and N. Tahara, "Microsphere synthesis by emulsion copolymerization of methyl methacrylate with poly(methacrylic acid) macromonomers," Polymer, vol. 37, no. 13, pp. 2853–2856, 1996.
12. N. B. Viswanthan, P. A. Thomas, J. K. Pandit, M. G. Kulkarni, and R. A. Mashelkar, "Preparation of non-porous microspheres with high entrapment efficiency of proteins by a (water-in-oil)-in-oil emulsion technique," Journal of Controlled Release, vol. 58, no. 1, pp. 9–20, 1999.
13. J. W. Vanderhoff and E. B. Brandford, Polymer Colloid I, Plenum Press, New York, NY, USA, 1971.
14. D.-H. Lee, J.-W. Kim, and K.-D. Suh, "Monodisperse micron-sized polymethylmethacrylate particles having a crosslinked network structure. V," Journal of Materials Science, vol. 35, no. 24, pp. 6181–6188, 2000.
15. M. Okubo, M. Shiozaki, M. Tsujihiro, and Y. Tsukuda, "Preparation of micron-size monodisperse polymer particles by seeded polymerization utilizing the dynamic monomer swelling method," Colloid & Polymer Science, vol. 269, no. 3, pp. 222–226, 1991.
16. J. Ugelstad, P. C. Mork, R. Schmid, T. Ellingsen, and A. Berge, "Preparation and biochemical and biomedical applications of new monosized polymer particles," Polymer International, vol. 30, no. 2, pp.

157–168, 1993.

17. C.-M. Tseng, M. S. El-Aasser, and J. W. Vanderhoff, "Optimization of polymer-solvent interaction in a multisolvent system by the unifac group-contribution method," Journal of Applied Polymer Science, vol. 32, no. 5, pp. 5007–5019, 1986.
18. H. Bamnolker, B. Nitzan, T. Tennenbaum, B. BarToov, M. Hinz, and H. Seliger, "Scientific and Clinical Applications of Magnetic Carriers," in Proceedings of the 1st International Conference on Scientific and Clinical Applications of Magnetic Carriers, pp. 37–51, Rostock, Germany, September 1996.
19. A. Tuncel, R. Kahraman, and E. Piskin, "Monosize polystyrene lattices carrying functional-groups on their surfaces," Journal of Applied Polymer Science, vol. 51, pp. 1485–1498, 1994.
20. K. Hu, V. L. Dimonie, E. D. Sudol, and M. S. El-Aasser, "Monodisperse poly(butadiene/styrene) particles by dispersion polymerization," Journal of Applied Polymer Science, vol. 55, no. 10, pp. 1411–1415, 1995. ·
21. B. Thomson, A. Rudin, and G. Lajoie, "Dispersion copolymerization of styrene and divinylbenzene: synthesis of monodisperse, uniformly crosslinked particles," Journal of Polymer Science A, vol. 33, no. 3, pp. 345–347, 1995.
22. B. Thomson, A. Rudin, and G. Lajoie, "Dispersion copolymerization of styrene and divinylbenzene. II: effect of crosslinker on particle morphology," Journal of Applied Polymer Science, vol. 59, no. 13, pp. 2009–2028, 1996.
23. M. Holderle, G. Bar, and R. Mulhaupt, "Synthesis and characterization of oxazoline-functional polymer particles by free radical nonaqueous dispersion polymerization," Journal of Polymer Science A, vol. 35, no. 13, pp. 2539–2548.
24. D. Horak, F. Svec, and J. M. Frechet, "Preparation and control of surface properties of monodisperse micrometer size beads by dispersion copolymerization of styrene and butyl methacrylate in polar media,"Journal of Polymer Science A, vol. 33, no. 14, pp. 2329–2338, 1995.
25. J.-S. Song, F. Tronc, and M. A. Winnik, "Two-stage dispersion polymerization toward monodisperse, controlled micrometer-sized copolymer particles," Journal of the American Chemical Society, vol. 126, no. 21, pp. 6562–6563, 2004.
26. H. Kurioka, H. Uyama, and S. Kobayashi, "Peroxidase-catalyzed dispersion polymerization of phenol derivatives," Polymer Journal, vol. 30, no. 6, pp. 526–529, 1998.
27. M. Okubo, T. Yamashita, and E. Ise, "Thermodynamics for preparation of micron-sized, monodispersed, monomer-adsorbed polymer particles having snowman shape by utilizing the dynamic swelling method and the seeded polymerization," Proceedings of the Japan Academy B, vol. 75, no. 7, pp. 195–200, 1999.
28. M. Okubo, Y. Konishi, M. Takebe, and H. Minami, "Preparation of micron-sized, monodispersed,anomalous polymer particles by utilizing the solvent-absorbing/ releasing method,"Colloid and Polymer Science, vol. 278, no. 10, pp. 919–926, 2000.

CHAPTER 8

SYNTHESIS AND CHARACTERIZATION OF [60] FULLERENE-GLYCIDYL AZIDE POLYMER AND ITS THERMAL DECOMPOSITION

Ting Huang[1], Bo Jin[1,2], Ru Fang Peng[1], Cong Di Chen[1], Rong Zong Zheng[1], Yi He[2] and Shi Jin Chu[1]

[1]State Key Laboratory Cultivation Base for Nonmetal Composites and Functional Materials, Southwest University of Science and Technology, Mianyang 621010, Sichuan, China

[2]School of National Defence Science and Technology, Southwest University of Science and Technology, Mianyang 621010, Sichuan, China

ABSTRACT

A new functionalized [60]fullerene-glycidyl azide polymer (C_{60}-GAP) was synthesized for the first time using a modified Bingel reaction of [60]fullerene (C_{60}) and bromomalonic acid glycidyl azide polymer ester (BM-GAP). The product was characterized by Fourier transform infrared (FTIR), ultraviolet-visible (UV-Vis), and nuclear magnetic resonance spectroscopy (NMR) analyses. Results confirmed the successful preparation of C_{60}-GAP. Moreover, the thermal decomposition of C_{60}-GAP was analyzed by differential scanning calorimetry (DSC), thermogravimetric analysis coupled with infrared spectroscopy (TGA-IR), and in situ **FTIR. C_{60}-GAP decomposition showed a three-step thermal process. The first step was due to the reaction of the azide group and fullerene at approximately 150 °C. The second step was ascribed to the remainder decomposition of the GAP main chain and** N-heterocyclic **at approximately 240 °C. The final step was attributed to the burning decomposition of amorphous carbon and carbon cage at around 600 °C.**

INTRODUCTION

Fullerene has been extensively developed with remarkable achievement since Kratschmer et al. first produced [60]fullerene (C_{60}) on a preparative scale [1,2]. C_{60} and its functionalized derivatives are important and can be remarkably applied in various studies because of their unique structure and physical properties [3,4,5]. As C_{60} has good features such as thermal stability, antioxidation, etching resistance, good compatibility with propellant components, and beneficial for preventing aging [6], C_{60} can replace carbon black as the solid rocket fuel additive that can increase the burning rate and combustion catalytic efficiency and decrease the amount of NO_x in the exhaust gas [7,8]. If some energetic groups, such as the nitro group, would be added into C_{60}, an enhanced fuel additive may be obtained [9]. Numerous nitration methods have been used to synthesize polynitro-fullerene derivatives under different

Citation: Huang, T.; Jin, B.; Peng, R.F.; Chen, C.D.; Zheng, R.Z.; He, Y.; Chu, S.J. Synthesis and Characterization of [60]Fullerene-Glycidyl Azide Polymer and Its Thermal Decomposition. Polymers 2015, 7, 896-908, doi:10.3390/polym7050896

conditions and nitration reagents, such as nitric acid [10], NO_2 [11,12,13], or N_2O_4 [14], in CS_2. Generally, the nitro group that directly bonds to fullerene reacts slowly with H_2O to afford partially hydroxylated products poly(hydroxynitro)-fullerenes [13]. Therefore, achieving a stable energetic fullerene derivative remains a challenge.

Energetic polymers with azide groups have recently attracted much attention, because they can be used in solid rocket propellant systems and composite explosives as plasticizers to reduce the vulnerability of storage and transport, impart additional energy, increase performance, improve mechanical properties, and enhance system stability [15,16,17]. Moreover, the terminal hydroxyl group of glycidyl azide polymer (GAP) can be easily modified through various reactions such as esterification, acetalization, etherification, etc. [18,19,20,21,22,23,24]. In this paper, GAP was further functionalized via esterification with malonyl dichloride and subsequent brominate reaction to afford bromomalonic acid GAP ester (BM-GAP), which can easily react with C_{60} through a modified Bingel reaction [25] to afford a new fullerene derivative, namely, [60]fullerene GAP (C_{60}-GAP), which perhaps could become a new material of energetic burning rate additive. The thermal decomposition performance and decomposition mechanism of C_{60}-GAP were also examined in detail.

EXPERIMENTAL SECTION

Materials and Methods

The reagents for the organic synthesis were pure commercial products from Aladdin Industrial Corporation (Shanghai, China). The solvents were purchased from Kelong Chemical Reagents Co. (Chengdu, China). The 300–400 mesh silica gel was purchased from Qingdao Hailang. GAP (M_n = 505, M_w/M_n = 1.39) were provided by the Liming Chemical Engineering Research and Design Institute of Luoyang. C_{60} was obtained from Puyang Yongxin Fullerene Technology Co., Ltd. (purity > 99.5%). All the chemicals were used as received.

Synthesis of Malonic Acid Glycidyl Azide Polymer Ester

Malonyl chloride (1.25 mL, 13 mmol) diluted with 30 mL of CH_2Cl_2 were added dropwise to a stirred solution of GAP (M_n = 505, M_w/M_n = 1.39) (10 g, 26 mmol) and N,N-dimethylformamide (0.76 mL, 13 mmol) as acid binding agent in redistilled CH_2Cl_2 (100 mL) under N_2 atmosphere in an ice bath. The mixture was stirred for 30 min and then overnight at room temperature. The reaction solution was repeatedly washed with deionized water to neutralize the pH. The organic phase was collected and dried by anhydrous Na_2SO_4. After filtration, the solution was dried at 45 °C in vacuum to achieve 9.34 g (yield of 86%) of malonic acid glycidyl azide polymer ester (M-GAP). The GPC analysis revealed a molecular weight (M_n) of 1197 g/mol and a molecular weight distribution (M_w/M_n) of 1.31. ^{1}H NMR (400 MHz, $CDCl_3$, δ): 5.12 (br, 2H, CH_2), 3.77–3.48 (m, 3H, CH_2CH), 3.45–3.30 (m, 2H, CH_2N_3), 1.29–1.12 (m, 3H, CH_3); FT-IR (KBr): υ = 2929 (–CH–), 2876 (–CH_2–), 2102, 1280 (–N_3), 1756, 1739 (C=O), 1118 (C–O–C), 986, 936, 849, 670, 628, 556 cm^{-1}.

Synthesis of Bromomalonic Acid Glycidyl Azide Polymer Ester

Bromine (0.31 mL) diluted in 10 mL of CH_2Cl_2 was slowly dropped in a stirred solution of 4.67 g M-GAP in 100 mL CH_2Cl_2, and the mixture was stirred at 30 °C for 30 min. The mixture color gradually changed from yellow to orange-red after bromine was added dropwise to the mixture, and after 6 h of stirring, the solution color became stable. The resulting solution was repeatedly washed by saturated NaBr solution and deionized water to neutralize the solution. The organic phase was dried with anhydrous Na_2SO_4. After filtration, the filtrate was rotary evaporated to remove the solvent and dried under 45 °C in vacuum to yield 4.36 g (yield of 85%) of bromomalonic acid glycidyl azide polymer ester (BM-GAP). The GPC analysis revealed a molecular weight (M_n) of 1268 g/mol and a molecular weight distribution (M_w/M_n) of 1.36. ^{1}H NMR (400 MHz, $CDCl_3$, δ): 5.13 (s, 1H, CHBr), 3.76–3.46 (m, 3H, CH_2CH), 3.43–3.26 (m, 2H, CH_2N_3), 1.25–1.11 (m, 3H, CH_3); FT-IR (KBr): υ= 2929 (–CH–), 2876 (–CH_2–), 2100, 1279 (–N_3), 1755, 1746 (C=O), 1115 (C–O–C), 989, 936, 850, 671, 630, 556 cm^{-1}.

Synthesis of [60]Fullerene Glycidyl Azide Polymer

0.1198 g C_{60} (0.17 mmol) and 0.1066 g glycine (0.17 mmol) were dissolved in 100 mL of chlorobenzene and 30 mL of DMSO by using ultrasonic irradiation method. Thereafter, 0.6755 g of BM-GAP was added to the above solution under mild stirring. The mixture was stirred at room temperature for 12 h, and then the resulting solution was repeatedly washed with deionized water to remove DMSO and sarcosine. Chlorobenzene was then removed under vacuum. The residue was separated on a silica gel column with carbon disulfide as the eluent to obtain 23.50 mg unreacted C_{60} and with toluene/ethyl acetate (1:2) as the eluent to afford the crude product of C_{60}-GAP. Then the crude product was washed with 150 mL methanol (3 × 50 mL) and dried in vacuo at 50 °C to afford 0.3048 g C_{60}-GAP, with a yield of 83%. The GPC analysis revealed a molecular weight (M_n) of 2176 g/mol and a molecular weight distribution (M_w/M_n) of 1.45. ^{1}H NMR (400 MHz, $CDCl_3$, δ): 3.74–3.43 (m, 3H, CH_2CH), 3.41–3.24 (m, 2H, CH_2N_3), 1.21–1.09 (m, 3H, CH_3); FT-IR (KBr): υ = 2920 (–CH–), 2867 (–CH_2–), 2198 (–N_3), 1744 (C=O), 1110 (C–O–C), 524, 1219,1456 (C_{60}), 945, 872, 663, 590 cm^{-1}.

Characterization

FTIR spectra were recorded on a Nicolet 6700 FTIR spectrometer (Thermo Fisher Scientific Co., Waltham, MA, USA) with a resolution of 4 cm^{-1} from 400 to 4000 cm^{-1} wavelength by being pressed with KBr tablet. UV-Vis spectra were collected on a UNICON UV-2102 PCS spectrometer (Shimadzu Co., Kyoto, Japan) with a rate of 1 nm/min from 200 to 600 nm wavelength and dichloromethane as solvent. ^{1}H NMR and ^{13}C NMR spectra were recorded at room temperature on a Bruker Advance DRX 400-MHz instrument (Bruker BioSpin AG, Fällanden, Switzerland) with $CDCl_3$ as the solvent and tetramethylsilane as internal standard. Gel permeation chromatography (GPC) was recorded on preplinc GPC spectrometer (J2 Scientific Co., Columbia, MO, USA) with an injection volume of 10 mL and tetrahydrofuran (THF) as the solvent. Differential scanning calorimetry (DSC) experiment was obtained in the 0–500 °C temperature range on a Q200 DSC instrument (TA Instruments Co., New Castle, DE, USA) at a heating rate of 10 °C/min under high-purity nitrogen flow (99.99%; flow rate, 50 mL/min). Thermogravimetry analysis (TGA) was performed in the 0–800 °C temperature range on a SDT Q600 TGA instrument (TA Instruments Co., New Castle, DE, USA) in air atmosphere with a heating rate of 10 °C/min.

RESULTS AND DISCUSSION

Synthesis and Characterization of C_{60}-GAP

C_{60} can replace carbon black as the solid rocket fuel additive that can increase the burning rate and combustion catalytic efficiency, but both C_{60} and carbon black are the inert catalyst, and adding C_{60} or carbon black into propellant will reduce the energy of propellant [8]. Therefore, obtaining a stable energetic fullerene derivative remains a challenge. Considering glycidyl azide polymer (GAP) is an azide energetic material and can be easily modified through various reactions, GAP was grafted on $C_{60}$to afford energetic fullerene derivatives C_{60}-PGN in three steps, as shown in Scheme 1. Firstly, M-GAP was synthesized through esterification of monomethyl-terminated GAP and malonyl chloride. Then, M-GAP was brominated using bromine as brominating agent and the bromosubstituted product BM-GAP was obtained. Subsequently, BM-GAP reacted with C_{60} through a modified Bingel reaction [25] to yield C_{60}-GAP.

The C_{60}-GAP structure was characterized by UV-vis, FTIR, and ^{1}H and ^{13}C NMR spectroscopy analyses. Figure 1 shows the FTIR spectrum of C_{60}-GAP. The peaks located at 524, 1219, and 1456 cm^{-1} are the characteristic absorption peaks of C_{60}, and the strong absorption peak at 2098 cm^{-1} corresponds to the $-N_3$ group of GAP. The peaks at 2920 and 2867 cm^{-1} are ascribed to the C–H symmetric and antisymmetric stretching vibrations of GAP, respectively. The broad intense stretching vibration absorption peak at 1110 cm^{-1} is attributed to the ester group C–O–C stretching vibration. In addition, a strong absorption band at 1744 cm^{-1} clearly indicates the carbonyl group (C=O) stretching vibration of the ester group [26].

The UV-Vis spectra of C_{60}, BM-GAP, and C_{60}-GAP are shown in Figure 2. The C_{60} UV-Vis spectrum shows the cage absorption peaks at approximately 231, 258, 331, and 405 nm. The BM-GAP UV-Vis spectrum shows an absorption peak at 296 nm. Compared with BM-GAP, three absorption peaks at 285, 314, and 427 nm appear in the C_{60}-GAP UV-Vis spectrum, which shows the GAP group in C_{60}, and the weak absorption band at 427 nm is typical of monoadducts at the closed 6,6-junction of C_{60} [27].

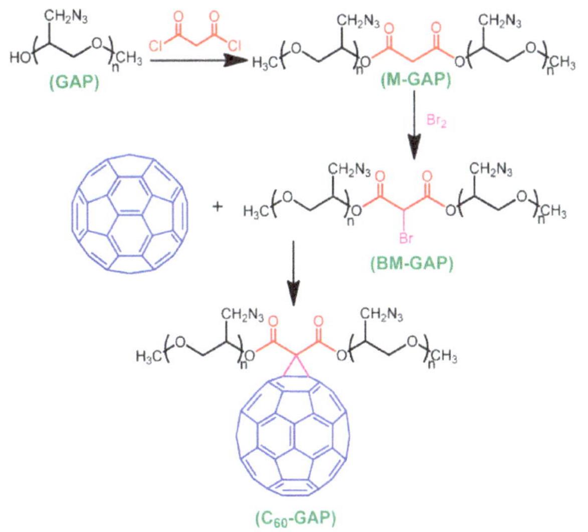

Scheme 1: Schematic representation of [60]fullerene-glycidyl azide polymer (C_{60}-GAP) synthesis.

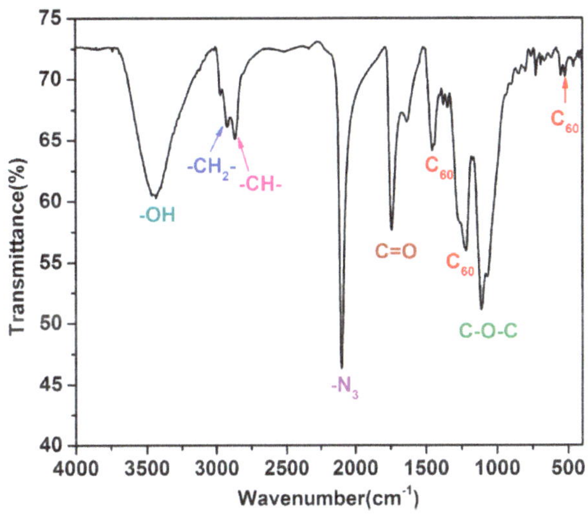

Figure 1: Fourier transform infrared (FTIR) spectrum of C_{60}-GAP.

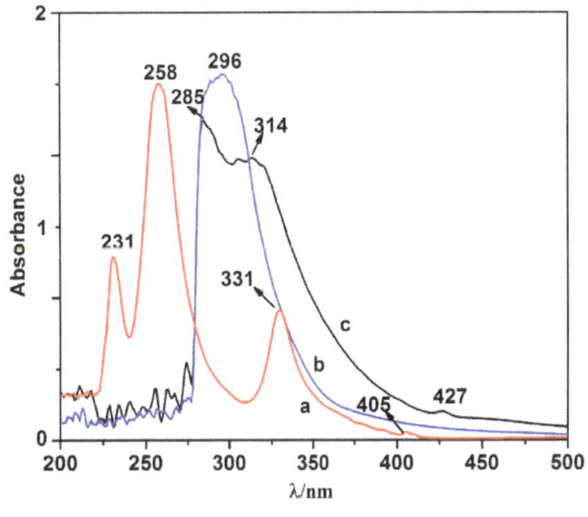

Figure 2: UV-Vis spectra of (**a**) C_{60}, (**b**) bromomalonic acid glycidyl azide polymer ester (BM-GAP) and (**c**) C_{60}-GAP.

Figure 3a,b represent the ^{1}H NMR spectra of BM-GAP and C_{60}-GAP, respectively. In Figure 3a, the peaks at around 3.76 and 3.45 ppm are attributed to three $-CH_2CH-$ protons (denoted b and c) of the polyether main chain of GAP, respectively. The peaks from 3.43 to 3.26 ppm are attributed to two $-CH_2N_3$ protons (denoted d) of the pendant group of GAP, and the peaks observed from 1.25 to 1.11 ppm are ascribed to the three methyl-terminated protons ($-CH_3$, denoted a) [28]. An evident peak at 5.13 ppm belongs to the –CHBr– proton (denoted e). In Figure 3b, the ^{1}H NMR spectrum of C_{60}-GAP is similar to that of BM-GAP, but lower signal less at approximately 5.13 ppm and the chemical shift of each proton moves upfield because of the reaction with C_{60}.

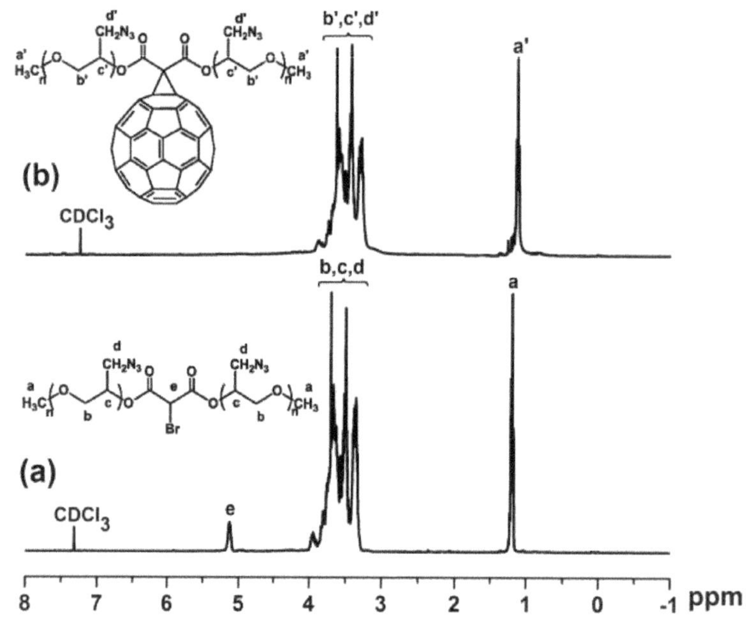

Figure 3: ^{1}H NMR spectra of (**a**) BM-GAP and (**b**) C_{60}-GAP.

The ^{13}C NMR spectra of BM-GAP and C_{60}-GAP are presented in Figure 4. In Figure 4a, the brominated active methylene carbon (–COCHBrCO–, denoted a) of BM-GAP appears at approximately 50.52 ppm. The signals from 52.12 to 51.34 ppm are attributed to the azidomethyl carbon of azidoacetate group ($-CH_2CH(CH_2N_3)O-$, denoted d). The peaks around 70.17 to 69.31 ppm result from the methylene carbon of the main chain ($-CH_2CH(CH_2N_3)O-$, denoted e). The peaks at approximately 79.57 to 78.43 ppm are attributed to the methine carbon of the main chain ($-CH_2CH(CH_2N_3)O-$, denoted c) [27], and the peak at 68.19 ppm is attributed to the methyl-terminated carbon ($-OCH_3$, denoted f) [29]. The carbonyl carbon (C=O, denoted b) of the main chain appears at around 165.58 ppm [30]. When BM-GAP reacts with C_{60}, the chemical shift of each carbon for C_{60}-GAP (Figure 4b) moves upfield. In Figure 4b, the two sp[3]-carbons of the C_{60} cage (denoted g›) appear at 73.34 ppm, and 58 sp[2]-carbons of fullerene cage (denoted h') overlap and appear from 153.31 to 130.34 ppm as a broad peak [31,32].

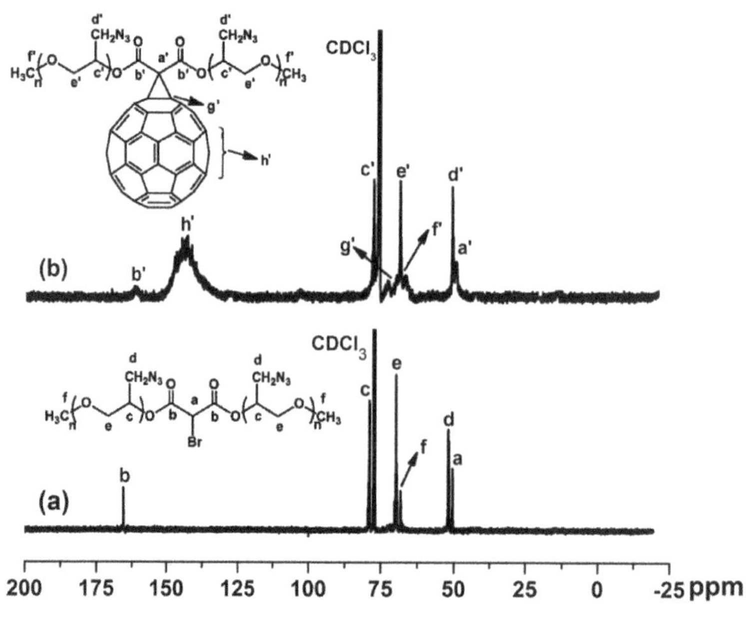

Figure 4: ^{13}C NMR spectra of (**a**) BM-GAP and (**b**) C_{60}-GAP.

Thermal Analysis

The thermal stability of energetic materials has an important effect on the preparation, storage, processing, and application properties of these materials. Thus, DSC and TGA methods were adopted to evaluate the decomposition behavior of C_{60}-GAP.

The DSC curve of C_{60}-GAP under N_2 atmosphere is shown in Figure 5. Only two exothermic peaks for C_{60}-GAP appear from 100 to 500 °C. The first exothermic peak at 155 °C is probably caused by the initial intramolecular or intermolecular cycloaddition of the $-N_3$ group and fullerene carbon cage and subsequent decomposition to release nitrogen [33,34,35,36,37]. The other exothermic peak at 242 °C is maybe due to the GAP main chain and N-heterocyclic decomposition.

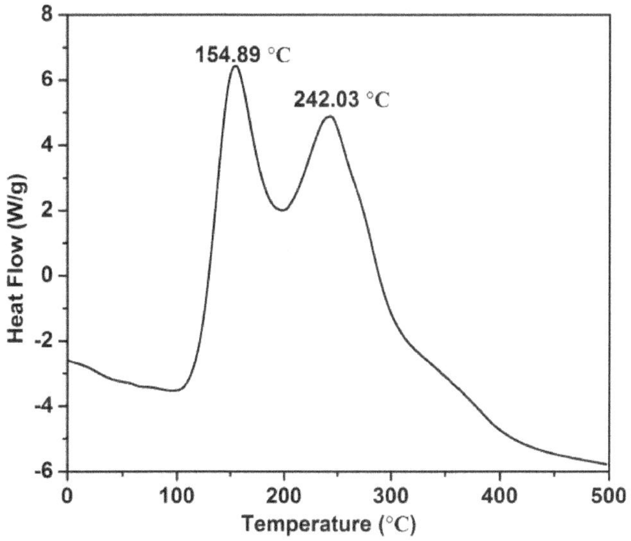

Figure 5: Differential scanning calorimetry (DSC) curve of C_{60}-GAP under N_2atmosphere.

TGA-IR spectroscopy was used to rapidly identify the constituents of the thermal decomposition gas to determine the thermal decomposition mechanism of C_{60}-GAP (Figure 6). In Figure 6a, the TGA curve shows the three-step thermal degradation of C_{60}-GAP under air atmosphere. The first stage of thermal degradation appears at 150 °C, with around 10.35% weight loss. The other two onsets thermal degradations appear at 300 °C and 600 °C, with approximately 22.89% and 64.18% weight losses, respectively. Figure 6c,d depict the IR spectra of the gas phase products of thermal decomposition of C_{60}-GAP at 300 and 580 °C, respectively. In Figure 6c, the decomposed products of C_{60}-GAP at 300 °C under air atmosphere are mainly CO_2 (2306 and 2348 cm^{-1}), NO (1829 cm^{-1}), NO_2 (1635 cm^{-1}), and H_2O (3448 cm^{-1} and 1400–1600 cm^{-1}). Figure 6d shows that the decomposed products of C_{60}-GAP at 580 °C under air atmosphere are mainly CO_2 (2306, 2333, 2358, and 2377 cm^{-1}), CO (2109 and 2176 cm^{-1}), and H_2O (3452 cm^{-1} and 1400–1600 cm^{-1}). By contrast, the decomposition temperature of GAP is at approximately 200 °C, and the main gas products are CO, HCN, HCHO, and NH_3 [38,39,40]. Thus, the results indicate that the first thermal degradation of C_{60}-GAP at 150 °C is probably due to the decomposition of the $-N_3$ group, and the decomposition of $-N_3$ at 150 °C is probably going through the initial intramolecular or intermolecular cycloaddition of $-N_3$ and fullerene carbon cage and subsequent decomposition to release nitrogen. The second thermal degradation at about 200~400 °C is probably due to the decomposition of the GAP main chain and N-heterocyclic decomposition; and the last step is maybe because of the thermal decomposition of fullerene cage (Scheme 2).

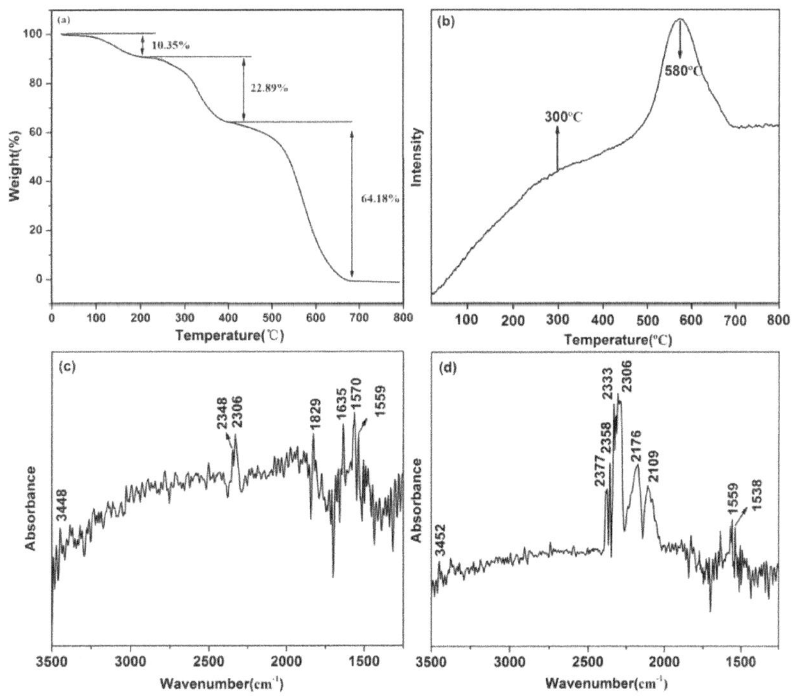

Figure 6: (**a**) TG curve of C_{60}-GAP under air atmosphere; (**b**) IR absorption intensity of the gas released from C_{60}-GAP by time diversification; FTIR spectra of the gaseous product released by C_{60}-GAP at decomposition temperatures of (**c**) 300 °C, and (**d**) 580 °C.

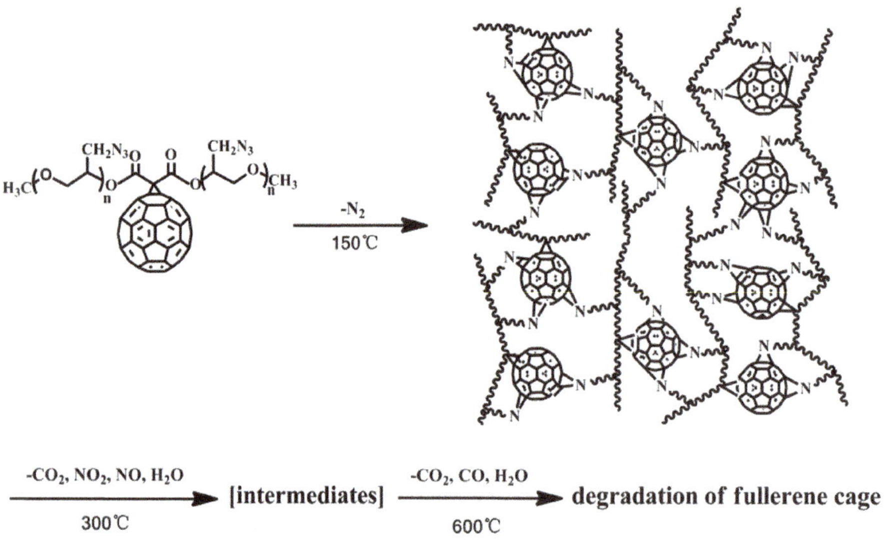

Scheme 2: Thermal decomposition mechanism of C_{60}-GAP under air atmosphere.

To prove the previous assumption, in situ FTIR was used to rapidly identify the constituents of decomposition condensed-phase residue. Figure 7a shows the IR spectra of C_{60}-GAP decomposition condensed-phase residue at 27 °C, 106 °C, 124 °C, 149 °C, 172 °C, 205 °C, 261 °C, and 305 °C under air atmosphere. Figure 7b shows the IR absorption intensity ratio of $-N_3$ ($\upsilon = 2098\ cm^{-1}$) and C=O ($\upsilon = 1735\ cm^{-1}$), as well as that of C_{60} ($\upsilon = 525\ cm^{-1}$) and C=O ($\upsilon = 1735\ cm^{-1}$), in C_{60}-GAP decomposition condensed-phase FTIR spectra at specified temperatures. The IR absorption intensity ratio of $-N_3$ ($\upsilon = 2098\ cm^{-1}$) and C=O ($\upsilon = 1735\ cm^{-1}$) ($A_{-N3}/A_{C=O}$), as well as that of C_{60} ($\upsilon = 525\ cm^{-1}$) and C=O ($\upsilon = 1735\ cm^{-1}$) ($A_{C60}/A_{C=O}$), rapidly decreased at 150 °C. The decrease in $A_{-N3}/A_{C=O}$ shows that the $-N_3$ started rapid decomposition at 150 °C, and the reduction in $A_{C60}/A_{C=O}$ at 150 °C indicates that the decomposition of the $-N_3$ group is accompanied with the damage of fullerene conjugated structure. Thus, the TGA-IR and in situ FTIR analytical results strictly corroborate that C_{60}-GAP decomposition at 150 °C is through the initial intramolecular or intermolecular reaction of $-N_3$ and the fullerene carbon cage.

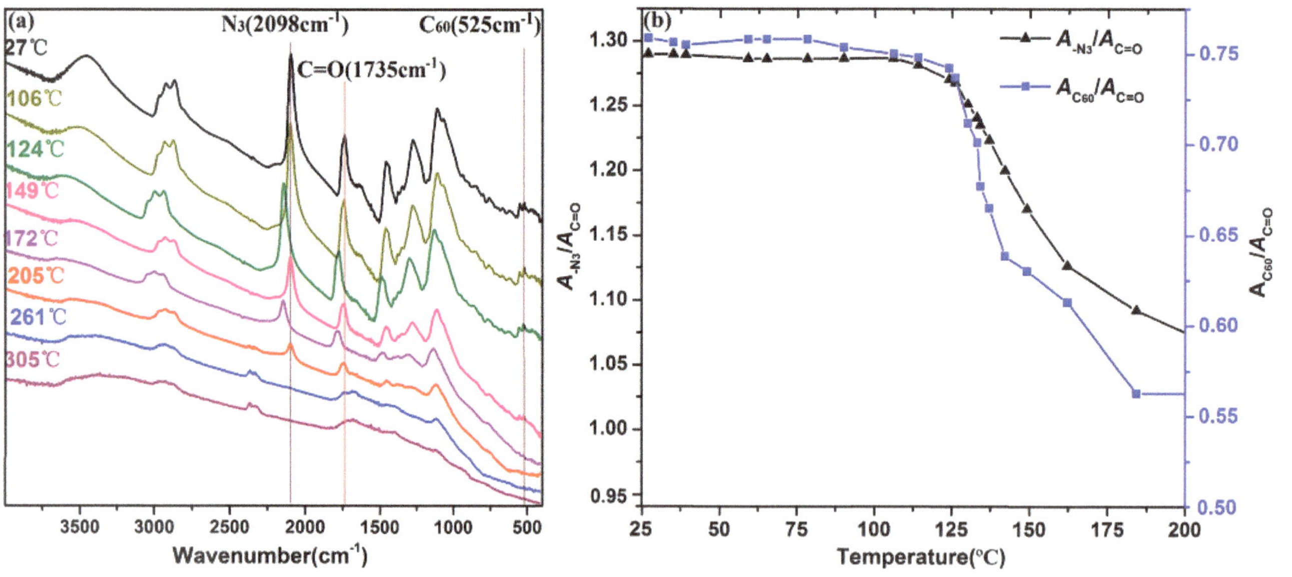

Figure 7: (**a**) FTIR spectra of C_{60}-GAP decomposition condensed-phase at different temperatures under air atmosphere; (**b**) Plots of the IR absorption intensity ratio of $-N_3$ ($\upsilon = 2098\ cm^{-1}$) and C=O ($\upsilon = 1735\ cm^{-1}$), C_{60} ($\upsilon = 525\ cm^{-1}$) and C=O ($\upsilon = 1735\ cm^{-1}$) in C_{60}-GAP decomposition condensed-phase FTIR spectra at different temperatures under air atmosphere.

CONCLUSIONS

A new energetic polymer fullerene derivative, C_{60}-GAP, has been successfully synthesized by BM-GAP and C_{60}. Its structure was systematically characterized by FTIR, UV-vis, and ^1H and ^{13}C NMR spectroscopy analyses. The thermal decomposition mechanism of C_{60}-GAP was analyzed by DSC, TGA-FTIR, and in situ FTIR. The results showed that the initial decomposition of C_{60}-GAP at 150 °C could be ascribed to the intramolecular or intermolecular reaction of the $-N_3$group and fullerene carbon cage. The second thermal degradation at about 200~400 °C is probably due to the decomposition of the GAP main chain and N-heterocyclic decomposition. And the last step is maybe because of the thermal decomposition of carbon cage.

ACKNOWLEDGMENTS

We are grateful for financial support from the National Natural Science Foundation of China (Project No. 21301142, 51372211), National Defense Fundamental Research Projects (Project No. A3120133002), Youth Innovation Research Team of Sichuan for Carbon Nanomaterials (2011JTD0017), Applied Basic Research Program of Sichuan Province (2014JY0170) and Southwest University of Science and Technology Researching Project (13zx9107, 13zxfk09).

REFERENCES

1. Hirsch, A. The Chemistry of the Fullerenes; Thieme Medical Publishers: Stuttgart, Germany, 1994.
2. Dresselhaus, M.S.; Dresselhaus, G.; Eklund, P.C. Science of Fullerenes and Carbon Nanotubes: Their Properties and Applications; Academic Press: Waltham, MA, USA, 1996.
3. Astefanei, A.; Nunez, O.; Galceran, M.T. Analysis of C_{60}-fullerene derivatives and pristine fullerenes in environmental samples by ultrahigh performance liquid chromatography-atmospheric pressure photoionization-mass spectrometry.J. Chromatogr. A 2014, 1365, 61–71. [PubMed]
4. Butts, C.P.; Jazdzyk, M. The preparation and structures of non-hydrocarbon functionalized fullerene-diamine adducts.Chem. Commun. 2003, 13, 1530–1531.
5. Chaves, C.W.; Moraes, D.E. Single or functionalized fullerenes interacting with heme group. AIP Adv. 2014, 9, 097119.
6. Tzirakis, M.D.; Orfanopoulos, M. Radical reactions of fullerenes: from synthetic organic chemistry to materials science and biology. Chem. Rev. 2013, 7, 5262–5321.
7. Han, X.; Wang, T.F.; Lin, Z.K.; Han, D.L.; Li, S.F.; Zhao, F.Q. RDX/AP CMDB propellants containing fullerenes and carbon black additives. Defence Sci. J. 2009, 3, 284–293.
8. Jin, B.; Peng, R.F.; Zhao, F.Q.; Yi, J.H.; Xu, S.Y.; Wang, S.B.; Chu, S.J. Combustion effects of nitrofulleropyrrolidine on RDX-CMDB propellants. Propellants Explos. Pyrotech. 2014, 39, 874–880.
9. Chiang, L.Y.; Bhonsle, J.B.; Wang, L.; Shu, S.F.; Chang, T.M.; Hwu, J.R. Efficient one-flask synthesis of water-soluble [60]fullerenols. Tetrahedron 1996, 14, 4936–4972.
10. Hamwi, A.; Marchand, V. Oxidized fullerene derivatives containing hydroxyl, nitro and fluorine groups. Fullerene Sci. Technol. 1996, 5, 835–851.
11. Franco, J.U.; Ell, J.R.; Hilton, A.K.; Hammons, J.C.; Olmstead, M.M. $C_{60}Cl_6$, $C_{60}Br_8$ and $C_{60}(NO_2)_6$ as selective tools in organic synthesis. Fuller. Nanotub. Carbon Nanostruct. 2009, 4, 349–360.
12. Cataldo, F. Preparation of polynitrofullerene by the action of dinitrogen tetroxide on C_{60}. Fullerene Sci. Technol. 1997, 1, 257–265.
13. Franco, C.; Ornella, U.; Giancarlo, A. Synthesis and explosive decomposition of polynitro[60] fullerene. Carbon 2013,62, 413–421.
14. Zhai, R.S.; Arindam, D.; Chien, K.H.; Chau, C.H.; Taizoon, C.; Long, Y.C. Polymeric fullerene oxide films produced by decomposition of hexanitro[60] fullerene. Carbon 2004, 2, 395–403.
15. Kumari, D.; Singh, H.; Patil, M.; Thiel, W.; Pant, C.S.; Banerjee, S. Synthesis, characterization, thermal and computational studies of novel tetra-azido esters as energetic plasticizers. Thermochim. Acta 2013, 562, 96–104.
16. Lee, D.H.; Kim, K.T.; Jung, H.; Kim, S.H.; Park, S.; Jeon, H.B.; Paik, H.; Min, B.S.; Kin, W. Characterization of 1,2,3-triazole crosslinked polymers based on azide chain-ends prepolymers and a dipolarophile curing agent as propellant binders: The effect of a plasticizer. J. Taiwan Inst. Chem. Eng. 2014, 6, 3110–3116.
17. Min, B.S.; Kim, C.K.; Ryoo, M.S.; Kim, S.Y. Azide-bearing polymer-based solid composite propellant prepared by a dual curing system consisting of a urethane-forming reaction and a dipolar addition reaction. Fuel 2014, 15, 165–171.
18. Ding, Y.Z.; Hu, C.; Guo, X.; Che, Y.Y.; Huang, J. Structure and mechanical properties of novel composites based on glycidyl azide polymer and propargyl-terminated polybutadiene as potential binder of solid propellant. J. Appl. Polym. Sci. 2014, 7, 40007.
19. Huang, T.; Jin, B.; Peng, R.F.; Chu, S.J. Synthesis and characterization of a new energetic plasticizer: Acyl-terminated GAP. Int. J. Polym. Anal. Charact. 2014, 6, 522–531.
20. Jin, B.; Dong, H.S.; Peng, R.F.; Shen, J.; Tan, B.S.; Chu, S.J. Synthesis and characterization of poly(vinyl 2,4,6-trinitrophenylacetal) as a new energetic binder. J. Appl. Polym. Sci. 2011, 3, 1643–1648.
21. Frankel, M.B.; Grant, L.R.; Flanagan, J.E. Historical development of glycidyl azide polymer. J. Propuls. Power 1992, 3, 560–563.
22. Kim, E.S.; Yang, V.; Lian, Y.C. Modeling of HMX/GAP pseudo-propellant combustion. Combust. Flame 2002, 3, 227–245.
23. Hu, C.; Guo, X.; Jing, Y.; Chen, J.; Zhang, C.; Huang, J. Structure and mechanical properties of crosslinked glycidyl azide polymers via click chemistry as potential binder of solid propellant. J. Appl. Polym. Sci. 2014, 16, 40636.
24. Li, N.; Xiong, X.F.; Mo, H.C.; Lu, X.M.; Liu, Y.J. Synthesis and properties of azide terminated and acetate terminated glycidyl azide polymer. J. Solid Rocket Technol. 2013, 2, 234–236.
25. Jin, B.; Shen, J.; Peng, R.F.; Zheng, R.Z.; Chu, S.J. Efficient cyclopropanation of [60]fullerene starting from bromo-substituted active methylene compounds without using a basic catalyst. Tetrahedron Lett. 2014, 36, 5007–5010.
26. Guerra, S.; Trinh, T.M.N.; Schillinger, F.; Muhlberger, L.; Sigwalt, D.; Holler, M. The di-t-butylsilylene protecting group as a bridging unit in linear and macrocyclic bis-malonates for the regioselective multifunctionalization of C_{60}.Tetrahedron Lett. 2013, 46, 6251–6257.

27. Zhao, Y.B.; Luo, Y.J.; Li, X.M. Synthesis and characterization of BAMO-r-GAP copolymer. Polym. Mater. Sci. Eng. 2012,9, 1–4.

28. Mohan, Y.M.; Raju, K.M. Synthesis and characterization of low molecular weight azido polymers as high energetic plasticizers. Int. J. Polym. Anal. Charact. 2004, 5, 289–304.

29. Sigwalt, D.; Schillinger, F.; Guerra, S.; Holler, M.; Berville, M.; Nierengarten, J.F. An expeditious regioselective synthesis of [60]fullerene e,e,e tris-adduct building blocks. Tetrahedron Lett. 2013, 32, 4241–4244.

30. Shirosaki, T.; Harisaki, R.; Horikawa, M.; Sakurai, H.; Nagaoka, S.; Ihara, H. Synthesis of a series of malonic diester-introduced fullerene derivatives. Synth. Commun. 2014, 2, 275–279.

31. Wang, G.W.; Li, J.X.; Xu, Y. Synthesis of C_{60}-fused tetrahydrothiophene derivatives via nucleophilic cycloaddition of thiocyanates. Org. Biomol. Chem. 2008, 16, 2995–2999.

32. Li, F.B.; You, X.; Wang, G.W. Synthesis of disubstituted [60]fullerene-fused lactones: Ferric perchlorate promoted reaction of [60]fullerene with malonate esters. Org. Lett. 2010, 21, 4896–4899.

33. Zhang, L.Y.; Liu, Z.R.; Wang, X.H.; Heng, S.Y.; Pan, Q.; Zhao, F.Q. Investigation on thermal decomposition of GAP by FTIR spectroanalysis. J. Solid Rocket Technol. 2010, 5, 549–553.

34. Korobeinichev, O.P.; Kuibida, L.V.; Volkov, E.N.; Shmakov, A.G. Mass spectrometric study of combustion and thermal decomposition of GAP. Combust. Flame 2002, 1, 136–150.

35. Sun, Y.; Li, S. The effect of nitrate esters on the thermal decomposition mechanism of GAP. J. Hhazard. Mater. 2008, 1, 112–117.

36. Kay, K.Y.; Han, K.J.; Yu, Y.J.; Park, Y.D. Dendritic fullerenes (C_{60}) with photoresponsive azobenzene groups.Tetrahedron Lett. 2002, 29, 5053–5056.

37. González, S.; Martín, N.; Swartz, A.; Guldi, D.M. Addition reaction of azido-exTTFs to C_{60}: Synthesis of fullerotriazoline and azafulleroid electroactive dyads. Org. Lett. 2003, 4, 557–560.

38. Fazhoğlu, H.; Hacaloğlu, J. The effect of triethanolamine on thermal decomposition of GAP. J. Macromol. Sci. A 2002, 7, 759–768.

39. Fazlıoğlu, H.; Hacaloğlu, J. Thermal decomposition of glycidyl azide polymer by direct insertion probe mass spectrometry. J. Anal. Appl. Pyrolysis 2002, 2, 327–338.

40. Jin, B.; Shen, J.; Peng, R.F.; Shu, Y.J.; Tan, B.S.; Chu, S.J.; Dong, H.S. Synthesis, characterization, thermal stability and sensitivity properties of the new energetic polymer through the azidoacetylation of poly(vinyl alcohol). Polym. Degrad. Stab. 2012, 4, 473–480.

CHAPTER 9

A REVIEW ON NATURAL FIBER REINFORCED POLYMER COMPOSITE AND ITS APPLICATIONS

Layth Mohammed,[1] M. N. M. Ansari,[1] Grace Pua,[1] Mohammad Jawaid,[2] and M. Saiful Islam[3]

[1]Centre for Advance Materials, Department of Mechanical Engineering, Universiti Tenaga Nasional, 43000 Kajang, Selangor, Malaysia

[2]Laboratory of Biocomposite Technology, Institute of Tropical Forestry and Forest Products (INTROP), Universiti Putra Malaysia (UPM), 43400 Serdang, Selangor, Malaysia

[3]Department of Chemistry, Faculty of Science, Universiti Putra Malaysia (UPM), 43400 Serdang, Selangor, Malaysia

ABSTRACT

Natural fibers are getting attention from researchers and academician to utilize in polymer composites due to their ecofriendly nature and sustainability. The aim of this review article is to provide a comprehensive review of the foremost appropriate as well as widely used natural fiber reinforced polymer composites (NFPCs) and their applications. In addition, it presents summary of various surface treatments applied to natural fibers and their effect on NFPCs properties. The properties of NFPCs vary with fiber type and fiber source as well as fiber structure. The effects of various chemical treatments on the mechanical and thermal properties of natural fibers reinforcements thermosetting and thermoplastics composites were studied. A number of drawbacks of NFPCs like higher water absorption, inferior fire resistance, and lower mechanical properties limited its applications. Impacts of chemical treatment on the water absorption, tribology, viscoelastic behavior, relaxation behavior, energy absorption flames retardancy, and biodegradability properties of NFPCs were also highlighted. The applications of NFPCs in automobile and construction industry and other applications are demonstrated. It concluded that chemical treatment of the natural fiber improved adhesion between the fiber surface and the polymer matrix which ultimately enhanced physicomechanical and thermochemical properties of the NFPCs.

INTRODUCTION

The increase in environmental consciousness and community interest, the new environmental regulations and unsustainable consumption of petroleum, led to thinking of the use of environmentally friendly materials. Natural fiber is considered one of the environmentally friendly materials which have good properties compared to synthetic fiber [1].

A late current industry research identified that the worldwide natural fiber reinforced polymer composites industry sector reached U$2.1 billion in 2010. Current pointers are that interest in NFPCs industry will keep on growing quickly around the world. The utilization of NFPCs has expanded considerably in the shopper merchandise as developing industry sectors throughout the last few years. As indicated by evaluations, over 5 years (2011–2016), the NFPCs industry is estimated to grow 10% worldwide [2].

Citation: Layth Mohammed, M. N. M. Ansari, Grace Pua, Mohammad Jawaid, and M. Saiful Islam, "A Review on Natural Fiber Reinforced Polymer Composite and Its Applications," International Journal of Polymer Science, vol. 2015, Article ID 243947, 15 pages, 2015. doi:10.1155/2015/243947

Natural fibers in simple definition are fibers that are not synthetic or manmade. They can be sourced from plants or animals [3]. The use of natural fiber from both resources, renewable and nonrenewable such as oil palm, sisal, flax, and jute to produce composite materials, gained considerable attention in the last decades, so far. The plants, which produce cellulose fibers can be classified into bast fibers (jute, flax, ramie, hemp, and kenaf), seed fibers (cotton, coir, and kapok), leaf fibers (sisal, pineapple, and abaca), grass and reed fibers (rice, corn, and wheat), and core fibers (hemp, kenaf, and jute) as well as all other kinds (wood and roots) [4]. The most common and commercially natural fibers in the world and world production have been shown in Table 1.

Table 1: Natural fibers in the world and their world production [4]

Fiber source	World production (10^3 ton)
Bamboo	30.000
Sugar cane bagasse	75.000
Jute	2300
Kenaf	970
Flax	830
Grass	700
Sisal	375
Hemp	214
Coir	100
Ramie	100
Abaca	70

Fiber reinforced polymer matrix got considerable attention in numerous applications because of the good properties and superior advantages of natural fiber over synthetic fibers in term of its relatively low weight, low cost, less damage to processing equipment, good relative mechanical properties such as tensile modulus and flexural modulus, improved surface finish of molded parts composite, renewable resources, being abundant [5], flexibility during processing, biodegradability, and minimal health hazards. NFPCs with a high specific stiffness and strength can be produced by adding the tough and light-weight natural fiber into polymer (thermoplastic and thermoset) [6]. On the other hand, natural fibers are not free from problems and they have notable deficits in properties. The natural fibers structure consists of (cellulose, hemicelluloses, lignin, pectin, and waxy substances) and permits moisture absorption from the surroundings which causes weak bindings between the fiber and polymer. Furthermore, the couplings between natural fiber and polymer are considered a challenge because the chemical structures of both fibers and matrix are various. These reasons for ineffectual stress transfer during the interface of the produced composites. Accordingly, natural fiber modifications using specific treatments are certainly necessary. These modifications are generally centered on the utilization of reagent functional groups which have ability for responding of the fiber structures and changing their composition. As a result, fiber modifications cause reduction of moisture absorption of the natural fibers which lead to an excellent enhancement incompatibility between the fiber and polymer matrix [7].

The wide applications of NFPCs are growing rapidly in numerous engineering fields. The different kinds of natural fibers reinforced polymer composite have received a great importance in different automotive applications by many automotive companies such as German auto companies (BMW, Audi Group, Ford, Opel, Volkswagen, Daimler Chrysler, and Mercedes), Proton company (Malaysian national carmaker), and Cambridge industry (an auto industry in USA). Beside the auto industry, the applications of natural fiber composites have also been found in building and construction industry, sports, aerospace, and others, for example, panels, window frame, decking, and bicycle frame [8].

In a review of chemical treatments of natural fibers, Kabir and coworkers [9] concurred that treatment is an important factor that has to be considered when processing natural fibers. They observed that fibers loose hydroxyl groups due to different chemical treatments, thereby reducing the hydrophilic behavior of the fibers and causing enhancement in mechanical strength as well as dimensional stability of natural fiber reinforced polymer composites. Their general conclusion was that chemical treatment of natural fibers results in a remarkable improvement of the NFPCs.

NATURAL FIBER REINFORCED COMPOSITES (NFPCS)

Natural fiber polymer composites (NFPC) are a composite material consisting of a polymer matrix embedded with high-strength natural fibers, like jute, oil palm, sisal, kenaf, and flax [10]. Usually, polymers can be categorized into two categories, thermoplastics and thermosets. The structure of thermoplastic matrix materials consists of one or two dimensional moleculars, so these polymers have a tendency to make softer at an raised heat range and roll back their properties throughout cooling. On the other hand, thermosets polymer can be defined as highly cross-linked polymers which cured using only heat, or using heat and pressure, and/or light irradiation. This structure gives to thermoset polymer good properties such as high flexibility for tailoring desired ultimate properties, great strength, and modulus [3, 4]. Thermoplastics widely used for biofibers are polyethylene [11], polypropylene (PP) [12], and poly vinyly chloride (PVC); hereas phenolic, polyester, and epoxy resins are mostly utilized thermosetting matrices [10]. Different factors can affect the characteristics and performance of NFPCs. The hydrophilic nature of the natural fiber [5] and the fiber loading also have impacts on the composite properties [13]. Usually, high fiber loading is needed to attain good properties of NFPCs [14]. Generally, notice that the rise in fiber content causes improving in the tensile properties of the composites [8]. Another vital factor that considerably impacts the properties and surface characteristics of the composites is the process parameters utilized. For that reason, appropriate process techniques and parameters should be rigorously chosen in order to get the best characteristics of producing composite [10]. The chemical composition of natural fibers also has a big effect on the characteristics of the composite represented by the percentage of cellulose, hemicellulose, lignin, and waxes. Table 2 shows the chemical composition of some common natural fibers [4].

Table 2: Chemical composition of some common natural fibers [4]

Fiber	Cellulose (wt%)	Hemicellulose (wt%)	Ligning (wt%)	Waxes (wt%)
Bagasse	55.2	16.8	25.3	—
Bamboo	26–43	30	21–31	—
Flax	71	18.6–20.6	2.2	1.5
Kenaf	72	20.3	9	—
Jute	61–71	14–20	12-13	0.5
Hemp	68	15	10	0.8
Ramie	68.6–76.2	13–16	0.6–0.7	0.3
Abaca	56–63	20–25	7–9	3
Sisal	65	12	9.9	2
Coir	32–43	0.15–0.25	40–45	
Oil palm	65	—	29	—
Pineapple	81	—	12.7	—
Curaua	73.6	9.9	7.5	—
Wheat straw	38–45	15–31	12–20	—
Rice husk	35–45	19–25	20	—
Rice straw	41–57	33	8–19	8–38

Many researchers [8, 11, 15–17] have examined and researched the suitability, competitiveness, and capabilities of natural fibers embedded in polymeric matrices. The researchers [4, 18, 19] concentrated on the effect of the fiber surface modifications as well as manufacturing processes in improving fiber/polymer compatibility. On the other hand, some researchers studied and compared between different natural fiber composites and their stability in various applications [20]. Al-Oqla and Sapuan [20] investigated the properties of jute/plastic composites such as crystallinity, fiber modification, thermal stability, weathering resistance, durability, in addition to their suitability to the automotive industry throughout ecodesign components. while Mohanty et al. [21] studied the effects of jute fiber on the mechanical properties of pure biodegradable polymer (Biopol), the mechanical properties of the resulted composites, impact strength, tensile strength, and bending strength, showed an increase when compared with pure Biopol. The tensile strength of jute Biopol was enhanced by 50%, while bending strength and impact strength of the composites were enhanced by 30% and 90% in comparison to pure Biopol.

GENERAL CHARACTERISTICS OF NFPCS

The properties of natural fiber composite are different to each other according to previous studies, because of different kinds of fibers, sources, and moisture conditions. The performance of NFPCs relies on some factors, like mechanical composition, microfibrillar angle [20], structure [10], defects [22], cell dimensions [23], physical properties [4], chemical properties [24], and also the interaction of a fiber with the matrix [25]. Since every product in market has drawbacks, similarly, natural fiber reinforced polymer composites also have drawbacks. The couplings between natural fiber and polymer matrix are problem taken into consideration, as a result of the difference in chemical structure between these two phases. This leads to ineffective stress transfer during the interface of the NFPCs. Thus, the chemical treatments for the natural fiber are necessary to achieve good interface properties. The reagent functional groups in the chemical treatments have ability to react on the fiber structures and alter the fiber composition [9]. Natural fibers include a functional group named as hydroxyl group which makes the fibers hydrophilic. During manufacturing of NFPCs, weaker interfacial bonding occurs between hydrophilic natural fibre and hydrophobic polymer matrices due to hydroxyl group in natural fibres. This could produce NFPCs with weak mechanical and physical properties [8].

Mechanical Properties of the NFPCs

There are considerable enhancement and suggestions for the natural fibers that can be implemented in order to enhance their mechanical properties resulting in high strength and structure. Once the base structures are made strong, the polymers can be easily strengthened and improved [26]. There are number of aspects that effects of composite are performance level or activities, of which to name a few are the following;(a) orientation of fiber [5],(b)strength of fibers [8],(c)physical properties of fibers [27],(d)interfacial adhesion property of fibers [28] and many more. NFPCs are such composites whose mechanical efficiency is dependent upon the interface provided by fiber-matrix along with the stress transfer function in which stress is transferred to fiber from matrix. This has been reported by many investigators in several researches [1, 23, 29]. Characteristic components of natural fibers such as orientation [30], moisture absorption [31], impurities [32], physical properties [33], and volume fraction [34] are such features that play a constitutive role in the determination of NFPCs mechanical properties. Mechanical properties of PLA, epoxy, PP, and polyester matrices can be affected by many types of natural fibers and to show some of them, Figure 1 is included.

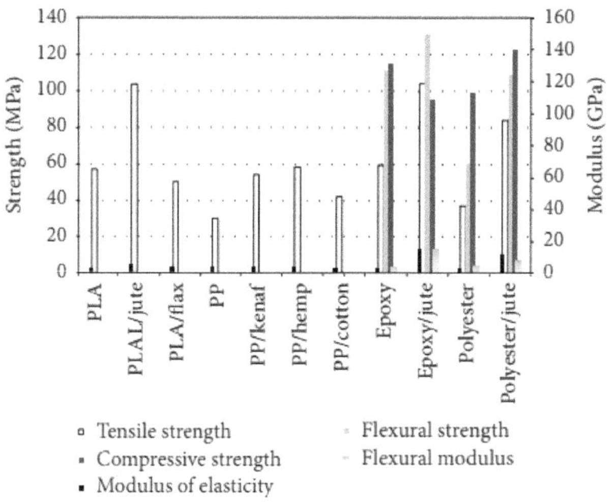

Figure 1: Some of mechanical properties of natural fiber reinforced polymer composite [5].

NFPCs show even better mechanical properties than a pure matrix in cases where jute fibers are added in PLA (polylactic-acid); in this case, 75.8% of PLA's tensile strength was improved; however, introduction or incorporation of flax fibers showed a negative impact on this addition. The addition of flax fibers resulted in 16% reduced tensile strength of the composites. Conversely, composites of PP were improved with the incorporation of hemp, kenaf, and cotton [5]. By far, maximum improvement is only seen in such composites where jute or polyester has been incorporated where a total of 121% improvement is evident compared to pure polyester [5]. However, due to the rubber phase present in gum compound, a greater range of flexibility is present in such materials which results in reduced stiffness and storage modulus.

It is also known that stiffness and stress transfer in composites increases with an increased or excessive addition of fiber which provides a better loss modulus and also a better storage modulus. The loss modulus is also considered to be increased with fiber addition up to 756 MPa at 50 phr fiber loading compared to the loss modulus of gum that is 415 MPa [8].

A group of researchers led by Ismail et al. [37] studied the effect of size and filler content on fiber characteristics that cures a wound or any part of the body. Along with this, mechanical properties of Oil Palm Wood Flour (OPWF) was also examined which is reinforced with (ENR) epoxidized natural rubber composites. When the fiber content is increased, torque of the fibers is also increased and with smallest possible particle size of OPWF, highest torque was noticed. However, increasing the factor of OPWF in ENR compounds showed reduced tensile strength and while reaching the break point a considerable elongation is evident. It also evident increase in elongation, tear strength, tensile modulus and hardness due to higher loading of OPWF. A higher tensile strength and tear strength as tensile modulus were identified in composites that were filled with even smallest proportion of OPWF [10]. The fracture behavior of composites is also affected due to the nonlinear mechanical behavior of natural fibers, under the influence of tensile-shear loads [1]. Table 3 shows the mechanical properties for common types of natural fiber in the world [38].

Table 3: Physicomechanical properties of natural fibers [38]

Fiber	Density (g/cm^3)	Tensile strength (MPa)	Young's modulus (GPa)	Elongation at break (%)
OPEFB	0.7–1.55	248	3.2	2.5
Flax	1.4	88–1500	60–80	1.2–1.6
Hemp	1.48	550–900	70	1.6
Jute	1.46	400–800	10–30	1.8
Ramie	1.5	500	44	2
Coir	1.25	220	6	15–25
Sisal	1.33	600–700	38	2-3
Abaca	1.5	980	—	—
Cotton	1.51	400	12	3–10
Kenaf (bast)	1.2	295	—	2.7–6.9
Kenaf (core)	0.21	—	—	—
Bagasse	1.2	20–290	19.7–27.1	1.1
Henequen	1.4	430–580	—	3–4.7
pineapple	1.5	170–1672	82	1–3
Banana	1.35	355	33.8	53

The bonding strength between fiber and polymer matrix in the composite is consider a major factor in order to get superior fiber reinforcement composites properties. Because of pendant hydroxyl and polar groups in fiber, this leads to extremely high moisture absorption of fiber, resulting in weak interfacial bonding between the fiber and the hydrophobic matrix polymers. To develop composites with good mechanical properties, chemical modification of fibre carried out to reduce the hydrophilic behavior of fibers and the absorption of moisture [15, 39].

The different surface treatments of advanced composites applications were reviewed by several researchers [40–42]. The effects of different chemical treatments on cellulosic fibers that were employed as reinforcements for thermoplastics and thermoset were also examined. For the treatments, the different kinds of chemical treatment include silane [43], alkali [44], acrylation [45], benzoylation [46], maleated coupling agents [47], permanganate [48], acrylonitrile and acetylation grafting [49], stearic acid [50], peroxide [51], isocyanate [52], triazine [53], fatty acid derivate (oleoyl chloride), sodium chloride, and fungal [9]. The main purpose of surface treatments of natural fibers to enhanced fibre/matrix interfacial bonding and stress transferability of the composites.

The impact of alkaline treatment on surface properties of Iranian cultivated natural fibers was studied by Cordeiroa et al. [54]. The research revealed that alkaline treatment gets rid of some chemical components on the surface of the fibers, comprising uranic acid (hemicellulose), aromatic moieties (extractives), and nonpolar molecules from the partial lignin depolymerisation. There is a stronger effect on chemical components of nonwood fibers. Improving the crystallinity of nonwood fibers, in the softwood fibers result in only a minor increase. Hence, alkaline treatment can result in a remarkable improvement in the specific interaction of the fibers as well as improving the fibers' wettability.

Le Troedec et al. [55] revealed the effects of some chemical treatments, comprising ethylenediaminetetraacetic acid (EDTA), NaOH, polyethylene imine (PEI), $CaCl_2$, and $Ca(OH)_2$. The effects were on the mechanical properties of the composite materials from mixtures of hemp fiber and lime by differential thermal evaluation and tests. The observation was that every treatment had a direct effect on the fiber surface. The treatment was with 6% NaOH and led to cleaning fibers by removing the amorphous compounds, and the increase of the crystallinity index of fiber bundles, while the EDTA treatment led to separates fibers and complex calcium ions related to pectins. In brief, PEI treatment shows all studied properties an intermediate character and lime water treatment depicts calcium ions' fixation at the surface of fibers in comparison to a calcium chloride treatment which does not.

May-Pat et al. [1] reported on the impact of the interphase properties from well controlled surface treatment in the case of natural fibers. The fracture behavior and the mechanical properties of a NFPC depend on the properties of constituents and region of the fiber surface, or interphase, where the stress transfer occurs. Furthermore, the tailoring of the interphase by different kinds of surface treatments and carefully characterizing it provides a better knowledge of the behavior of the NFPCs. Moreover, different fiber surface treatments modify the natural fiber microstructure specifically under high loading rates.

Venkateshwaran et al. [35] studied the effect of alkali (NaOH) treatments of various concentrations (0.5%, 1%, 2%, 5%, 10%, 15%, and 20%) on the mechanical properties of banana/epoxy composite. The results reported that as compared to other treated and untreated fiber composites, 1% NaOH treated fiber reinforced composites have a better properties. The alkali concentration on the fiber surfaces results in better mechanical properties of the resulted composite. However, the rising of alkali concentration maybe causes fiber surface damage, leading to a decrease of mechanical properties. The effect of different chemical treatment on the mechanical properties and characteristic of sisal-oil palm hybrid fiber reinforced natural rubber composites have been studied by John et al. [56]. With chemical treatment, the torque values increased which lead to greater crosslinking. Similarly, alkali treatment showed a rise in the composites' tensile strength in comparison to untreated composites and with 4% NaOH treated fibers, optimum tensile strength was seen for resulted composites. In contrast, for composites treated with 4% NaOH a strong interface is apparent because of a more superior adhesion between rubber and fiber is present which avoids solvent entry and a little swelling occurs.

Van de Weyenberg et al. [24] examined the impact of flax processing parameters, and the fiber treatment, on the mechanical properties of flax fiber reinforced epoxy composites. It was discovered that the employment of long flax slivers may not necessarily lead to more superior composite properties. The highest enhancement of the flexural properties of the flax fiber reinforced epoxy composites can be gotten by chemical treatments. There was an increase of transverse strength of up to 250% and transverse modulus, of up to 500%. In addition, the longitudinal properties of the UD composites (both modulus and strength) showed improvements with 40% or more.

Some modifications in the chemical and physical properties of the lignocellulosic fibers can be observed after the treatment of the fiber of rubber wood with laccase enzymes. These chemical treatments lead to the amorphous lignin content, changing the hemicellulose content and ultimately the natural crystallinity [57]. The fiber has treatment effect on morphological and single fiber tensile strength of EFB fiber. The EFB fiber treated with boiling water, 2% sodium hydroxide (NaOH), and mixing of boiling water and NaOH was examined by Norul Izani et al. [13]. It was revealed that it changed the properties of the fiber surface topography after the treatment. Compared to untreated EFB fiber, the treated EFB fibers with the two types of treatment were more thermal stability. On the other hand, the tensile strength and Young's modulus of the treated fiber showed a rise in comparison to untreated fibers. For tensile modulus, the alkaline treatment has enhanced the tensile properties of sugar palm fiber reinforced epoxy composites at better soaking times and concentrations of alkaline. On the other hand, the increase the alkaline concentration may lead to fiber damage [58].

Acrylonitrile Butadiene Styrene (ABS) with the coating effect of OPEFB fibers was tested BT. The coated process enhanced the mechanical and physical properties and also improved the fiber performance. The ABS treatment led to reducing the water absorption and also decreased the biodegradation potential of the fiber in contact with soil. With the coating, the tensile strength and elasticity moduli of the OPEFB fibers became better than what they were in the past. The surface area between fiber and soil particles increased by coating fiber, which led to improving the shear strength parameters of the fiber reinforced soils [66].

Flame Retardant Properties of the NFPCs

Due to eco-friendly and sustainability nature, natural fiber composites prefers as compared to conventional synthetic fibre based composites. They are applied in diversified domains [9, 18, 20, 38, 67] such as building materials [68], aerospace industry, and automotive industry [69]. Natural fibers and polymers are organic materials and are very sensitive to alter any features if flame is introduced to them. Flame retardancy is another aspect that has become greatly significant in order to fulfill safety measures taken while developing natural fiber composites.

In the presence of flame, burning of composites takes place in five different steps as shown below:

- Heating.
- Decomposition.
- Ignition.
- Combustion.
- Propagation [70].

If flame retardancy has been achieved in the aforementioned steps, no matter whether ignition step has been conducted or not, the procedure will be terminated before an actual ignition is set up. There are two forms of products that are obtained upon burning of composites; these two include high cellulose content and high lignin content. High cellulose provides chances of higher flammability whereas higher values of lignin show there is a greater chance of char formation [71]. Thermal resistance is provided by flax fibers [72]; however, silica or ash is another important feature that is helpful in extinguishing fire [73].

In order to enhance fire resistance of various NFPCs, different procedures are undertaken. Fire barriers are kind of barriers that can be applied to phenolics, ceramics, intumescents, glass mats, silicone, ablatives, and chemical additives too. Coatings and additives used in the system of intumescent are found to be very promising fire barrier treatments in which these barriers are expanded upon heating resulting in a cellular surface that is charred even. However, with the help of this charred surface, internal or underlying components and protected against flux and heat.

One of the well-known or profound flame retardants for reinforced polymers (natural fibers) is used with the combination of char developing cellulose material [74]. The only method of reducing combustion in this scenario is through increasing stability and char formation in the polymer. This will result in reduced flammability, decrease visible smoke, and restrict the volume of products produced due to combustions [72]. Fire retardant coating is another method that helps in enhancement of fire resistance property of composites. This coating is done at the end or finishing stage or impregnation. Due to changes in the fibers and lingo-cellulosic particles, fire resistance is altered during the process of manufacturing [75].

The two most widely used metal hydroxide flame retardants are known to be aluminum hydroxide [$Al(OH_3)$] and magnesium hydroxide [$Mg(OH)_2$] which are purposefully added to polymers. The chemical reaction through which these two flame retardants decompose are as follows:

$$2Al(OH)_3 \longrightarrow Al_2O_3 + 3H_2O \quad \Delta H = 20\,cal/g \tag{1}$$

$$Mg(OH)_2 \longrightarrow MgO + H_2O \quad \Delta H = 328\,cal/g \tag{2}$$

Out of these two flame retardants, magnesium hydroxide display better thermal stability as compared to aluminum hydroxide since the temperature range given off by the decomposition of magnesium hydroxide is nearly 300–320 degrees centigrade (°C) which is much higher than the temperature range offered by aluminum hydroxide that is only 200°C. For this reason, aluminum hydroxide is not considered to be thermally stable as it cannot be used for polyamides, polypropylene, or others whereas magnesium hydroxide can be used.

It was also shown in a research that the addition of expandable graphite (EG) and ammonium polyphosphate (APP) in composite polymer as a source of flame retardant (FR) helps in enhancing flax fiber reinforced PP composite's property of fire retardancy. It was also shown that heat release rate (HRR) in a composite additive of 30 wt% of flax fiber and 25 wt% of EG (expandable graphite) was decreased to 35 from 167 kW/m^2 [76].

Spirocyclic pentaerythritol bisphosphorate disphosphoryl melamine (SPDPM) is a type of intumescent flame retardant for PLA that was studied by Zhan et al. [77]. Char formation that is a result of spirocyclic pentaerythritol bisphosphorate disphosphoryl melamine (SPDPM), an active component of flame retardancy, enhances antidripping performance of PLA and flame retardancy too with an addition of 25 wt%. Flame retardancy is not an easy thing to be imparted and it is only possible if there is an extensive high loading of inorganic filler.

Hapuarachchi and Peijs [78] studied the development of fully bio-based natural fiber composite that has enhanced features of fire or flame retardancy. This natural fiber composite was developed with the help of PLA polymers that were derived from crops accompanied with 2 kinds of nanofillers which are able to produce synergy corresponding to flame retardancy. Upon analysis, after incorporating hemp fiber mat in PLA resin, decrease in PHRR (peak heat release) in calorimeter will be shown.

Biodegradability of the NFPCs

High strength composites are resultant products of natural fiber reinforcement in polymers which also provide extra or improved biodegradability, low cost, light weight, and enhanced properties related to mechanical structure [29]. At temperatures as high as 240°C, natural fibers start degrading whereas constituents of fiber, such as hemicelluloses, cellulose, lignin, and others, start degrading at different levels of temperature; for example, at 200°C lignin starts to decompose whereas at temperatures higher than this other constituents will also degrade [9].

Since thermal stability of the fibers is dependent on the structural constituents of fibers, it can be improved if the concentration levels or the structural constituents are completely removed, such as lignin and hemicelluloses. This can be achieved with the help of chemical treatments. Development of fibers and materials that provide services are two important aspects which should be considered while degradating natural fibers [9]. Natural fibers have a short lifetime with minimum environmental damage upon degradation whereas synthetic ones affect environment due to pollution caused by degradation. Lignin, hemicelluloses, and cellulose concentrations or composition affect thermal degradation features of lingo-cellulosic materials [13]. More than fifty percent weight of jute or Biopol composite is lost after exactly 1500 days of burial [21].

Energy Absorption of the NFPCs

High strength, energy absorption, and stiffness are obtained by composite materials which are widely used in automotive and motorsport sectors of industry mainly due to the property of mass reduction [79]. Enhanced energy absorption is evident from the increased volume fraction that is only possible in the presence of low speed such as 2.5 m/s [80]. On the other hand, at high speeds, such as 300 m/s, similar performance is shown by flax, jute, and hemp, but jute showed brittleness and low strength of fibers [81]. Potential of NFPCs that is required for application in providing sustainable energy absorption was investigated by Meredith et al. While keeping focus on motorsport [80]. Vacuum Assisted Resin Transfer Molding (VARTM) technique is used to test conical specimens of flax, jute, and hemp for their properties and features. Various values exhibited by different kinds of materials were recorded to analyze specific energy absorption (SEA).

Tribology Properties of NFPCs

Since every material has some wear and friction properties that degrade with respect to time, the tribological loadings are important to consider for an improved mechanical part design [5]. Around 90% failure is obtained due to differences in tribological loading conditions which alters their wear and friction properties [82]. Reinforcement is a method with which fiber's or polymers tribological properties are altered (either positively or negatively) [83]. Studies on different kinds of tribological analysis have been conducted on fibers including kenaf/epoxy [84], betelnut fiber reinforced polyester [85], sisal/phenolic resin [86], sugarcane fiber reinforced polyester (SCRP) [87], and cotton/polyester [88]. Improvement in wear performance of PLA was evident due to the addition of natural fibers in which wear rate of composites was quite low in comparison to wear rate at higher loads on neat PLA [82].

Kenaf fibers reinforced with epoxy composite were used by Chin and Yousif [84] for a kind of bearing application, in which they showed an 85% enhancement of wear performance and normal orientation in composites. Wear and friction features of glass fiber/polyester (GRP) and sugarcane fiber/polyester (SCRP) were studied by El-Tayeb with different parameters including speed, the time taken for test, and load [87]. The research results concluded SCRP as a competitor of GRP composite. The same characteristics were studied by Xin et al. [86] for sisal fiber reinforced resin brake composites which showed sisal fiber can be used instead of asbestos in brake pads [89, 90].

Laminated composites were developed with the help of three different natural fibers such as grewia, nettle, optiva, and sisal. This connection of natural fibers was studied by Bajpai et al. in which a hot compression procedure was used to incorporate three different materials in a PLA polymer [82]. Friction and wear features of composites were examined under different situations such as dry contact condition with varying operating parameters. Due to an option of variable operating parameters, applied load was varied in between range of 10 to 30 N with a speed ranging from 1 to 3 m/s and sliding distance of thousand to three thousand meters. The research results showed infusion of natural fiber mats, in PLA matrix, is capable of enhancing wear and frictional behavior of neat polymers. An approximate reduction of 10–44% in the coefficient of friction with a greater reduction of 70% seen in developed composites for specific wear rate is visualized in comparison to neat PLA [82].

Water Absorption Characteristics of the NFPCs

Natural fibers work well as reinforcement in polymers. However, the main weakness of the application of natural fibers is their susceptibility to moisture [91]. Mechanical properties of polymeric composites have a strong dependence on the interface adhesion between the fiber and the polymer matrix [15]. The natural fibers are rich of cellulose, hemicelluloses, lignin, and pectins, all of which are hydroxy 1 groups; that is, they are usually hydrophilic sources and strong polar whilst polymers show considerable hydrophobicity. Thus, there are major challenges of suitability between the matrix and fiber that weakens interface region between matrices and natural fibers [5]. At the composite materials' outer layers, water absorption happens and decreases gradually into the bulk of the matrix. A generally high water intake by composite materials results in an increased weight of wet profiles, a conceivable decline in their strength, and increment in their deflection, swelling, and causing pressure on nearby structures. These can cause warping, buckling, bigger possibility of their microbial inhabitation, freeze, and unfreeze caused destruction of mechanical characteristics of composite materials [92].

Oil palm fiber natural rubber (OPF-NR) composites increased in the water absorption percentage corresponding to an increase in fiber loading because of the fibers' hydrophilicity. The absorption behavior of NR showed modifications from Fickian to non-Fickian with additional OPF because of the microcracks and the viscoelastic nature of the polymer [8]. In the woven pandanus/banana fabric composites tests, woven pandanus fabric composites increased the water uptake compared to woven banana fabric composites because of higher lignin content and hemicellulose, as well as the presence of defects in the composite system [93]. Furthermore, temperature can affect the percentage of water absorption of the composites. At 65% humidity at 21°C, the equilibrium moisture content of some natural fiber can be observed in Table 4 [4]. It was revealed that the water absorption of OPF-NR composite was less than of OPF-sisal fiber-NR hybrid biocomposite. The incorporation of sisal fiber that contains relatively more holocellulose (23%), which is exceptionally hydrophilic brought on more water intake. Besides, the lignin content material of OPF (19%) was bigger than sisal fiber (9%). Lignin being hydrophobic reduces the water absorption [8].

Table 4: The equilibrium moisture content of different natural fiber at 65% relative humidity (RH) and 21°C [4]

Fiber	Equilibrium moisture content (%)
Sisal	11
Hemp	9.0
Jute	12
Flax	7
Abaca	15
Ramie	9
Pineapple	13
Coir	10
Bagasse	8.8
Bamboo	8.9

Many researchers [8, 56] showed the effect of coupling agent such as maleic anhydride polyethylene and chemical treatments such as bleaching, acetylation, and alkali treatment on reduction moisture absorption of NFPCs. The surface of the fibers is cleaned during the chemical treatments to ensure there are no impurities which increases the fiber surface roughness and preventing the moisture absorption via the removal of the coat of OH groups of fiber as seen in equation below [5, 9]:

$$\text{Fiber-OH} + \text{NaOH} \longrightarrow \text{Fiber-O}^-\text{Na}^+ + H_2O \quad (3)$$

Sreekala and Thomas [91] investigated the moisture absorption properties of OPEFB fiber in different temperature condition. They also studied the effect of different modification on OPEFB fiber, such as silane treatment, gamma irradiation, latex coating, mercerization, acetylation, peroxide treatment, and isocyanate treatment on moisture absorption properties. They concluded that all the treatment causes the seduction in moisture absorption properties in all temperatures.

The mercerization of OPF-sisal fiber-NR hybrid composites led to reducing water adsorption of the composite by improving the adhesive characteristics of fiber surface and also providing a large surface area which causes better mechanical interlocking [94]. Shinoj et al. [8] conducted the effect of chemical treatment for agave fibers on moisture adsorption before blending in a polymer matrix. The chemical treatment is achieved using four types of reagents, acetic anhydride (Ac), maleic anhydride (MA), styrene (S), and acrylic acid (AA). This paper concludes that the chemical treatment causes decrease in the global diffusivity of water and also concludes that the water mobility in the fiber core is more than at the surface.

Viscoelastic Behavior of the NFPCs

Dynamic mechanical measurements or the viscoelastic behavior over a range of temperatures provides valuable insight into the structure, morphology, and determination of the interface characteristics of natural fiber composite materials [95]. Storage modulus gives an insight into the load bearing capability and the stiffness behavior of natural fiber composite materials. Storage modulus is a measure of the maximum energy stored in the material during one cycle of oscillation. The mechanical damping coefficient is the ratio of the loss modulus to the storage modulus and is related to the degree of molecular mobility in the polymeric material. On the other hand, loss modulus is proportional to the amount of energy dissipated as heat by the sample [96].

The viscoelastic behavior of novel commingled biocomposites based on polypropylene/jute yarns is reported by [96]. By using commingling method, jute yarn reinforced polypropylene commingled composites were prepared. The viscoelastic behavior or dynamic mechanical properties of the result commingled composites were studied according to the fiber content and different kinds of chemical treatments such as potassium permanganate ($KMnO_4$), maleic anhydride modified polypropylene (MAPP), toluene diisocyanate (TDI), and stearic acid (ST). This study concluded that increasing of fiber content leads to increasing the storage and loss modulus of the composite. On the other hand, at all temperatures, the chemical treatment by $KMnO_4$ and MAPP leads to increasing the storage modulus and loss modulus of the respective treated composites than the untreated ones. By using the dynamic mechanical analyzer, Venkateshwaran et al. [35] studied the impact of alkali treatments on the viscoelastic behavior of natural fiber composite. The corresponding viscoelastic properties were determined as a function of temperature and frequency when the measurements were executed in the tensile mode of the used equipment. At the temperature range of 30–140°C at 0.1, 1, and 10 Hz frequencies, respectively, the experiments executed the graphs plotted, as storage modulus (E^-) versus temperature and $\tan\delta$ versus temperature, as shown in Figures 2 and 3 [35].

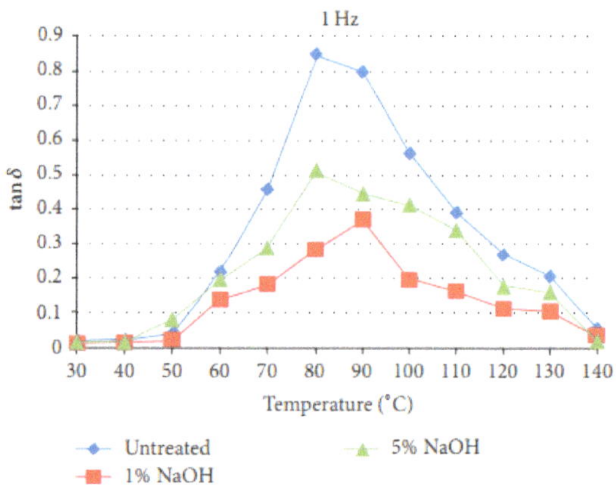

Figure 2: $\tan\delta$ versus temperature curves of the alkali treated and untreated composites at 1 Hz frequency [35].

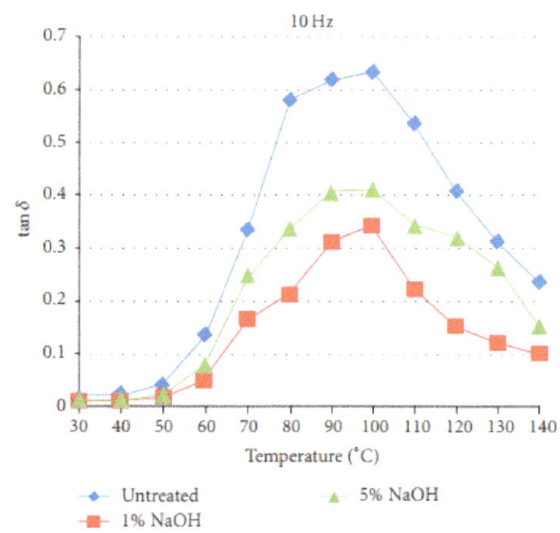

Figure 3: $\tan\delta$ versus temperature curves of alkali treated and untreated composites [35].

Relaxation Behavior of the NFPCs

Natural fiber possesses an intrinsic relaxation behavior which has a main function in the stress relaxation of NFPCs. Thus, the tensile stress relaxation of the reinforcing fiber needs to be studied in detail [38]. Sreekala et al. [29], for instance, did such a study focusing on the character of individual OPEFB fiber and, in addition, examined the fiber surface modification effects, ageing, and strain level on the fiber relaxation behavior. The fibers stress relaxation was lessened significantly with surface treatments, like latex modification, and thus decreasing resulting physical interaction between the latex particles and fiber surface. Also, water and thermal ageing reduce the rate of relaxation for the oil palm fiber; the rate of stress relaxation of the OPEFB fiber was optimized at 10% strain level which is shown in Figure 4, while the relaxation modulus values for the fiber show similar trends as in the case of stress relaxation as shown in Figure 5. On the other hand, the stress relaxation rate of OPF-sisal fiber-NR hybrid composites showed a reduction with an increase in fiber percentage [97].

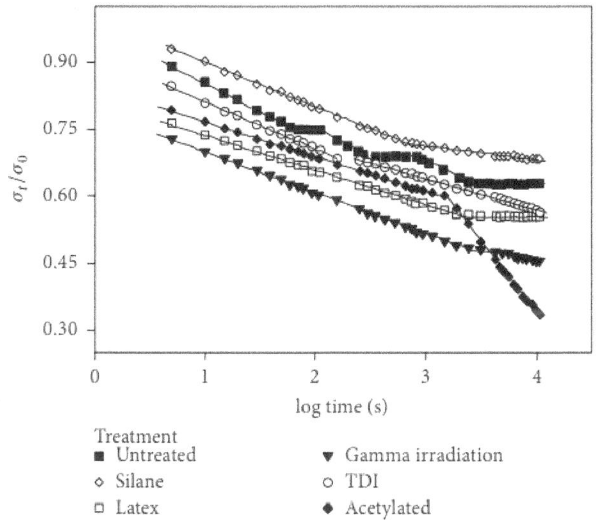

Figure 4: Stress relaxation curves of untreated and treated OPEFB fiber at 10% strain [29].

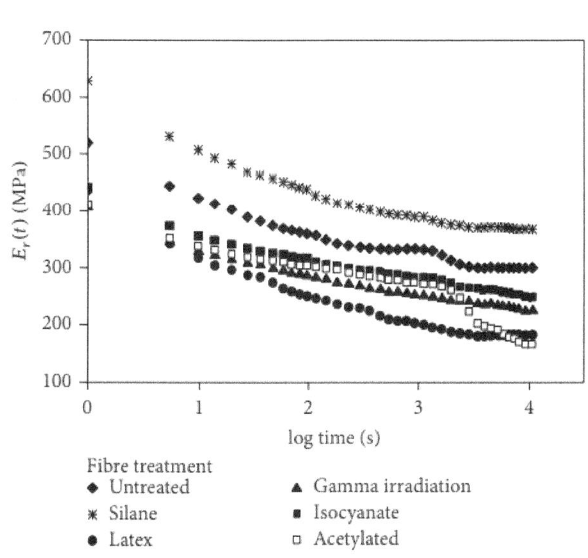

Figure 5: Relaxation modulus curves of untreated and treated OPEFB fiber at 10% strain [29].

Thermal Properties of NFPCs

The untreated OPFs are thermally more stable when compared with treated ones, while OPF is considered thermally more stable compared to flax fibers and hemp fibers. Increase of temperatures from 20°C to 150°C causes increase of the heat capability of OPFs particularly from 1.083 J g^{-1} °C^{-1} to 3.317 J g^{-1} °C^{-1} [98]. The thermal diffusivity, thermal conductivity, and specific heat of the flax/HDPE composites lessened with a rise in fiber composition. However, the thermal conductivity and thermal diffusivity showed no significant changes in the range of 170–200°C. The biocomposites' specific heat showed gradual increase with temperature [4]. By using polycarbonate to generate functional composites, pineapple leaf fiber was the support. The changed pineapple leaf fibers composite, treated with silane, showed the impact strengths and highest tensile. The composites' thermal stability is lower than the neat polycarbonate resin according to the thermogravimetric examination. Besides that, the thermal stability lessened with a rise in pineapple leaf fiber composition [4].

Enzymatic treatment for many natural fibers such as flax and hemp often natural fibers can lead to improvement in surface and thermal properties [56]. Hemicellulose and pectinase are the treatments which can improve the thermal properties of the fibers mentioned. The enzymes possess an attractive state for the improvement of the natural fibers' surfaces for natural fiber composite application [67].

Norul Izani et al. [13] studied the effect of chemical treatment on the morphological and tensile strength of the EFB fiber. The treatments were by the types of treatments, 2% sodium hydroxide (NaOH) and combination of both NaOH and boiling water. The chemical treatment by NaOH led to enhancing the fiber surface topography, thermal stability, and tensile strength of the fiber, while the chemical treatment using NaOH and water boiling caused the higher thermal properties of the EFB fibers compared to untreated fibers.

NATURAL FIBER POLYMER COMPOSITES APPLICATION

The applications of NFPCs are growing rapidly in numerous engineering fields. The different kinds of natural fibers such as jute, hemp, kenaf, oil palm, and bamboo reinforced polymer composite have received a great importance in different automotive applications, structural components, packing, and construction [5, 99]. NFPCs are finding in electrical and electronic industries, aerospace, sports, recreation equipment, boats, machinery office products, and so forth. The widespread application of NFPCs in polymer composites due to its low specific weight, relatively high strength, relatively low production cost, resistance to corrosion and fatigue, totally biodegradable, improving the surface finish of molded part composites, relatively good mechanical properties, available and renewable sources as compared to synthetic fibers [5, 98]. On the other hand, there is a physical disadvantage of the NFPCs such as moisture absorption, restricted processing temperature, and variable quality and this disadvantage led to limiting their performance [73].

Natural Fiber Composites Applications in the Interior Car

Most of the car companies in the world have done a lot of investigation in order to insert the NFPCs in their products. The car manufacture in Europe has done various researches to increase the applications of NFPCs in automotive industry, especially in car interior such as seat backs, parcel shelves, boot linens, front and rear door linens, truck linens, and door-trim panels [89]. Beside the use for car interior parts in automobile industry, natural fiber embedded in polymers has been used for high requirement applications for exterior auto body components, such as the middle section between the headlights above the fender of a passenger bus [18].

German auto companies (BMW, Audi Group, Ford, Opel, Volkswagen, Daimler Chrysler, and Mercedes) utilize the cellulose fibers composites in various automobile part, shown in Figure 6, such as using coco nut fibers rubber latex composites for the seats of the Mercedes Benz A-class model and using fax-sisal fiber mat reinforced epoxy door panels of Mercedes Benz E-class model [8]. Audi company uses flax/ sisal mat reinforced polyurethane composite with a mix to make door trim panels [60]. Ford is using kenaf fibers imported from Bangladesh in their "Mondeo" model and the door panels of the Mondeo are manufactured from kenaf reinforced PP composites while using flax in floor trays [61]. Kenaf and flax mixture has gone into package trays and door panel inserts for Opel Vectras. Volkswagen company used cellulose fiber to make Seatback, door panel, boot-lid finish panel, boot-liner in Passat Variant, Golf, A4, and Bora model.

Figure 6: Automobile components made of natural fiber composites [36].

BMW Group has a lot of NFPCs into its automobiles. BMW Group used about 10 000 tonnes of natural fiber in 2004 [100]. Each BMW 7 series car boats 24 kg of renewable raw materials, with flax and sisal in the interior door lining panels. Also use cotton in the soundproofing, wool in the upholstery, and wood fiber in the seat back. Daimler-Benz in Germany is also working with a range of natural fibers ± sisal, jute, coconut, European hemp, and flax ± as reinforcing fibers in high-quality polypropylene components in order to replace glass fibers. Daimler-Benz has developed the dashboards and center armrest consoles along with seat shells and paneling on seat backs. Moreover, it increased the utilization of NFPCs in some automobiles by approximately 98% over earlier models by utilizing natural fibers, for example, abaca and flax. On the other hand, the Cambridge industry made rear shelf trim panels of the 2000 model Chevrolet Impala using flax fiber polypropylene composite [8, 101]. Toyota, Proton, Volvo, and other automobile companies used cellulose fiber to make car parts as shown in Table 5.

Table 5: The application of natural fiber composites in automotive industry [59–62]

Manufacturer	Model	Application
Rover	2000 and others	Rear storage shelf/panel, and insulations
Opel	Vectra, Astra, Zafira	Door panels, pillar cover panel, head-liner panel, and instrumental panel
Volkswagen	Passat Variant, Golf, A4, Bora	Seat back, door panel, boot-lid finish panel, and boot-liner
Audi	A2, A3, A4, A4 Avant, A6, A8, Roadstar, Coupe	Boot-liner, spare tire-lining, side and back door panel, seat back, and hat rack
Daimler Chrysler	A, C, E, and S class, EvoBus (exterior)	Pillar cover panel, door panels, car windshield/car dashboard, and business table
BMW	3, 5 and 7 series and other Pilot	Seat back, headliner panel, boot-lining, door panels, noise insulation panels, and moulded foot well linings
Peugeot	406	Front and rear door panels, seat backs, and parcel shelf
Fiat	Punto, Brava, Marea, Alfa Romeo 146, 156, 159	Door panel
General Motors	Cadillac De Ville, Chevrolet Trail Blazer	Seat backs, cargo area floor mat
Toyota	ES3	Pillar garnish and other interior parts
Saturn	L300	Package trays and door panel
Volvo	V70, C70	Seat padding, natural foams, and cargo floor tray
Ford	Mondeo CD 162, Focus	Floor trays, door inserts, door panels, B-pillar, and boot-liner
Saab	9S	Door panels
Renault	Clio, Twingo	Rear parcel shelf
Toyota	Raum, Brevis, Harrier, Celsior,	Floor mats, spare tire cover, door panels, and seat backs
Mitsubishi		Cargo area floor, door panels, and instrumental panel
Mercedes Benz	C, S, E, and A classes	Door panels (flax/sisal/wood fibers with epoxy resin/UP matrix), glove box (cotton fibers/wood molded, flax/sisal), instrument panel support, insulation (cotton fiber), molding rod/apertures, seat backrest panel (cotton fiber), trunk panel (cotton with PP/PET fibers), and seat surface/backrest (coconut fiber/natural rubber)
	Trucks	Internal engine cover, engine insulation, sun visor, interior insulation, bumper, wheel box, and roof cover
Citroen	C5	Interior door panelling
Lotus	Eco Elise (July 2008)	Body panels, spoiler, seats, and interior carpets
Rover	2000 and others	Insulation, rear storage shelf/panel
VAUXHALL	Corsa, Astra, Vectra, Zafira	Headliner panel, interior door panels, pillar cover panel, and instrument panel

The Natural Fiber Applications in the Industry

Other than the car industry, the applications of NFPCs are found in building and construction, aerospace, sports, and more, such as partition boards, ceilings, boats, office products, and machinery. The most applications of NFPCs are concentrated on nonload bearing indoor components in civil engineering because of their vulnerability to environmental attack [72]. Green buildings are wanted to be ecologically mindful, suitable, and healthy place to live and work. Biocomposites is consider one of the major materials utilized as a part of green materials at this time. It could be categorized, with regard to their application in the building market, into two principle products: firstly, structural biocomposites, which include bridge as well as roof structure, and secondly, nonstructural biocomposites which include window, exterior construction, composites panels, and door frame [2].

The wide advantages of natural fibers reinforced composites such as high stiffness to weight ratio, lightweight, and biodegradability give them suitability in different application in building industries [102]. Van de Weyenberg et al. [24] have shown that good properties of thin walled elements such as high strength in tension and compression, made of sisal fiber reinforced composite, give it a wide area of application, for instance, structural building members, permanent formwork, tanks, facades, long span roofing elements, and pipes strengthening of existing structures. On the other hand, bamboo fiber can be used in structural concrete elements as reinforcement, while sisal fiber and coir fiber composites have been used in roofing components in order to replace asbestos [15]. Natural fiber reinforced concretes products in construction applications like sheets (both plain and corrugated) and boards are light in weight and are ideal for use in roofing, ceiling, and walling for the construction of low cost houses [103]. Table 6 shows the various applications of cellulose fiber in industry, construction, and other industries.

Table 6: Natural fiber composite applications in industry [3, 63–65]

Fiber	Application in building, construction, and others
Hemp fiber	Construction products, textiles, cordage, geotextiles, paper & packaging, furniture, electrical, manufacture bank notes, and manufacture of pipes
Oil palm fiber	Building materials such as windows, door frames, structural insulated panel building systems, siding, fencing, roofing, decking, and other building materials [14]
Wood fiber	Window frame, panels, door shutters, decking, railing systems, and fencing
Flax fiber	Window frame, panels, decking, railing systems, fencing, tennis racket, bicycle frame, fork, seat post, snowboarding, and laptop cases
Rice husk fiber	Building materials such as building panels, bricks, window frame, panels, decking, railing systems, and fencing
Bagasse fiber	Window frame, panels, decking, railing systems, and fencing
Sisal fiber	In construction industry such as panels, doors, shutting plate, and roofing sheets; also, manufacturing of paper and pulp
Stalk fiber	Building panel, furniture panels, bricks, and constructing drains and pipelines
Kenaf fiber	Packing material, mobile cases, bags, insulations, clothing-grade cloth, soilless potting mixes, animal bedding, and material that absorbs oil and liquids
Cotton fiber	Furniture industry, textile and yarn, goods, and cordage
Coir fibers	Building panels, flush door shutters, roofing sheets, storage tank, packing material, helmets and postboxes, mirror casing, paper weights, projector cover, voltage stabilizer cover, a filling material for the seat upholstery, brushes and brooms, ropes and yarns for nets, bags, and mats, as well as padding for mattresses, seat cushions
Ramie fiber	Use in products as industrial sewing thread, packing materials, fishing nets, and filter cloths. It is also made into fabrics for household furnishings (upholstery, canvas) and clothing, paper manufacture.
Jute fiber	Building panels, roofing sheets, door frames, door shutters, transport, packaging, geotextiles, and chip boards.

CONCLUSIONS

Natural fiber reinforced polymer composites have beneficial properties such as low density, less expensive, and reduced solidity when compared to synthetic composite products, thus providing advantages for utilization in commercial applications (automotive industry, buildings, and constructions). Using natural fibers as reinforcement for polymeric composites introduces positive effect on the mechanical behavior of polymers. This paper evaluates the characteristics and properties of natural fiber reinforced polymer composites: mechanical, thermal, energy absorption, moisture absorption, biodegradability, flame retardancy, tribology properties. Viscoelastic behavior and relaxation behavior of NFPCs are researched. Also the application of NFPCs in automobile and industry is reported. The effects of chemical treatment of the natural fiber properties were also addressed. The physical and mechanical properties of these NFPCs can be further enhanced through the chemical treatment, while moisture absorption of the NFPCs can be reduced through surface modification of fibers such as alkalization and addition of coupling agents.

REFERENCES

1. A. May-Pat, A. Valadez-González, and P. J. Herrera-Franco, "Effect of fiber surface treatments on the essential work of fracture of HDPE-continuous henequen fiber-reinforced composites," Polymer Testing, vol. 32, no. 6, pp. 1114–1122, 2013.
2. N. Uddin, Ed., Developments in Fiber-Reinforced Polymer (FRP) Composites for Civil Engineering, Elsevier, 2013.
3. A. Ticoalu, T. Aravinthan, and F. Cardona, "A review of current development in natural fiber composites for structural and infrastructure applications," in Proceedings of the Southern Region Engineering Conference (SREC '10), pp. 113–117, Toowoomba, Australia, November 2010.
4. O. Faruk, A. K. Bledzki, H.-P. Fink, and M. Sain, "Biocomposites reinforced with natural fibers: 2000–2010," Progress in Polymer Science, vol. 37, no. 11, pp. 1552–1596, 2012.
5. A. Shalwan and B. F. Yousif, "In state of art: mechanical and tribological behaviour of polymeric composites based on natural fibres," Materials & Design, vol. 48, pp. 14–24, 2013.
6. Y. Xie, C. A. S. Hill, Z. Xiao, H. Militz, and C. Mai, "Silane coupling agents used for natural fiber/polymer composites: a review," Composites Part A: Applied Science and Manufacturing, vol. 41, no. 7, pp. 806–819, 2010.
7. S. S. Ray and M. Bousmina, "Biodegradable polymers and their layered silicate nanocomposites: in greening the 21st century materials world," Progress in Materials Science, vol. 50, no. 8, pp. 962–1079, 2005.
8. S. Shinoj, R. Visvanathan, S. Panigrahi, and M. Kochubabu, "Oil palm fiber (OPF) and its composites: a review," Industrial Crops and Products, vol. 33, no. 1, pp. 7–22, 2011.
9. M.M.Kabir,H.Wang,K.T.Lau,andF.Cardona,"Chemical treatments on plant-based natural fibre reinforced polymer composites: an overview," Composites Part B: Engineering, vol. 43, no. 7, pp. 2883–2892, 2012.
10. H. Ku, H. Wang, N. Pattarachaiyakoop, and M. Trada, "A review on the tensile properties of natural fiber reinforced polymer composites," Composites Part B: Engineering, vol. 42, no. 4, pp. 856–873, 2011.
11. F. Z. Arrakhiz, M. El Achaby, M. Malha et al., "Mechanical and thermal properties of natural fibers reinforced polymer composites: doum/low density

polyethylene," Materials & Design, vol. 43, pp. 200–205, 2013.

12. G. Di Bella, V. Fiore, G. Galtieri, C. Borsellino, and A. Valenza, "Effects of natural fibres reinforcement in lime plasters (kenaf and sisal vs. Polypropylene)," Construction and Building Materials, vol. 58, pp. 159–165, 2014.
13. M. A. Norul Izani, M. T. Paridah, U. M. K. Anwar, M. Y. Mohd Nor, and P. S. H'Ng, "Effects of fiber treatment on morphology, tensile and thermogravimetric analysis of oil palm empty fruit bunches fibers," Composites Part B: Engineering, vol. 45, no. 1, pp. 1251–1257, 2013.
14. I. S. M. A. Tawakkal, M. J. Cran, and S. W. Bigger, "Effect of kenaf fibre loading and thymol concentration on the mechanical and thermal properties of PLA/kenaf/thymol composites," Industrial Crops and Products, vol. 61, pp. 74–83, 2014.
15. M. J. John and S. Thomas, "Biofibres and biocomposites," Carbohydrate Polymers, vol. 71, no. 3, pp. 343–364, 2008.
16. E. Jayamani, S. Hamdan, M. R. Rahman, and M. K. B. Bakri, "Comparative study of dielectric properties of hybrid natural fiber composites," Procedia Engineering, vol. 97, pp. 536–544, 2014. ·
17. C. C. Eng, N. A. Ibrahim, N. Zainuddin, H. Ariffin, and W. M. Z. W. Yunus, "Impact strength and flexural properties enhancement of methacrylate silane treated oil palm mesocarp fiber reinforced biodegradable hybrid composites," The Scientific World Journal, vol. 2014, Article ID 213180, 8 pages, 2014. ·
18. N. Graupner, A. S. Herrmann, and J. Müssig, "Natural and man-made cellulose fibre-reinforced poly (lactic acid)(PLA) composites: an overview about mechanical characteristics and application areas,"Composites Part A: Applied Science and Manufacturing, vol. 40, no. 6-7, pp. 810–821, 2009.
19. K. Bocz, B. Szolnoki, A. Marosi, T. Tábi, M. Wladyka-Przybylak, and G. Marosi, "Flax fibre reinforced PLA/TPS biocomposites flame retarded with multifunctional additive system," Polymer Degradation and Stability, vol. 106, pp. 63–73, 2014.
20. F. M. Al-Oqla and S. M. Sapuan, "Natural fiber reinforced polymer composites in industrial applications: feasibility of date palm fibers for sustainable automotive industry," Journal of Cleaner Production, vol. 66, pp. 347–354, 2014.
21. A. K. Mohanty, M. A. Khan, and G. Hinrichsen, "Surface modification of jute and its influence on performance of biodegradable jute-fabric/Biopol composites," Composites Science and Technology, vol. 60, no. 7, pp. 1115–1124, 2000.
22. T. Hänninen, A. Thygesen, S. Mehmood, B. Madsen, and M. Hughes, "Mechanical processing of bast fibres: the occurrence of damage and its effect on fibre structure," Industrial Crops and Products, vol. 39, no. 1, pp. 7–11, 2012.
23. V. K. Thakur and M. K. Thakur, "Processing and characterization of natural cellulose fibers/thermoset polymer composites," Carbohydrate Polymers, vol. 109, pp. 102–117, 2014.
24. I. Van de Weyenberg, J. Ivens, A. De Coster, B. Kino, E. Baetens, and I. Verpoest, "Influence of processing and chemical treatment of flax fibres on their composites," Composites Science and Technology, vol. 63, no. 9, pp. 1241–1246, 2003.
25. D. Dai and M. Fan, "Wood fibres as reinforcements in natural fibre composites: structure, properties, processing and applications," in Natural Fibre Composites: Materials, Processes and Properties, chapter 1, pp. 3–65, Woodhead Publishing, 2014. ·
26. V. S. Srinivasan, S. R. Boopathy, D. Sangeetha, and B. V. Ramnath, "Evaluation of mechanical and thermal properties of banana-flax based natural fibre composite," Materials & Design, vol. 60, pp. 620–627, 2014.
27. J.-C. Bénézet, A. Stanojlovic-Davidovic, A. Bergeret, L. Ferry, and A. Crespy, "Mechanical and physical properties of expanded starch, reinforced by natural fibres," Industrial Crops and Products, vol. 37, no. 1, pp. 435–440, 2012.
28. A. R. Kakroodi, S. Cheng, M. Sain, and A. Asiri, "Mechanical, thermal, and morphological properties of nanocomposites based on polyvinyl alcohol and cellulose nanofiber from Aloe vera rind," Journal of Nanomaterials, vol. 2014, Article ID 903498, 7 pages, 2014. ·
29. M. S. Sreekala, M. G. Kumaran, and S. Thomas, "Stress relaxation behaviour in oil palm fibres," Materials Letters, vol. 50, no. 4, pp. 263–273, 2001.
30. B. Ren, T. Mizue, K. Goda, and J. Noda, "Effects of fluctuation of fibre orientation on tensile properties of flax sliver-reinforced green composites," Composite Structures, vol. 94, no. 12, pp. 3457–3464, 2012.
31. Y. Pan and Z. Zhong, "A micromechanical model for the mechanical degradation of natural fiber reinforced composites induced by moisture absorption," Mechanics of Materials, vol. 85, pp. 7–15, 2015. ·
32. E. Jayamani, S. Hamdan, M. R. Rahman, and M. K. B. Bakri, "Investigation of fiber surface treatment on mechanical, acoustical and thermal properties of betelnut fiber polyester composites," Procedia Engineering, vol. 97, pp. 545–554, 2014.
33. L. Boopathi, P. S. Sampath, and K. Mylsamy, "Investigation of physical, chemical and mechanical properties of raw and alkali treated Borassus fruit fiber," Composites Part B: Engineering, vol. 43, no. 8, pp. 3044–3052, 2012.
34. M. Ramesh, T. S. A. Atreya, U. S. Aswin, H. Eashwar, and C. Deepa, "Processing and mechanical property evaluation of banana fiber reinforced polymer composites," Procedia Engineering, vol. 97, pp. 563–572, 2014. ·
35. N. Venkateshwaran, A. Elaya Perumal, and D. Arunsundaranayagam, "Fiber surface treatment and its effect on mechanical and visco-elastic behaviour of banana/epoxy composite," Materials & Design, vol. 47, pp. 151–159, 2013.
36. S. N. Monteiro, K. G. Satyanarayana, A. S. Ferreira et al., "Selection of high strength natural fibers,"Matéria,

vol. 15, no. 4, pp. 488–505, 2010.

37. H. Ismail, H. D. Rozman, R. M. Jaffri, and Z. A. Mohd Ishak, "Oil palm wood flour reinforced epoxidized natural rubber composites: the effect of filler content and size," European Polymer Journal, vol. 33, no. 10–12, pp. 1627–1632, 1997.

38. M. Jawaid and H. P. S. Abdul Khalil, "Cellulosic/synthetic fibre reinforced polymer hybrid composites: a review," Carbohydrate Polymers, vol. 86, no. 1, pp. 1–18, 2011.

39. M. George, P. G. Mussone, Z. Abboud, and D. C. Bressler, "Characterization of chemically and enzymatically treated hemp fibres using atomic force microscopy and spectroscopy," Applied Surface Science, vol. 314, pp. 1019–1025, 2014. ·

40. P. Wongsriraksa, K. Togashi, A. Nakai, and H. Hamada, "Continuous natural fiber reinforced thermoplastic composites by fiber surface modification," Advances in Mechanical Engineering, vol. 5, Article ID 685104, 2013.

41. J. Gassan and A. K. Bledzki, "Possibilities for improving the mechanical properties of jute/epoxy composites by alkali treatment of fibres," Composites Science and Technology, vol. 59, no. 9, pp. 1303–1309, 1999.

42. M. Z. Rong, M. Q. Zhang, Y. Liu, G. C. Yang, and H. M. Zeng, "The effect of fiber treatment on the mechanical properties of unidirectional sisal-reinforced epoxy composites," Composites Science and Technology, vol. 61, no. 10, pp. 1437–1447, 2001.

43. T. P. T. Tran, J.-C. Bénézet, and A. Bergeret, "Rice and Einkorn wheat husks reinforced poly(lactic acid) (PLA) biocomposites: effects of alkaline and silane surface treatments of husks," Industrial Crops and Products, vol. 58, pp. 111–124, 2014.

44. S. I. Hossain, M. Hasan, M. N. Hasan, and A. Hassan, "Effect of chemical treatment on physical, mechanical and thermal properties of ladies finger natural fiber," Advances in Materials Science and Engineering, vol. 2013, Article ID 824274, 6 pages, 2013.

45. A. O›Donnell, M. A. Dweib, and R. P. Wool, "Natural fiber composites with plant oil-based resin,"Composites Science and Technology, vol. 64, no. 9, pp. 1135–1145, 2004.

46. H. Luo, G. Xiong, C. Ma et al., "Mechanical and thermo-mechanical behaviors of sizing-treated corn fiber/polylactide composites," Polymer Testing, vol. 39, pp. 45–52, 2014.

47. H. Ismail, A. Rusli, and A. A. Rashid, "Maleated natural rubber as a coupling agent for paper sludge filled natural rubber composites," Polymer Testing, vol. 24, no. 7, pp. 856–862, 2005.

48. A. Paul, K. Joseph, and S. Thomas, "Effect of surface treatments on the electrical properties of low-density polyethylene composites reinforced with short sisal fibers," Composites Science and Technology, vol. 57, no. 1, pp. 67–79, 1997.

49. F. Corrales, F. Vilaseca, M. Llop, J. Gironès, J. A. Méndez, and P. Mutjè, "Chemical modification of jute fibers for the production of green-composites," Journal of Hazardous Materials, vol. 144, no. 3, pp. 730–735, 2007.

50. F. G. Torres and M. L. Cubillas, "Study of the interfacial properties of natural fibre reinforced polyethylene," Polymer Testing, vol. 24, no. 6, pp. 694–698, 2005.

51. A. Hidayat and S. Tachibana, "Characterization of polylactic acid (PLA)/kenaf composite degradation by immobilizedmyceliaofPleurotusostreatus,"International Biodeterioration & Biodegradation, vol. 71, pp. 50–54, 2012. ·

52. L. He, X. Li, W. Li, J. Yuan, and H. Zhou, "A method for determining reactive hydroxyl groups in natural fibers: application to ramie fiber and its modification," Carbohydrate Research, vol. 348, pp. 95–98, 2012.

53. K. Xie, H. Liu, and X. Wang, "Surface modification of cellulose with triazine derivative to improve printability with reactive dyes," Carbohydrate Polymers, vol. 78, no. 3, pp. 538–542, 2009.

54. N. Cordeiroa, M. Ornelasa, A. Ashorib, S. Sheshmanic, and H. Norouzic, "Investigation on the surface properties of chemically modified natural fibers using inverse gas chromatography," Carbohydrate Polymers, vol. 87, no. 4, pp. 2367–2375, 2012.

55. M. Le Troedec, D. Sedan, C. Peyratout et al., "Influence of various chemical treatments on the composition and structure of hemp fibres," Composites Part A: Applied Science and Manufacturing, vol. 39, no. 3, pp. 514–522, 2008.

56. M. J. John, B. Francis, K. T. Varughese, and S. Thomas, "Effect of chemical modification on properties of hybrid fiber biocomposites," Composites Part A: Applied Science and Manufacturing, vol. 39, no. 2, pp. 352–363, 2008.

57. M. Nasir, A. Gupta, M. D. H. Beg, G. K. Chua, and A. Kumar, "Fabrication of medium density fibreboard from enzyme treated rubber wood (Hevea brasiliensis) fibre and modified organosolv lignin,"International Journal of Adhesion and Adhesives, vol. 44, pp. 99–104, 2013.

58. M. Rokbi, H. Osmani, A. Imad, and N. Benseddiq, "Effect of chemical treatment on flexure properties of natural fiber-reinforced polyester composite," Procedia Engineering, vol. 10, pp. 2092–2097, 2011. ·

59. H. L. Bos, The potential of flax fibres as reinforcement for composite materials [Ph.D. thesis], Technische Universiteit Eindhoven, Eindhoven, The Netherlands, 2004.

60. B. C. Suddell, "Industrial fibres: recent and current developments," in Proceedings of the Symposium on Natural Fibres, vol. 20, pp. 71–82, FAO, CFC, Rome, Italy, October 2008.

61. K. Pickering, Ed., Properties and Performance of Natural-Fibre Composites, Woodhead Publishing, Cambridge, UK, 2008.

62. A. K. Mohanty, M. Misra, and L. T. Drzal, Eds., Natural Fibers, Biopolymers, and Biocomposites, Taylor &

Francis, 2005.

63. T. Sen and H. N. Reddy, "Various industrial applications of hemp, kinaf, flax and ramie natural fibres,"International Journal of Innovation, Management and Technology, vol. 2, pp. 192–198, 2011.

64. U. S. Bongarde and V. D. Shinde, "Review on natural fiber reinforcement polymer composite,"International Journal of Engineering Science and Innovative Technology, vol. 3, no. 2, pp. 431–436, 2014.

65. L. Mwaikambo, "Review of the history, properties and application of plant fibres," African Journal of Science and Technology, vol. 7, no. 2, p. 121, 2006.

66. F. Bateni, F. Ahmad, A. S. Yahya, and M. Azmi, "Performance of oil palm empty fruit bunch fibres coated with acrylonitrile butadiene styrene," Construction and Building Materials, vol. 25, no. 4, pp. 1824–1829, 2011.

67. M. George, P. G. Mussone, and D. C. Bressler, "Surface and thermal characterization of natural fibres treated with enzymes," Industrial Crops and Products, vol. 53, pp. 365–373, 2014.

68. K. Joseph, R. D. T. Filho, B. James, S. Thomas, and L. H. de Carvalho, "A review on sisal fiber reinforced polymer composites," Revista Brasileira de Engenharia Agrícola e Ambiental, vol. 3, no. 3, pp. 367–379, 1999.

69. C. Alves, P. M. C. Ferrão, A. J. Silva et al., "Ecodesign of automotive components making use of natural jute fiber composites," Journal of Cleaner Production, vol. 18, no. 4, pp. 313–327, 2010.

70. M. Sain, S. H. Park, F. Suhara, and S. Law, "Flame retardant and mechanical properties of natural fibre-PP composites containing magnesium hydroxide," Polymer Degradation and Stability, vol. 83, no. 2, pp. 363–367, 2004.

71. A. Alhuthali, I. M. Low, and C. Dong, "Characterisation of the water absorption, mechanical and thermal properties of recycled cellulose fibre reinforced vinyl-ester eco-nanocomposites," Composites Part B: Engineering, vol. 43, no. 7, pp. 2772–2781, 2012.

72. Z. N. Azwa, B. F. Yousif, A. C. Manalo, and W. Karunasena, "A review on the degradability of polymeric composites based on natural fibres," Materials & Design, vol. 47, pp. 424–442, 2013.

73. E. Gallo, B. Schartel, D. Acierno, F. Cimino, and P. Russo, "Tailoring the flame retardant and mechanical performances of natural fiber-reinforced biopolymer by multi-component laminate," Composites Part B: Engineering, vol. 44, no. 1, pp. 112–119, 2013.

74. S. Fatima and A. R. Mohanty, "Acoustical and fire-retardant properties of jute composite materials,"Applied Acoustics, vol. 72, no. 2-3, pp. 108–114, 2011.

75. N. P. G. Suardana, M. S. Ku, and J. K. Lim, "Effects of diammonium phosphate on the flammability and mechanical properties of bio-composites," Materials & Design, vol. 32, no. 4, pp. 1990–1999, 2011.

76. B. Schartel, U. Braun, U. Knoll et al., "Mechanical, thermal, and fire behavior of bisphenol a polycarbonate/multiwall carbon nanotube nanocomposites," Polymer Engineering & Science, vol. 48, no. 1, pp. 149–158, 2008.

77. J. Zhan, L. Song, S. Nie, and Y. Hu, "Combustion properties and thermal degradation behavior of polylactide with an effective intumescent flame retardant," Polymer Degradation and Stability, vol. 94, no. 3, pp. 291–296, 2009.

78. T. D. Hapuarachchi and T. Peijs, "Multiwalled carbon nanotubes and sepiolite nanoclays as flame retardants for polylactide and its natural fibre reinforced composites," Composites Part A: Applied Science and Manufacturing, vol. 41, no. 8, pp. 954–963, 2010.

79. G. Savage, "Development of penetration resistance in the survival cell of a Formula 1 racing car,"Engineering Failure Analysis, vol. 17, no. 1, pp. 116–127, 2010.

80. J. Meredith, R. Ebsworth, S. R. Coles, B. M. Wood, and K. Kirwan, "Natural fibre composite energy absorption structures," Composites Science and Technology, vol. 72, no. 2, pp. 211–217, 2012.

81. P. Wambua, B. Vangrimde, S. Lomov, and I. Verpoest, "The response of natural fibre composites to ballistic impact by fragment simulating projectiles," Composite Structures, vol. 77, no. 2, pp. 232–240, 2007.

82. P. K. Bajpai, I. Singh, and J. Madaan, "Tribological behavior of natural fiber reinforced PLA composites,"Wear, vol. 297, no. 1-2, pp. 829–840, 2013.

83. N. S. M. El-Tayeb, B. F. Yousif, and T. C. Yap, "Tribological studies of polyester reinforced with CSM 450-R-glass fiber sliding against smooth stainless steel counterface," Wear, vol. 261, no. 3-4, pp. 443–452, 2006.

84. C. W. Chin and B. F. Yousif, "Potential of kenaf fibres as reinforcement for tribological applications,"Wear, vol. 267, no. 9-10, pp. 1550–1557, 2009.

85. U. Nirmal, B. F. Yousif, D. Rilling, and P. V. Brevern, "Effect of betelnut fibres treatment and contact conditions on adhesive wear and frictional performance of polyester composites," Wear, vol. 268, no. 11-12, pp. 1354–1370, 2010.

86. X. Xin, C. G. Xu, and L. F. Qing, "Friction properties of sisal fibre reinforced resin brake composites,"Wear, vol. 262, no. 5-6, pp. 736–741, 2007.

87. N. S. M. El-Tayeb, "A study on the potential of sugarcane fibers/polyester composite for tribological applications," Wear, vol. 265, no. 1-2, pp. 223–235, 2008.

88. S. A. R. Hashmi, U. K. Dwivedi, and N. Chand, "Graphite modified cotton fibre reinforced polyester composites under sliding wear conditions," Wear, vol. 262, no. 11-12, pp. 1426–1432, 2007.

89. M. M. Davoodi, S. M. Sapuan, D. Ahmad, A. Aidy, A. Khalina, and M. Jonoobi, "Concept selection of car bumper beam with developed hybrid bio-composite material," Materials & Design, vol. 32, no. 10, pp. 4857–4865, 2011.

90. N. Banthia and R. Gupta, "Influence of polypropylene fiber geometry on plastic shrinkage cracking in concrete," Cement and Concrete Research, vol. 36, no. 7, pp. 1263–1267, 2006.

91. M. S. Sreekala and S. Thomas, "Effect of fibre surface modification on water-sorption characteristics of oil palm fibres," Composites Science and Technology, vol. 63, no. 6, pp. 861–869, 2003.

92. M. H. Ab Ghani and S. Ahmad, "The comparison of water absorption analysis between counterrotating and corotatingtwin-screwextruderswithdifferentantioxidants content in wood plastic composites,"Advances in Materials Science and Engineering, vol. 2011, Article ID 406284, 4 pages, 2011.

93. M. Mariatti, M. Jannah, A. A. Bakar, and H. P. S. A. Khalil, "Properties of banana and pandanus woven fabric reinforced unsaturated polyester composites," Journal of Composite Materials, vol. 42, no. 9, pp. 931–941, 2008.

94. S. N. Monteiro, V. Calado, R. J. S. Rodriguez, and F. M. Margem, "Thermogravimetric behavior of natural fibers reinforced polymer composites—an overview," Materials Science and Engineering A, vol. 557, pp. 17–28, 2012.

95. O. Hassan, T. P. Ling, M. Y. Maskat et al., "Optimization of pretreatments for the hydrolysis of oil palm empty fruit bunch fiber (EFBF) using enzyme mixtures," Biomass and Bioenergy, vol. 56, pp. 137–146, 2013.

96. G. George, E. T. Jose, D. Åkesson, M. Skrifvars, E. R. Nagarajan, and K. Joseph, "Viscoelastic behaviour of novel commingled biocomposites based on polypropylene/jute yarns," Composites Part A: Applied Science and Manufacturing, vol. 43, no. 6, pp. 893–902, 2012.

97. V. Fiore, T. Scalici, and A. Valenza, "Characterization of a new natural fiber from Arundo donax L. as potential reinforcement of polymer composites," Carbohydrate Polymers, vol. 106, no. 1, pp. 77–83, 2014.

98. S. Shinoj, R. Visvanathan, and S. Panigrahi, "Towards industrial utilization of oil palm fibre: physical and dielectric characterization of linear low density polyethylene composites and comparison with other fibre sources," Biosystems Engineering, vol. 106, no. 4, pp. 378–388, 2010.

99. E. Sassoni, S. Manzi, A. Motori, M. Montecchi, and M. Canti, "Novel sustainable hemp-based composites for application in the building industry: physical, thermal and mechanical characterization," Energy and Buildings, vol. 77, pp. 219–226, 2014.

100. J. Holbery and D. Houston, "Natural-fiber-reinforced polymer composites in automotive applications,"JOM, vol. 58, no. 11, pp. 80–86, 2006.

101. S. H. Shuit, K. T. Tan, K. T. Lee, and A. H. Kamaruddin, "Oil palm biomass as a sustainable energy source: a Malaysian case study," Energy, vol. 34, no. 9, pp. 1225–1235, 2009.

102. A. R. Kakroodi, Y. Kazemi, and D. Rodrigue, "Mechanical, rheological, morphological and water absorption properties of maleated polyethylene/hemp composites: effect of ground tire rubber addition,"Composites Part B: Engineering, vol. 51, pp. 337–344, 2013.

103. S. Kalia and B. S. Kaith, Cellulose Fibers: Bio- and Nano-Polymer Composites, Springer, 2011.

CHAPTER 10

ENZYMATIC SYNTHESIS AND CHEMICAL RECYCLING OF NOVEL POLYESTER-TYPE THERMOPLASTIC ELASTOMERS

Tsukuru Yagihara and Shuichi Matsumura

Department of Applied Chemistry, Faculty of Science and Technology, Keio University, 3-14-1 Hiyoshi, Kohoku-ku, Yokohama 223-8522, Japan;

ABSTRACT

Novel polyester-type thermoplastic elastomers based on poly(alkylene succinate)s were synthesized by the lipase-catalyzed copolymerization of cyclic diol/succinate oligomer and cyclic diol/alkylthiosuccinate oligomer. These copolymers exhibited biodegradabilities by activated sludge and a wide range of thermal and mechanical properties that were dependent on the molecular structure and the content of side alkylthio groups. The degree of crystallinity of the copolymer decreased with increasing content of alkylthio groups, which were introduced into the polymer chain as a soft segment. Furthermore, lipase-catalyzed depolymerization of these copolymers into cyclic oligomers and repolymerization of the oligomers was carried out. A repolymerized copolymer having the same M_w and monomer composition as the initial copolymer was obtained, indicating the chemical recyclability of the copolymer.

INTRODUCTION

Thermoplastic elastomers (TPE) are widely used as soft materials for industrial and medical applications. Unlike crosslinked-type elastomers, TPEs show thermoplasticity and can be easily recycled with remolding of the parent material. The typical example of a TPE is the ethylene-α-olefin copolymer, consisting of a polyethylene main chain and an alkyl side chain unit. The main chain acts as the hard segment because of its high crystallinity, while the alkyl side chain unit acts as the soft segment due to its steric hindrance to crystallization. The degree of crystallinity decreases with increasing content of soft segments; thus, the polymer changes from a rigid plastic to a soft elastomer [1,2,3].

In the field of elastomers, environmentally benign properties become more important [4,5].Examples of such properties include the replacement of petrochemical resources by bio-based resources, biodegradability and chemical recyclability. However, environmentally benign elastomers have not been developed yet. Currently available TPEs, such as ethylene-α-olefin copolymer, are only produced from petrochemicals and are highly resistant to biodegradation.

Thus, for the next generation of elastomers, establishment of an environmentally benign production method is an important issue. In recent years, enzymatic synthesis has

Citation: Yagihara, T.; Matsumura, S. Enzymatic Synthesis and Chemical Recycling of Novel Polyester-Type Thermoplastic Elastomers. Polymers 2012, 4, 1259-1277, doi: 10.3390/polym4021259

been studied as a novel and green methodology for polymer production. Enzymes are renewable and nontoxic catalysts with high catalytic activities. For example, lipases can act as powerful catalysts for polyesterification reactions under mild anhydrous conditions. The syntheses of high molecular weight polyesters by ring-opening polymerization (ROP) of cyclic oligomers have also been reported [6,7,8,9]. In addition, using a lipase as the catalyst in anhydrous dilute solution, polyesters can be easily depolymerized into repolymerizable cyclic oligomers. In such a way, lipases may also contribute to the establishment of sustainable polymer production and chemical recycling systems [10,11,12].

Diol/diacid-type aliphatic polyesters are typical polymers that can be readily produced and chemically recycled by lipases. It is reported that poly(butylene succinate) (PBS) and poly(hexamethylene succinate) (PHS) are produced by the lipase-catalyzed polymerization of succinic acid with butane-1,4-diol and hexane-1,6-diol, respectively [13,14]. The polymers exhibit good biodegradability and can be considered to be bio-based plastics, because succinic acid can be produced by fermentation, and butane-1,4-diol and hexane-1,6-diol can be produced by the reduction of bio-based succinic acid and adipic acid, respectively. However, the impact strength of PBS is relatively low, preventing its use in various applications.

One effective way to resolve this problem is to form PBS-based thermoplastic elastomers by the introduction of soft segments into the polymer chain [15,16,17,18]. The impact strength may be increased, and also such elastomers are expected to be biodegradable and chemically recyclable. In this study, a series of polyester-type elastomers consisting of PBS and PHS moieties as the hard segment and an alkyl side chain unit as the soft segment were prepared by the lipase-catalyzed copolymerization of cyclic oligomers as shown in Scheme 1. Their thermal and mechanical properties, biodegradabilities and chemical recyclabilities were evaluated.

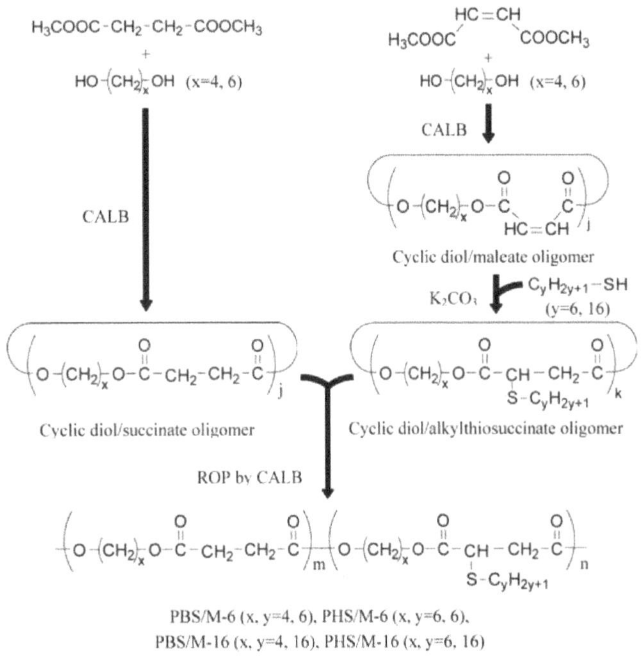

Scheme 1: Enzymatic synthesis of novel thermoplastic elastomers.

EXPERIMENTAL SECTION

Materials

Dimethyl succinate, dimethyl maleate, butane-1,4-diol, hexane-1,6-diol and 1-hexanethiol were purchased from Wako Pure Chemical Industries, Ltd. (Osaka, Japan). 1-Hexadecanethiol was purchased from TCI Co., Inc. (Tokyo, Japan). Immobilized lipase from Candida antarctica was kindly supplied by Novozymes Japan, Ltd. [Chiba, Japan; CALB: Novozym 435, a lipase (lipase B) from C. antarctica produced by submerged fermentation of a genetically modified Aspergillus oryzaemicroorganism and absorbed on a macroporous acrylic resin (10000 propyl laurate units per gram; lipase activity based on ester synthesis)]. The enzyme was dried under vacuum (3 mmHg) over P_2O_5 at 25 °C for 2 h before use. Molecular sieves 4A were purchased from Sigma-Aldrich Co. (St. Louis, MO, USA) and were dried at 120 °C for 2 h before use.

Measurements

The weight-average (M_w) and number-average (M_n) molecular weight and the molecular weight distribution (M_w/M_n) of the polymer were determined by size exclusion chromatography (SEC) using SEC columns (Shodex K-805L + K-800D, Showa Denko Co., Ltd., Tokyo, Japan) with a refractive index detector; chloroform was used as the eluent at 1.0 mL·min^{-1}. The SEC system was calibrated with polystyrene standards of a narrow molecular weight distribution. The molecular weight of the oligomers was measured by matrix-assisted laser desorption ionization time-of-flight mass spectra (MALDI-TOF MS) using a Bruker Ultraflex mass spectrometer equipped with a nitrogen laser. The detection was performed in the refraction mode, a cobalt matrix was used for oligomers, sodium bromide was used as a cation source, and positive ionization was used. ^{1}H NMR spectra were recorded on a Varian 300 Fourier transform spectrometer (Varian Inc., CA, USA) operating at 300 MHz.

The crystallization temperature (T_c), melting temperature (T_m) and glass transition temperature (T_g) of the polymer were determined by differential scanning calorimetry (DSC-60, Shimadzu Co., Kyoto, Japan). The measurements were made with a sample (10 mg) on a DSC plate. The polymer samples were heated at a rate of 10 °C·min^{-1} from 30 °C to 120 °C (first scan), cooled at a rate of −30 °C·min^{-1} to −80 °C and then scanned with heating at a rate of 10 °C·min^{-1} to 120 °C. Wide angle X-ray diffraction (WAXD) patterns were obtained using Ni-filtered Cu-Kα radiation at room temperature. A voltage and current of X-ray source were set to be 40 kV and 40 mA. Crystallinity (χ_c) of the films was calculated from diffracted X-ray intensity data in the 2θ range of 5–40° using the calculation program (DISCOVER, Bruker AXS K.K., Yokohama, Japan). The mechanical properties of the film samples were determined by an Autograph instrument (Shimadzu Co., Kyoto, Japan) at the cross-head speed of 30 mm·min^{-1}.

The biodegradability of the copolymers was evaluated by biochemical oxygen demand (BOD) measurements. BOD was determined with a BOD tester (VELP Scientifica s.r.l., Usmate,

MI, Italy) by the oxygen consumption method, according to the modified MITI test. The activated sludge was obtained from a municipal sewage plant in Yokohama City. BOD biodegradation (BOD/ThOD) was calculated from the BOD values and the theoretical oxygen demand (ThOD).

Enzymatic Synthesis of Novel Thermoplastic Elastomer

Preparation of Cyclic Diol/Succinate Oligomer

The cyclization was carried out in a round-bottomed flask with molecular sieves 4A placed at the top of the flask in the vapor phase to absorb the byproducts. In a typical procedure, dimethyl succinate (73 mg; 0.50 mmol) and butane-1,4-diol (48 mg; 0.53 mmol) were reacted by CALB (121 mg; 100 wt.%) as the catalyst in toluene (24 mL; 5.0 mg·mL^{-1}) under a nitrogen atmosphere at 90 °C for 48 h.

The reaction mixture was diluted with chloroform, and the CALB was removed by filtration. The solvent was then evaporated under reduced pressure to obtain cyclic butylene/succinate (BS) oligomer. The crude product was purified by column chromatography (silica gel 60N, 40 g) using chloroform-ethyl acetate (6:1 v/v, R_f = 0.40) as an eluent to obtain cyclic BS dimer as white crystals in 44% yield. In a similar procedure, cyclic hexamethylene/succinate (HS) dimer was produced by the reaction of dimethyl succinate and hexane-1,6-diol in 36% yield.

Cyclic BS dimer: ^{1}H-NMR (300 MHz, $CDCl_3$): δ 1.66–1.76 [4H, m, $-CH_2(CH_2)_2CH_2-$], 2.60–2.69 [4H, m, $-(C{=}O)(CH_2)_2(C{=}O)-$], 4.05–4.20 [4H, m, $-OCH_2(CH_2)_2CH_2O-$]. m.p.: 114–116 °C.

Cyclic HS dimer: ^{1}H-NMR (300 MHz, $CDCl_3$): δ 1.31–1.45 [4H, m, $-(CH_2)_2(CH_2)_2(CH_2)_2-$], 1.54–1.71 [4H, m, $-CH_2CH_2(CH_2)_2CH_2CH_2-$], 2.59–2.67 [4H, m, $-(C{=}O)(CH_2)_2(C{=}O)-$], 4.10 [4H, t, J = 6.6 Hz, $-OCH_2(CH_2)_4CH_2O-$]. m.p.: 57–59 °C.

Preparation of Cyclic Diol/Maleate Oligomer

In a typical procedure, dimethyl maleate (72 mg; 0.50 mmol) and butane-1,4-diol (48 mg; 0.53 mmol) were reacted by CALB (120 mg; 100 wt.%) as the catalyst in toluene (24 mL; 5.0 mg·mL^{-1}) under a nitrogen atmosphere at 90 °C for 48 h. The solvent was evaporated under reduced pressure. The reaction mixture was diluted with chloroform, and the CALB was removed by filtration to obtain cyclic butylene/maleate (BM) oligomer.

In a similar procedure, cyclic hexamethylene/maleate (HM) oligomer was produced by the reaction of dimethyl maleate and hexane-1,6-diol.

Preparation of Cyclic Diol/Alkylthiosuccinate Oligomer

The addition of thiol was carried out in a round-bottomed flask. In a typical procedure, a mixture of cyclic BM oligomer (170 mg; 1.0 mmol), 1-hexanethiol (118 mg; 1.0 mmol) and K_2CO_3 (51 mg; 30 wt.%) as the catalyst was stirred in dry acetone (2.9 mL; 100 mg·mL^{-1}) and reacted under an argon atmosphere at 25 °C for 24 h.

The reaction mixture was diluted with chloroform, and K_2CO_3 was removed by filtration. The solvent was evaporated under reduced pressure to obtain cyclic butylene/hexylthiosuccinate (BM-6) oligomer. In a similar procedure, cyclic hexamethylene/hexylthiosuccinate (HM-6), cyclic butylene/hexadecylthiosuccinate (BM-16) and hexamethylene/hexadecylthiosuccinate (HM-16) oligomers were obtained.

The crude product was purified by column chromatography (silica gel 60N, 40 g) using hexane-ethyl acetate (6:1 v/v, R_f = 0.38) as the eluent to obtain cyclic BM-6 dimer and cyclic HM-6 monomer as liquids in 38% and 37% yields, respectively. In the case of the addition of 1-hexadecanethiol, a mixed solvent of hexane/chloroform/acetone (6:1:1 v/v/v, R_f = 0.53) was used as the eluent to obtain cyclic BM-16 dimer and HM-16 monomer as white crystals in 36% and 34% yields, respectively.

Cyclic BM-6 dimer: ^{1}H-NMR (300 MHz, $CDCl_3$): δ 0.88 (3H, t, J = 6.9 Hz, CH_3CH_2-), 1.20–1.43 [6H, m, $CH_3(CH_2)_3CH_2-$], 1.50–1.68 (2H, m, $-CH_2CH_2CH_2S-$), 1.76–1.96 [4H, m, $-OCH_2(CH_2)_2CH_2O-$], 2.52–2.92 (2H, m, $-SCH_2CH_2-$), 2.71 [1H, dd, J = 11.4, 14.1 Hz, $-C({=}O)C(H)HC(S-)H-$], 2.83 [1H, dd, J = 6.0, 14.1 Hz, $-C({=}O)C(H)HC(S-)H-$], 3.72 [1H, dd, J = 6.0, 11.4 Hz, $-CH_2C(S-)H(C{=}O)-$], 4.12–4.38 [4H, m, $-OCH_2(CH_2)_2CH_2O-$]. $C_{14}H_{24}O_4S$ (288.41): Calcd. C 58.30, H 8.39, S 11.12; Found C 58.37, H 8.58, S 11.14.

Cyclic HM-6 monomer: ^{1}H-NMR (300 MHz, $CDCl_3$): δ 0.88 (3H, t, J = 6.9 Hz, CH_3CH_2-), 1.20–1.43 [6H, m, $CH_3(CH_2)_3CH_2-$], 1.44–1.84 [10H, m, $-CH_2CH_2CH_2S-$, $-OCH_2(CH_2)_4CH_2O-$], 2.57–2.74 (2H, m, $-SCH_2CH_2-$), 2.70 [1H, dd, J= 3.6, 16.2 Hz, $-C({=}O)C(H)HC(S-)H-$], 2.99 [1H, dd, J = 12.3, 16.2 Hz, $-C({=}O)C(H)HC(S-)H-$], 3.67 [1H, dd, J = 3.6, 12.3 Hz, $-CH_2C(S-)H(C{=}O)-$], 4.10–4.26 [3H, m, $-OCH_2(CH_2)_4C(H)HO-$], 4.43 [1H, ddd, J = 3.3, 7.8, 11.7, $-OCH_2(CH_2)_4C(H)HO-$]. $C_{16}H_{28}O_4S$ (316.46): Calcd. C 60.73, H 8.92, S 10.13; Found C 60.65, H 9.07, S 10.40.

Cyclic BM-16 dimer: ^{1}H-NMR (300 MHz, $CDCl_3$): δ 0.88 (3H, t, J = 6.6 Hz, CH_3CH_2-), 1.16–1.42 [26H, m, $CH_3(CH_2)_{13}CH_2-$], 1.52–1.65 (2H, m, $-CH_2CH_2CH_2S-$), 1.78–1.93 [4H, m, $-OCH_2(CH_2)_2CH_2O-$], 2.55–2.66 (2H, m, $-SCH_2CH_2-$), 2.69 [1H, dd, J = 11.4, 14.4 Hz, $-C({=}O)C(H)HC(S-)H-$], 2.82 [1H, dd, J = 6.0, 14.4 Hz, $-C({=}O)C(H)HC(S-)H-$], 3.72 [1H, dd, J = 6.0, 11.4 Hz, $-CH_2C(S-)H(C{=}O)-$], 4.14–4.33 [4H, m, $-OCH_2(CH_2)_2CH_2O-$]. $C_{24}H_{44}O_4S$ (428.68): Calcd. C 67.24, H 10.35, S 7.48; Found C 67.58, H 10.43, S 7.44. m.p.: 62–64 °C.

Cyclic HM-16 monomer: ^{1}H-NMR (300 MHz, $CDCl_3$): δ 0.88 (3H, t, J = 6.9 Hz, CH_3CH_2-), 1.18–1.42 [26H, m, $CH_3(CH_2)_{13}CH_2-$], 1.45–1.54 (2H, m, $-CH_2CH_2CH_2S-$), 1.54–1.84 [6H, m, $-OCH_2(CH_2)_4CH_2O-$], 2.58–2.70 (2H, m, $-SCH_2CH_2-$), 2.70 [1H, dd, J = 3.6, 15.9 Hz, $-C({=}O)C(H)HC(S-)H-$], 2.99 [1H, dd, J = 12.6, 15.9 Hz, $-C({=}O)C(H)HC(S-)H-$], 3.67 [1H, dd, J = 3.6, 12.6 Hz, $-CH_2C(S-)H(C{=}O)-$], 4.15–4.35 [3H, m, $-OCH_2(CH_2)_4C(H)HO-$], 4.43 [1H, ddd, J = 3.3, 7.5, 10.8, $-OCH_2(CH_2)_4C(H)HO-$]. $C_{26}H_{48}O_4S$ (456.73): Calcd. C 68.37, H 10.59, S 7.02; Found C 68.13, H 10.86, S 6.94. m.p.: 40–42 °C.

Lipase-Catalyzed ROP of Cyclic Oligomers

The ROP was carried out in a screw-capped vial with molecular sieves 4A placed at the top of the vial in the vapor phase. In a typical procedure, cyclic BS dimer (29.1 mg) and cyclic BM-6 dimer (20.9 mg) were copolymerized in the presence of CALB (25 mg; 50 wt.%) under a nitrogen atmosphere at 120 °C for 24 h. The polymerization mixture was diluted with chloroform, and the CALB was removed by filtration. The purification was carried out by reprecipitation using chloroform (good solvent)/1-propanol (poor solvent) to obtain PBS/M-6 in 68% yield.

PBS/M-6: ^{1}H-NMR (300 MHz, $CDCl_3$): δ 0.89 (t, J = 5.7 Hz, CH_3CH_2–), 1.19–1.42 [m, $CH_3(CH_2)_3CH_2$–], 1.50–1.64 (m, –$CH_2CH_2CH_2S$–), 1.64–1.82 [m, –$OCH_2(CH_2)_2CH_2O$–], 2.53–2.75 [m, –SCH_2CH_2–, –(C=O)$(CH_2)_2$(C=O)–], 2.67 [dd, J = 5.7, 17.1 Hz, –C(=O)C(H)HC(S–)H–], 2.99 [dd, J = 9.6, 17.1 Hz, –C(=O)C(H)HC(S–)H–], 3.64 [dd, J = 5.7, 9.6 Hz, –CH_2C(S–)H(C=O)–], 4.04–4.24 [m, –$OCH_2(CH_2)_2CH_2O$–].

PHS/M-6: ^{1}H-NMR (300 MHz, $CDCl_3$): δ 0.88 (t, J = 6.9 Hz, CH_3CH_2–), 1.21–1.49 [m, $CH_3(CH_2)_3CH_2$–, –$(CH_2)_2(CH_2)_2(CH_2)_2$–], 1.52–1.74 [m, –$CH_2CH_2(CH_2)_2CH_2CH_2$–,–$CH_2CH_2CH_2S$–], 2.55–2.76 [m, –SCH_2CH_2–, –(C=O)$(CH_2)_2$(C=O)–], 2.67 [dd, J = 5.4, 10.8 Hz, –C(=O)C(H)HC(S–)H–], 2.99 [dd, J = 9.6, 10.8 Hz, –C(=O)C(H)HC(S–)H–], 3.64 [dd, J= 5.4, 9.6 Hz, –CH_2C(S–)H(C=O)–], 4.08–4.20 [br t, –$OCH_2(CH_2)_4CH_2O$–].

PBS/M-16: ^{1}H-NMR (300 MHz, $CDCl_3$): δ 0.88 (t, J = 6.6 Hz, CH_3CH_2–), 1.15–1.42 [m, $CH_3(CH_2)_{13}CH_2$–], 1.46–1.64 (m, –$CH_2CH_2CH_2S$–), 1.64–1.78 [m, –$OCH_2(CH_2)_2CH_2O$–], 2.55–2.67 [m, –SCH_2CH_2–, –(C=O)$(CH_2)_2$(C=O)–], 2.67 [dd, J = 5.7, 16.8 Hz, –C(=O)C(H)HC(S–)H–], 2.99 [dd, J = 9.9, 16.8 Hz, –C(=O)C(H)HC(S–)H–], 3.63 [dd, J = 5.7, 9.9 Hz, –CH_2C(S–)H(C=O)–], 4.02–4.23 [m, –$OCH_2(CH_2)_2CH_2O$–].

PHS/M-16: ^{1}H-NMR (300 MHz, $CDCl_3$): δ 0.88 (t, J = 6.9 Hz, CH_3CH_2–), 1.15–1.43 [m, $CH_3(CH_2)_{13}CH_2$–, –$(CH_2)_2(CH_2)_2(CH_2)_2$–], 1.55–1.78 [m, –$CH_2CH_2CH_2S$–, –$CH_2CH_2(CH_2)_2CH_2CH_2$–], 2.52–2.72 [m, –SCH_2CH_2–, –(C=O)$(CH_2)_2$(C=O)–], 2.67 [dd, J = 5.7, 17.1 Hz, –C(=O)C(H)HC(S–)H–], 2.99 [dd, J = 9.9, 17.1 Hz, –C(=O)C(H)HC(S–)H–], 3.63 [dd, J= 5.7, 9.9 Hz, –CH_2C(S–)H(C=O)–], 3.90–4.23 [br t, –$OCH_2(CH_2)_4CH_2O$–].

Chemical Recycling of Polymers

The degradation was carried out in a round-bottomed flask with molecular sieves 4A placed at the top of the flask in the vapor phase to absorb the byproducts. In a typical procedure, PBS/M (100 mg) and CALB (100 mg; 100 wt.%) were stirred with toluene (20 mL; 5 mg·mL^{-1}) and reacted under a nitrogen atmosphere at 100 °C for 72 h. The reaction mixture was diluted with chloroform, and the CALB was removed by filtration to obtain degradation products consisting of cyclic oligomers.

The degradation products were subjected to repolymerization without any purification. The reaction was carried out by a method similar to the ROP of cyclic oligomers noted above.

RESULTS AND DISCUSSION

Preparation of Cyclic Oligomers

Cyclic diol/succinate oligomers (cyclic BS and HS) as the hard segment were prepared by the cyclization of dimethyl succinate and diols with CALB in dilute toluene solution. A slight excess of diols was required in order to prevent production of species having terminal succinate ends, which were difficult to isolate from the cyclic oligomers. Cyclic diol/alkylthiosuccinate oligomers (cyclic BM-6, HM-6, BM-16 and HM-16) as the soft segment were prepared by the cyclization of dimethyl maleate and diols, followed by the addition of thiols in the presence of K_2CO_3 in acetone solution.

These cyclic oligomers were successfully isolated by silica gel column chromatography. The ^{1}H-NMR spectra supported the cyclic structures in that the terminal methyl ester and hydroxyl group disappeared after the reaction. Ring size of the cyclic oligomers was determined by MALDI-TOF MS spectrometry. As shown in Table 1, the main components of the cyclic oligomers were dimers or monomers. Relatively higher melting points of cyclic BS and HS suggest that the diol/succinate oligomers have crystallinity; therefore, they are expected to act as the hard segment of the copolymer. Also, cyclic BM-16 and HM-16 were crystals, while cyclic HM-6 and BM-6 were liquids at room temperature as shown in Table 1. These results indicate that hexyl side chains disturb the crystallization process; however, hexadecyl side chains might crystallize with their packing effect.

Table 1: Main components, yields and melting points of cyclic oligomers

Cyclic Oligomer	Ring Size	Yield (%)	m.p. (°C)
BS	Dimer	44	114–116
HS	Dimer	36	57–59
BM-6	Dimer	38	–
HM-6	Monomer	37	–
BM-16	Dimer	36	62–64
HM-16	Monomer	34	40–42

Lipase-Catalyzed ROP of Cyclic Oligomers

The isolated main components of the cyclic oligomers as shown in Table 1 were subjected to the lipase-catalyzed ROP. Samples of PBS/M-6, PHS/M-6, PBS/M-16 and PHS/M-16 with various monomer ratios were prepared by the copolymerization of cyclic diol/succinate and cyclic diol/alkylthiosuccinate using CALB in bulk. A higher polymerization temperature of 120 °C was needed, because the melting point of the cyclic BS dimer was 115 °C.

Figure 1 shows a time course of molecular weight changes of the PHS/M-6 and compositional changes of the HM-6 unit in the PHS/M-6. The copolymerization was carried out for 72 h with an initial molar ratio of cyclic HS/HM-6 = 75/25 (mol/mol). The molecular weight was measured by SEC, and the HM-6 content was determined by ^{1}H-NMR spectroscopy. It was found that the M_w of the copolymer reached the highest value of 86000 g·mol^{-1} after 24 h of reaction. In addition, the composition of HM-6 reached the highest of 14 mol%, although this value was lower than the feed ratio. These results indicate that cyclic diol/alkylthiosuccinate oligomers have lower polymerizability than cyclic diol/succinate oligomers due to steric hindrance from the alkyl side chain.

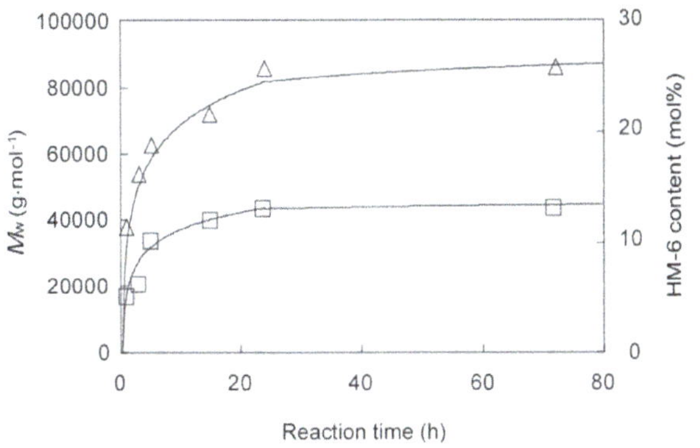

Figure 1: Time course of M_w (△) and HM-6 content (□) of PHS/M-6 (HM-6 feed ratio: 25 mol%).

The M_w and the yield of the copolymers having various monomer ratios were obtained by the lipase-catalyzed polymerization for 24 h and the results are shown in Table 2. It was found that high molecular weight copolymers were produced. The contents of side chain units (M-6 and M-16) in all copolymers were lower than the initial feed ratio. Copolymers containing hexyl side chain units (PBS/M-6 and PHS/M-6) changed from a rigid plastic to a liquid polymer with increasing content of the alkyl side chain unit. The homopolymer with hexyl side chain units (PHM-6) was a viscous liquid below room temperature. Although copolymers containing hexadecyl side chain units (PBS/M-16 and PHS/M-16) also changed from a rigid plastic to a liquid polymer with increasing content of side chain units, the homopolymer of the hexadecyl side chain unit (PBS/M-16 and PHM-16) was a solid polymer. This indicated that the hexadecyl side chain units tended to crystallize with each other in addition to disturbing the crystallization of the main chain.

Table 2: M_w and M_w/M_n of polymers with various contents of the side chain unit.

Polymer	Feed Ratio of Side Chain Unit (mol%)	Content of Side Chain Unit in Copolymer (mol%)	M_w (g·mol^{-1})	M_w/M_n	Yield (%)
PBS	0	0	82000	1.47	78
PBS/M-6	10	7	75000	1.54	70
	15	12	77000	1.55	72
	30	22	85000	1.61	68
	40	33	84000	1.58	65
PBM-6	100	100	86000	1.60	58
PBS/M-16	7	5	64000	2.34	70
	15	11	65000	2.05	63
	25	17	63000	2.21	52
	30	23	59000	1.93	57
PBM-16	100	100	54000	1.96	53
PHS	0	0	90000	1.89	75
PHS/M-6	10	6	84000	1.72	68
	25	15	88000	1.85	57
	30	17	85000	1.88	59
	33	19	95000	1.93	60
PHM-6	100	100	87000	1.87	51
PHS/M-16	10	6	72000	2.24	67
	15	10	66000	1.85	57
	20	13	69000	2.31	51
	25	15	64000	2.07	54
PHM-16	100	100	71000	2.54	42

Crystallinities

Crystallinities of the polymer films were measured by wide angle X-ray diffraction (WAXD). The degree of crystallinity (χ_c) was calculated using the ratio of the peak area corresponding to the crystalline structure and the total area of all peaks in the WAXD pattern. The obtained χ_c values of the copolymers are shown in Figure 2. The χ_c of the copolymer decreased with increasing content of the alkyl side chain unit. This suggests that the alkyl side chain units prevent the crystallization of the main chain, and create an amorphous area.

Figure 2 also shows that the copolymers containing butane-1,4-diol (PBS/M-6 and PBS/M-16) have higher χ_c than those of hexane-1,6-diol (PHS/M-6 and PHS/M-16). Figure 3 shows the WAXD profiles of the BS homopolymer (PBS) and HS homopolymer (PHS). The difference in profiles indicates that the crystallization processes of PBS and PHS are different. This can be explained by the ratio of ester bonds in the main chains [19]. Since PBS has a higher ratio of ester bonds, PBS can adopt a stiff crystalline structure. According to the study of Ihn et al., PBS crystallizes in a monoclinic system [20]. In contrast, the crystallization process of PHS was scarcely affected by the ester bonds due to the lower ratio of the ester bonds in the main chain. The two sharp peaks detected in the WAXD patterns of PHS (2θ = 21° and 24°) are similar to those of polyethylene [21]. This suggests that PHS possesses a crystalline structure mainly consisting of a zigzag structure caused by the packing of the hexamethylene units in the main chain.

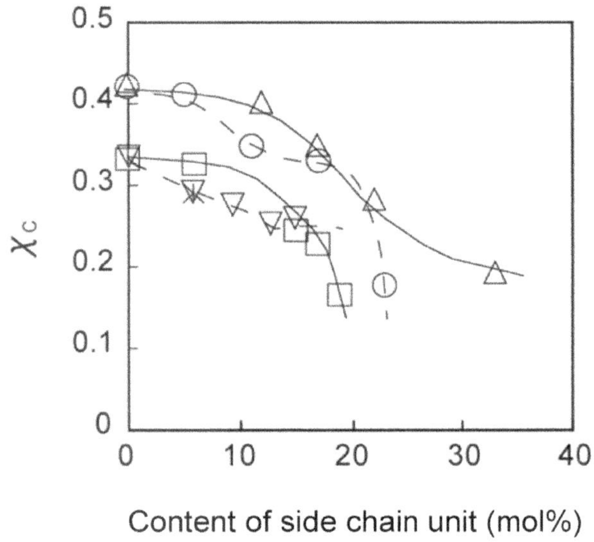

Figure 2: The degree of crystallinity (χ_c) of PBS/M-6 (△), PHS/M-6 (□),PBS/M-16 (○)and PHS/M-16 (▽) with various monomer contents.

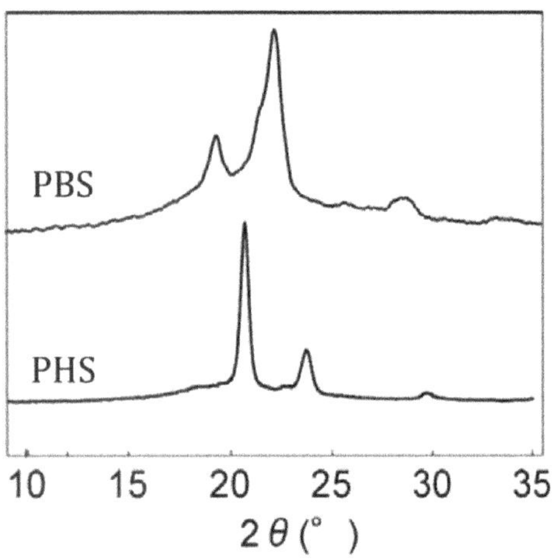

Figure 3: Wide angle X-ray diffraction spectra of PBS and PHS films.

Additionally, the alkyl side chain also affected the crystallinity of the copolymer. In order to evaluate the difference between hexyl and hexadecyl side chain units, WAXD profiles of PHS/M-6 and PHS/M-16 with various contents of side chain units are shown in Figure 4. Two crystalline peaks corresponding to methylene packing were detected around 2θ of 21° and 24° in all copolymers. Figure 4a shows that both of the crystalline peaks in PHS/M-6 decreased with increasing HM-6 content. This suggests that the hexyl side chain unit simply disturbs the crystallization process of the main chains. In contrast, Figure 4b shows that the crystalline peak at 21° in PHS/M-16 decreased with increasing HM-16 content, while the crystalline peak at 24° increased from 10 to 15 mol% with increasing HM-16 content. These results indicate that the hexadecyl side chain units create a new crystalline structure themselves, in addition to disturbing the crystallization process of the main chains.

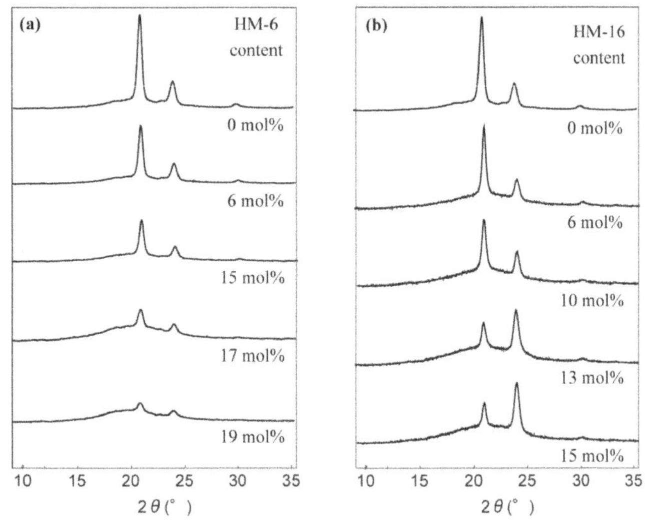

Figure 4: Wide angle X-ray diffraction spectra of (**a**) PHS/M-6 and (**b**) PHS/M-16films with various monomer contents.

Thermal Properties

The thermal properties of the copolymers were measured by DSC. The T_c was detected at a cooling rate of −30 °C·min^{-1}, and the T_m was detected at a heating rate of 10 °C·min^{-1}. The T_g could not be obtained by DSC measurement. The results are shown in Figure 5 and Figure 6. Both T_c and T_m of all copolymers decreased with increasing content of alkyl side chain units. This is due to the decrease in the χ_c caused by the steric hindrance of the alkyl side chain units. Also, PBS/M-6 and PBS/M-16 having higher χ_c showed higher T_c and T_m relative to PHS/M-6 and PHS/M-16.

It was found that the thermal properties of the copolymer were dependent on the alkyl side chain. Figure 7 shows the melting peaks of PHS/M-6 (a) and PHS/M-16 (b) with various contents of the alkyl side chain unit obtained by DSC measurement. Figure 7a shows that the melting peaks of PHS/M-6 broadened with the increasing content of the alkyl side chain unit, and the copolymer changes to an amorphous polymer with no melting peak. This suggests that the crystalline area of the copolymer simply decreased with increasing content of HM-6. On the other hand, new melting peaks appeared in the temperature range of 0–20 °C with increasing HM-16 content above 23 mol% as shown in Figure 7b. These new peaks must correspond to the new crystalline structure created by the hexadecyl side chain units. Sharp decreases in T_m values of PBS/M-16 and PHS/M-16 as shown in Figure 6 are due to a shift of the crystalline structure from the main chains to the hexadecyl side chain units.

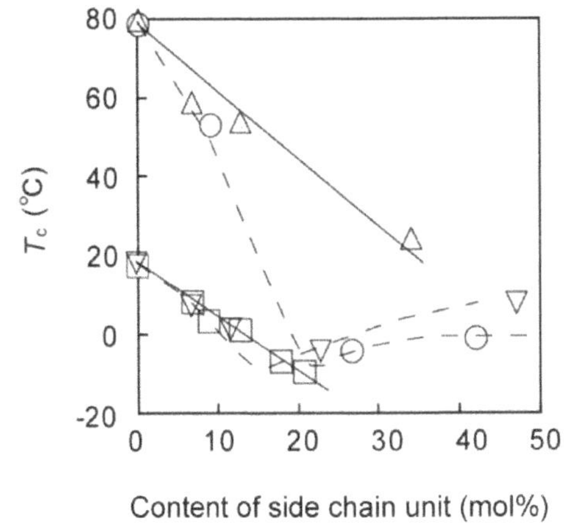

Figure 5: T_c of PBS/M-6 (Δ), PHS/M-6 (□), PBS/M-16 (○) and PHS/M-16 (▽)with various monomer contents.

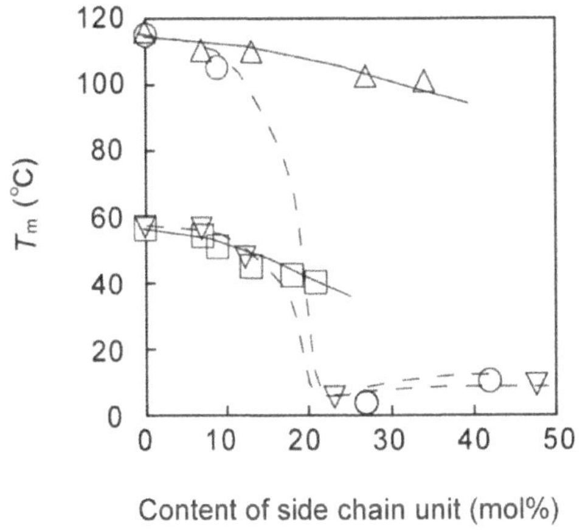

Figure 6: T_m of PBS/M-6 (Δ), PHS/M-6 (□), PBS/M-16 (○) and PHS/M-16 (▽) with various monomer contents.

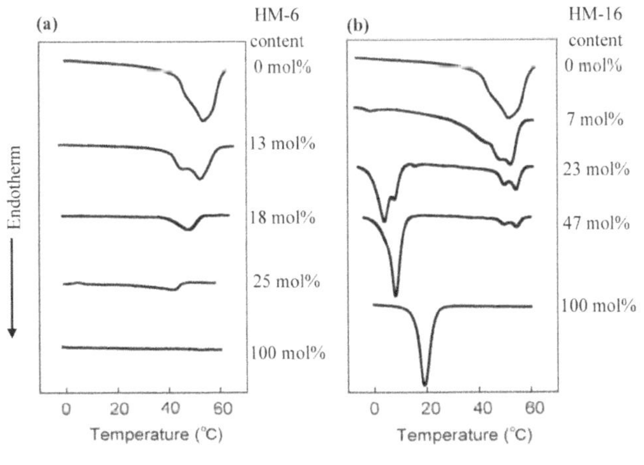

Figure 7: Melting peaks of (**a**) PHS/M-6 and (**b**) PHS/M-16 with various monomer contents.

Mechanical Properties

The mechanical properties of the copolymers were measured by tensile tests at 25 °C. Stress-strain curves of the copolymers were obtained and values of Young's modulus, tensile strength and elongation at break were calculated. As a typical example, the stress-strain curves of PHS/M-6 with various HM-6 contents are shown in Figure 8, and the values of the mechanical properties are summarized in Table 3. These properties were strongly dependent on the content of the alkyl side chain unit in the copolymer. With increasing content of the alkyl side chain unit, the Young's modulus and tensile strength simply decreased, although the elongation at break first increased and then decreased. The yield points of the copolymer disappeared with increasing content of the alkyl side chain unit as shown in Figure 8. These results indicate that a certain content of the alkyl side chain unit imparts a flexibility to the copolymer, while an excess amount of the alkyl side chain unit imparts weakness and flexibility. These tendencies were observed for all of the copolymers; thus the alkyl side chain unit was confirmed to act as a soft segment.

Cyclic stress-strain curves were obtained from cyclic loading tests with 5-cycles load-unload to the maximum elongation of 100% [22]. As a typical example, cyclic stress-strain curves of PHS/M-6 (HM-6 content: 19 mol%) are shown in Figure 9, and values of elastic recovery in the first cycle are shown in Table 3. It was found that considerable strain remains after the first cycle, and similar values of elastic recovery were obtained in subsequent cycles. The remaining strain of the copolymer may be due to the orientation of the polymer chains and the restructuring of a crystalline area with elongation. The copolymer films that underwent the elongation maintain the oriented structure of the polymer chains; therefore, further orientation is unlikely to occur in subsequent cycles. In addition, the elastic recovery increased with increasing content of the alkyl side chain unit. It is suggested that in copolymers with lower χ_c, amorphous areas were mainly stretched, and as a result, prevented the crystalline areas from deformation [23].

Table 3: Mechanical properties of copolymers with various monomer contents

Polymer	Content of Side Chain Unit (mol%)	Degree of Crystallinity (χ_c)	Young's Modulus (MPa)	Tensile Strength (MPa)	Elongation at Break (%)	Elastic Recovery (%)
PBS	0	0.42	292	29.8	18	–
PBS/M-6	7	0.40	195	18.9	20	–
	12	0.35	148	13.5	40	–
	22	0.28	118	12.5	485	29
	33	0.19	21.8	4.4	310	40
PBS/M-16	5	0.41	263	24.1	20	–
	11	0.35	192	22.0	25	–
	17	0.33	116	12.5	54	–
	23	0.18	12.3	2.7	10	–
PHS	0	0.33	188	16.9	392	17
PHS/M-6	6	0.32	156	15.5	594	22
	15	0.25	90.3	8.8	520	28
	17	0.23	46.3	7.6	660	39
	19	0.17	7.9	2.6	338	51
PHS/M-16	6	0.29	111	18.0	838	21
	10	0.27	57.3	10.1	582	30
	13	0.25	27.7	4.6	243	39
	15	0.26	13.5	2.6	59	–

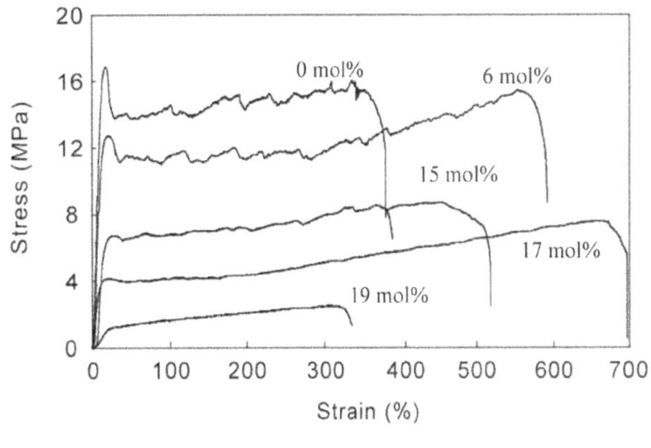

Figure 8: Stress-strain curves of PHS/M-6 with various HM-6 contents.

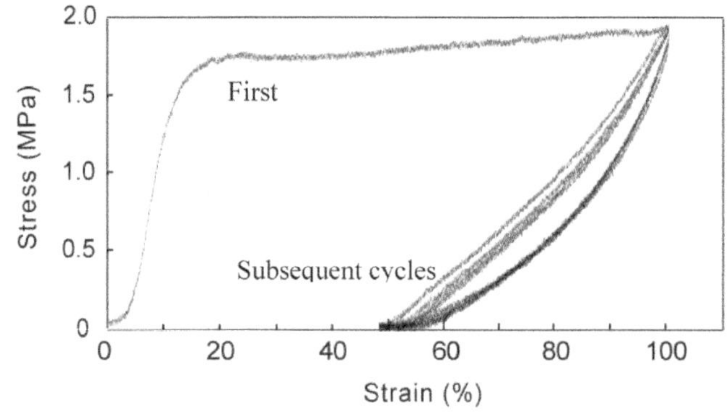

Figure 9: Cyclic stress-strain curves of PHS/M-6 (HM-6 content: 19 mol%).

These mechanical properties were also dependent on the methylene length of both the main chain and side chain. PBS/M-6 and PBS/M-16 showed relatively higher Young's modulus and tensile strength and lower elongation at break when compared to PHS/M-6 and PHS/M-16. This indicates that the crystalline structure, which is mainly affected by the ester bonds, has rigid and inflexible properties. Additionally, PHS/M-16 showed a relatively sharper decrease in the Young's modulus and an increase in the elongation at break when compared to PHS/M-6. This is due to the extension of the alkyl side chain that leads to an increase in the steric hindrance. In contrast, the tensile strength decreased similarly between PHS/M-16 and PHS/M-6. In order to investigate this tendency, a relationship between tensile stress at 200% elongation and Young's modulus is shown in Figure 10. It was found that PHS/M-16 exhibited a higher tensile stress relative to PHS/M-6. This is due to the difference in packing effect of the alkyl side chains. Upon elongation of the copolymers, both main chains and side chains were oriented and the crystalline areas became tougher. In this case, longer alkyl side chains may produce bigger packing effects that might contribute to reinforce the crystalline area [24].

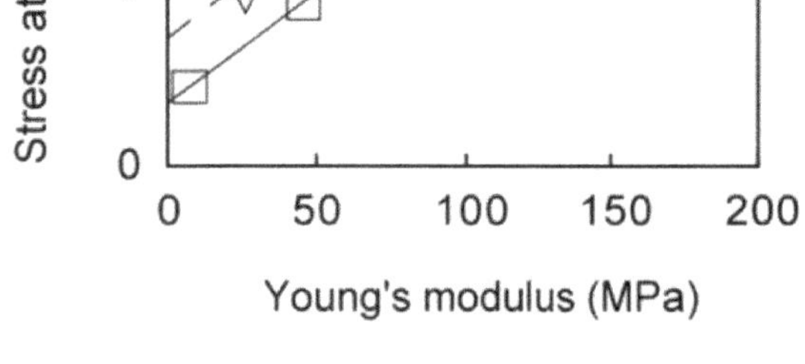

Figure 10: Relationship between Young's modulus and stress at 200% strainof PHS/M-6 (□) and PHS/M-16 (▽) with various monomer contents.

Biodegradabilities

Biodegradabilities of PHS/M-6 with various monomer contents were evaluated using activated sludge according to the BOD tests at 25 °C for 38 days. Figure 11 shows the time course of biodegradation of PHS/M-6 as a typical example. Aniline was used as a reference for the biodegradation test. It was found that biodegradation rates of PHS/M-6 increased with increasing HM-6 content. This is due to the decrease in χ_c of PHS/M-6, as shown in Table 2. The amorphous unit increased upon increasing HM-6 content, which promoted the hydrolysis of the ester bonds in the copolymer [22].

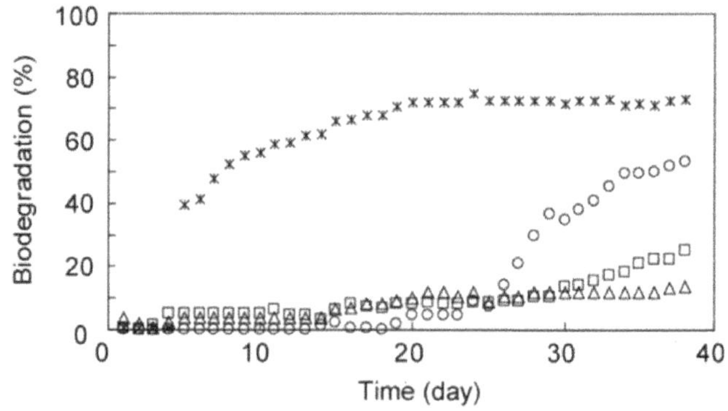

Figure 11: BOD-biodegradation of PHS/M-6 with various HM-6 contents [0 mol% (Δ), 6 mol% (□), 17 mol% (○)] and aniline (*) using activatedsludge at 25 °C.

Chemical Recyclabilities

The chemical recyclabilities of the copolymers were evaluated by lipase-catalyzed degradation into cyclic oligomers and repolymerization of the resulting cyclic oligomers. Figure 12 shows SEC profile changes of PHS/M-6 (HM-6 content: 17 mol%) by the degradation and repolymerization processes. A copolymer having a M_w of 76000 g·mol^{-1} was completely degraded into oligomers having a M_w of 800 g·mol^{-1} by CALB after a 3-day reaction as shown in Figure 12a,b. No terminal carboxyl or hydroxyl groups were detected in the ^{1}H-NMR spectrum, thus confirming cyclic structures of the degradation products. Also, molecular masses corresponding to the cyclic oligomers with several ring sizes (z = 2–8) were observed in the MALDI-TOF MS spectrum of the degradation products, while no linear oligomers of any molecular mass could be detected. These results suggested that the degradation products exclusively consisted of the repolymerized cyclic oligomers. In order to establish a convenient method of chemical recycling, the crude product was subjected to ROP without further purification.

The lipase-catalyzed repolymerization and reprecipitation were carried out using the same method as the usual ROP. Figure 12b,c shows that the degradation products readily repolymerized to a copolymer having a M_w of 70000 g·mol^{-1} and HM-6 content of 16 mol%, almost the same as those of the initial copolymer. These results confirmed the chemical recyclabilities of the copolymers by a lipase. A slight decrease in the molecular weight might be caused by a small amount of impurities.

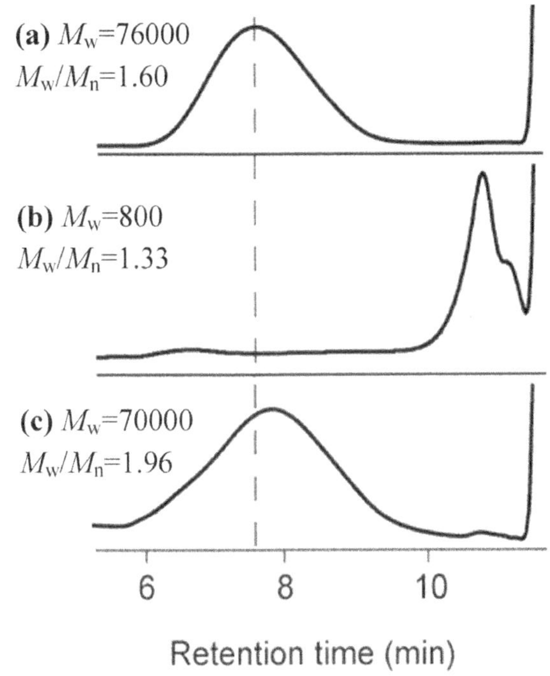

Figure 12: (**a**) SEC profiles of PHS/M-6, (**b**) degradation products and (**c**) repolymerizedPHS/M-6.

CONCLUSIONS

Novel thermoplastic elastomers, PBS/M-6, PHS/M-6, PBS/M-16 and PHS/M-16, were prepared by the lipase-catalyzed polymerization of bio-based diols and dicarboxylic acids via cyclic oligomers. These copolymers exhibited biodegradability, chemical recyclability and a wide range of mechanical properties such as Young's modulus of 8–292 MPa, tensile strength of 2.6–29.8 MPa and elongation at break of 10–838%. These values were dependent on the content of alkyl side chain units and the methylene units in both the main and side chains. Based on these results, these novel polyesters have potential for applications as environmentally benign thermoplastic elastomers in the field of polymeric materials.

ACKNOWLEDGMENTS

Immobilized lipase from Candida antarctica (CALB, Novozym 435) was kindly supplied by Novozymes Japan Ltd. (Chiba, Japan). This work was supported by a Grant-in-Aid for General Scientific Research from the Ministry of Education, Culture, Sports, Science, and Technology, Japan.

REFERENCES

1. Wang, H.P.; Khariwala, D.U.; Cheung, W.; Chum, S.P.; Hiltner, A.; Baer, E. Characterization of some new olefinic block copolymers. Macromolecules 2007, 40, 2852–2862.
2. Poon, B.C.; Dias, P.; Ansems, P.; Chum, S.P.; Hiltner, A.; Baer, E. Structure and deformation of an elastomeric propylene-ethylene copolymer. J. Appl. Polym. Sci. 2007, 104, 489–499.
3. Simanke, A.G.; Galland, G.B.; Baumhardt Neto, R.; Quijada, R.; Mauler, R.S. Influence of the type and the comonomer contents on the mechanical behavior of ethylene/α-olefin copolymers. J. Appl. Polym. Sci. 1999, 74, 1194–1200.
4. Ebata, H.; Toshima, K.; Matsumura, S. Lipase-catalyzed synthesis and properties of poly[(12-hydroxydodecanoate)-co-(12-hydroxystearate)] directed towards novel green and sustainable elastomers. Macromol Biosci. 2008, 8, 38–45.
5. Wanamaker, C.L.; O'Leary, L.E.; Lynd, N.A.; Hillmyer, M.A.; Tolman, W.B. Renewable-resource thermoplastic elastomers based on polylactide and polymenthide. Biomacromolecules 2007, 8, 3634–3640.
6. Matsumura, S. Enzymatic synthesis ofpolyesters via ring-opening polymerization. Adv. Polym. Sci. 2006, 194, 95–132.
7. Sugihara, S.; Toshima, K.; Matsumura, S. New strategy for enzymatic synthesis of high-molecular-weight poly(butylene succinate) via cyclic oligomers. Macromol. Rapid Commun. 2006, 27, 203–207.
8. Uyama, H.; Takeya, K.; Hoshi, N.; Kobayashi, S. Lipase-catalyzed ring-opening polymerization of 12-dodecanolide.Macromolecules 1995, 28, 7046–7050.
9. Bisht, K.S.; Henderson, L.A.; Gross, R.A. Enzyme-catalyzed ring-opening polymerization ofω-pentadecalactone.Macromolecules 1997, 30, 2705–2711.
10. Kondo, A.; Sugihara, S.; Kuwahara, M.; Toshima, K.; Matsumura, S. Lipase-catalyzed ring-opening polymerizationof molecularly pure cyclic oligomers for usein synthesis and chemical recycling of aliphatic polyesters. Macromol. Biosci.2008, 8, 533–539.
11. Tsuneizumi, Y.; Kuwahara, M.; Okamoto, K.; Matsumura, S. Chemical recycling of poly(lactic acid)-based polymer blendsusing environmentally benign catalysts. Polym. Degrad. Stab. 2010, 95, 1387–1393.
12. Shimokawa, K.; Kato, M.; Matsumura, S. Enzymatic synthesis and chemical recyclingof polythiocaprolactone. Macromol. Chem. Phys. 2011, 212, 150–158.
13. Uyama, H.; Inada, K.; Kobayashi, S. Lipase-catalyzed synthesis of aliphatic polyesters by polycondensation of dicarboxylic acids and glycols in solvent-free system. Polym. J. 2000, 32, 440–443.
14. Berkane, C.; Mezoul, G.; Lalot, T.; Brigodiot, M.; Marechal, E. Lipase-catalyzed polyester synthesis in organic medium.study of ring-chain equilibrium. Macromolecules 1997, 30, 7729–7734.
15. Ba, C.; Yang, J.; Hao, Q.; Liu, X.; Cao, A. Syntheses and physical characterization of new aliphatic triblock poly(L-lactide-b-butylene succinate-b-L-lactide) s bearing soft and hard biodegradable building blocks. Biomacromolecules2003, 4, 1827–1834.
16. Wang, G.; Gao, B.; Ye, H.; Xu, J.; Guo, B. Synthesis and characterizations of branched poly(butylenesuccinate) copolymers with 1,2-octanediol segments. J. Appl. Polym. Sci. 2010, 117, 2538–2544.
17. Cao, A.; Okamura, T.; Ishiguro, C.; Nakayama, K.; Inoue, Y.; Masuda, T. Studies on syntheses and physical characterization of biodegradable aliphatic poly(butylene succinate-co-ε-caprolactone)s. Polymer 2002, 43, 671–679.

18. Huang, X.; Li, C.; Zheng, L.; Zhang, D.; Guan, G.; Xiao, Y. Synthesis, characterization and properties of biodegradable poly(butylene succinate)-block-poly(propylene glycol) segmented copolyesters. Polym. Int. 2009, 58, 893–899.
19. Kojio, K.; Furukawa, M.; Motokucho, S.; Shimada, M.; Sakai, M. Structure-mechanical property relationships for poly(carbonate urethane) elastomers with novel soft segments. Macromolecules 2009, 42, 8322–8327.
20. Ihn, K.J.; Yoo, E.S.; Im, S.S. Structure and morphology of poly(tetramethylene succinate) crystals. Macromolecules 1995,28, 2460–2464.
21. Caminiti, R.; Pandolfi, L.; Ballirano, P. Structure of polyethylene from X-ray powder diffraction: Influence of the amorphous fraction on data analysis. J. Macromol. Sci. B 2000, 39, 481–492.
22. Kobayashi, T.; Matsumura, S. Enzymatic synthesis and properties of novel biodegradable and biobased thermoplastic elastomers. Polym. Degrad. Stab. 2011, 96, 2071–2079.
23. Kelterborn, J.C.; Soong, D.S. Time-dependent morphologies and viscoelastic properties of block copolymers. Polym. Eng. Sci. 1982, 22, 654–672.
24. Deimede, V.; Andrikopoulos, K.S.; Voyiatzis, G.A.; Konstandakopoulou, F.; Kallitsis, J.K. Molecular orientation of blue luminescent rigid-flexible polymers. Macromolecules 1999, 32, 8848–8856.

CHAPTER 11

IMPROVED BIODEGRADABLE RADIATION CURED POLYMERIC FILM PREPARED FROM CHITOSAN-GELATIN BLEND

Zinia Nasreen,[1] Mubarak A. Khan,[2] and A. I. Mustafa[3]

[1]Department of Chemistry, Dhaka University of Engineering and Technology, Gazipur 1700, Bangladesh

[2]Radiation and Polymer Chemistry Laboratory, Institute of Nuclear Science and Technology, Bangladesh Atomic Energy Commission, P.O. Box 3787, Dhaka 1000, Bangladesh

[3]Department of Applied Chemistry and Chemical Engineering, University of Dhaka, Dhaka 1000, Bangladesh

ABSTRACT

The mechanical, thermal, swelling, and release properties of chitosan-gelatin (CG) films have been investigated in order to verify the influence of UV and gamma radiation on the stability of the films. Thin films of chitosan and gelatin (1 : 3, w/w) that were radiated with 100 krad of gamma dose showed the best performance and the TS values reached 25, 45, and 49 MPa, respectively, for chitosan, gelatin, and blend. The corresponding highest TS values were 23, 42, and 45 MPa, respectively, for 10 passes of UV radiation. The effect of radiation over gelatin, chitosan, and CG blend caused modification in the arrangement of molecules in the crystal lattice that is significant by XRD analysis. Surfaces of the films were also investigated by scanning electron microscope (SEM). Fourier transform infrared spectroscopy (FTIR) studies further revealed structural changes of the films. These changes were attributed to understanding the behavior of the irradiated chitosan, gelatin, and CG blend on application of thermal energy using DSC and TGA studies, water uptake of the films in aqueous medium, and soil degradation properties to observe the best possibility for its application.

INTRODUCTION

Biopolymers attract a significant attention to the scientists because they are abundantly found worldwide. Most of the biopolymers can be used as biomaterial. Chitosan and gelatin are natural biopolymers and have the film forming properties [1–4]. Extensive research works are going on chitosan and also its property modifications for suitable applications. It is now matured that chitosan can suitably be applied in biomedical, pharmaceutical, and agricultural applications [5–7]. Chitosan is 2-amino-2-deoxy-β-D-glucose unit. The advantage of chitosan is its biodegradability and its antibacterial activity. It is reported that chitosan grafted copolymer nanoparticles were obtained by free radical polymerization with different comonomers; these new polymers allow encapsulation of proteins and sustained drug release [8–10].

Gelatin is a hydrocolloid natural biopolymer. It dissolves in hot distilled water and forms jelly when cooled. Gelatin is obtained by partial hydrolysis of collagen derived from natural sources such as skin, connective tissue, and bones of animals involving denaturation and physical and chemical degradation. Gelatin is a protein derivative obtained by boiling

Citation: Zinia Nasreen, Mubarak A. Khan, and A. I. Mustafa, "Improved Biodegradable Radiation Cured Polymeric Film Prepared from Chitosan-Gelatin Blend," Journal of Applied Chemistry, vol. 2016, Article ID 5373670, 11 pages, 2016. doi:10.1155/2016/5373670

collagen. Commercial gelatin is a vitreous solid having a faint yellow color. The uses of gelatin depend on its characteristic properties. Its proteinaceous nature makes it used in food, abrasive, gummed tape, and adhesives and also as a protective colloid for the prevention of coalescence [11–15]. On the other hand, gelatin is widely utilized in the manufacture of various articles and materials as it is a polymeric material. The number and the variety of reactive groups along the chain enhance the possibilities of modifying gelatin using various potential reagents. Addition of monomers to gelatin was found to be similar to vinyl polymerization to give an extended carbon chain attached to the side group of the gelatin molecules. Blends of biopolymer with common polymers are of great significance for expanding application to biochemical and biodegradable materials [10, 16–18]. Gelatin is biodegradable and has good biocompatibility, which makes it used as tissue replacement material. Chitosan was graft copolymerized to enhance the property and also combined with gelatin to be used as wound dressing materials. Permanent cross-linking of gelatin can be achieved by reaction with aldehydes such as formaldehyde, glyoxal, or glutaraldehyde [19–21]. Chitosan-gelatin surgical wound dressing is a kind of excellent wound healing material. In order to testify the safe reliability, antibacterial property, and wound healing ability, a lot of experiments were executed. The results showed its safe reliability and good antibacterial property. The film casting solution including pectin and gelatin was irradiated at 0, 10, 20, and 30 KGY to investigate the irradiation effect on the mechanical properties of the film and tensile strength of the 10 KGY irradiated film was the highest among the treatments.

Aqueous solutions of carboxymethylated chitosan (CM chitosan) were radiated with γ-ray in various conditions. The radiation chemical yields of CM chitosan degradation were found to decrease at lower pH in which the polymer chains tend to coiled conformation. FTIR and UV spectra showed that main chain structures of CM chitosan remained and some carbonyl/carbonyl groups were formed during the degradation.

Use of UV radiation source has some advantages like low energy consumption, ambient temperature operations, no emission of solvents, reduced reaction time, enhanced product quality, and so forth [22]. Photo-initiated graft polymerization is one of the fasted and most efficient methods to improve physicomechanical properties of natural polymers. UV radiation is weak type of nonionizing radiation source [23]. The application of gamma radiation is increasing. Over the past few decades there has been a continuous and significant growth in the development and application of radiation techniques, primarily in coating and adhesive industry. As this technique continues to develop innovative new products based on high efficiency and easy process control a logical extension for this technology is in the field of composite materials. The use of gamma radiation to natural biopolymers offers several advantages, such as continuous operation, minimum time requirement, less atmospheric pollution, curing at ambient temperature, and increased design flexibility through process control [9, 24]. The aim of the present investigation was to prepare and characterize thin films of chitosan, gelatin, and their blends. The effect of gamma and UV radiations on the mechanical and thermal properties of the prepared films was also found out.

EXPERIMENTAL

Materials

Pharmaceutical grade gelatin (bloom number 185) was collected from Global Capsules Ltd., Barisal, Bangladesh. Chitosan was extracted from Prawn (Penaeus monodon) shell [8]. Sodium hydroxide (NaOH), hydrochloric acid (HCl), and ethanoic acid (CH_3COOH) were supplied by Merck, Germany.

Preparation of Thin Film of Chitosan, Gelatin, and Blends

2% percent chitosan solution (w/w) was prepared in aqueous solution of acetic acid (2%). The films of chitosan (thickness was about 0.1 mm) were made by casting. The dried films were preserved in polyethylene bags and stored in desiccators for further analysis. Gelatin films were also prepared by casting from its aqueous solution using 2% gelatin solution. For the preparation of blend (chitosan and gelatin ratio was 1 : 3 w/w) gelatin was dissolved in hot water by continuous stirring with magnetic stirrer for about an hour; then aqueous chitosan solution was mixed and properly stirred [25]. The solution was cast onto the silicon cloth frames mounted on flat glass plate for film formation and dried in an oven at 70°C for 12 hours. The dried films (about 0.20 mm thickness) thus prepared were peeled off and cut into small pieces for further tests.

Use of Gamma Irradiation

Films of chitosan, gelatin, and blend were irradiated using a Co-60 Gamma source (25 kci) of the Bangladesh Atomic Energy Commission (BAEC), Saver, Dhaka. Dose rate was 350 krad/hour.

UV Curing of the Films

Samples were irradiated under a UV lamp (254–313 nm, model UV-200, 1ST-Technik, Germany) with 2 kW power capacity at various intensities expressed in number of passes. The conveyor speed of the UV manicure was 4 m/min. The samples were subjected to various characterization processes after 24 hours of radiation.

Mechanical Tests of the Films

Tensile strength (TS), tensile modulus (TM), and elongation at break (Eb%) of the film samples were determined using a universal testing machine (Houston S testing series, UK). The load range was 500 N with 10 mm/min crosshead speed and 20 mm gauge length. All the results were taken as the average values of 6 samples. Relative humidity was at least 65%.

Water Uptake of the Films

Film samples were immersed in the static water beaker at 25°C for different time periods (10–90 seconds). Before immersion in the water, the weight of the samples was measured. After certain period of time, samples were taken out from the beaker, wiped using tissue papers, and then weighed again. The water uptake (mass gained) was calculated by the subtraction of initial weight from the final weight.

Thermal Analysis

The thermal test of the films was taken using computer controlled TG/DTA 6300 system controlled to an EXSTAR 6000 STATION, Seiko Instrument, Inc., Japan. The TG/DTA module uses a horizontal system balance mechanism. All the experiments were performed in the nitrogen atmosphere. Samples weights were 10–15 mg, and heating rate was 10°C/min.

X-Ray Diffraction Studies

A PHILIPS PW3040 Pert PRO X-Ray Diffractometer was used. The films were exposed to CuK_α radiation with a primary beam of 40 kV and 30 mA with a sampling pitch of 0.02° and time for each step data collection was 0.1 sec. A 2θ scan was taken from 15° to 35° to get possible fundamental peaks where Ni filter was used to reduce CuK_β radiation. All the data of the samples were analyzed using computer software "X'PERT HIGHSCOPE." The integrated intensity of the blend films was investigated by using the following equation:

$$\text{Integrated Intensity } (I) = \text{Peak Height } (h) \times \text{FWHM}, \quad (1)$$

where FWHM is Full Width at Half Maximum.

Fourier Transform Infrared Spectroscopy (FTIR)

The FTIR spectroscopy of chitosan-gelatin blend films was performed by FTIR Spectrophotometer (Paragon 500 Model, PerkinElmer, Beaconsfield, Buckinghamshire, UK) in the wave number range 400–3500 cm^{-1} with resolution of 4 cm^{-1}. The FTIR spectrum was taken in a transmittance mode.

Scanning Electron Microscopic (SEM) Study

The morphological study of the chitosan-gelatin blend film was done using a JEOL 6400 SEM at an accelerating voltage of 10 kV. The SEM specimens were sputter-coated with gold.

RESULTS AND DISCUSSION

Mechanical Properties of the Films of Chitosan, Gelatin, and Blend

Thin films of chitosan, gelatin, and a blend of chitosan and gelatin (1 : 3, w/w) were prepared by casting. Tensile strength (TS), tensile modulus (TM), and percent elongation at break (Eb%) were evaluated and are given in Table 1. It was found that TS, TM, and Eb of the blend are higher than those of the chitosan or gelatin films. The TS of the blend film was found to be 43 MPa, that is, 139 and 10% higher over chitosan and gelatin films, respectively [26].

Table 1: Mechanical properties of the thin films of chitosan, gelatin, and blend

Film	Tensile strength (MPa)	Tensile modulus (MPa)	Elongation at break (%)
Chitosan	18 ± 0.5	445 ± 35	10 ± 1
Gelatin	39 ± 0.8	656 ± 44	6 ± 0.4
Blend	43 ± 1	782 ± 40	18 ± 1.5

Effect of Gamma Radiation on the Films of Chitosan, Gelatin, and Their Blend

Thin films of chitosan, gelatin, and blend were exposed to gamma radiation varying total doses at a rate of 350 krad/hour to find out the effect of gamma radiation on the films. Tensile strength (TS), tensile modulus (TM), and percent elongation at break (Eb%) of the films were evaluated. The results are presented in Figures 1–3. The results of TS values for the films of chitosan, gelatin, and blend against total gamma dose are shown in Figure 1. It was found that, at 25 krad of total dose, TS of chitosan, gelatin, and blend was decreased and then moved upward and the highest values of TS were found at 100 krad. After 100 krad, TS was found to decrease slowly and at 500 krad of total dose, TS decreased more than that of the original values. From this investigation it is clearly found that gamma radiation has a strong effect on chitosan, gelatin, and their blend. It can be concluded that, at 100 krad of exposure of gamma radiation, chitosan, gelatin, and blend gained 39, 15, and 14% increment of TS over their nonirradiated counterparts (Figure 15).

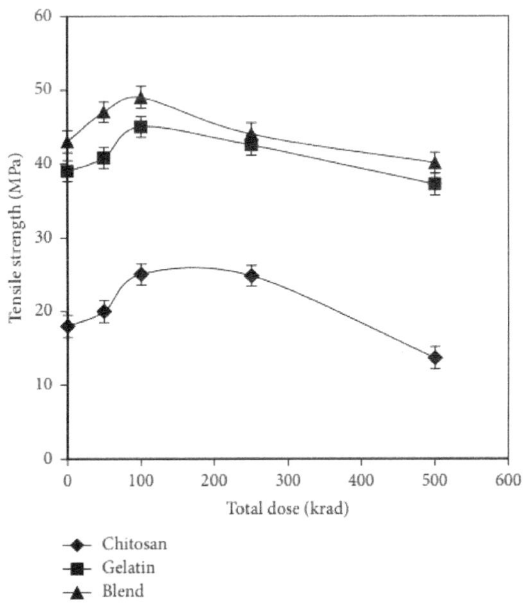

Figure 1: Tensile strength against gamma radiation dose (krad) for chitosan, gelatin, and blend.

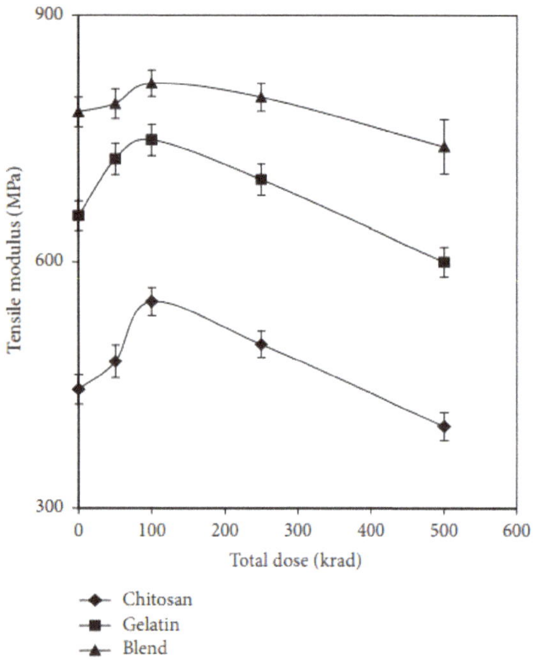

Figure 2: Tensile modulus against gamma radiation dose (krad) for chitosan, gelatin, and blend.

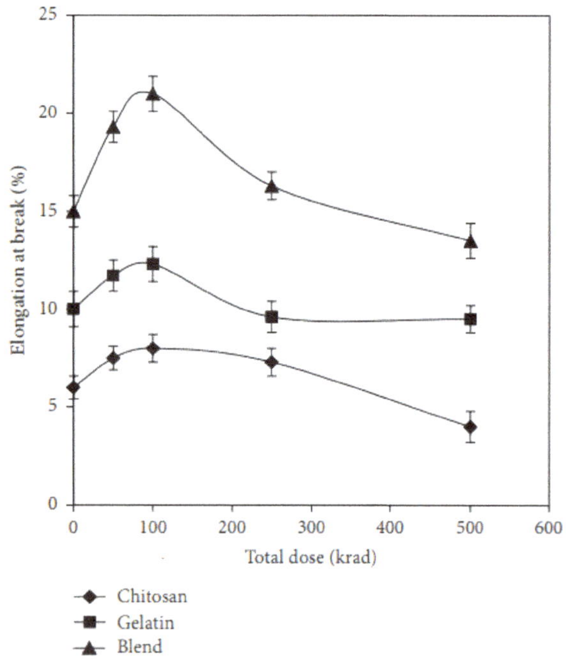

Figure 3: Elongation at break against gamma radiation dose for chitosan, gelatin, and blend.

In Figure 2, tensile modulus (TM) is plotted against gamma radiation dose of the films of chitosan, gelatin, and their blend. Almost similar results were found like TS of the films of chitosan, gelatin, and blend that is explained above. At low dose (25 krad), TM was decreased and after that TM increased and we found the highest values at 100 krad. High (500 krad) gamma dose caused low TM values of all the films. Figure 3explains the elongation at break against gamma radiation dose for the films of chitosan, gelatin, and blend. This result indicated that at 100 krad of total dose onto the films Eb (%) values decreased and then moved upward. These findings supported the results for TS and TM of the films of chitosan, gelatin, and blend. From the above discussion, it is quite clear that, at 100 krad of total dose, TS and TM increased but Eb decreased because, at this dose, films became strong and elongation became low, as expected. Actually gamma radiation can play a significant role for the modification of polymers. Gamma radiation is ionizing radiation and has strong effect on natural polymers [27]. Use of gamma radiation is attributed to higher mechanical properties; gamma radiation produces three types of reactive species in polymer (chitosan, gelatin) (Schemes 1 and 2). These are ionic, radical, and peroxide species. The peroxide species are produced because the polymers were irradiated in presence of oxygen. When chitosan or gelatin or blend is subjected to gamma radiation, some radicals are produced. Gamma radiation also ruptures some carbon-carbon bonds and produces radicals. Chain scission may also take place to form other radicals. Cross-linking and chain scission occurred when polymers are exposed to gamma radiation. Polysaccharides and other natural polymers generally degrade by breaking of glycosidic linkage under gamma radiation. Gelatin is natural biopolymer that consists of protein molecules which can easily fragment due to gamma radiation. The nature of gelatin chain scission is like where some of the free radicals generated. The generated free radicals may cross-link each other and form more cross-linked gelatin molecules with the exposure of gamma ray; as a result tensile properties might be increased. It is reported in the literature that chitosan molecules also form free radicals in a similar nature when irradiated to gamma radiation.

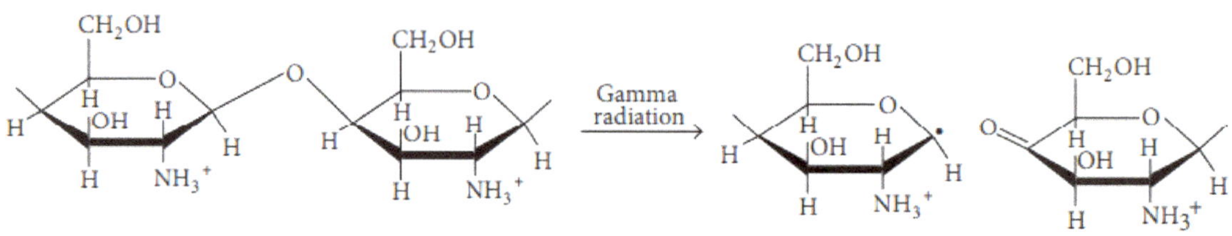

Scheme 1: Degradation mechanism of chitosan under gamma source.

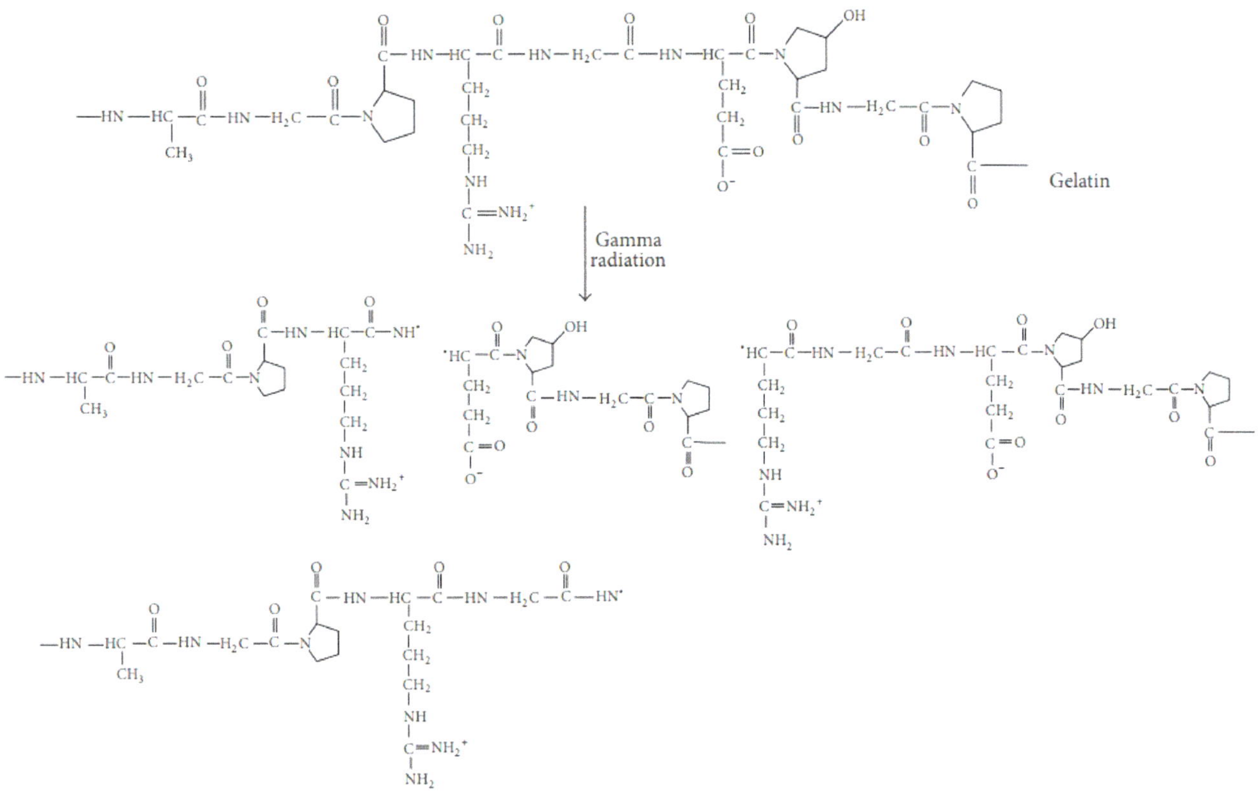

Scheme 2: Degradation mechanism of gelatin under gamma source.

Effect of UV Radiation on the Films of Chitosan, Gelatin, and Their Blend

To investigate the effect of UV radiation on chitosan and gelatin, thus films were exposed to the UV source. The films of chitosan, gelatin, and their blend were treated with UV radiation varying intensities. The UV radiation source was conveyor type and the intensity is termed as the number of UV passes. One pass of UV indicated that the sample was exposed to UV under 60 cm length of conveyor belt. The details of the UV source are mentioned in the experimental section. Tensile strength (TS), tensile modulus (TM), and percent elongation at break (Eb%) of the films of chitosan, gelatin, and blend were evaluated by universal testing machine. The results are presented in Figures 4–6. The results of TS values for the films of chitosan, gelatin, and blend against UV radiation intensities (termed as the number of passes) are shown in Figure 4. It was found that, at 5 passes of UV, TS of the films of chitosan, gelatin, and blend slightly decreased and then moved upward; the highest values of TS were found at 10 passes of UV. It was also noticed that TS was again decreased slowly at 25 passes of UV dose. At 25 passes, TS was found to decrease more than that of the original values. From this investigation it is clearly found that UV radiation has a weak effect on chitosan, gelatin, and their blend. It can be concluded that, at 10 passes of UV radiation, chitosan, gelatin, and blend gained 28, 7.6, and 4.6% increment of TS over their nonirradiated counterparts. The effect of UV radiation on the TM of the films of chitosan, gelatin, and blend was studied and the results are depicted in Figure 5. It was found that, at 5 passes of UV, TM of the films decreased slightly but after that dose, TM gradually increased and the highest TM was observed at 10 passes of UV. At higher dose of UV, TM decreased again. Elongation at break (Eb%) of the chitosan, gelatin, and blend versus number of UV passes is presented in Figure 6. This figure represents that, at 5 passes of UV, Eb% increased and after that dose it decreased and the nature is quite similar to that mentioned in Figure 3. TS and TM increase with increasing irradiation doses up to a certain limit but after that, they decrease due to opposite phenomenon such as photo cross-linking and photodegradation that were taking place simultaneously under UV radiation. In lower dose, free radicals are stabilized by combination reactions; as a result, photo cross-linking or intercross-linking occurs between the neighboring polymeric molecules. The higher is the number of active sites generated on the polymeric substrate, the greater is the cross-linking efficiency. At higher UV radiation (above 10 passes), TS and TM values decrease, which may be associated with the photodegradation of polymer backbone at higher UV does. At higher radiation, main chain may be broken and polymer may degrade into fragments. As a result, tensile properties decreased with increasing higher UV doses. During photodegradation, there will be loss in strength due to primary bond breakage in the polymer constituents [28, 29].

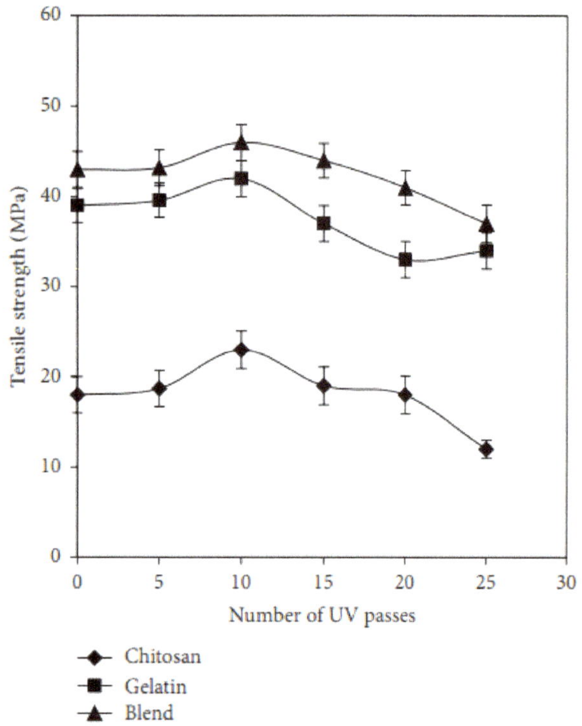

Figure 4: Tensile strength versus number of UV passes for chitosan, gelatin, and blend.

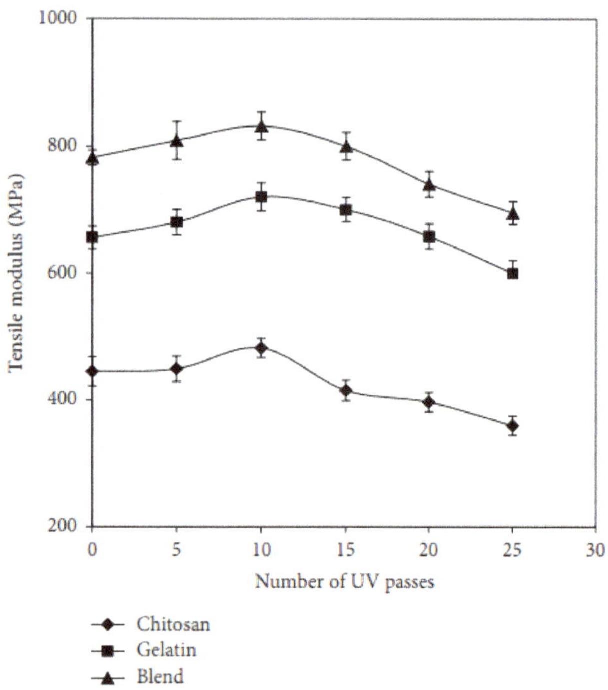

Figure 5: Tensile modulus versus number of UV passes for chitosan, gelatin, and blend.

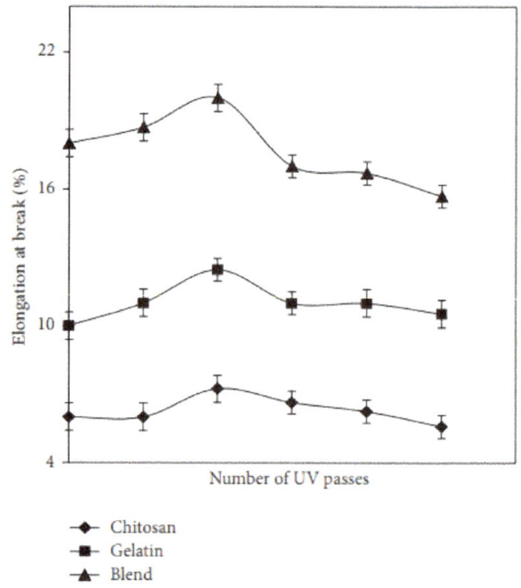

Figure 6: Elongation at break versus number of UV passes for gelatin, chitosan, and blends.

Water Uptake of the Films

Gamma treated (100 krad) films of chitosan, gelatin, and blend were immersed in the static water bath for 90 sec at room temperature (25°C). It is to be noted here that this dose of gamma radiation performed the best mechanical properties over the other gamma doses or UV treated films. Percentages of water uptake of the treated films of chitosan, gelatin, and blend are plotted in Figure 7. It was observed that all samples absorbed a significant amount of water up to 10 sec, then continue to absorb, and became almost to plateau with time. For 90 sec duration, films of chitosan, gelatin, and blend absorbed 22, 24, and 28% of water, respectively. It is clearly mentioned that blend absorbed the highest amount of water over the chitosan or gelatin films. On the other hand, chitosan absorbed the lowest amount of water during aqueous immersion of the films.

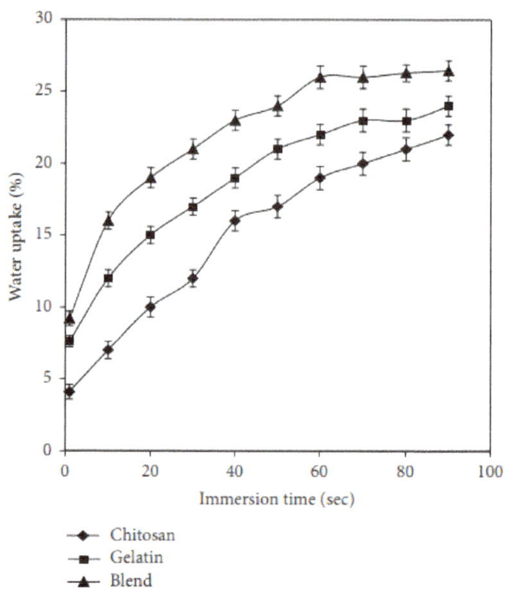

Figure 7: Water uptake of the films of gamma treated (100 krad) chitosan, gelatin, and blend.

Scanning Electron Microscopic (SEM) Analysis of the Films

Surfaces of the prepared films of chitosan, gelatin, and blend were studied by SEM and the images are given in Figure 8. From the investigation it was found that some rough surfaces appeared for chitosan which may be attributed from microbubbles formed during casting. The gelatin surface is quite good compared to that of the chitosan surface. In the blend surface, few pits are clearly evidenced and this is generated from bubbles of gelatin during casting. Gelatin has a strong tendency to form gels with a lot of bubbles and during experiment it was tried to remove all the bubbles. But some microbubbles still remained and it was not possible to remove all of the microbubbles. From the SEM of blend, no boundary line is found between gelatin and chitosan and it is indicated that blend between chitosan and gelatin mixed successfully.

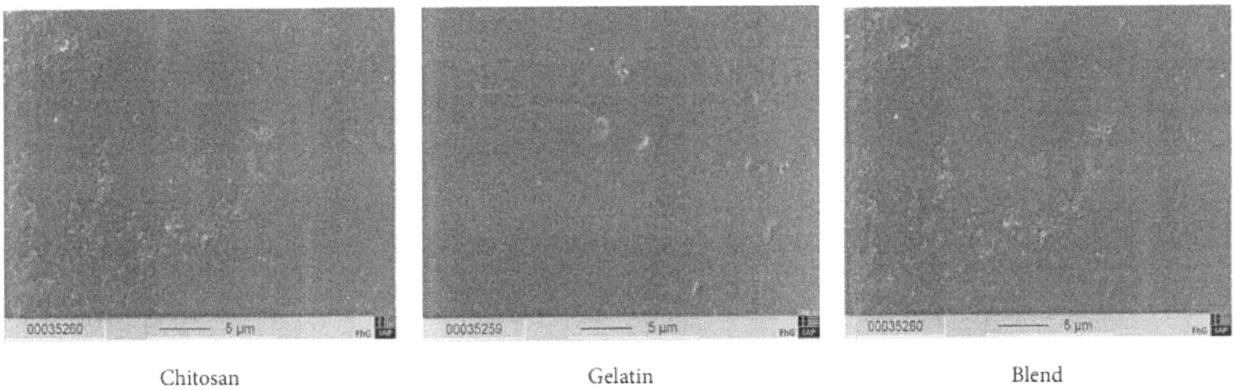

Figure 8: SEM images of the surface of the films of chitosan, gelatin, and blend.

Differential Scanning Calorimetric (DSC) Analysis of the Films

The glass transition temperature, melting temperature, and melting enthalpy of the chitosan, gelatin, and their blend were measured by using DTA technique. Figure 9(a) shows the first heating scan programmed from 0 to 600° C at a heating rate of 10° C/min. The midpoint of the specific heat change associated with the glass transition was taken as the glass transition temperature (T_g) and the peak temperature of the melting endotherm was taken as the melting temperature (T_m) of the transition. The area underneath the endotherm divided by the total sample weight (ΔH/gelatin, j/g) indicates the percentage of structural order in the gelatin.

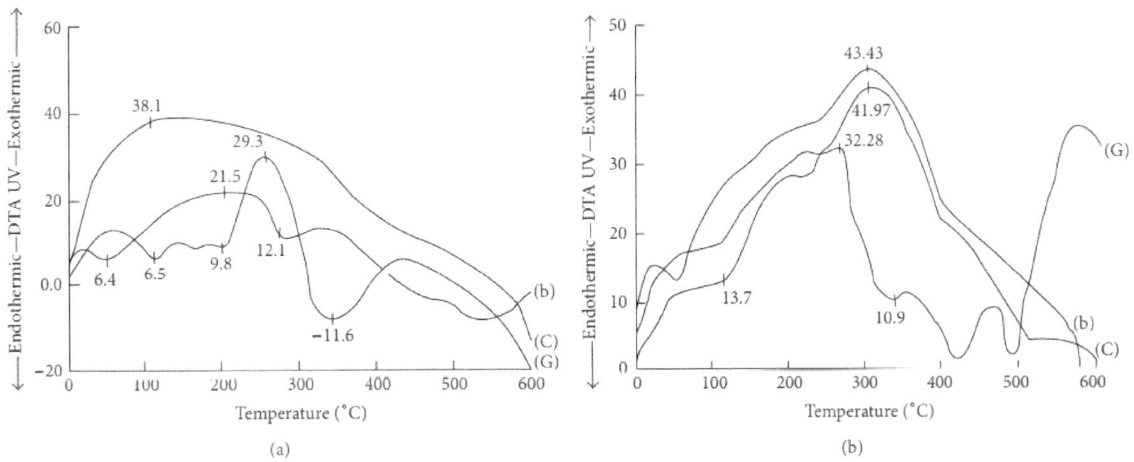

Figure 9: (a) DTA studies of the raw chitosan (C), gelatin (G), and their blend (b). (b) DTA studies of the irradiated chitosan (C), gelatin (G), and their blend (b).

From Figure 9(a), for raw gelatin (G), T_m value is 29.3, at a temperature range between 250° C and 300° C, for raw chitosan film (C), Tm value is 38.1, and for blend (b), T_m value is 21.5, where, from Figure 9(b), it is clearly shown that T_m value increased in every case, whereas, from Figure 10(b), it is clearly shown that Tm value increased in every case. Gamma irradiated gelatin shows endothermic peak at 250° C–300° C and Tm value is 32.28, 41.97 for irradiated chitosan film (C), and 43.43 for irradiated blend.

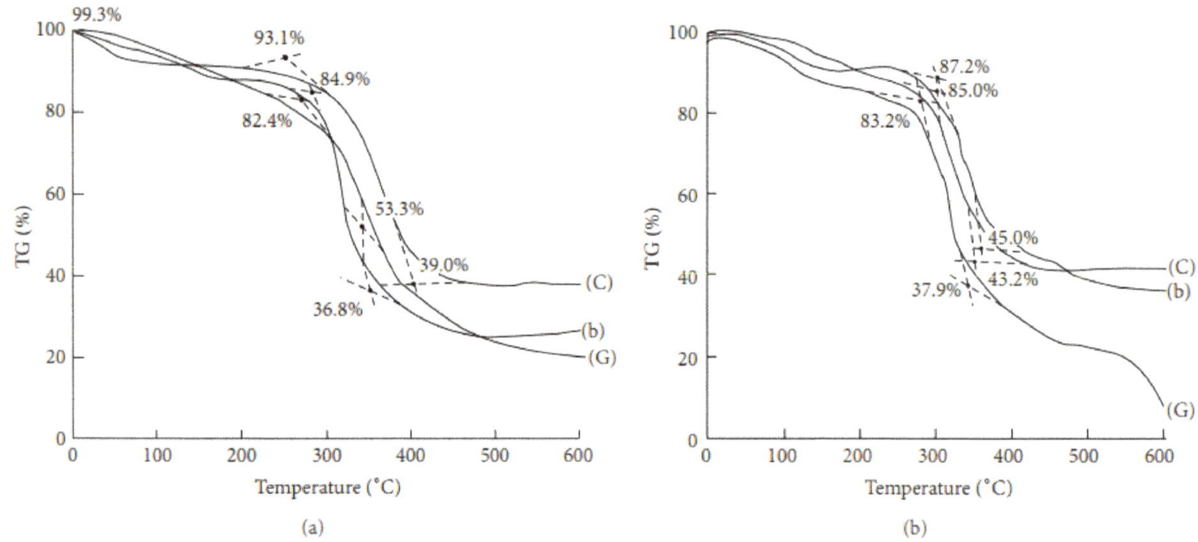

Figure 10: (a) TG% studies of the raw chitosan (C), gelatin (G), and their blend (b). (b) TG% studies of the irradiated chitosan (C), gelatin (G), and their blend (b).

Thermogravimetric Analysis (TGA Analysis)

Figures 10(a) and 10(b) show the TGA traces for raw samples and irradiated samples. From Figure 10(a), it is clearly observed that for all three samples, raw gelatin (G), raw chitosan (C), and raw chitosan-gelatin blend (b), the water content decreases continuously in two zones. The first weight loss approximately at 50°C–100°C was due to the loss of water; the second weight loss of water started at about 260°C, showing that different extent of thermal degradation of gelatin took place. From Figure 10(b), for irradiated gelatin, chitosan, and blend of chitosan-gelatin films, water content decreases in two zones. The first weight loss was at 70°C–130°C and second loss was at 270°C–350°C.

X-Ray Diffraction (XRD) Studies

Figure 11 showed the XRD spectra of raw and irradiated gelatin films. Raw gelatin (A) sample showed critical change in crystallinity in comparison with the irradiated gelatin (B) sample of the same ratios. However, it is clear that sharper peaks (increased crystal size) were obtained from the gamma irradiated gelatin sample at 10° (2θ) and 30° (2θ). Figure 12 showed XRD spectra of chitosan with more than one crystalline peak, at 10° (2θ) and 30° (2θ). In the case of gamma irradiated chitosan, there was significant decrease in the intensity of characteristic peaks of chitosan [30]. The distinct differences in the diffraction patterns of raw chitosan (A) and irradiated chitosan (B) could be attributed to modification in the arrangement of molecules in the crystal lattice. In irradiated chitosan more suppressed peaks from 10° (2θ) and 30° (2θ) were observed, which might be due to amorphization.

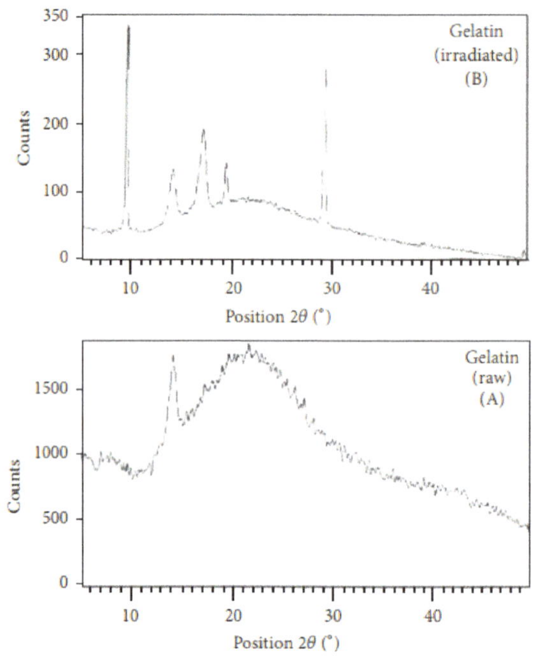

Figure 11: XRD studies of the raw (A) and irradiated (B) gelatin (G).

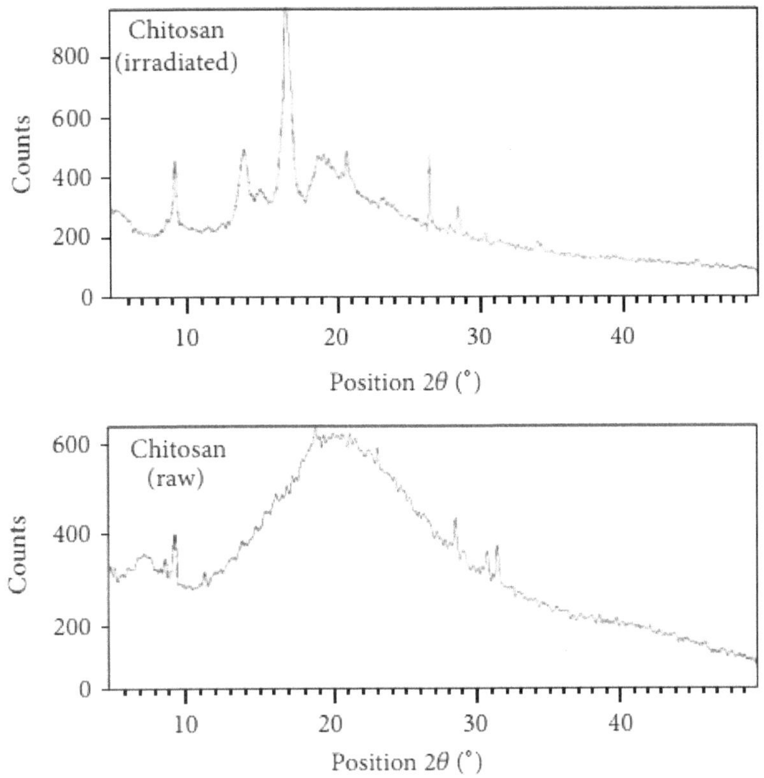

Figure 12: XRD studies of the raw (A) and irradiated (B) chitosan (C).

Figure 13 showed the XRD spectra of raw and irradiated blends of chitosan and gelatin film. In case of raw blend film (A), it showed two prominent crystalline peaks at 10° (2θ) and 20° (2θ). But in case of irradiated blend film (B), there was a significant decrease in the intensity of characteristic peaks at 10° (2θ) and 20° (2θ). But a single peak was found at 18° (2θ) and then suppressed peaks were observed, which might be due to amorphization.

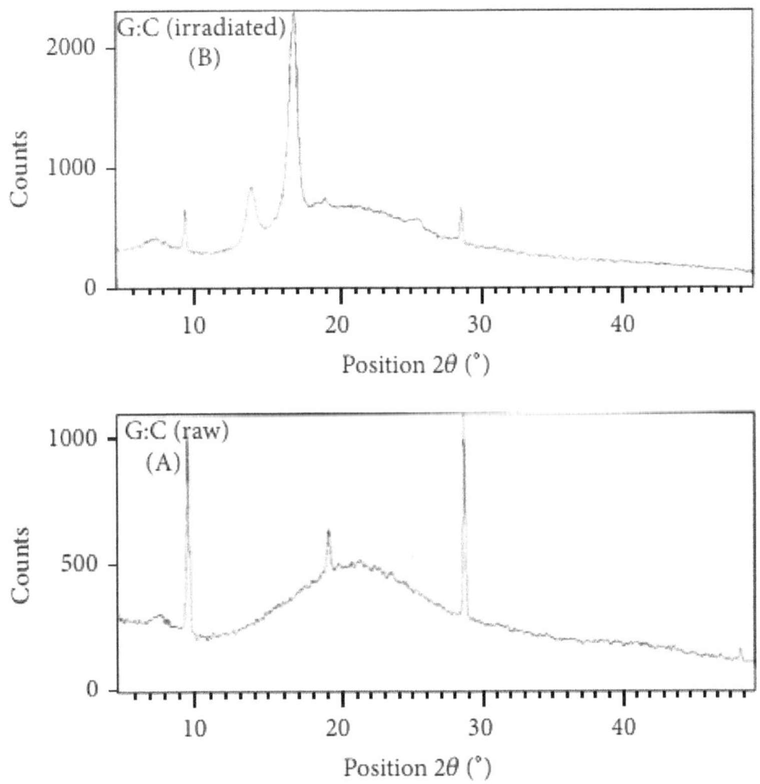

Figure 13: XRD studies of the raw (A) and irradiated (B) chitosan-gelatin blend (b).

Fourier Transform Infrared (FTIR) Study

FTIR is of importance in the study of the molecular structure. The width and intensity of the spectrum bands, as well as the position of the peaks, are all sensitive to environment changes and to the conformation to the macromolecules on the molecular level. Intermolecular interaction occurs when different polymers are compatible, so that the FTIR spectrum of the blend is different from those of pure polymers, which is advantageous to the study extent of polymer compatibility. The FTIR spectrum of gamma irradiated films of chitosan-gelatin blend is shown in Figure 14. A characteristic band at 4383.9 cm^{-1} is attributed to $-NH_2$ and $-OH$ groups stretching vibration and the band for amine group at 2669.3 cm^{-1} is seen in the infrared spectrum of chitosan-gelatin blend.

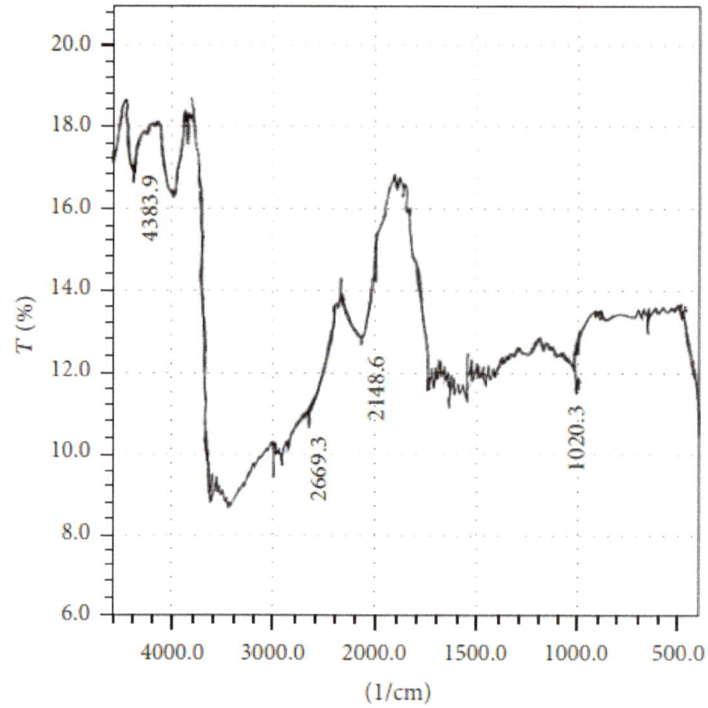

Figure 14: FTIR studies of the irradiated chitosan-gelatin blend (b).

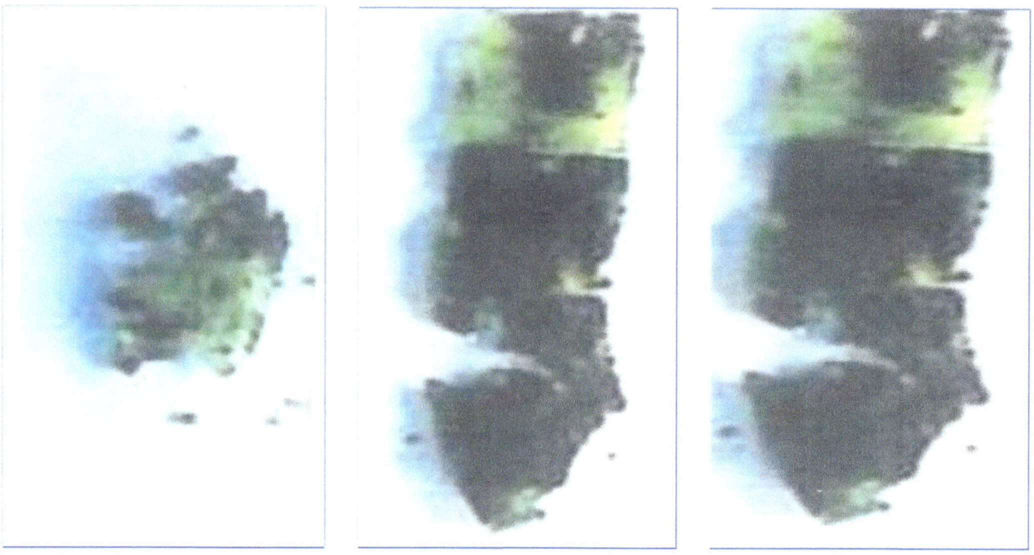

(a) (b) (c)

Figure 15: Biodegradability tests (after 24 hours in soil) of the nonirradiated gelatin (a), gamma treated (100 krad) gelatin (b), and UV treated (10 passes) gelatin films (c).

Another major absorption band observed between 1220 cm^{-1} and 1020 cm^{-1} represents the free amino group (NH_3^+) at C2 of glucosamine, a major group present in chitosan [31].

CONCLUSIONS

Films of chitosan, gelatin, and their blend (chitosan and gelatin ratio was 1 : 3, w/w) were prepared by casting and mechanical properties were evaluated. It was investigated to find out the effect of UV and gamma radiation on the mechanical properties of chitosan, gelatin, and their blend. From this investigation, it can be concluded that both UV and gamma radiations are effective in improving the mechanical properties of chitosan, gelatin, and their blend. Comparing both types of radiation, gamma radiation found more potential than that of UV radiation. TS of gamma treated blend is 49 Krad and Eb% is 21%. The thermomechanical properties have been drastically improved for gamma treated (100 Krad) films. Water uptake values became almost static after 70 sec of immersing in aqueous media. Scanning electron microscopic analysis revealed that the blend was mixed homogenously. DSC and TGA studies were performed to understand the behavior of both of raw and irradiated films on application of thermal energy. Characterization of irradiated chitosan film, gelatin film, and blend of chitosan-gelatin film by XRD provided the evidence of reduction in hydrophilicity. FTIR of chitosan-gelatin blend is also studied. X-ray diffractograms of irradiated blend films showed the highest integrated intensity. The experimental results showed that the blend films exhibit higher thermal stability and improved mechanical properties in dry state, which suggests the occurrence of interaction detected by XRD. The experimental results showed that the blend films (100 Krad) exhibit higher thermal stability and improved mechanical properties.

REFERENCES

1. F. Qian, F. Cui, J. Ding, C. Tang, and C. Yin, "Chitosan graft copolymer nanoparticles for oral protein drug delivery: preparation and characterization," Biomacromolecules, vol. 7, no. 10, pp. 2722–2727, 2006.
2. Y. Hong, Z. Mao, H. Wang, C. Gao, and J. Shen, "Covalently crosslinked chitosan hydrogel formed at neutral pH and body temperature," Journal of Biomedical Materials Research Part A, vol. 79, no. 4, pp. 913–922, 2006.
3. R. A. A. Muzzarelli, Chitin, Pergamon Press, New York, NY, USA, 1977.
4. G. A. F. Roberts, Chitin Chemistry, MacMillan Press, Houndmills, UK, 1992.
5. M. Sugimoto, M. Morimoto, H. Sashiwa, H. Saimoto, and Y. Shigemasa, "Preparation and characterization of water-soluble chitin and chitosan derivatives," Carbohydrate Polymers, vol. 36, no. 1, pp. 49–59, 1998.
6. L. Zhao, H. Mitomo, N. Nagasawa, F. Yoshii, and K. Tmikazu, "Radiation induced degradation of carboxymethylated chitosan on aqueous solution," Carbohydrate Polymers, vol. 51, pp. 169–175, 2003.
7. R. Jayakumar, M. Prabaharan, R. L. Reis, and J. F. Mano, "Graft copolymerized chitosan—present status and applications," Carbohydrate Polymers, vol. 62, no. 2, pp. 142–158, 2005.
8. M. A. Khan, S. Ferdous, and A. I. Mustafa, "Improvement of physico-mechanical properties of chitosan films by photocuring with acrylic monomers," Journal of Polymers and the Environment, vol. 13, no. 2, pp. 193–201, 2005.
9. S. Ferdous, A. I. Mustafa, and M. A. Khan, "Study on mechanical properties of photocured films of chitosan/PVA and PEO/PVA blend with acrylic monomers," Journal of Macromolecular Science Part A: Pure and Applied Chemistry, vol. 40, no. 8, pp. 817–832, 2003.
10. L. Huang, M. Zhai, J. Peng, J. Li, and G. Wei, "Radiation-induced degradation of carboxymethylated chitosan in aqueous solution," Carbohydrate Polymers, vol. 67, no. 3, pp. 305–312, 2007.
11. I. Kaur, R. Barsola, A. Gupta, and B. N. Misra, "Graft copolymerization of acrylonitrile and methacrylonitrile onto gelatin by mutual irradiation method," Journal of Applied Polymer Science, vol. 54, no. 8, pp. 1131–1139, 1994.
12. M. Sivakumar, P. R. G. Radhakrishnan, and H. Kothandaraman, "Grafting of glycidyl methacrylate onto gelatin," Journal of Applied Polymer Science, vol. 43, no. 10, pp. 1789–1794, 1991.
13. S. Matsuda, H. Iwata, N. Se, and Y. Ikada, "Bioadhesion of gelatin films crosslinked with glutaraldehyde,"Journal of Biomedical Materials Research, vol. 45, no. 1, pp. 20–27, 1999.
14. I. Arvanitoyannis, A. Nakayama, and S.-I. Aiba, "Edible films made from hydroxypropyl starch and gelatin and plasticized by polyols and water," Carbohydrate Polymers, vol. 36, no. 2-3, pp. 105–119, 1998.
15. T. Scherzer and R. Schubert, "Oxygen permeability of electron beam cured gelatin methacrylate layers,"Polymers for Advanced Technologies, vol. 9, no. 10-11, pp. 777–785, 1998.
16. C.-M. Deng, L.-Z. He, M. Zhao, D. Yang, and Y. Liu, "Biological properties of the chitosan-gelatin sponge wound dressing," Carbohydrate Polymers, vol. 69, no. 3, pp. 583–589, 2007.
17. C. Jo, H. Kang, N. Y. Lee, J. H. Kwon, and M. W. Byun, "Pectin- and gelatin-based film: effect of gamma irradiation on the mechanical properties and biodegradation," Radiation Physics and Chemistry, vol. 72, no. 6, pp. 745–750, 2005.
18. T. Kojima, M. Bessho, M. Furuta, S. Okuda, and M. Hara, "Characterization of biopolymer hydrogels produced by γ-ray irradiation," Radiation Physics and Chemistry, vol. 71, no. 1-2, pp. 233–236, 2004.
19. C.-Y. Lii, P. Tomasik, H. Zaleska, S.-C. Liaw, and V. M.-F. Lai, "Carboxymethyl cellulose-gelatin complexes," Carbohydrate Polymers, vol. 50, no. 1, pp. 19–26, 2002.
20. M. Tamada, N. Seko, and F. Yoshii, "Application of radiation-graft material for metal adsorbent and crosslinked natural polymer for healthcare product," Radiation Physics and Chemistry, vol. 71, no. 1-2, pp. 221–225, 2004.
21. S. Kapoor and C. Gopinathan, "Reduction and aggregation of silver, copper and Cadmium ions in aqueous solutions of gelatin and carboxymethyl

cellulose," Radiation Physics and Chemistry, vol. 53, no. 2, pp. 165–170, 1998.

22. J. P. Fouassier, Photoinitiation, Photopolymerization, and Photocuring: Fundamentals and Applications, Hanser Gardner Publications, Cincinnati, Germany, 1995.

23. A. Lagos and J. Reyes, "Grafting onto chitosan. I. Graft copolymerization of methyl methacrylate onto chitosan with Fenton's reagent (Fe^{2+}–H_2O_2) as a redox initiator," Journal of Polymer Science Part A: Polymer Chemistry, vol. 26, no. 4, pp. 985–991, 1988.

24. B. Jansen and G. Ellinghorst, "Radiation-induced modification of polyurethane-elastomers with hydroxyethyl methacrylate," Journal of Polymer Science: Polymer Symposia, vol. 66, no. 1, pp. 465–473, 1979.

25. B. G. Sparks and D. G. Murry, United States Patent 4,572, 906, 1986.

26. P. Haque, A. I. Mustafa, and M. A. Khan, "Effect of crosslinking monomers on the physico-mechanical and degradation properties of photografted chitosan film," Carbohydrate Polymers, vol. 68, no. 1, pp. 109–115, 2007.

27. P. Haque, A. I. Mustafa, and M. A. Khan, "Development and modification of ethylene glycol grafted chitosan films by photocuring," Nuclear Instruments and Methods in Physics Research Section B: Beam Interactions with Materials and Atoms, vol. 236, no. 1–4, pp. 314–317, 2005.

28. F. Khan and S. R. Ahmad, "Graft copolymerization reaction of water-emulsified methyl methacrylate with preirradiated jute fiber," Journal of Applied Polymer Science, vol. 65, no. 3, pp. 459–468, 1997.

29. M. M. I. Khalil, N. M. El-Sawy, and G. A. El-Shobaky, "Gamma-irradiation effects on the thermal and structural characteristics of modified, grafted polypropylene," Journal of Applied Polymer Science, vol. 102, no. 1, pp. 506–515, 2006.

30. Y. Wan, K. A. M. Creber, B. Peppley, and V. T. Bui, "Synthesis, characterization and ionic conductive properties of phosphorylated chitosan membranes," Macromolecular Chemistry and Physics, vol. 204, no. 5-6, pp. 850–858, 2003.

31. K. Singh, A. K. Tiwary, and V. Rana, "Ethylenediaminediacetic acid bis(carbido amide chitosan): synthesis and evaluation as solid carrier to fabricate nanoemulsion," Carbohydrate Polymers, vol. 95, no. 1, pp. 303–314, 2013.

CHAPTER 12

INTRINSIC EMBEDDED SENSORS FOR POLYMERIC MECHATRONICS: FLEXURE AND FORCE SENSING

Leif P. Jentoft[1], Aaron M. Dollar, Christopher R. Wagner[3] and Robert D. Howe[1]

[1]Harvard School of Engineering and Applied Sciences, Cambridge, MA 02138, USA

[2]Department of Mechanical Engineering and Materials Science, Yale School of Engineering and Applied Science, New Haven, CT 06511, USA

[3]Formerly with the Harvard School of Engineering and Applied Sciences, Cambridge, MA 02138, USA

ABSTRACT

While polymeric fabrication processes, including recent advances in additive manufacturing, have revolutionized manufacturing, little work has been done on effective sensing elements compatible with and embedded within polymeric structures. In this paper, we describe the development and evaluation of two important sensing modalities for embedding in polymeric mechatronic and robotic mechanisms: multi-axis flexure joint angle sensing utilizing IR phototransistors, and a small (12 mm), three-axis force sensing via embedded silicon strain gages with similar performance characteristics as an equally sized metal element based sensor.

INTRODUCTION

In order to become commercially viable, the majority of robotic and mechatronic systems must eventually become compatible with inexpensive, mass-manufacturing processes. In addition to challenges associated with fabricating structures with intrinsic sensors, a number of properties of polymers, such as multi-axial compliance, thermal insulation, and viscoelasticity/creep, make developing useful compatible transducers challenging. However, aspects such as the kinematic configuration of the structure as well as the state of the loading condition must frequently be sensed, even in the presence of those challenging material properties, in order to allow for the desired performance of the robotic or mechatronic system. In this paper, we explore two types of polymeric transducer configurations designed to give satisfactory performance in spite of these properties: a multi-axis flexural joint sensor for compliant structures, and a three-axis load sensor designed to be miniaturizable and incorporated into the structure of small polymeric devices such as surgical instruments.

Fabrication processes such as multi-material molding and insert molding allow for some expansion of the types of systems that can be easily fabricated with modern processes, but have not yet produced fully-integrated sensorized commercial systems with intrinsic transducers. On the scale of small-batch fabrication of research hardware, a popular polymer-based processes is Shape Deposition Manufacturing (SDM) [1,2], which can allow for the fabrication of compliant mechanisms that are very difficult to fabricate with traditional techniques. Complex mechanisms with embedded components can be created as a single part, eliminating the need for fasteners, and

Citation: Jentoft, L.P.; Dollar, A.M.; Wagner, C.R.; Howe, R.D. Intrinsic Embedded Sensors for Polymeric Mechatronics: Flexure and Force Sensing. Sensors 2014, 14, 3861-3870, doi:10.3390/s140303861

reducing the likelihood of damage to fragile components by encasing them within the part structure. To this point, however, SDM structures have been almost purely passive or open-loop mechanisms, devoid of sensing and feedback control [2–5].

In order to expand the utility of polymeric robotics, sensors must be developed that both integrate with and exploit the characteristics of plastics and rubbers [6–9]. Integrating these sensors with the robot structure promises to greatly reduce fabrication costs and complexity, and to increase the robustness of the resulting robotic systems—a crucial feature for commercially-viable systems. In addition to durability and ease of manufacture, polymeric transducers are desirable for MRI-compatibility, and a number of sensor designs have been explored for this application [9–11]. Additionally, considerable progress has been made towards creating soft sensors that stretch like their biological counterparts [12,13].

In this paper, we describe the development and evaluation of two types of sensors that can be utilized to give good sensing performance despite challenging aspects of polymer material properties. The first of these is an infrared phototransistor-based sensor that provides three axes of deflection sensing for compliant flexure joints. Flexures are a frequently-used and simple way of achieving polymeric joints, but are inherently more complex than simple rotational joints and require special sensing to estimate their multi-degree-of-freedom (DOF) state. The second transducer presented is a three-axis stain-gage force sensor. Force sensing is a useful and often necessary capability for a number of applications in which contact with external object must be controlled. However, traditional designs, while accurate and repeatable, are expensive, heavy, and fragile. By embedding strain gages into the structure of the mechanism, we create lightweight, durable, and inexpensive force transducers that are simple to fabricate.

PHOTOTRANSISTOR FLEXURE SENSING

Flexures are a simple way to add articulation to polymeric mechanisms. They are low-cost, easy to fabricate, frictionless (but not zero stiffness), and robust—even to off-axis loads. Furthermore, they are an easy way to incorporate compliance, and a single flexure can allow deflection around multiple axes, which provides advantages for grasping [3], proprioception, and to detect contact with the environment by the deflection of the joint [14]. Examples of flexure-based joints in robotic systems include the Sprawl series of legged robots [2], the SDM Hand [3], the UB Hand [15], and Compliant Framed Modular Robots [16], among others.

However, while flexures pose many advantages over traditional revolute joints, they are not compatible with standard approaches to measuring joint position, such as potentiometers or encoders. The pose of single-DOF flexure joints has been measured with piezoresistive bend sensors [15], hall-effect sensors [17], and optoelectronic sensors [18], the latter of which we utilize in this paper, but for a multi-DOF sensor. The only literature on measuring the pose of multi-DOF flexure joints (beyond our preliminary design presented in [19]) utilizes strain gauges distributed along the length of a long flexure joint [16] and achieves fairly good performance, albeit with significant mechanical and computational complexity to do so. It is therefore desirable to develop methods to measure the configuration of multi-DOF flexure joints with simpler methods and without needing to instrument the flexure itself.

Design

The sensor consists of a single infrared LED (VLMD3100-GS08, Vishay Semiconductor, Malvern, PA, USA) shining on to two pairs of phototransistors (four total—OP501DA, Optek Technology, Carrollton, TX, USA), as shown in Figure 1. The two phototransistors in each pair are mounted at different angles, so that as the finger bends, the LED moves from shining directly on one towards shining directly on the other. As the finger twists, the LED moves from one pair to the other, generating approximately 1 V response from each phototransistor (configured as a photodarlington) over a 220-ohm pull-up resistor To calibrate the design, a first-order polynomial approximation is used to map sensor readings to Euler-angle representation of orientation:

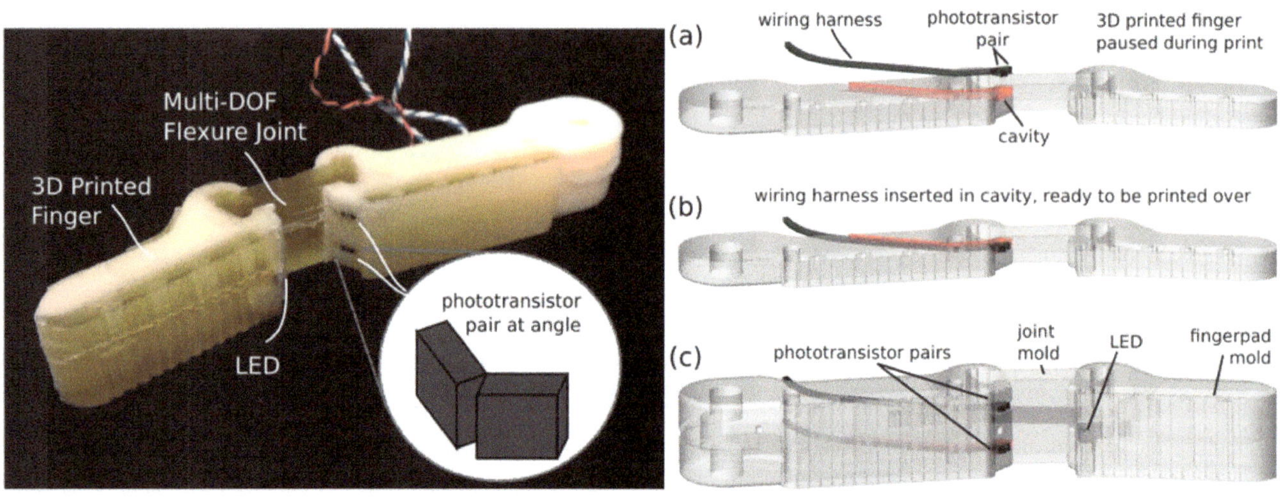

Figure 1: Joint-angle sensor design. An infrared LED shines across the joint onto two angled pairs of phototransistors (**left**). An embedded fused-deposition manufacturing method is used to integrate sensing into the finger design (**right**): (a) print is paused; (b) wired sensor is inserted into cavity designed to hold it; (c) whole assembly is printed over.

$$\begin{bmatrix} \theta_x \\ \theta_y \\ \theta_z \end{bmatrix} = \begin{bmatrix} a_{1,1} & \cdots & a_{1,5} \\ a_{2,1} & \cdots & a_{2,5} \\ a_{3,1} & \cdots & a_{3,5} \end{bmatrix} \begin{bmatrix} 1 \\ v_1 \\ v_2 \\ v_3 \\ v_4 \end{bmatrix}$$

To fabricate the sensor, a wiring harness is created with the phototransistors and LED. This is laid into the plastic finger which is printed by a fused-deposition manufacturing process (3D printer). After a cavity for the sensor is printed, the printer is paused and the harness is laid inside as shown in Figure 1. Printing then resumes, and as plastic is extruded over the sensor it fixes it in place. This process both provides a cavity to align the sensor, and removes the need for later assembly. The finger design includes cavities for flexure joints (16 mm × 6 mm × 17 mm) and finger pads, which are then filled with two-part urethane rubber (PMC 780, Smooth-On Inc., Easton, PA, USA—Shore-A durometer 80). The walls of the cavities are then peeled off, leaving the flexure joint as shown in Figure 1.

Methods and Results

To test the response of the finger, the orientation of the distal link is measured with an electromagnetic tracker (TrakSTAR, Ascension Technologies, Shelburne, VT, USA) at 50 Hz; voltages are measured at 10 bit resolution with an Arduino Micro (Arduino, Ivrea, Italy) at 50 Hz, and interpolated in MatLab. The finger is loaded from the tip using a string (simulating fingertip contact), and the results are plotted in Figure 2. The respective performance of the sensor for each degree of freedom is shown in Table 1. Note that the varying stiffness of the joint in different degrees of freedom results in differing magnitudes of deflection.

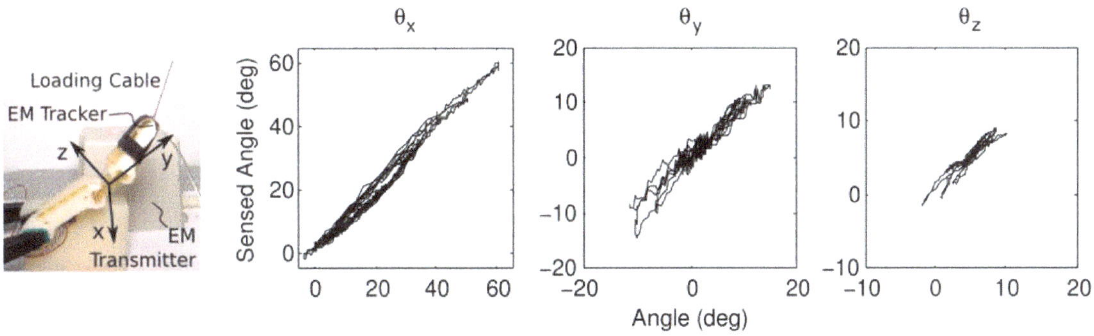

Figure 2: Sensor response: (**left**) experimental setup and (**right**) sensed angles across joint *vs.* actual angles measured with electromagnetic tracker.

Table 1: Joint sensor performance

Angle	*Range*	*Max Error*	*RMS Error*
θ_x	[−4, 61]	5.2°	1.7°
θ_y	[−11, 15]	5.0°	1.3°
θ_z	[−2, 10]	2.0°	0.60°

Discussion

The results show that the design is capable of measuring the deflection of multi-DOF flexure joints, and demonstrate a new method to integrate sensors into polymeric devices. While the errors are higher than seen in typical rotational encoders, there is no other published work measuring 3-DoF flexure deformations to the authors knowledge, and compliant fingers that adapt to the shape of object require less precise information regarding finger placement than do stiff pin-jointed fingers.

The primary source of error comes from the simple calibration between sensor values and flexure deformation (note that only first-order calibration terms are used to avoid over fitting). The flexure is able to deflect in all six degrees of freedom (translation and rotation), but only rotations are measured (the most significant deflection modes—flexion about *x*, *y*, and *z*in Figure 2). However, the flexure is significantly stiffer in translational degrees of freedom due to the joint and finger geometry (the flexure and distal link are roughly ten times longer than the flexure thickness) so these other modes play a less significant role in finger behavior.

Embedding the sensors during the printing process provides a number of advantages. The printed device itself serves as an alignment jig, enabling faster assembly and tighter tolerances. The printed material also provides protection for fragile wires (and, for some sensor types, the sensors themselves).

STRAIN-BASED FORCE SENSING

Multi-axis force sensing provides essential information about the interaction between the robot and the environment. Strain gages are often used in high quality force transducers due to their high sensitivity. However, the process of bonding strain gages to the transducer substrate is complicated and time-consuming, and often fails if any of the steps are not performed exactly to specification. Prior work by the authors has found that embedded strain gages can bond well to the structural material without the need for adhesives and give accurate measures of the structural strain in the part [6]. Integrating strain gages into polymeric structures via molding therefore may allow devices to be created with the sensitivity of strain gauges combined with the high durability and ease of construction of polymeric molding. In the following sections we present the design and evaluation of a miniature three-axis force transducer created using the SDM process, consisting of multi-stage molding steps.

Miniature Three-Axis Force Sensor

We integrated six strain gages to create a miniature three-axis force sensor (5 mm × 6.5 mm × 14.5 mm). Our primary motivation for targeting a design of this size was to meet the stringent design requirements of a force-sensing grasper jaw for use with minimally invasive surgical instruments [20,21], and as such, must fit through a small port (5 mm to 12 mm in diameter).

In the following section, we describe the design, construction, and evaluation of our current prototype, designed to form a pair of force-sensing gripper jaws that fit through a 12 mm port. The same technology can be readily applied to generate 5–10 mm pairs of sensors. This prototype is meant as an evaluation platform for an embedded strain-gage based force transducer on a small size scale, and does not incorporate the ideal jaw geometry for surgical grasping tasks.

Sensor Design and Construction

The force sensor is a dual-serial-beam configuration. One strain gage is located on each of the four sides of the proximal bending beam, sensing two bending moments (corresponding to X and Z forces in Figure 3). Two strain gages are located in the distal bending beam to sense the final axis (Y) of force (Figure 3). The sensor is integrated with an aluminum base allowing simple attachment to manipulators. Because of the SDM process, wires to the individual gages are integrated into the sensor core, removing the need for strain relief. A further advantage is that all wires leave the sensor at the same point providing straightforward and robust wire management.

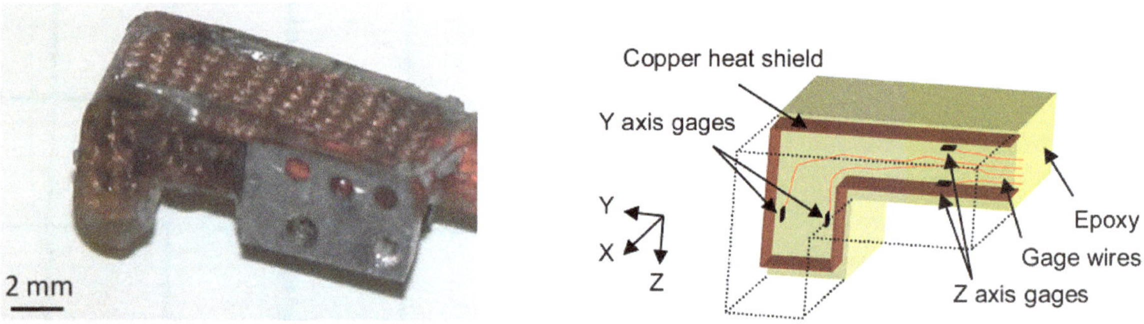

Figure 3: Photograph and cross-sectional diagram of three-axis force sensor. Note copper heat shield.

Silicon strain gages (1 mm × 0.25 mm, SS-037-022-500P, Micron Instruments, Simi Valley, CA, USA) were used to achieve high sensitivity in a small package. Individual strain gage resistances were sensed through a Wheatstone bridge, followed by an instrumentation amplifier circuit with a gain of 100.

The sensor construction is carried out in a two-pour casting process, each with a separate mold cavity. The first pour of liquid polymer embeds the strain gages and wires in an epoxy core (approximately half of the size of the final sensor) and attaches the core to the aluminum base. During this first stage, the gages are suspended in the mold via their lead wires, and positioned by hand to be close to their desired position and orientation. The second pour into a mold of the shape of the final sensor part embeds a copper braid (Ungar-Wick #4, Ungar Products, Apex, North Carolina, CA, USA) heat shield around the core. The epoxy used (Resin 105 Fast Cure, West System, Bay City, MI, USA) was chosen for its stiffness and low creep properties, low mixing viscosity (to fill all parts of the mold), and low curing temperature (to cure inside a wax mold).

A key component of our sensor design is the heat shield to equalize temperature variation between gages. Epoxy is an insulator, which negates the standard temperature differential rejection scheme of examining the difference in strains on opposite sides of a thermally-conducting bending beam. With the heat shield in place, the temperature differences are small and resistance differences between the two gages depends principally on strain.

Sensor Evaluation

To calibrate the SDM sensor, known loads were applied along each coordinate axis. The calibration matrix was found using a linear least squares method on the gage voltages and known loads. Temperature effects were minimized by allowing the temperature of the sensor to vary during calibration. Estimated forces *versus* known forces are shown in Figure 4 to demonstrate the calibration as well as the linearity (average R2 of 0.998). The force sensor has a range greater than 2 N with an RMS error in calibration of 0.15 N. This range is 40% of the desired full range for a wide range of surgical applications [22], but is appropriate for many tasks.

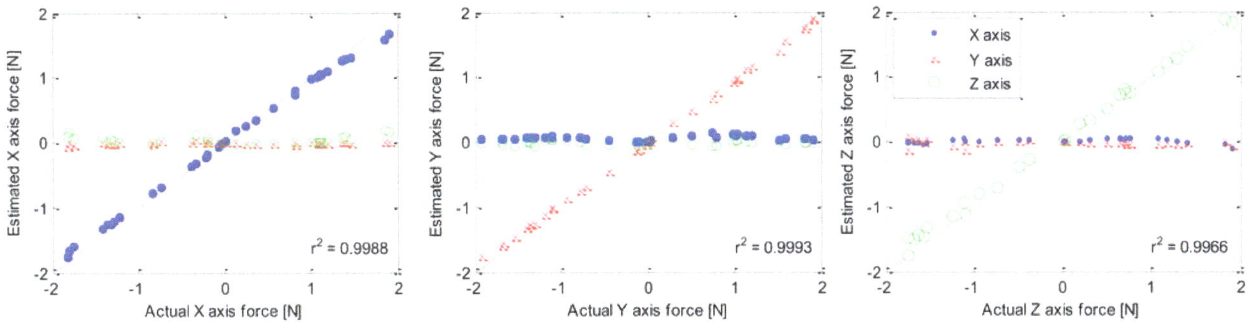

Figure 4: Sensor response to loads showing linearity.

To characterize noise and resolution, the unloaded sensors were placed in an enclosed container and allowed to thermally equilibrate. Data was then taken for 20 s. Because the major component of noise is at high frequencies, resolution is coupled with sampling frequency. At 1,000 Hz sampling, the RMS noise was approximately 0.1 N.

To examine the benefit of the copper braid heat shield, a sensor was fabricated without the inclusion of a heat shield. After calibration, sensors were allowed to thermally equilibrate in open air. With the non-heat shielded sensor, we measured a drift of 1.2 N over 5 min. The sensor with the heat shield drifts only 0.15 N in 5 min, almost an order of magnitude improvement.

A metal-based force-sensing jaw of the same size scale, using an aluminum element, was also fabricated and is reported in [21]. The resulting average RMS error in calibration was 0.07 N and average R2 was 0.930. Accordingly, the SDM sensor described above has approximately twice the calibration error as the more traditional, metal-based design. However, the ease of fabrication and acceptable performance of the molded sensor make it more appealing for many applications, particularly those requiring "disposable" hardware.

Discussion

Due to thermally-insulating and viscoelastic properties of most polymers, force transducers constructed with polymeric support structures will likely never perform as well as metal-based sensors. The benefits of the SDM approach, however, are many. The force sensitivity of strain gages is retained without the complex bonding process normally associated with strain gages. The sensor is robust due to its monolithic construction. The force ranges can be easily adjusted by depth of gage placement within the element. Wire management is straightforward, with all wires exiting the sensor at the same point, and intrinsic strain relief is provided by the epoxy. Finally, a construction process in which a single pour would encapsulate the strain gages and the heat shield could lead to a straightforward mass production scheme. In mass production the sensors could be low cost and potentially disposable, thus removing the issue of repeated sterilization for surgery applications.

An advantage of traditional metal-based sensors not reflected in the above SDM design is that they are relatively moment insensitive due to moment rejecting flexure design (binocular-shaped cutouts) where applied forces cause the flexure to translate instead of bending [23]. An SDM based sensor could take advantage of this design principle as well, at the cost of a more intricate mold design.

The viscoelasticity of polymeric structures typically causes undesirable creep effects, which would cause a slowly changing force signal under a constant load. We specifically chose a low-creep epoxy to minimize these effects, but we would expect to observe some creep for loads of long duration (hundreds of seconds). Further work will include investigation of other polymers and inclusion of other materials within the polymer to reduce creep, as well as targeting specific applications that are insensitive to creep, such as short time-duration load sensing.

CONCLUSIONS/OUTLOOK

The design approaches and fabrication techniques presented here demonstrate that sophisticated sensors can be readily incorporated into polymeric structures. A central advantage is that the fabrication process can enable the creation of highly effective sensors by embedding inexpensive, prepackaged transducers to create specialized sensing structures. These sensors are part of the robot structure and are created using the same tools and forming techniques as the mechanical structure, requiring minimal additional effort. This also permits optimization of the overall mechanical properties of the system as well as facilitates cable routing. In the joint-angle sensor presented above, phototransistors and LEDs are molded into a finger during a fused-deposition manufacturing printing operation. This approach is readily extensible to other sensors such as hall-effect sensors and allows easy alignment of the sensors to the device.

In the multi-axis force sensor presented above, the strain gauge wiring is molded within the sensor element, providing strain relief and enhanced robustness. This approach can be applied to the entire robot, with all cabling for sensors and actuators molded into the robot structure. While SDM-like processes were used to fabricate the prototypes presented in this paper, a number of established industrial manufacturing processes exist that can enable similar functionality. Multi-shot injection molding is used to create parts such as tool handles with integrated soft grips and parts with integral o-rings and gaskets. Overmolding and insert molding allow for the embedding of prefabricated components and are used to encase electronic components within polymer housings. These processes and others might be used directly or slightly modified in order to mass-produce inexpensive commercial polymeric robotic and mechatronic devices incorporating the sensor designs presented here.

ACKNOWLEDGMENTS

A portion of this work was published in a preliminary conference paper version [6]. This work was supported in part by US Office of Naval Research grant number N00014-98-1-0669, US National Science Foundation grants EEC-9731748 and IIS-0905180, US National Institutes of Health grant R01 HL073647-01, and US Army contract DAMD 17-01-1-0677. The authors would like to thank Lael Odhner and Alejandro Carrillo for assistance fabricating the flexure sensor prototype.

REFERENCES

1. Merz, R.; Prinz, F.B.; Ramaswami, K.; Terk, M.; Weiss, L. Shape Deposition Manufacturing. Proceedings of the Solid Freeform Fabrication Symposium, Austin, TX, USA, 8–10 August 1994.
2. Clark, J.E.; Cham, J.G.; Bailey, S.A.; Froehlich, E.M.; Nahata, P.K.; Full, R.J.; Cutkosky, M.R. Biomimetic Design and Fabrication of a Hexapedal Running Robot. Proceedings of the 2001 IEEE International Conference on Robotics and Automation (ICRA'01), Seoul, Korea, 21–26 May 2001; pp. 3643–3649.
3. Dollar, A.M.; Howe, R.D. The highly adaptive SDM hand: Design and performance evaluation. *Int. J. Robot. Res.*2010, *29*, 585–597.
4. Asbeck, A.T.; Kim, S.; Cutkosky, M.R.; Provancher, W.R.; Lanzetta, M. Scaling hard vertical surfaces with compliant microspine arrays. *Int. J. Robot. Res.* 2006, *25*, 1165–1179.
5. Cutkosky, M.; Dario, P.; Stefanini, C. A High Force Miniature Gripper Fabricated via Shape Deposition Manufacturing. Proceedings of the 2003 IEEE International Conference on Robotics and Automation (ICRA' 03), Taipei, Taiwan, 15–18 September 2003; pp. 1836–1841.
6. Dollar, A.M.; Wagner, C.R.; Howe, R.D. Embedded Sensors for Biomimetic Robotics via Shape Deposition Manufacturing. Proceedings of the IEEE/RAS-EMBS International Conference on Biomedical Robotics and Biomechatronics (BioRob 2006), Pisa, Italy, 20–22 February 2006; pp. 763–768.
7. Park, Y.-L.; Ryu, S.C.; Black, R.J.; Chau, K.K.; Moslehi, B.; Cutkosky, M.R. Exoskeletal force-sensing end-effectors with embedded optical fiber-Bragg-grating sensors. *IEEE Trans. Robot.* 2009, *25*, 1319–1331.
8. Lincoln, L.S.; Bamberg, S.J.M.; Parsons, E.; Salisbury, C.; Wheeler, J. An Elastomeric Insole for 3-Axis Ground Reaction Force Measurement. Proceedings of the 4th IEEE RAS & EMBS International Conference on Biomedical Robotics and Biomechatronics (BioRob 2012), Rome, Italy, 24–27 June 2012; pp. 1512–1517.
9. Yaul, F.M.; Bulovic, V.; Lang, J.H. A flexible underwater pressure sensor array using a conductive elastomer strain gauge. *J. Microelectromech. Syst.* 2012, *21*, 897–907.
10. Polygerinos, P.; Puangmali, P.; Schaeffter, T.; Razavi, R.; Seneviratne, L.D.; Althoefer, K. Novel Miniature MRI-Compatible Fiber-Optic Force Sensor for Cardiac Catheterization Procedures. Proceedings of the 2010 IEEE International Conference on Robotics and Automation (ICRA'10), Anchorage, AK, USA, 3–7 May 2010; pp. 2598–2603.
11. Tan, U.-X.; Yang, B.; Gullapalli, R.; Desai, J.P. Design and Development of a 3-Axis MRI-Compatible Force Sensor. Proceedings of the 2010 IEEE International Conference on Robotics and Automation (ICRA'10), Anchorage, AK, USA, 3–7 May 2010; pp. 2586–2591.
12. Park, Y.-L.; Chen, B.-R.; Wood, R.J. Design and fabrication of soft artificial skin using embedded microchannels and liquid conductors. *IEEE Sens. J.* 2012, *12*, 2711–2718.
13. Vogt, D.M.; Park, Y.-L.; Wood, R.J. Design and Characterization of a Soft Multi-Axis Force Sensor Using Embedded Microfluidic Channels. *IEEE Sens. J.* 2013, *13*, 4056–4064.
14. Jentoft, L.P.; Howe, R.D. Determining Object Geometry with Compliance and Simple Sensors. Proceedings of the 2011 IEEE International Conference on Intelligent Robots and Systems (IROS'11), San Francisco, CA, USA, 25–30 September 2011; pp. 3468–3473.
15. Lotti, F.; Tiezzi, P.; Vassura, G.; Biagiotti, L.; Palli, G.; Melchiorri, C. Development of UB Hand 3: Early Results. Proceedings of the 2005 IEEE International Conference on Robotics and Automation (ICRA'05), Barcelona, Spain, 18–22 April 2005; pp. 4488–4493.
16. Minor, M.A.; Merrell, R. Instrumentation and algorithms for posture estimation in compliant framed modular mobile robots. *Int. J. Robot. Res.* 2007, *26*, 491–512.
17. Dollar, A.M.; Howe, R.D. A robust compliant grasper via shape deposition manufacturing. *IEEE/ASME Trans. Mechatron.* 2006, *11*, 154–161.
18. Cavallo, A.; De Maria, G.; Natale, C.; Pirozzi, S. Optoelectronic joint angular sensor for robotic fingers. *Sens. Actuators A Phys.* 2009, *152*, 203–210.
19. Jentoft, L.P.; Howe, R.D. Compliant Fingers Make Simple Sensors Smart. Proceedings of the 2010 IFToMM/ASME Workshop on Underactuated Grasping (UG2010), Montreal, Canada, 19 August 2010.
20. Wagner, C.R.; Howe, R.D. Force Feedback Benefit Depends on Experience in Multiple Degree of Freedom Robotic Surgery Task. *IEEE Trans. Robot.* 2007, *23*, 1235–1240.
21. Wagner, C.R. Force Feedback in Surgery: Physical Constraints and Haptic Information. Ph.D. Thesis, Harvard University, Cambridge, MA, USA, 2006.
22. Peirs, J.; Clijnen, J.; Reynaerts, D.; Brussel, H.V.; Herijgers, P.; Corteville, B.; Boone, S. A micro optical force sensor for force feedback during minimally invasive robotic surgery. *Sens. Actuators A Phys.* 2004, *115*, 447–455.
23. *Strain Gauge Based Transducers: Their Design and Construction*; Technical Report; Measurements Group Inc.: Raleigh, NC, USA, 1988.

CHAPTER 13

OPTIMIZATION OF ENVIRONMENTALLY BENIGN POLYMERS BASED ON THYMINE AND POLYVINYL SULFONATE USING PLACKETT-BURMAN DESIGN AND SURFACE RESPONSE

Julieta Ledesma,[1] Santiago A. Bortolato,[2] Carlos E. Boschetti,[1] and Débora M. Martino[2,3]

[1]Área Tecnología Química, Facultad de Ciencias Bioquímicas y Farmacéuticas, Universidad Nacional de Rosario, Suipacha 531, S2002LRK Rosario, Argentina

[2]Instituto de Desarrollo Tecnológico para la Industria Química (INTEC), Güemes 3450, S3000GLN Santa Fe, Argentina

[3]Facultad de Humanidades y Ciencias, Universidad Nacional del Litoral, Paraje El Pozo, S3080HOF Santa Fe, Argentina

ABSTRACT

Traditional approaches to the development of integrated circuits involve the use and/or manufacture of toxic materials that have a potential environmental impact. An extensive research has been done to design environmentally benign synthetic polymers containing nucleic acid bases, which can be used to enhance the photoresistor technologies. Water soluble, environmentally benign photopolymers of 1-(4-vinylbenzyl) thymine (VBT) and vinylphenyl sufonate (VPS) undergo a photodimerization reaction when exposed to low levels of ultraviolet irradiation leading to an immobilization of the copolymer on a variety of substrates. Plackett-Burman design (PBD) and central composite design (CCD) were applied to identify the significant factors influencing the polymer crosslinking and dye adsorption processes, which are relevant in the fabrication of copolymer films for potential photoresist use. The PBD results assigned a maximum absorption signal of 0.67, while optimal conditions obtained in this experiment following the CCD method predictions provided a response of 0.83 ± 0.03, being a solid foundation for further use of this methodology in the production of potential photoresistors. The pH effect was relevant for low concentrations but not significant for higher concentrations. To the best of our knowledge, this was the first report applying statistical experimental designs to optimize the crosslinking of thymine-based polymers.

INTRODUCTION

The success of semiconductor technologies in recent decades has been based primarily on what is called the photolithography technique, often used to make integrated circuits [1]. Photolithography is the process by which optical methods are used to transfer the circuit patterns from master images (masks) to substrates, typically silicon wafers. The traditional approaches to the development of integrated circuits involve obtaining small pieces at low cost, without considering the toxicity of the materials used and/or manufactured, and their potential environmental impact. During the last decades, extensive research has been done to design environmentally benign synthetic polymers containing nucleic acid bases, which can be used to enhance the photoresistor technologies [2–5]. The need for an inexpensive, stable, and versatile system which incorporates thymine moieties in the polymer backbone leads to the design of a new synthetic monomer,

Citation: Julieta Ledesma, Santiago A. Bortolato, Carlos E. Boschetti, and Débora M. Martino, "Optimization of Environmentally Benign Polymers Based on Thymine and Polyvinyl Sulfonate Using Plackett-Burman Design and Surface Response," Journal of Chemistry, vol. 2013, Article ID 947137, 9 pages, 2013. doi:10.1155/2013/947137

the 1-(4-vinylbenzyl) thymine (VBT) [6], that has the ability to photocrosslink upon irradiation with short-wavelength ultraviolet (UV) component of sunlight [6–8]. This process was bioinspired on the photodimerization of thymine, one of the DNA bases, both in vivo and in vitro under irradiation with ultraviolet light, which disrupts the helical structure of DNA [4, 5].

The versatility of VBT makes it an attractive research platform, since the balance between photoreactivity, solubility, and noncovalent interactions can be fine-tuned for a variety of applications. Since the VBT homopolymer is water insoluble due to strong intermolecular interactions, the copolymerization of VBT with charged functional groups can improve its solubility. Among them, substituted styrenes either cationic such as vinylbenzyl triethylammonium chloride (VBA) and N-butyl-N,N-dimethyl-(4-vinylbenzyl)ammonium chloride (BDMQ), or anionic such as vinylphenyl sufonate (VPS), allow for obtaining a fully water-processable photoresist, eliminating the need of toxic solvents [9, 10]. The resulting photopolymers of VBT and VPS or VBA are water soluble, environmentally benign, and undergo a photodimerization reaction of the pendant thymine units when exposed to low levels of ultraviolet irradiation. This process leads to an immobilization of the copolymer on a variety of substrates in response to irradiation and allows the removal of unexposed regions by aqueous wash (photoresist), having potential applications in a variety of fields [9–19]. Trakhtenberg et al. introduced the photoresistive capacity of VBT-VPS copolymers and demonstrated that they can be used as a viable template for enzymatic synthesis of conductive polyaniline, allowing for surface photopatterning with a conductive polymer under mild conditions [15]. The same copolymer was also used in photoresist applications [9, 20]. In those works, the authors immobilized the copolymers to form thin films on polyethylene terephthalate (PET) substrate and demonstrated that the process is potentially useful for the advancement of photoresistors [20].

In photolithography technique, in order to create the patterns, the wafers are coated with a photosensitive material called photoresist, which changes its solubility upon exposure to light through an appropriate mask of UV opaque material. This allows the selective irradiation of only the desired areas of the photoresist-coated wafer, thus transferring the pattern from the mask onto the wafer. After being exposed to light the photoresist-coated wafer goes into a developing solution that selectively removes the resist layer from unexposed areas, resulting in a wafer with the semiconductor surface partially protected by the resist [21]. In subsequent steps, the unprotected areas are modified, for example, by chemical etching, while protected areas remain intact. Finally the remaining photoresist is stripped off the wafer, which now features the pattern identical to the one on the mask. This process can be repeated several times, using different masks [1]. Figure 1 schematically presents the photolithography process.

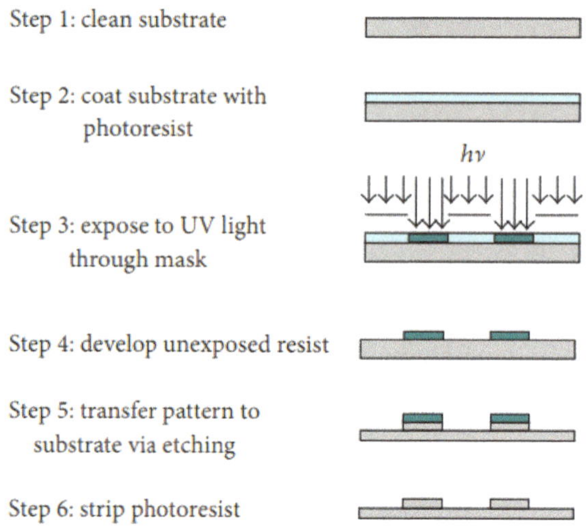

Figure 1: Description of photolitography process.

Taking into account that the photoimmobilized copolymer film is very thin (<1 μm) and translucent, the immobilized copolymer coatings are typically visualized by dipping the resulting patterned films into a solution of an oppositely charged dye. The dye chosen for VPS was methylene blue (MB) given that it has a net positive charge that favors the physisorption process on the polymer, as it was shown [9]. The dye selectively adsorbs on the charges of the photocrosslinked copolymer, providing clear contrast from the unreacted regions. The amount of adsorbed dye can be monitored by UV-VIS spectroscopy, measuring the absorbance at 610 nm (MB absorbance maximum), and therefore a correspondence among the measured absorbance and the amount of immobilized charged groups on the substrate surface can be made [9].

The technological applications of these polymers were seen earlier than the necessary basic studies, which could help to understand their behavior and improve their service. Therefore, more studies need to be done in order to optimize the proposed photocrosslinking method due to the large amount of variables that have effect on the processing and development of these copolymers. To begin with, thymine is a compound having acid-base behavior with a pKa ~9, and it has been demonstrated that at pH greater than pKa thymine exists mostly in its anionic form, in an equilibrium between two tautomers (ratio 1 : 1), one of which cannot crosslink [22]. It was found experimentally that increasing the proportion of VPS in its VBT-VPS copolymer, the pH of the solution increases to values close to 11, which could produce a significant reduction in the efficiency of the crosslinking process. However, the efficiency and reproducibility of the method used to prepare VBT-VPS coatings may depend on other factors, including the coating thickness and the copolymer composition [20]. Although the works cited before [9, 20] are very helpful to screen important factors influencing film preparation, they are insufficient to establish a general rule for manufacturing certain photoresist polymers with reliable properties. In general, it is difficult to obtain the comprehensive effects of some factors through a reasonable number of experiments. Solving this problem in a rational way involves performing an experimental design [23]. This procedure can be summarized in three stages: (1) identification of factors that can influence the results of the experiments, (2) use of statistical analysis to separate and evaluate the effects of the various factors involved "screening phase"; and (3) with the significant factors, get the best response possible for the developed experiment "response surface phase" [23]. The primordial difference between the classical one-variable-at-a-time method and the experimental design is that, in the latter case, the values of all the factors are varied in each experiment in a programmed and rational way. Some of the studied factors will probably have no influence on the experiment, and only few will act upon the response; therefore, the essential factors can be detected while keeping the number of trials to a minimum.

Plackett-Burman design (PBD) is a method that can inspect up to N-1 factors in N experiments and is commonly used in the screening phase of a wide range of processes [24–26]. It has been proven to be an efficient way of evaluating a large number of variables and identifying the significant ones from a statistical point of view [23–26]. Therefore, in this study, PBD method was applied to identify the significant factors influencing the polymer crosslinking and dye adsorption processes that are relevant in the fabrication of copolymer films for potential photoresist use. The effects of seven controllable factors on the process performances were studied. Selected operational variables were pH, concentration, dyeing time, solution stability, coating thickness, comonomer ratios, and film drying time. The absorbance corresponding to the methylene blue dye at 610 nm, as an indication of VBT-VPS crosslinking efficiency, was chosen as the response value of the PBD experiments.

Once the various factors affecting the results of an experiment have been identified through PBD, a separate method is needed to determine the combination of factor levels, which will provide the optimum response. First it is necessary to carefully define what is meant by "optimum response" in a given experimental procedure; in our case, the optimum outcome of the experiment would be the maximum signal-to-noise ratio in the absorbance measurements. With that in mind, the response surface methodology (RSM) is an efficient experimental tool with which the optimal conditions of a multivariable system may be determined. This strategy includes a large number of statistical techniques [23, 27], in particular the Central Composite Design (CCD) which was used to explore the relationships between several descriptive variables and one response variable.

In this work, the optimization of the main factors affecting the polymer photocrosslinking and subsequent dye adsortion processes was carried out by means of an experimental design. The goal of the present research is to quantitatively assess the effects of numerous variables over the photocrosslinking process and to propose the best experimental conditions for the improvement of photoresistors based on VBT and VPS copolymers in a rational way and at low cost of experiments.

MATERIALS AND METHODS

Materials

All reagents were purchased in the purest available form and were used as received. Sodium hydroxide, isopropanol, and acetone were purchased from Cicarelli (Buenos Aires, Argentina). 4-Vinylbenzyl chloride, 2,2′-azobis-2-methylpropionitrile (AIBN), 2,6-di-tert-butyl-4-methylphenol, thymine, and methylene blue (MB) were purchased from Sigma-Aldrich (SA) (Buenos Aires, Argentina). VBT was synthesized from thymine and vinylbenzyl chloride as described previously [6]. VPS salt was purchased from Fluka (Buenos Aires, Argentina). Based on ^{1}H NMR spectra and melting point results, the monomeric products were deemed pure enough for the synthesis of the polymers.

Hydrophilic polyethylenterephthalate film (PET-X4C1, Dupont) was used as substrate without previous treatment. Coatings were done using wire-round milled coating rods purchased from RDS Corp. (Webster, NY, USA). Irradiations were performed using a UV hand lamp (Spectroline UL, Model ENF 260c, Spectronics Corporation Westbury, NY, USA). Absorbance measurements were taken with Spectronic Genesys spectrophotometer (Thermo Electron Corporation, USA), UV-VIS spectra were collected on Jasco V-630 UV-VIS spectrophotometer (Jasco Inc., USA), and ^{1}H NMR spectra were taken on Bruker 300 MHz NMR spectrometer.

Copolymer Synthesis and Characterization

To produce water-soluble polymers, VBT was copolymerized in a free radical process with the anionic monomer, VPS. The ratio of VBT: VPS comonomers influences the behavior of the VBT polymeric system and varies depending on the application. Therefore, VBT_n: VPS_m have been synthesized with n=1 and m=1, 4, 8, 16, 32. VBT: VPS 1:1 copolymer: to a 300 mL, 3-neck, roundbottomed flask containing 250 mL of a 50% (v/v) aqueous solution of isopropanol, VBT (11.9 g, 0.049 mol) and VPS (10.1 g, 0.049 mol) were added. The solution was heated to 65° C while stirring and 0.22 g of AIBN was added. Stirring was continued for 16 h while the temperature was maintained at 65° C. The reaction mixture was cooled to room temperature and rotary evaporated to concentrate to 125 mL in vacuo. Adding the aqueous solution to 1 L of cold acetone precipitated the polymer. Subsequently, the white solid precipitate was filtered and then dried under vacuum for 2 days. To verify the absence of unreacted monomers, the precipitated polymer was analyzed by 1 H NMR spectroscopy and the typical vinyl group signal at chemical shifts between 5 and 6 ppm was not observed in the spectra. Additionally, elemental analysis was used to confirm copolymer ratios [9]. Scheme 1 presents the structure of the VBT-VPS copolymers. VBT: VPS 1 : 4, 1 : 8, 1 : 16, and 1 : 32 copolymers were synthesized with identical procedures only varying the corresponding ratios of starting monomers.

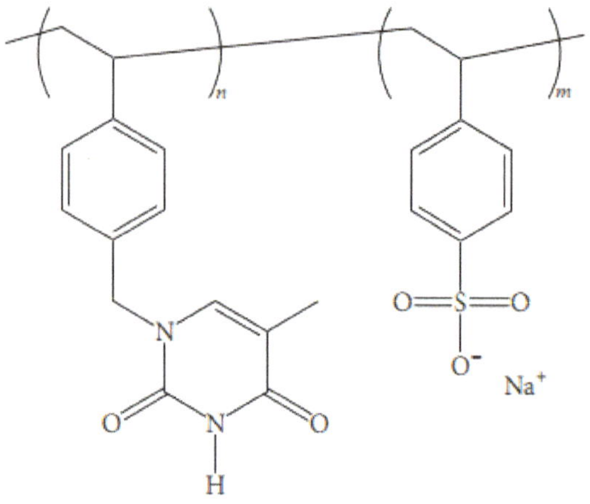

Scheme 1: Structure of VBT-VPS copolymers (n=1 and m=1, 4, 8, 16, or 32).

Coating Preparation and Film Irradiation

Aqueous solutions of each VBT : VPS_m (m = 1, 4, 8, 16, 32) copolymers were distributed homogeneously on PET film using either a no. 3 or no. 6 wire-round milled coating rod. The films were dried at room temperature for one hour and in a vacuum oven (at 80°C ~20 torr) for another hour to give a uniform wet thickness of 6.8 µm or 13.6 µm according to the coating rod specifications [28]. Films were protected from the environmental light throughout the process.

All copolymer-coated films were irradiated with a UV hand lamp at 254 nm from a distance of 1 cm at different times (from 10 to 60 seconds) for the preliminary studies and one minute for the experimental design. This process leads to the immobilization of the polymer in response to the irradiation (photoresist material). The curing reaction due to radiation was achieved at room temperature; thus these coatings can be prepared on heat sensitive materials.

The samples were then immersed into an aqueous solution of MB dye (1×10^{-3} M) for 30 or 120 s. Next, they were dried by hot air or at room temperature and visually examined for uniform coloring. The photocrosslinking was monitored indirectly by UV-VIS spectroscopy measuring the efficiency of dye adsorption.

Plackett-Burman Design

Plackett-Burman design (PBD) [29] was used in the present study to screen the essential variables that significantly influence the photocrosslinking process. In this study, a 12-run PBD, including a replicated one to analyze the model lack of fit, was applied to evaluate seven factors: five numerical and two categorical. Each variable was examined at two levels, according to the guidelines of PBD [29]. Table 1 illustrates the seven variables and their corresponding levels (low/high for numerical variables and one/two for categorical variables) used in the experimental design. The values of the levels were set according to our preliminary experimental results.

Table 1: Experimental variables used in the Plackett-Burman design

Variable	Type	Units	Symbol	Low level	High level
pH	Numerical	[a]	A	6	12
Concentration	Numerical	% g L^{-1}	B	2	10
Dyeing time	Numerical	Sec	C	30	120
Stability	Numerical	Day	D	1	30
Coating thickness	Numerical	μm	E	6.8	13.6
				Level one	Level two
Drying film[b]	Categorical	[a]	F	Yes	No
Comonomer ratio[c]	Categorical	[a]	G	1:1	1:4

[a]Not applicable. [b]Drying as explained in Section 2.3. [c]1:1 = VBT : VPS_1, 1:4 = VBT : VPS_4.

The PBD and the response values of absorbance corresponding to MB dye adsorption on photocrosslinked copolymer, both the observed and predicted by the PBD model, are shown in Table 2.

Table 2: Twelve-trial PBD matrix for seven variables (A to G): experimental input values, and observed and predicted absorbance[a].

Run order	Experimental values							Absorbance[b]	
	A	B	C	D	E	F	G	Observed	Predicted
1	12.0	2	30	1	13.6	Yes	1:4	0.18	0.17
2	6.00	10	30	30	13.6	No	1:4	0.54	0.52
3	12.0	10	120	1	13.6	No	1:1	0.58	0.61
4	12.0	2	120	30	13.6	Yes	1:4	0.10	0.11
5	12.0	2	30	30	6.8	No	1:1	0.44	0.37
6	6.00	2	30	1	6.8	Yes	1:1	0.39	0.43
7	6.00	10	30	30	13.6	Yes	1:1	0.67	0.70
8	6.00	2	120	1	13.6	No	1:1	0.61	0.58
9	6.00	2	120	30	6.8	No	1:4	0.15	0.19
10	12.0	10	30	1	6.8	No	1:4	0.24	0.28
11	6.00	10	120	1	6.8	Yes	1:4	0.36	0.29
12	6.00	2	120	30	6.8	No	1:4	0.17	0.19

[a]A–G are symbols shown in Table 1. [b]Absorbance measured at 610 nm.

Response Surface Methodology

The optimal levels of the significant factors and the interactions of these variables in a photocrosslinking process were analyzed by Central Composite Design (CCD). In this study, a two-factor, five-level CCD with 10 runs was employed. The tested variables (pH and polymer concentration) were denoted as X_1 and X_2, respectively. The numerical variables were assessed at five different levels, combining factorial, axial, and central points, according to the guidelines of CCD [27]. The values for the three tested variables are shown in Table 3.

Table 3: Levels of the variables tested in the CCD

Variable	Type	Symbol	Factorial levels		Axial levels		Central level
pH	Numerical	X_1	7.50	10.50	6.00	12.00	9.00
Concentration	Numerical	X_2	4.00	8.00	2.00	10.00	6.00

Statistical Analysis and Software

The analysis of variance (ANOVA) was conducted to determine the importance of the models and the regression coefficients, both in PBD and CCD assays. The statistical significance of the employed models, their terms, and their lack of fits were checked by Fisher's F-test [30]. This test is used for comparing the model variance with the residual (error) variance: if the variances are close to the same value, their ratio will be close to unity, and it is less likely that any of the factors, terms, or lack of fits will have a significant effect on the response.

In CCD, the quality of the polynomial equation was judged from the determination coefficient (R^2), the predicted R^2 (pred-R^2), which measures how good the model predicts a response value, the R^2 adjusted for the number of parameters in the model relative to the number of points in the design (Adj-R^2), which estimates the amount of variation around the mean explained by the model, and its statistical significances were checked also by Fisher's F-test [30]. The surface response and the contour plots of the predicted model responses were used to assess the interactive relationships between the significant variables.

Design-Expert software [27] was used for designing the experiments as well as for the regression and graphical analysis of the experimental data obtained.

RESULTS AND DISCUSSION

Preliminary Step

The variables that could influence the polymer photocrosslinking and subsequent dye adsorption process were defined as follows. It was demonstrated that the irradiation time is positively related to the degree of photocrosslinking and how this relationship is modified when changing the VBT : VPS copolymer ratio up to a proportion 1 : 16 [9]. However, such analysis was carried out qualitatively, and therefore it is not possible to predict accurately the behavior of the photocrosslinking process.

Figure 2 shows the total intensity of MB absorption signal as function of irradiation time for polymer films at the same concentration (2%), with different VBT to VPS ratios (1 : 1 to 1 : 32). It can clearly be observed that the greater the proportion of VPS to VBT, the larger the irradiation time required to immobilize the polymer on the substrate surface and the lower the amount of dye adsorbed on the polymer after saturation. In general, these results are in good agreement with previous reported data [9]. However, from the potential technological-utility point of view, it is clear that mixtures with higher proportion of VPS (1 : 8, 1 : 16, and 1 : 32) are not convenient since the sensitivity of the absorption signal is poor at 60 seconds compared with the VBT : VPS (1 : 1). Based on these preliminary results, the analysis of the experiments was completed only with the polymers of the lowest ratio of VPS (1 : 1 and 1 : 4).

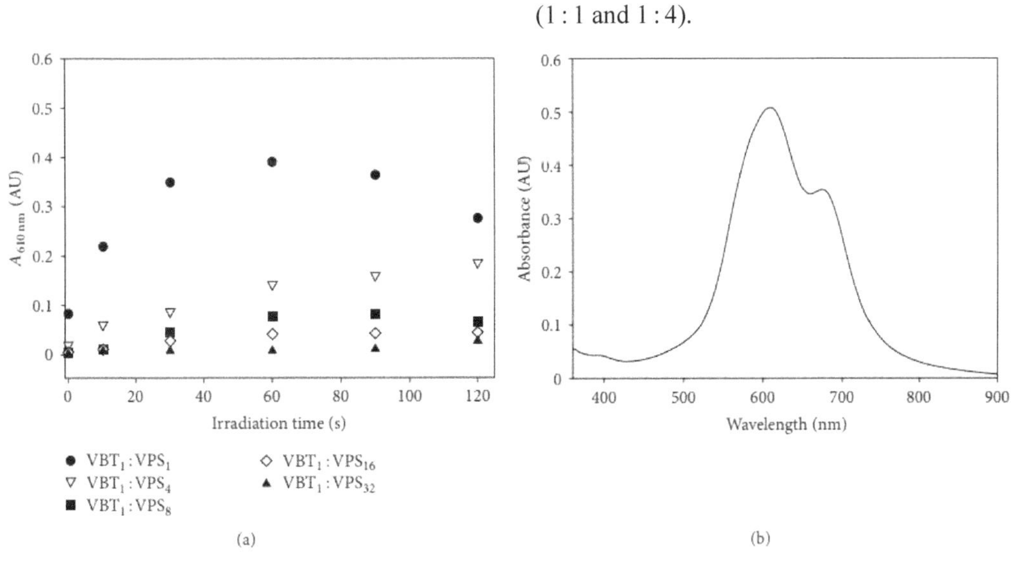

Figure 2: (a) Time evolution of MB absorption signal for polymer films at the same concentration (2%), with different VBT to VPS ratios (1 : 1 to 1 : 32). (b) UV-VIS absorption spectra of polymer film of VBT : VPS_1 at 2% and 60 seconds irradiation.

On the other hand, it is reasonable that the dyeing time significantly influences the observed signal as this process favors the physisorption, although previous studies did not explain in detail the effect of this factor [9]. Therefore, dyeing time is defined as a potential influential variable in the process.

Experimentally it was found that the pH value of the $VBT_1 : VPS_{32}$ solution increases to 11. Furthermore, the pH of the solution containing fewer amounts of VPS is around 6.5. On this basis, the pH influence on the photocrosslinking process was analyzed, since it can affect the acid-base behavior of thymine [22] and therefore reduce the efficiency of photocrosslinking process.

In addition, it was found experimentally that the pH of the solutions with greater VPS percentage (1 : 8 to 1 : 32) was modified over time, and after 30 days of preparation the pH stabilized in a value close to 6. While the explanation for this phenomenon is not completely clear, the stability of the solutions was considered as a potential influential variable in the process.

It has been shown that the film thickness controls the crosslinking process, being more effective as the thickness reduces [31]. Therefore two coating thicknesses were compared to see how this variable could influence the process. Similarly, it is likely that increasing the polymer concentration in the solution influences the process. Previous works used 10% aqueous solution of VBT-VPS [9, 20]. From an environmentally perspective, it seemed interesting to verify whether or not lower concentration values give good responses, as well.

Finally, it was reported that the retained water in the coated film could modify the process [31, 32]. Two alternative drying methods were compared to test the statistical significance of the drying method: (a) one hour at room temperature and (b) an additional hour in oven at 100°C.

In summary the seven experimental variables used in the PBD are shown in Table 1.

Screening Step

The data listed in Table 2 indicates a wide variation in the absorption signal corresponding to methylene blue dye adsorbed on the photocrosslinked polymer, from 0.10 to 0.67, in the 12 trials. This variation suggests that the optimization process is crucial for improving and controlling the dye adsorption after the photocrosslinking process. The analysis of the regression coefficients of the seven factors presented in Table 4 showed that factors B, C, and F had positive effects on dye adsortion, whereas factors A, D, E, and G had negative effects.

Table 4: Effects of the variables and statistical analysis of the PBD

Source	Coefficient estimate[a]	Sum of squares	df[b]	Mean square	*F* value	*P* value Prob > *F*
Model		0.4215968	7	0.06022811	13.1201459	0.0128
A	−0.0630	0.03967072	1	0.03967072	8.64190442	**0.0424**
B	0.079	0.06239922	1	0.06239922	13.593102	**0.0211**
C	−0.02242	0.00503972	1	0.00503972	1.09785677	0.3539
D	−0.0058	0.00034116	1	0.00034116	0.07431762	0.7986
E	0.0600	0.03602371	1	0.03602371	7.84743665	**0.0487**
F	0.0368	0.01359198	1	0.01359198	2.96088871	0.1604
G	−0.1287	0.1657404	1	0.1657404	36.1050351	**0.0039**
Residual		0.01836203	4	0.00459051		
Lack of fit		0.01812003	3	0.00604001	24.9587134	0.1458
Pure error		0.000242	1	0.000242		
Cor Total[c]		0.43995883	11			

[a]In terms of coded factors. For the intercept, the estimated coefficient is 0.39. [b]Degree of freedom. [c]Sum of total squares corrected for the mean.
The significant factors (at 5% significance level) are shown in bold type.

A Model F value of 13.12 (see Table 4) indicates that the model is significant; there is only a 1.28% chance that a "Model F Value" this large could occur due to noise [30]. The variable with "Prob > F" value lower than 0.0500 points out that model terms are significant, while values greater than 0.1000 indicate that the model terms are insignificant [30]. It was clear that variables A, B, E, and G were the significant factors, while variables C, D, and F were considered insignificant and were not included in the next step of the CCD experiments (see Table 4).

Figure 3 shows the time evolution of the MB adsorption signal for copolymer films of $VBT : VPS_1$ and $VBT : VPS_4$ at 2% concentration. It can clearly be seen that the time evolution of 1 : 1 is faster than 1 : 4, reaching a similar maximum absorption value at 60 sec and 450 sec, respectively. Therefore, at the irradiation time used in the experiments (60 sec) the signal intensity of the MB dye adsorbed on the crosslinked polymers is about 65% lower for 1 : 4 than 1 : 1, being the former less energy efficient. These results are consistent with the fact that decreasing the amount of the crosslinkable monomer, VBT, also slows down the curing process and consequently the amount of dye adsorbed after washing. For that reason, the comonomer ratio 1 : 4 was not included in the CCD model. On the other hand, as it was shown in a reported work [31], films made using coating rods no. 6 give rise to thicker films (13.6 μm) that need longer irradiation times to crosslink and produce comparable signal intensities of the adsorbed dye. Therefore, the coating thickness was also excluded in the CCD model.

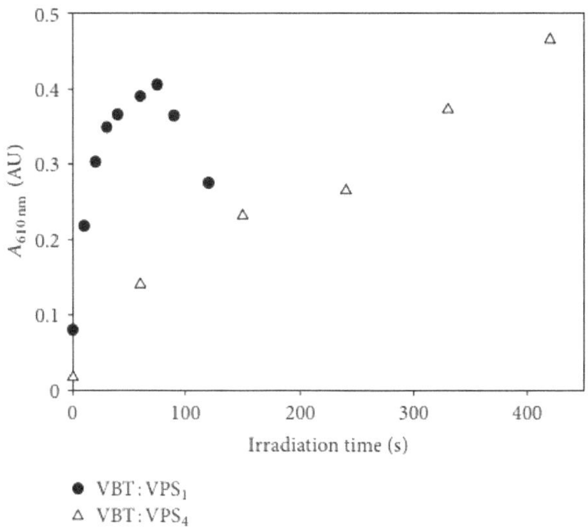

Figure 3: Time evolution of MB adsorption signal for the copolymers VBT : VPS_1 (•) and VBT : VPS_4 (Δ) at the same concentration (2%).

Finally, the inclusion of a duplicate as mentioned in the experimental section allowed the model fit determination. The "lack of fit F value" of 24.96 suggests that it is not significant relative to the pure error: there is a 14.58% chance that a "Lack of Fit F-value" this large could occur due to experimental noise [30].

Optimization Step

CCD was used to study the correlations between the significant factors and to determine their optimal levels. The central composite design matrix of tested variables and the experimental results are represented in Table 5.

Table 5: Central composite design matrix and the corresponding experimental and predicted responses for dye adsorption on crosslinked polymer process

Run order	pH	Concentration	Absorbance[a]	
			Observed	Predicted
1	9.0	6	0.63	0.57
2	7.5	8	0.60	0.58
3	7.5	4	0.49	0.50
4	9.0	2	0.41	0.41
5	10.5	4	0.46	0.48
6	6.0	6	0.48	0.50
7	10.5	8	0.75	0.73
8	9.0	6	0.57	0.57
9	9.0	10	0.71	0.74
10	12.0	6	0.63	0.64

[a] Absorbance measured at 610 nm.

The competence of the model was checked using ANOVA, as shown in Table 6. The "F value" of the model was 34.4, and the P value of "Prob > F" was <0.05, suggesting that the model proposed for the raised CCD (two-factor interaction) was significant. Linear terms of X_1, X_2 and interaction term of X_1X_2 were significant for dye adsorption process.

Table 6: ANOVA for RSM two factor interaction model

Source	Coefficient estimate[a]	Sum of squares	df[b]	Mean square	F value	P value Prob > F
Model		0.10592423	3	0.03530808	34.3997072	0.0004
X_1-pH	−0.069	0.01380408	1	0.01380408	13.4489461	0.0105
X_2-Conc.	−0.095	0.08377837	1	0.08377837	81.6230066	0.0001
X_1X_2	0.015	0.00834178	1	0.00834178	8.12716909	0.0292
Residual		0.00615844	6	0.00102641		
Lack of Fit		0.00441794	5	0.00088359	0.50766308	0.7806
Pure Error		0.0017405	1	0.0017405	34.3997072	0.0004
Cor Total[c]		0.11208267	9			

[a] In terms of actual factors. For the intercept, the coefficient estimate is 0.9406. [b] Degree of freedom. [c] Sum of total squares corrected for the mean.

From ANOVA, the experimental results of RSM fitted via the response surface regression procedure is a polynomial equation with the following expression:

$$Y = \beta_0 + \beta_1 X_1 + \beta_2 X_2 + \beta_{12} X_1 X_2, \qquad (1)$$

in which Y is the predicted response, X_i are the independent variables with i=1 and 2, β_0 is the offset term, β_i is the ith linear coefficient, and β_{12} is the X_1 and X_2 interaction coefficient.

The coefficient of determination (R^2) was calculated as 0.9451 for dye adsortion according to (1), indicating a good agreement between the experimental and the predicted values. The pred-R^2 of 0.8627 was in reasonable agreement with adj-R^2 of 0.9176. The value of "adequate precision," which measures the signal-to-noise ratio, resulted in a value of 16.5 suggesting an adequate signal [27]. The "lack of fit F value" of 0.51 implied that the "lack of fit" was not significantly relative to the pure error. The model was found to be adequate for prediction within the range of variables employed.

The graphical representations of the regression model (see (1)) and their corresponding contour plot were obtained using Design-Expert software and are presented in Figure 4 for comonomer ratio 1 : 1. The surface response plot represents the effect of two independent numerical variables, that is, pH and polymer concentration. The shape of the corresponding contour plot indicates whether or not the mutual interaction between the independent variables is significant. As shown in Figure 4, the surface response of absorbance indicates the presence of two regions of maximum response, one at high concentration and pH and the other at low pH. The optimum value found for the numerical variables for this copolymer composition using the CCD model equation is 9% for concentration and 10.7% for pH.

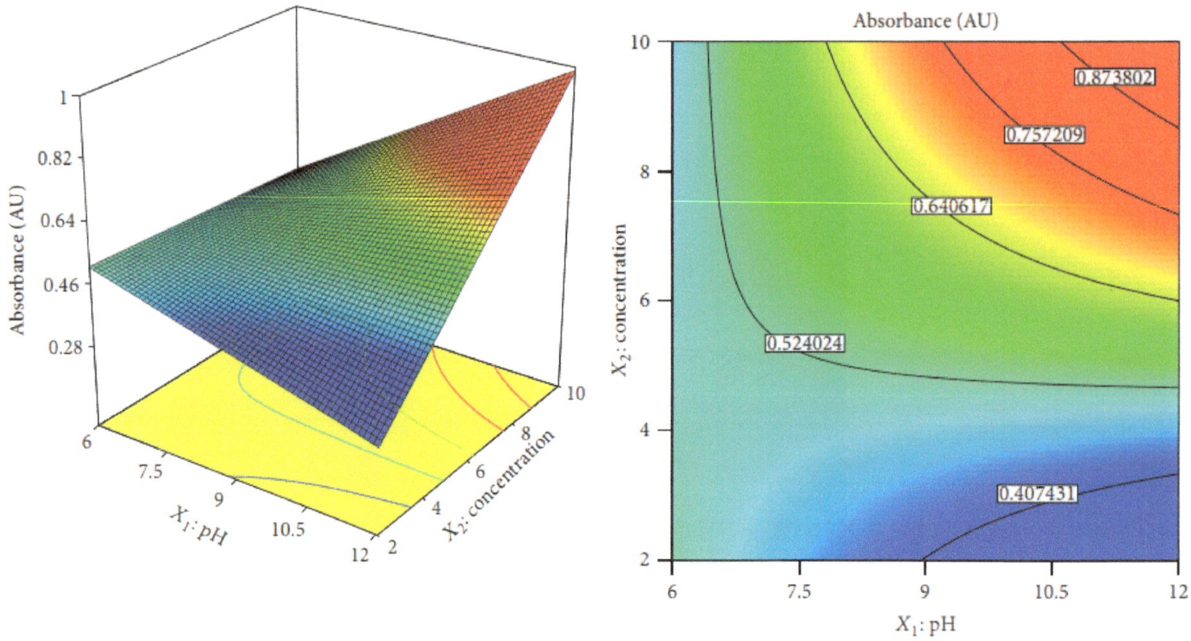

Figure 4: Three-dimensional response surface plots and two-dimensional contour plots for dye adsorption process on photocrosslinking polymer showing the interactions between pH and concentration variables for VBT : VPS_1.

A decrease in absorbance happens when the pH value is increased from 6 to 12, for concentrations up to 6%, giving rise to a relative maximum in the surface. The observed decrease of the absorbance at low concentration suggested that at high pH value the thymine moiety is present in the anionic form producing a strong reduction of the crosslinking process, as it was mentioned above.

At concentrations higher than 6% the tendency is inverse, so that the absorbance enlarges when pH varies from 6 to 12, showing an absolute maximum at 10% concentration. Therefore, at higher concentrations that effect of pH is diminished, and the absorbance turns out to be in direct relation to the polymer concentration.

The nature of the contour plot indicates that the mutual interaction between the two independent variables (X_1, X_2) was significant, which means that the effect of pH on the response was dependent on the level of concentration used, being negligible at concentrations larger than 6%.

Validation of the RSM Model

The validation of the statistical model and the regression equation were conducted by taking into account the values of the variables corresponding to maximum response calculated from (1). Under these optimized conditions, the predicted response for dye adsortion was 0.82, and the observed experimental value was 0.83 ± 0.03, an average of triplicate measurements. These results confirmed the validity of the model, and the experimental values were regarded to be quite close to the predicted values.

CONCLUSIONS

Results suggested that statistical strategy is an effective and important tool for the optimization of experimental parameters involved in dye adsorption on crosslinked polymer processes. In addition, to the best of our knowledge, this was the first report applying statistical experimental designs to optimize the crosslinking of thymine-based polymers. The PBD results assigned a maximum absorption signal of 0.67, while following the CCD method predictions, a response of 0.83 was obtained. The pH effect was relevant for low concentrations (less than 6%) but, as it was shown, not important for higher concentrations. Optimal conditions obtained in this experiment lead to a solid foundation for further use of this methodology in the production of potential photoresistors.

ACKNOWLEDGMENTS

Débora M. Martino, Carlos E. Boschetti, and Santiago A. Bortolato are members of the Research Council of CONICET. The authors would like to thank Universidad Nacional del Litoral (CAI+D Tipo II PI 11-57), CONICET (PIP 112-200801-01079 and D-1280/2011), and the NSF OISE (Grant no. 1031394) for the financial support.

REFERENCES

1. H. J. Levinson, Principles of Lithography, SPIE The International Society for Optical Engineering, 2001.
2. K. Takemoto, "Functional monomers and polymers containing nucleic acid bases," Journal of Polymer Science: Polymer Symposia, vol. 55, pp. 105–125, 1976.
3. Y. Inaki, "Thymine polymers as high resolution photoresists and reversible photo-recording materials,"Polymer News, vol. 17, pp. 367–371, 1992.
4. G. M. Blackburn and R. J. H. Davies, "The structure of thymine photo-dimer," Journal of the Chemical Society C, vol. 23, pp. 2239–2244, 1966.

5. A. A. Lamola and J. P. Mittal, "Solution photochemistry of thymine and uracil," Science, vol. 154, no. 3756, pp. 1560–1561, 1966.
6. C. M. Cheng, M. J. Egbe, M. J. Grasshoff et al., "Synthesis of 1-(vinylbenzyl)thymine, a novel, versatile multi-functional monomer," Journal of Polymer Science A, vol. 33, pp. 2515–2519, 1995.
7. J. M. Grasshoff, L. D. Taylor, and J. C. Warner, "Copolymeric mordants and photographic products and processes containing same," U.S. Patent 5,395,731, March 1995.
8. J. M. Grasshoff, L. D. Taylor, and J. C. Warner, "Vinylbenzyl thymine monomers," U.S. Patent 5,455,349, October 1995.
9. C. Kiarie, J. Bianchini, S. Trakhtenberg, and J. C. Warner, "Methylene blue adsorption on thymine based polyvinylphenylsulfonate films," Journal of Macromolecular Science A, vol. 42, no. 11, pp. 1489–1496, 2005.
10. J. R. Bianchini, K. Saito, T. B. Balin, V. Dua, and J. C. Warner, "Thymine-based, water-soluble phototerpolymers: their preparation and synthesis," Journal of Polymer Science A, vol. 45, no. 7, pp. 1296–1303, 2007.
11. A. S. Cannon, J. Raudys, A. Undurti, and J. C. Warner, "Photoreactive Polymers and Devices for use in Hair Treatments," PCT Int. Appl., WO 2004058187, p. 23, 2004.
12. J. M. Grasshoff, L. D. Taylor, and J. C. Warner, "Method of imaging using a polymeric photoresist having pendant vinylbenzyl thymine groups," U.S. Patent 5,616,451, April 1997.
13. J. M. Grasshoff, L. D. Taylor, and J. C. Warner, "Copolymers having pendant functional thymine groups," U.S. Patent 5,708,106, January 1998.
14. L. Lloyd-Kindstrand and J. C. Warner, "Thymine-containing styrene polymers as environmentally benign photoresists," in Biopolymers, S. Matsumura and A. Steinbüchel, Eds., vol. 9, pp. 165–174, John Wiley & Sons, Weinheim, Germany, 2003.
15. S. Trakhtenberg, Y. Hangun-Balkir, J. C. Warner et al., "Photo-cross-linked immobilization of polyelectrolytes for enzymatic construction of conductive nanocomposites," Journal of the American Chemical Society, vol. 127, no. 25, pp. 9100–9104, 2005. ·
16. K. Saito, L. R. Ingalls, J. Lee, and J. C. Warner, "Core-bound polymeric micellar system based on photocrosslinking of thymine," Chemical Communications, no. 24, pp. 2503–2505, 2007.
17. K. Saito and J. C. Warner, "Core-shell thymine containing polymeric micelle system: study of controlled release of riboflavin," Green Chemistry Letters and Reviews, vol. 2, no. 2, pp. 71–76, 2009.
18. G. Kaur, S. L. Y. Chang, T. D. M. Bell, M. T. W. Hearn, and K. J. Saito, "Bioinspired core-crosslinked micelles from thymine-functionalized amphiphilic block copolymers: hydrogen bonding and photo-crosslinking study," Journal of Polymer Science A, vol. 49, pp. 4121–4128, 2011.
19. S. A. Bortolato, A. L. Barbarini, R. M. Benitez, D. A. Estenoz, and D. M. Martino, "Photo-induced curing of thymine-based bioinspired polymers. A chemometric analysis," Latin American Applied Research. In press.
20. S. Trakhtenberg, J. C. Warner, R. Nagarajan, F. F. Bruno, L. A. Samuelson, and J. Kumar, "Spectroscopic and microscopic analysis of photo-cross-linked vinylbenzylthymine copolymers for photoresist applications," Chemistry of Materials, vol. 18, no. 12, pp. 2873–2881, 2006.
21. M. J. Bowden, "A perspective on resist materials for fine-line lithography," in Materials For Microlithography. Radiation-Sensitive Polymers, L. F. Thompson, C. G. Willson, and J. M. J. Frechet, Eds., ACS, Washington, DC, USA, 1984.
22. M. A. Morsy, A. M. Ai-Somali, and A. Suwaiyan, "Fluorescence of thymine tautomers at room temperature in aqueous solutions," Journal of Physical Chemistry B, vol. 103, no. 50, pp. 11205–11210, 1999.
23. N. J. Miller and N. C. Miller, Statistics and Chemometrics For Analytical Chemistry, chapter 7, Prentice Hall, 5th edition, 2005.
24. D. B. Hibbert, "Experimental design in chromatography: a tutorial review," Journal of Chromatography A, vol. 910, pp. 2–13, 2012.
25. P. Wang, Z. Wang, and Z. Wu, "Insights into the effect of preparation variables on morphology and performance of polyacrylonitrile membranes using Plackett-Burman design experiments," Chemical Engineering Journal, vol. 193, pp. 50–58, 2012.
26. J. Zhou, X. Yu, C. Ding et al., "Optimization of phenol degradation by Candida tropicalis Z-04 using Plackett-Burman design and response surface methodology," Journal of Environmental Sciences, vol. 23, no. 1, pp. 22–30, 2011.
27. Design Expert Version 7, Stat-Ease Inc., Minneapolis, Minn, USA.
28. D. M. MacLeod, "Wire-Wound rod coating," in Coating Technology Handbook, D. Satas, Ed., Marcel Dekker, New York, NY, USA, 1991.
29. R. L. Plackett and J. P. Burman, "The design of optimum multifactorial experiments," Biometrika, vol. 33, pp. 305–325, 1946.
30. W. P. Gardiner, Statistical Analysis Methods For Chemists. A Software-Based Approach, The Royal Society of Chemistry, London, UK, 1997.
31. A. L. Barbarini, D. L. Reyna, and D. M. Martino, "The effect of light intensity, film thickness, and monomer composition in styrene-based bioinspired polymers," Green Chemistry Letters and Reviews, vol. 3, no. 3, pp. 231–237, 2010.
32. A. L. Barbarini, D. M. Martino, and D. A. Estenoz, "Synthesis, characterization and curing of bioinspired polymers based on vinyl benzyl thymine and triethyl ammonium chloride," Macromolecular Reaction Engineering, vol. 4, pp. 453–460, 2010.

CHAPTER 14

AN IONIC-POLYMER-METALLIC COMPOSITE ACTUATOR FOR RECONFIGURABLE ANTENNAS IN MOBILE DEVICES

Yi-Chen Lin[1], Chung-Yi Yu[1], Chung-Min Li[1], Chin-Heng Liu[1], Jiun-Peng Chen[2], Tah-Hsiung Chu[2] and Guo-Dung John Su[1]

[1]Graduate Institute of Photonics and Optoelectronics, National Taiwan University, No. 1, Sec. 4, Roosevelt Rd., Taipei 106, Taiwan

[2]Graduate Institute of Communication Engineering, National Taiwan University, No. 1, Sec. 4, Roosevelt Rd., Taipei 106, Taiwan

ABSTRACT

In this paper, a new application of an electro-active-polymer for a radio frequency (RF) switch is presented. We used an ionic polymer metallic composite (IPMC) switch to change the operating frequency of an inverted-F antenna. This switch is light in weight, small in volume, and low in cost. In addition, the IPMC is suitable for mobile devices because of its driving voltage of 3 volts and thickness of 200 μm. The IPMC acts as a normally-on switch to control the operating frequency of a reconfigurable antenna in mobile phones. We experimentally demonstrated by network analysis that the IPMC switch could shift its operating frequency from 1.1 to 2.1 GHz, with return losses of than −10 dB at both frequencies. To minimize electrolysis and maximize the operation time in air, propylene carbonate electrolyte with lithium perchlorate ($LiClO_4$) was applied inside the IPMC. The results showed that when the IPMC was actuated over three months at 3.5 V, the tip displacement fell by less than 10%. Therefore, an IPMC actuator is a promising choice for application to a reconfigurable antenna.

INTRODUCTION

Polymers have attractive advantages; they are soft, flexible, and lightweight. Some polymers can even change in size or shape under electrical stimulation. These are called electro-active polymers (EAPs). EAPs can provide large deformation with low driving voltage and produce force as actuators. Compared to conventional methods such as switches and mechanical actuators, EAPs have some attractive advantages like small electrical energy consumption, light weight and compliant properties, biocompatibility, ability to operate in air and aquatic media, insensitivity to magnetic fields and simple fabrication processes, making them suitable for actuators and sensors [1]. Based on the driving mechanism, EAPs can be classified into two major types: electronic EAPs and ionic EAPs [2]. Electronic EAPs can generate greater force than ionic EAPs and hold induced displacement under an electric field, but they require a high driving voltage. The deformation mechanism of electronic EAPs is the Coulomb force, including dielectric, electrostatic, piezoelectric and ferroelectric forces. Ionic EAPs generate deformation by movement or diffusion of ions between two electrodes. They can induce bending

Citation: Lin, Y.-C.; Yu, C.-Y.; Li, C.-M.; Liu, C.-H.; Chen, J.-P.; Chu, T.-H.; Su, G.-D.J. An Ionic-Polymer-Metallic Composite Actuator for Reconfigurable Antennas in Mobile Devices. Sensors 2014, 14, 834-847, doi:10.3390/s140100834

displacement at relatively low voltage, but at such a voltage, it is hard to make them maintain the same position. Popular ionic EAPs include ionic polymer gels, ionic polymer-metal composites, conductive polymers, and carbon nanotubes. Generally speaking, ionic EAPs can deform greatly when they are stimulated by electric signals.

An ionic polymer metallic composite (IPMC) consists of an ion-exchange polymer membrane, two coated electrodes, and mobile cations. An IPMC can bend gradually with increasing voltage. When the applied voltage is off, the IPMC is flat. When the voltage is on, the hydrated positive ions (Na^+, Li^+, H^+) carry water and move toward the anode. This mechanism of ionic motion causes the IPMC to bend [3]. An IPMC can be made into an actuator, and the actuator can be used in several applications, such as biomechanics, grabbers, and robotics.

The antenna design of portable devices is important. Antennas receive different signals according to their lengths in order to ensure good communication quality. Long antennas are used for low frequency bands. A cell phone is usually designed to receive many bands, such as GSM (Global System for Mobile Communications) 900, GSM 1800, Bluetooth, and local wireless. However, cell phones must also be small to satisfy the consumer. Thus, it is challenging for designers to integrate various antennas. We propose a concept to connect short antennas into one long antenna, or a reconfigurable antenna. A reconfigurable antenna has controllable parameters, such as the radiation pattern or operating frequency. Using this type of antenna allows effective use of the inner space of the cell phone. It is well known that changing the length of an antenna can shift the operating frequency. In order to design an antenna that can change size, RF switches have been introduced. An RF switch is used to change the radiation pattern or operating frequency [4–8]. Generally speaking, RF switches use PIN (positive-intrinsic-negative) diodes and GaAs transistors to control the signal, but those switches need relatively high power and provide low isolation. RF MEMS (micro-electro-mechanical systems) switches have also been proposed [9,10]. RF MEMS switches have low power consumption, low insertion loss, and high isolation, and they play an important role in RF devices. Although RF MEMS switches can shift the frequency of the antenna, the required driving voltage, at least nine volts [11], is considered high for portable devices. A common portable device can offer about 3 volts for switching, so the high driving voltage of MEMS switches is not ideal for portable devices. In order to overcome this problem, this paper introduces an IPMC actuator used as a switch. The low driving voltage and large tip displacement of IPMC actuators result in high isolation.

The operating frequency of an antenna is dependent on its physical length. To design a switch that can change the band of an antenna, we propose a bridge structure. The concept is to use an IPMC to connect two antenna paths, which can extend the physical length of the antenna and change the operating frequency between the high and low bands. This device consists of one antenna, an extension path, and a movable bridge controlled by the IPMC actuator, as shown in Figure 1. In this paper, we first show how to make IPMC switches. Then, we discuss the design concept of a reconfigurable antenna device using an IPMC switch. The simulation and experimental results show that a small driving voltage can be used to shift the operating frequency of the IPMC switch. Regarding the reliability, we know that the IPMC must work in wet conditions because it needs water molecules to bend. When an IPMC is used in mobile devices, however, it is actuated by voltage and operates in the air, where electrolysis and evaporation could occur. To solve these problems, a water solution was replaced in this study with propylene carbonate. We demonstrate herein that this is a promising solution to designing a reconfigurable antenna. Section 2 and Section 4 are related to characteristics and reliability of IPMC. While the antenna design and simulations are given Section 3. Finally, the paper is concluded in Section 5.

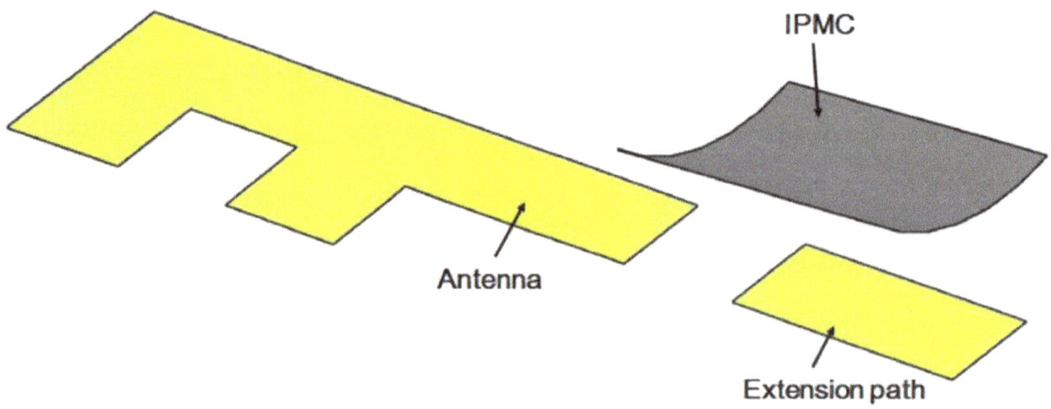

Figure 1: Schematic representation of one IPMC switch moved to control the operating band of an antenna.

IONIC POLYMER METALLIC COMPOSITE SWITCH

Fabrication Process

To fabricate an IPMC actuator, chemical decomposition, also known as electroless-plating, is required [12]. It is a method that utilizes reduction from a metal complex solution. This process has three major steps: surface treatment, first reduction, and second reduction. We used Nafion™-117 as an ion-exchange membrane and a platinum layer as electrodes. The fabrication flow is illustrated in Figure 2.

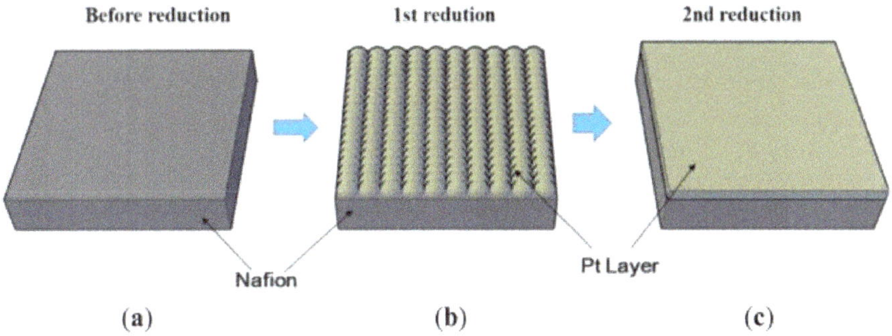

Figure 2: IPMC fabrication process. (**a**) Step 1: surface treatment; (**b**) Step 2: first plating and (**c**) Step 3: second plating.

Step 1: The first step of the IPMC fabrication was to roughen the surface of the ion-exchange membrane by sand blasting to increase the interfacial area density between the polymer membrane and metal composites. To remove any particles and impurities, we used ultrasonic washing and chemical treatment with boiled hydrochloric acid (low concentration) and boiled de-ionized water (DI water), respectively. Ion-adsorption was accomplished by immersing the membrane in platinum complex solution [$Pt(NH_3)_4Cl_2$], and ammonium hydroxide solution (NH_4OH) was introduced to neutralize the solution before the membrane was left to stand at room temperature for at least one night.

Step 2: Before the first plating, we immersed the membrane in DI water at 40 °C and added 2 mL of the aqueous sodium borohydride solution ($NaBH_4$, 5 wt%) and 20 mL of the $NaBH_4$. We added $NaBH_4$ every 30 min while gradually raising the temperature to 60 °C. As a result, a dark layer of Pt particles was deposited on the surface of the membrane, as shown in Figure 2b. Then the membrane was removed and rinsed with water, after which it was immersed in dilute HCl (0.1 M) for at least eight hours. The primary reaction for this process is:

$$NaBH_4 + 4[Pt(NH_3)]^{2+} + 8OH^- \rightarrow 4Pt + 16NH_3 + NaBO_2 + 6H_2O \quad (1)$$

Step 3: In the first plating, only a very thin platinum (Pt) layer was deposited. Therefore, the next step was to improve the electroplating process by plating a thicker layer of platinum by secondary reduction. Aqueous solution of the complex [$Pt(NH_3)_4$]Cl_2 was prepared, and 0.6 mL of the 25% ammonium hydroxide solution was added. Also, a 5 wt% aqueous solution of hydroxylamine hydrochloride ($H_2NOH{\cdot}HCl$) and 20% solution of hydrazine ($H_2NNH_2{\cdot}H_2O$) were prepared. The membrane was placed into the Pt solution at 40 °C, and 3 mL of hydroxylamine hydrochloride solution and 1.5 mL of the hydrazine solution were added every 30 min for 4 h as the temperature was gradually increased to 60 °C. During this process, the surface electrodes of metallic particles grew smoothly on both surfaces of the IPMC.

Step 4: After the IPMC was rinsed with water and boiled in dilute HCl (0.1 M), it was stored in sodium hydroxide solution (0.1 M) for 3 days. Na^+ in the composite can be exchanged for any cation by immersing the IPMC in a solution of another cation, such as LiOH, which can be used to replace the Na^+ with lithium cations.

Step 5: Propylene carbonate (99.5% w/w) 25 mL was prepared as solvent and 0.6 M lithium perchlorate ($LiClO_4$) was added. Then the IPMC was immersed in the propylene carbonate solution for 3 days.

Morphology Measurement

With the naked eye, it was not easy to recognize the differences between the electrode surfaces after the first plating and the second plating. Both electrodes were silver in color. Figure 3 shows top-view SEM micrographs (JSM-7600F, JEOL, Tokyo, Japan) of the IPMC surface electrodes after the initial plating and the second plating. The magnification is 30,000 times at 15 kV. In Figure 3a, it can be seen that the platinum particles were lumps and the cracks were deep. After the second reduction step, however, the platinum particles were small and the cracks were fine, as shown in Figure 3b. The samples were cut into pieces of 10 mm × 5 mm and the surface resistance was measured on the diagonal. The surface resistances with the initial plating and with the second plating were about 22.6 and 6.4 Ω, respectively. These results showed that after the second chemical plating, the surface electrodes were smoother and surface resistance was reduced. These conditions reduce evaporation of the solution in the ion exchange membrane.

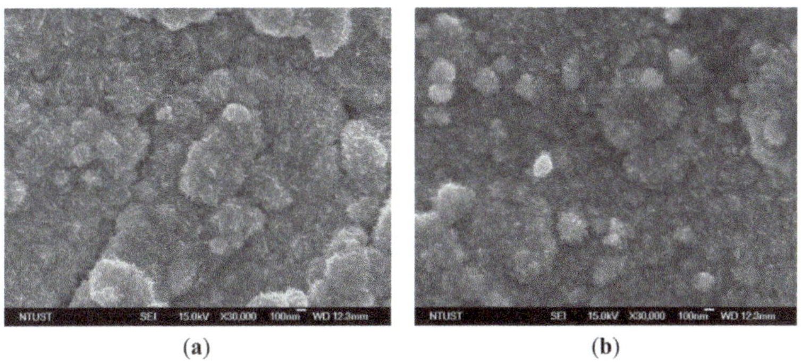

Figure 3: Top-view SEM micrographs of surface electrodes after (**a**) initial plating (magnification 30,000×) and (**b**) second plating (magnification 30,000×).

Tip Displacement of IPMC

Tip displacement was measured with a model number LK-G80 laser displacement sensor (Keyence, Itasca, IL, USA). The setup is illustrated in Figure 4a. The IPMC was actuated by a voltage source, and the laser sensor detected the position of the end of the IPMC during the operating period. Via a controller, the position data was sent to a computer for recording and analysis. Figure 4b shows the tip displacement of the IPMC switch *versus* time. The applied voltage was 3 V. We found that tip displacement was close to 2 mm at 1 s and reached the maximum value of about 6 mm in 6 s; the corresponding velocity was 1 mm/s. This result ensured that the IPMC switch would have two antenna paths of over 1 mm for changing of the operating frequency, as discussed in the next section.

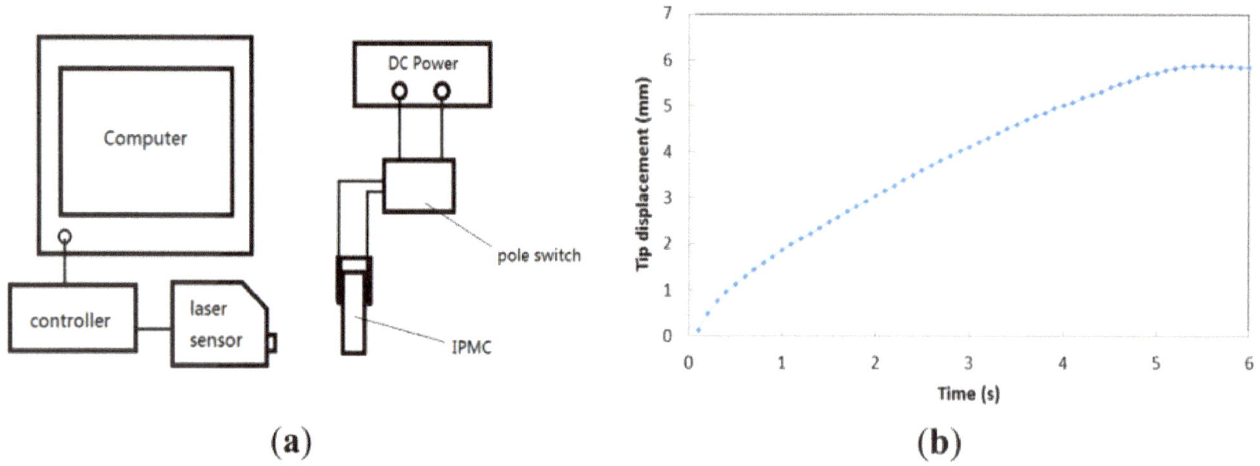

Figure 4: (**a**) Schematic diagram of tip displacement measured by laser sensor; (**b**) Diagram of tip displacement of IPMC switch *versus* response time at 3 V.

RECONFIGURABLE INVERTED-F ANTENNA

Design Concept

In order to prevent electric current from being drawn directly into the antenna, the IPMC could not be used directly as a switch. Indeed, the metal electrodes on the IPMC could affect antenna performance due to the metal shielding effect. We used an electrical-isolation material, a thin piece of glass, to separate the IPMC and copper sheet, as shown in Figure 5a. The thickness of the glass with the copper sheet was 1.145 mm, considering the weight of glass and cooper sheet, IPMC should be large enough to offer needed force to handle the structure. At the same time, the space was limited in a mobile phone. We chose a 7 mm × 10 mm IPMC to make a reconfigurable antenna. If we decreased the size of the IPMC, the force might be too small to contact the antenna properly.

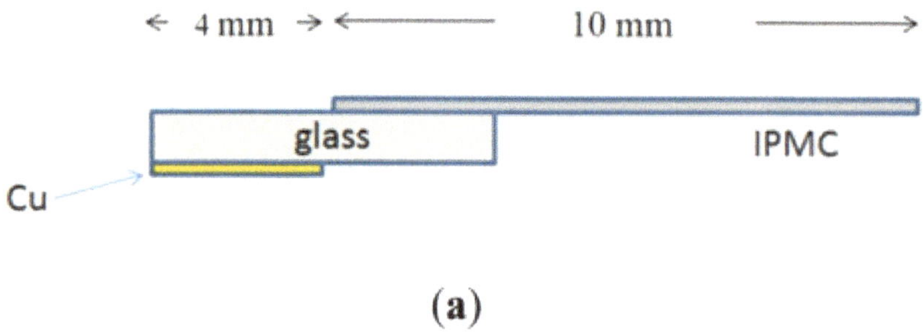

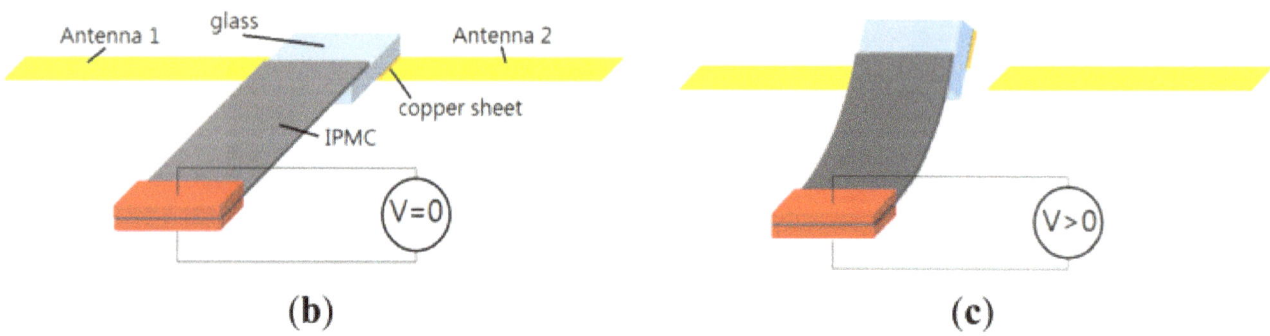

Figure 5: The illustration of normally-on IPMC switch design: (**a**) cross-section; (**b**) the on state ($V = 0$) and (**c**) the off state ($V > 0$).

We know that the IPMC will bend toward the positive electrode, so it is suitable for use as either a normally-on switch or a normally-off switch. In our proposed configuration, the IPMC-bridge was a normally-on switching device. When there was no applied voltage, the IPMC switch touched the antenna and was pressed by a load to avoid backward relaxation issue. The extended path was in a conducting state, shown in Figure 5b as on-state. When the voltage was applied, the IPMC switch bent away from the device and the entire system was an open circuit, shown inFigure 5c as off-state. Due to this feature, we were able to design an extra extension path for the antenna and use this IPMC switch to connect it. The backward relaxation appeared in off-state and it didn›t affect our control to change frequency because the distance between the actuated IPMC and the antenna was more than 1 mm.

Simulation Results

The antenna simulations were performed with the high frequency structure simulator (HFSS) developed by Ansoft® (Canonsburg, PA, USA). We used a popular antenna style, the inverted-F antenna, as our major design. The inverted-F antenna is widely used in wireless communication because of its low profile, low weight, and easy fabrication. We can pre-calculate the operating frequency by empirical formula. The three main parameters that decide the operating frequency are illustrated in Figure 6a. We can vary the combination of the three parameters to select the desired operating frequency and adjust the input impedance of the antenna to about 50 Ω of pure resistance. This removes the need for another impedance-matching circuit to match the microwave transmission line. To install the inverted-F antenna in a mobile phone model, we folded the upper arm of the antenna and assumed the parameter L to be the length of the new upper arm, as in Figure 6b. The structural parameter S in our design was 9 mm, L was 21 mm, and H was 2 mm. $L + H$ were roughly a quarter of the wavelength, which provided an operating frequency of 2.1 GHz and characteristic impedance near 50 Ω. The material of the antenna, feed line, ground line, and ground plane were copper with very high electrical conductivity, and the dielectric substrate was FR-4 epoxy, which is often used in printed circuit boards.

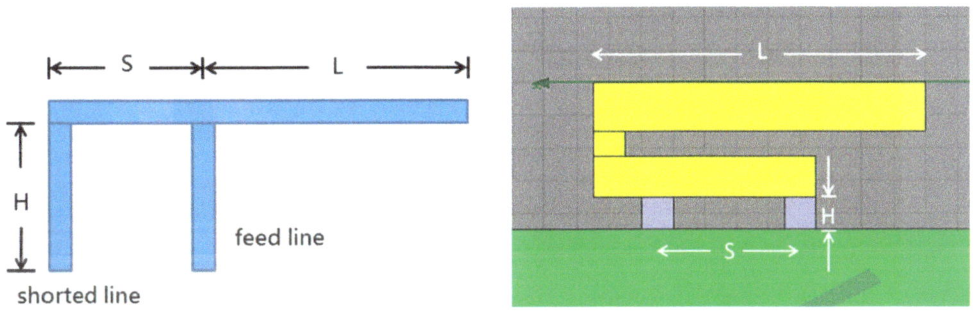

(a) (b)

Figure 6: (**a**) Structure of inverted-F antenna; and (**b**) Simulation layout with substrate board.

The purpose of our design concept, which is illustrated in Figure 1, is to extend the length of an antenna by using a switch. Therefore, it is necessary to set an extended path for connection. In order to design the extended path of an antenna, we used simulations to check the gap size, as shown in Figure 7. When there was no gap, the extended path was connected with the inverted-F antenna and the antenna was 40 mm long. The operating frequency was 1.1 GHz. When the gap was greater than 1 mm, the antenna was shorter and the operating frequency was around 2.1 GHz. Thus, we set the gap at 1 mm. Please note that we chose these frequencies for demonstration purposes; operating frequencies of 900 MHz and 1800 MHz can be achieved by adjusting the length of the antenna.

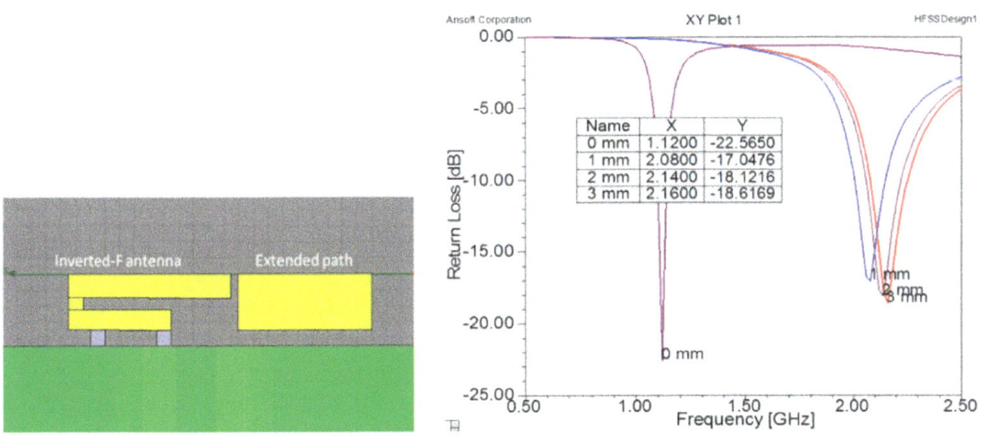

(a) (b)

Figure 7: (**a**) Design of extended path; (**b**) Optimization of gap between two antenna strips. The horizontal axis represents frequency and the vertical axis represents return loss.

Next, we created a small piece of copper to serve as a switch. The piece of copper was 7 mm × 4 mm. Figure 8 a shows the copper sheet set above the middle of the gap. If the copper sheet adhered to the ends of the two antenna strips, the strips were connected together and the effective length of the antenna was increased. If the copper sheet was separated from the two antenna strips, they were disconnected and the effective length of the antenna was decreased. The results of height variation of the switch are shown in Figure 8b. That figure shows that if the copper sheet left the plane of the antenna, the operating the frequency increased radically to 2.1 GHz. From the above results, we believe that using an IPMC actuator to control the height of the copper sheet is a proper method to achieve a dual-band inverted-F antenna.

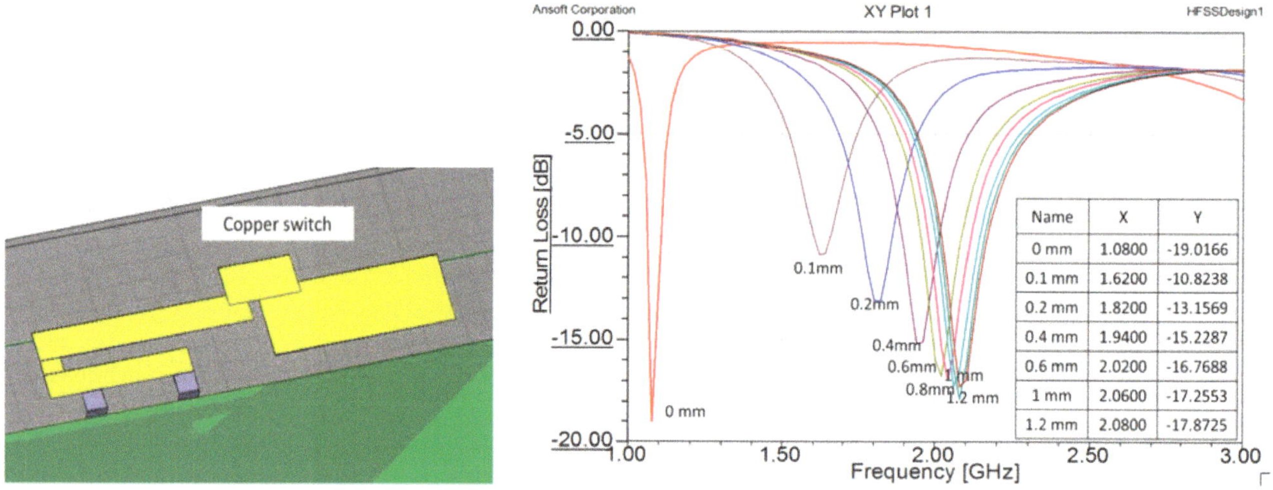

(a) (b)

Figure 8: (**a**) Design model of copper sheet switch; (**b**) Result of height variation of switch. The horizontal axis represents frequency and the vertical axis represents return loss.

Fabrication and Measurement of IPMC inside a Mobile Device

After simulation, we fabricated a similar antenna in a mobile phone. We pasted two strips of copper tape as an inverted-F antenna and extended the path on the shell, as shown in Figure 9a. Semi-rigid coaxial cable with a Sub Miniature version A (SMA) connector was used. When the back cover and the main body of the cell phone were combined, the antenna could touch the two metal contact points, one of which served as the grounded point and the other as the feed point, which was connected to the semi rigid coaxial cable with the SMA connector. We used the IPMC actuator to connect the antennas and extend the path, as shown in Figure 9b.

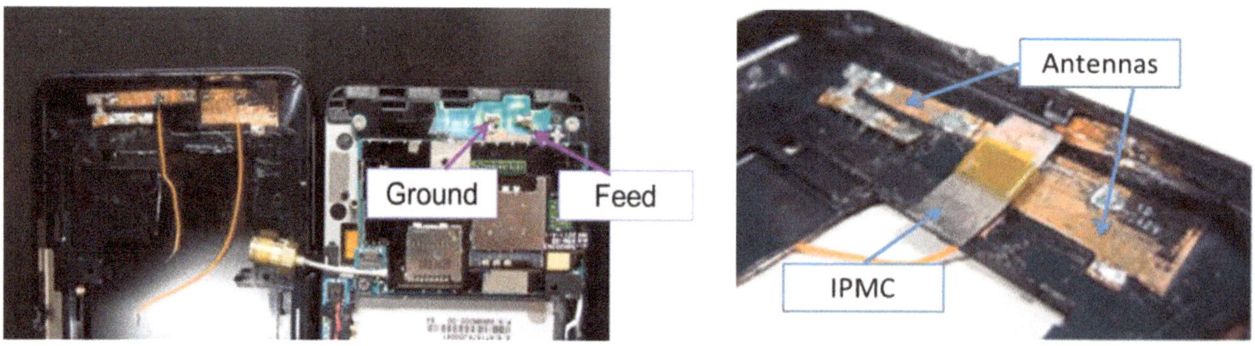

(a) (b)

Figure 9: (**a**) Structure of antennas in mobile phone model; (**b**) IPMC switch connects two antenna strips.

We measured the operating frequency *versus* return loss with a model number E5061B network analyzer (Agilent, Santa Clara, CA, USA). The network analyzer was set to output impulse signals at a range of frequencies while measuring the reflected amplitude at the same time. Thus, the resonant frequency of the antenna could be determined. As illustrated in Figure 10, we used a switch to open or close the voltage applied on the IPMC actuator. This device was connected to the Agilent network analyzer to measure the operation frequency and the return loss. The DC power supply was used to control the IPMC switch.

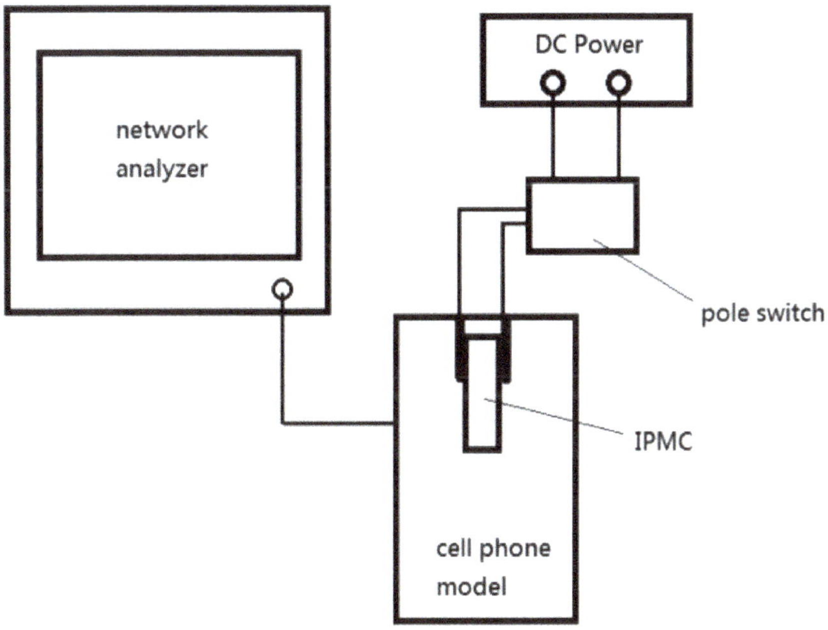

Figure 10: Schematic diagram of operation frequency and return loss measured by network analyzer.

The inverted-F antenna we designed was measured to have a best response of −14.92 dB at 2.14 GHz. In comparison, the previous simulation result was −18.26 dB at 2.12 GHz. A line chart of the experimental results and simulation are shown in Figure 11. The operating frequency peak values of both the experiment and the simulation were near 2.1 GHz, but the return loss peak value of the experiment was slightly higher than that of the simulation. Also, the bandwidth of the experiment was slightly wider than that of the simulation. This could be due to the antenna fabrication tolerance.

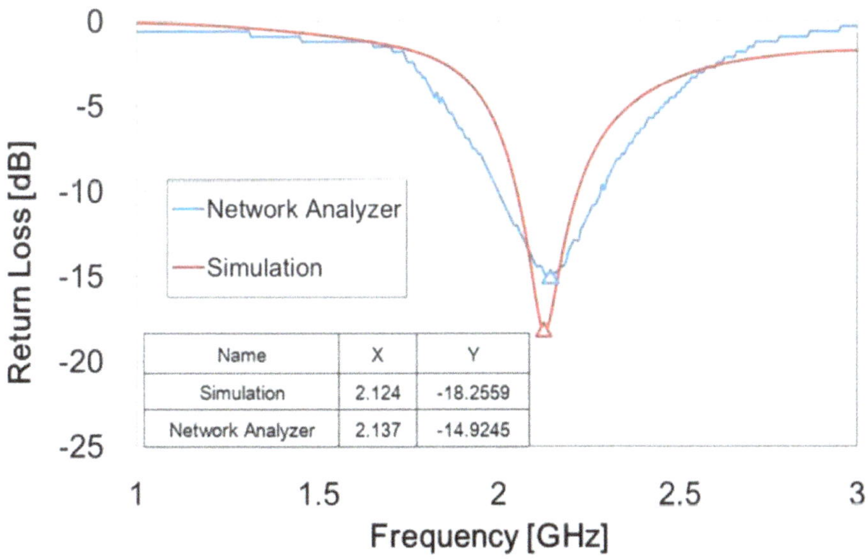

Figure 11: Line chart of experimental results compared with simulation when IPMC switch operates. The horizontal axis represents frequency and the vertical axis represents return loss.

Figure 12 shows the results of extending the antenna path when the IPMC switch contacted the inner ends of the two antenna strips. The operating frequency shifted to 1.06 GHz and the return loss was −10.37 dB, which means our design was able to attain the goal of changing the operating frequency to 1.1 GHz. The simulation result was −13.26 dB at 1.09 GHz. Although the return loss was higher, it was still within an acceptable range, below −10 dB. When the voltage was switched back to 3 V, the condition returned to the off state and the operating frequency returned to 2.1 GHz. Table 1 compares the measurement results. The operating frequency clearly changed between the two switch states, and both return losses were less than −10 dB. We operated this IPMC switch several times, and the response time results are presented in Table 2. The average response time from the off state to the on state (from 2.1 to 1.1 GHz) was 1.14 s, and the response time from the on state to the off state (from 1.1 to 2.1 GHz) was about 1.34 s. These response times demonstrate that the operating frequency could be changed easily and fast. We feel confident in stating that the effective length of the antenna was successfully changed by the IPMC switch.

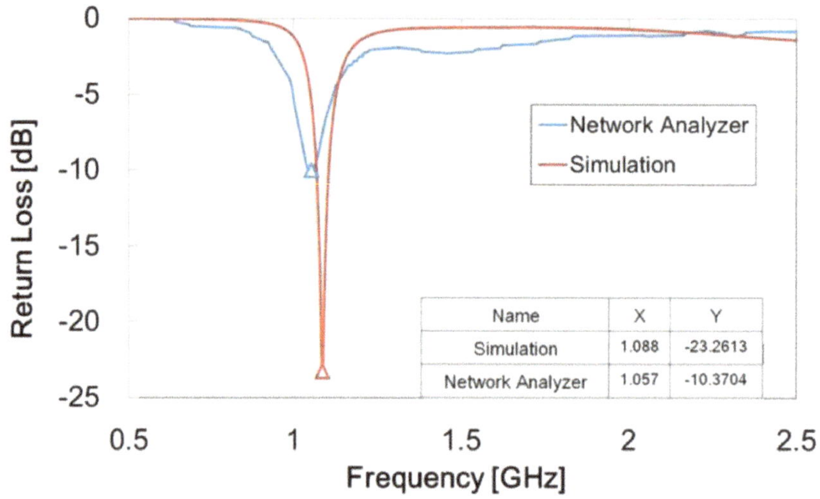

Figure 12: Line chart of experimental results compared with simulation when IPMC switch is off. The horizontal axis represents frequency and the vertical axis represents return loss.

Table 1: Result of operating frequency and return loss of reconfigurable antenna in mobile device

Voltage(V)	Switch state	Operating frequency(GHz)	Return loss(dB)
0	on	1.06	−10.4
3	off	2.13	−14.9

Table 2: Response time with changing frequency in a reconfigurable antenna

Switch Status	1st (s)	2nd (s)	3rd (s)	4th (s)	Average (s)
Off to On	1.12	1.24	1.19	1.01	1.14
On to Off	1.37	1.52	1.19	1.29	1.3425

RELIABILITY

As previously mentioned, the IPMC bends because of the hydrated positive ions carrying water that moves toward the anode. If the IPMC is exposed to air without the water supply, the water can evaporate and the water uptake can be decreased. Furthermore, we usually operated the IPMC actuator at 3 V, at which electrolysis of the water is inevitable. These defects degrade the bending performance and increase the difficulty of designing the mechanism.

To solve the above-mentioned questions, we used propylene carbonate electrolyte with lithium perchlorate ($LiClO_4$) for the IPMCs. The electrolyte uses propylene carbonate as its solvent and Li cations as its free ions. Propylene carbonate is a polar solvent that has very low vapor pressure and does not evaporate easily at room temperature. It also has a high dielectric permittivity that allows larger IPMC voltages without the problem of electrolysis.

In this work, we prepared propylene carbonate (99.5% w/w) 25 mL as solvent and added lithium perchlorate ($LiClO_4$) 0.6 M. The IPMCs were immersed into the propylene carbonate solution for 3 days. The IPMCs were 10 mm × 6 mm (sample A) and 10 mm × 5 mm (sample B), and we measured the tip displacement and weight of the IPMCs operated at 3.5 Volts DC each day. After each daily measurement, we stored the IPMC samples at room temperature.

Figure 13a is a statistical chart of variation in tip displacement of the IPMC after it was soaked in propylene carbonate. The tip displacements of both sample A (10 mm × 6 mm) and sample B (10 mm × 5 mm) were about 6 mm over three months. There was a slight drop after day 90. These results show that the performance of the IPMCs with propylene carbonate was steady, without evaporation or electrolysis. Figure 13b shows the variation in weight of the IPMC after being soaked in propylene carbonate. The weights of sample A and sample B were about 0.026 g and 0.02 g on the first day, respectively. After being actuated several times for several days, their weights fell to 0.022 g and 0.018 g on day 8, respectively. Thereafter, the weights remained stable. We think that the weight loss in the beginning was due to evaporation of the residual water content or gradual electrolysis under the applied voltage. After that, only propylene carbonate remained.

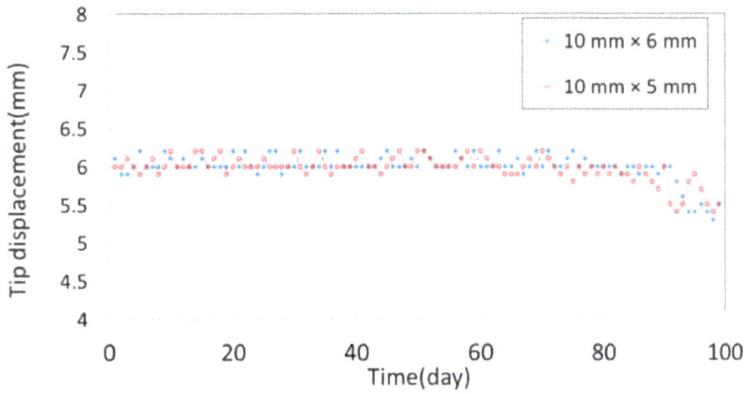

(a)

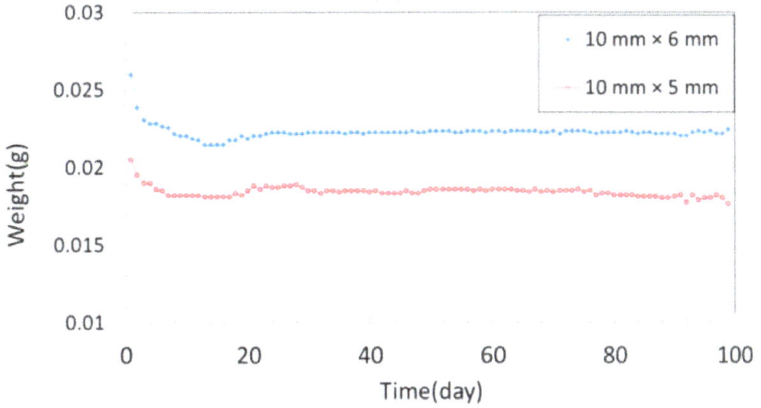

(b)

Figure 13: (**a**) Variation in tip displacement of IPMC with propylene carbonate solution. (**b**) Variation in weight of IPMC with propylene carbonate solution.

We normalized the data of the tip displacement of the IPMC with propylene carbonate solution and compared it to that with lithium hydroxide water solution, as shown in Figure 14. As time went by, the tip displacement of the IPMC with propylene carbonate solution remained almost the same. In contrast, with water solution, it dropped to less than 40% after day 20. In the future, attempts should be made to control the deposition density of the metal electrodes to minimize solvent loss. Coating other metals on the surface of the IPMC to make the electrodes denser could reduce the surface resistance and evaporation. Another way to eliminate solvent loss would be to encapsulate the IPMC. We feel confident that the lifetime of the IPMC in air can be effectively prolonged. From these results, we believe that propylene carbonate is a promising replacement for the water solution and is possible to make fast-speed polymer actuators [13] in future.

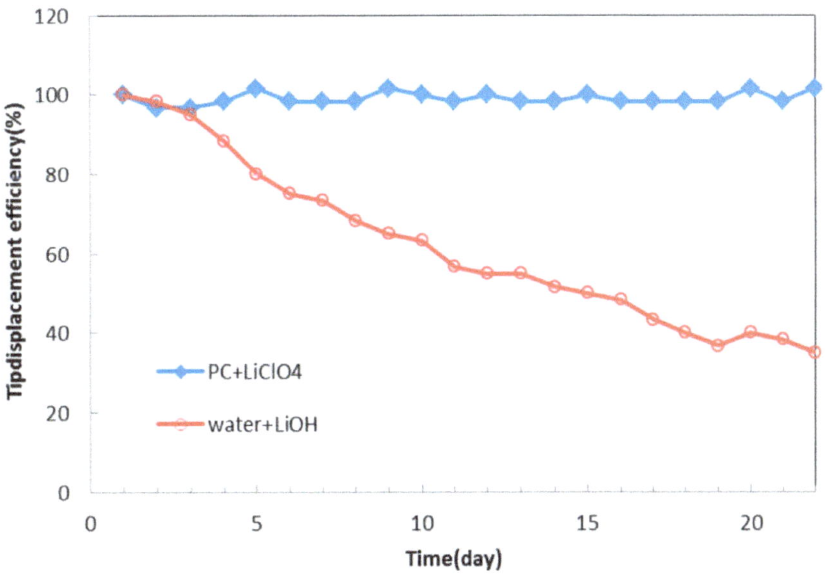

Figure 14: Variation of normalized tip displacement of IPMC with propylene carbonate and water solution.

CONCLUSIONS

In this paper, we have demonstrated a new RF switch that is light in weight and has a low driving power, large deformation, and frequency-shifting capability. After our experiments, we investigated a setup with a bridge style switch. This switch used the IPMC as an actuator to move a copper sheet upward and downward. When the IPMC-bridge was turned off, the antenna could be considered to be longer due to the copper sheet connecting the antennas and thereby extending the path. Simulation results showed that the operating frequency changed from 1.09 to 2.12 GHz and that the return losses at both frequencies were both less than −10 dB. Using a network analysis system, we determined that the original operating frequency of the antenna, 1.07 GHz, could be changed to 2.14 GHz when the IPMC switch was on. It has been demonstrated experimentally that the operating frequency can be effectively changed from a low band to a high band.

To prolong the lifetime of the IPMC in air, we used propylene carbonate electrolyte with $LiClO_4$. The IPMCs were still functional over three months after being actuated daily at 3.5 Volts DC voltage. Our device has lower driving voltage, which is suitable for application in cell phones. The IPMC switch is a potential solution to integrating antenna systems on mobile devices.

ACKNOWLEDGMENTS

This project was financially supported by National Taiwan University under contract number 102R7747.

REFERENCES

1. Malone, E.; Lipson, H. Freeform fabrication of electroactive polymer actuators and electromechanical devices.*Rapid Prototyp. J.* 2006, *12*, 244–253.
2. Bar-Cohen, Y.; Zhang, Q. Electroactive polymer actuators and sensors. *MRS Bull.* 2008, *33*, 173–181.
3. Bonomo, C.; Fortuna, L.; Giannone, P.; Graziani, S. A method to characterize the deformation of an IPMC sensing membrane. *Sens. Actuator A Phys.* 2005, *123–124*, 146–154.
4. Daneshmand, M.; Mansour, R.R. Multiport MEMS-based waveguide and coaxial switches. *IEEE Trans. Microw. Theory Tech.* 2005, *53*, 3531–3537.
5. Dail, C.-L.; Peng, H.-J.; Liu, M.-C.; Wu, C.-C.; Yang, L.-J. Design and fabrication of RF MEMS switch by the CMOS process. *Tamkang J. Sci. Eng.* 2005, *8*, 197–202.
6. Vaughan, R.G. Switched parasitic elements for antenna diversity. *IEEE Trans. Antennas Propag.* 1999, *47*, 399–405.
7. Zhang, S.; Huff, G.H.; Feng, J.; Bernhard, J.T. A pattern reconfigurable microstrip parasitic array. *IEEE Trans. Microw. Antennas Propag.* 2004, *52*, 2773–2776.
8. Huff, G.H.; Feng, J.; Zhang, S.; Bernhard, J.T. A novel radiation pattern and frequency reconfigurable single turn square spiral microstrip antenna. *IEEE Microw. Wirel. Compon. Lett.* 2003, *13*, 57–59.
9. Ativanichayaphong, T.; Cai, Y.; Wang, J.; Chiaoand, M.; Chiao, J.-C. Design Considerations of Reconfigurable Antennas Using MEMS Switches. Proceedings of the SPIE 2005 Microelectronics, MEMS, and Nanotechnology Symposium, Microelectronics: Design, Technology, and Packaging Conference, Brisbane, Australia, 11–15 December 2005.
10. Rebeiz, G.M.; Muldavin, J.B. RF MEMS switches and switch circuits. *IEEE Microw. Mag.* 2001, *2*, 59–71.
11. Pacheco, S.P.; Katehi, L.P.B.; Nguyen, C.T.-C. Design of low actuation voltage RF MEMS switch. *IEEE Microw. Symp. Dig.* 2000, *1*, 165–168.
12. Kim, K.J.; Shahinpoor, M. Ionic polymer–metal composites: II. Manufacturing techniques. *Smart Mater. Struct.*2003, *12*, 65–79.
13. Wu, Y.; Alici, G.; Spinks, G.M.; Wallace, G.G. Fast trilayer polypyrrole bending actuators for high speed applications. *Synth. Met.* 2006, *156*, 1017–1022.

CHAPTER 15

SYNTHESIS AND ANTIBACTERIAL TESTING OF SILVER/POLY (ETHER AMIDE) COMPOSITE NANOFIBERS WITH ULTRALOW SILVER CONTENT

Shuai Liang,[1] Geng Zhang,[2] Jianzhong Min,[1] Junqiao Ding,[3] and Xingmao Jiang[1,4]

[1]Key Laboratory of Fine Petrochemical Engineering, Changzhou University, Changzhou 213164, China

[2]Department of Chemistry, College of Science, Huazhong Agricultural University, Wuhan 430070, China

[3]State Key Laboratory of Polymer Physics and Chemistry, Changchun Institute of Applied Chemistry, Chinese Academy of Sciences, Changchun 130022, China

[4]Changzhou Yingzhong Nano Technology Company Limited, Changzhou 213023, China

ABSTRACT

Antimicrobial materials have attracted much attention all over the world. Herein, a new kind of antimicrobial material, poly (ether amide) (Pebax®) nanofibers containing Ag nanoparticles, was prepared by electrospinning method. The Ag/Pebax® composites were characterized by scanning electron microscopy (SEM), transmission electron microscopy (TEM), X-ray diffraction (XRD), X-ray photoelectron spectroscopy (XPS), differential scanning calorimetry (DSC), and thermogravimetric analyzer (TGA) measurements. The antimicrobial properties of Ag/Pebax® composites against Escherichia coli (E. coli; ATCC25922 and avirulent) and Staphylococcus aureus (S. aureus; ATCC6538 and avirulent) were evaluated by membrane adhering method. It was found that the Ag content played an important part in the antimicrobial ability of Ag/Pebax® composites. When the mass ratio of $AgNO_3$ to Pebax® in the precursor was 0.15%, the inhibition rate can reach >99.9% and antimicrobial activity against E. coli and S. aureus was 5.8 and 5.6, respectively, exceeding the antimicrobial testing standards JIS Z 2801. The above results indicated that the Ag/Pebax® composite was a promising antimicrobial material that can be used in many applications.

INTRODUCTION

Nowadays, human beings are facing a variety of infectious diseases, not only conventional sicknesses (e.g.,malaria) but also emerging viral influenza (such as SARS and avian flu), all of which were spread by microorganism. Therefore, in the past few years, antimicrobial agents have been paid more and more attention, in which silver and its compounds were proved to possess potent antimicrobial efficiency against microorganism [1, 2]. It is generally believed that the bacterial cell will be killed upon the contact with silver, because the positively charged Ag cations strongly bind to electron donor groups containing sulphur, oxygen, or nitrogen in the cell membrane and enzymes, resulting in structural and functional changes in the cell and finally death of the microorganism [3–7]. Moreover, multiplication of bacteria may be prevented owing to the interaction between the Ag^+ and DNA structure [4, 5]. In addition, because of immediate death of microorganisms upon contact with Ag^+, there are minimal chances for development of bacterial resistance [8]. In comparison with Ag ions, Ag

Citation: Shuai Liang, Geng Zhang, Jianzhong Min, Junqiao Ding, and Xingmao Jiang, "Synthesis and Antibacterial Testing of Silver/Poly (Ether Amide) Composite Nanofibers with Ultralow Silver Content," Journal of Nanomaterials, vol. 2014, Article ID 684251, 10 pages, 2014. doi:10.1155/2014/684251

nanoparticles are superior in heat resistance, chemical stability, release rate of Ag^+, and service time, bringing about more opportunities for antimicrobial applications [9].

On the basis of the aforementioned advantages, combining Ag nanoparticles with polymers to form Ag/polymer composites has attracted much interest of many researchers, because the Ag/polymer composite can be applied in diverse fields, such as medical use, daily necessities, and clothing. Recently, Ag/polymers composites with different polymer matrixes, for example, cellulose acetate [10], polyurethane [8], poly (ethylene oxide) [11], poly (vinyl alcohol) [1], and poly (lactic-co-glycolic acid) [12], have been fabricated by electrospinning. The electrospinning is a popular technique to prepare high-performance polymer products due to its cost effectiveness, versatility, and simplicity. Briefly, electrospinning is a process by which a polymer solution is ejected under high electric fields to produce polymeric fibers [13]. The fibers are collected as a nonwoven and porous web with a high surface to volume ratio, small pore sizes, and high porosity [1, 8, 14], which is suitable for biomedical use, disinfection, water purifying systems, and so on.

Pebax is a kind of copolymer, and its structure unit consists of dodecyl amide and ether components, which offers a perfect combination of mechanical strength, breathability, water barrier property, gas barrier property, flexibility, chemical resistance, no skin allergy, and ease of processing. So it can be used to make urinary catheter, arterial catheter, PTCA (percutaneous transluminal coronary angioplasty) catheter device, infusion bags, and so forth [15–18]. Furthermore, Pebax is a high-performance thermoplastic polyamide elastomer with excellent durability and workability, which can be widely used as high-performance sports materials [19–21]. If the antibacterial ability can be rendered to the Pebax by embedding Ag nanoparticles, the performance of Pebax will be dramatically improved with certainty. However, to the best of our knowledge, the preparation of Ag/Pebax composite has not been reported so far.

Herein, we employed electrospinning technique and ultraviolet treatment to prepare Ag/Pebax composites without any other reducing agent. The size-uniform Ag nanoparticles were well dispersed in the Pebax matrix. The antimicrobial properties of this Ag/Pebax composite against Escherichia coli and Staphylococcus aureuswere studied via membrane adhering method. Study results showed that Ag/Pebax presented excellent antimicrobial properties at an ultralow $AgNO_3$ content in the precursor (0.15%), which was much lower than that of conventional antimicrobial Ag/polymer composites reported in the literature [1, 8–10, 12], making Ag/Pebax composite a promising antimicrobial material with low-cost and excellent antimicrobial properties, which could be widely used in many fields in the near future.

EXPERIMENTAL SECTION

Preparation of Ag/Pebax Composites

First, Pebax 2533 (ARKEMA Firm) was added into isopropanol (Shanghai Lingfeng, >99%) and the mixture was magnetically stirred for 5 h at 60°C to obtain Pebax solution with a concentration of 66.7 mg mL^{-1}. Meanwhile, 6.7 mg mL^{-1} $AgNO_3$ (Shanghai Lingfeng, 99.99%) in isopropanol solution was prepared in dark. Then a given amount of $AgNO_3$ solution was added into the Pebax solution and the mass ratio of $AgNO_3$ to Pebax was controlled from 0.05% to 0.25%. After sonication, the homogeneous solution was fed into a positively charged spinneret with a diameter of 0.6 mm attached to an electrospinning apparatus [22]. The voltage used for electrospinning was 13 kV and the collection distance was 10 cm. The solution was spun under its own gravity without flow control equipment and the product obtained ($AgNO_3$/Pebax) was dried freely in air. Finally, the as-prepared samples were treated under UV light (254 nm, 14 W) for 6 h to obtain Ag/Pebax composites, which was denoted as Ag/Pebax-1, Ag/Pebax-2, Ag/Pebax-3, Ag/Pebax-4, and Ag/Pebax-5 for the sample with the mass ratio of $AgNO_3$ to Pebax at 0.05%, 0.10%, 0.15%, 0.20%, and 0.25%, respectively.

Characterizations

The morphology of the resulting electrospun nonwoven fiber mat was evaluated using scanning electron microscopy (SEM, Zeiss, SUPRA 55, Germany). The transmission electron microscopy (TEM) image was obtained on a JEOL JEM-2100 microscope operating at 200 kV equipped with energy dispersive spectrometer (EDS). The crystalline structure of the sample was identified by X-ray diffraction (XRD, Rigaku D/max 2500PC) using Cu Kα radiation. A simultaneous differential scanning calorimeter (PerkinElmer, DSC 8500, USA) was operated in the temperature range from −40°C to 80°C with a heating rate of 10°C min^{-1} under a nitrogen atmosphere. The thermal degradation behaviors of Ag/Pebax nanofibers were studied on a thermogravimetric analyzer (PerkinElmer, Pyris 1 TGA, USA) from 30°C to 800°C at a heating rate of 10°C min^{-1} under nitrogen at a flow rate of 20 mL min^{-1}. The chemical state of the sample was examined by X-ray photoelectron spectroscopy (XPS) on a VG Scientific ESCALABMKLL spectrometer using Al Kα X-ray source (10 mA, 15 kV).

Bacterial Culture and Test for Antibacterial Activity

The bacteria used for the test were Escherichia coli (E. coli: ATCC25922) and Staphylococcus aureus (S. aureus: ATCC6538). Any water used was distilled and all reagents were of a grade appropriate for microbiological purposes. The stock culture was stored at 5°C to 10°C on an appropriate medium and the bacteria were transferred from the stock culture to a slant culture medium and incubated at (35 ± 1)°C for 24 h. Then bacteria were transferred onto another slant culture medium and incubated for another 20 h. After that, the bacteria were transferred into 1/500 NB (nutrient broth) prepared in advance, obtaining a bacterial concentration between 2.5×10^5 and 10×10^5 cells mL^{-1}, which was used as the test inoculum. Then 0.4 mL of the test inoculum was pipetted onto the test surface and covered by a piece of square Ag/Pebax film with side length of 40 mm and a thickness of ~1 mm. The test specimen was incubated at (35 ± 1)°C with a relative humidity of 90% for 24 h. After incubation, 20 mL of neutralizer was added to wash the specimen and then the viable bacteria count was determined by pour-plate culture method. For the testing, all petri dishes should be inverted and incubated at (35 ± 1)°C for 40 h, and then the number of colonies in petri dishes was counted [22].

The number of viable bacteria was calculated by

$$N = C \times D \times V, \quad (1)$$

Where N is the number of viable bacteria recovered per test specimen; C is the average plate count for the duplicate plates; V is the dilution factor for the plate counted; and is the volume in mL of neutralizer added to the specimen.

The antibacterial activity was calculated by

$$R = U_t - A_t, \quad (2)$$

where R is the antibacterial activity; U_t is the average of the common logarithm of the number of viable bacteria recovered from the untreated test specimens after 24 h; and A_t is the average of the common logarithm of the number of viable bacteria recovered from the treated specimens after 24 h [23, 24].

RESULTS AND DISCUSSION

The electrospun Pebax was tough and flexible, and it is white in color (Figure 1(a)). The SEM and TEM images (Figure 2) indicated that the electrospun Pebax was mesh on a microlevel formed by the connection between Pebax nanofibers, proving that the electrospinning technique can be applied successfully on Pebax.

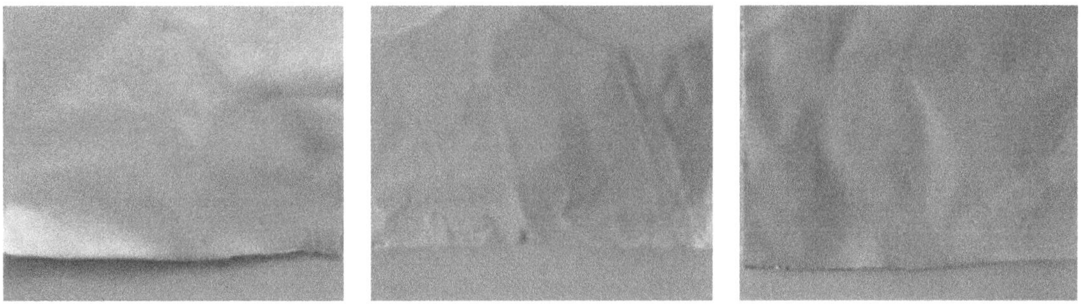

(a) (b) (c)

Figure 1: Optical photographs of an electrospun nanofiber web: (a) Pebax, (b) $AgNO_3$/Pebax-3, and (c) Ag/Pebax-3.

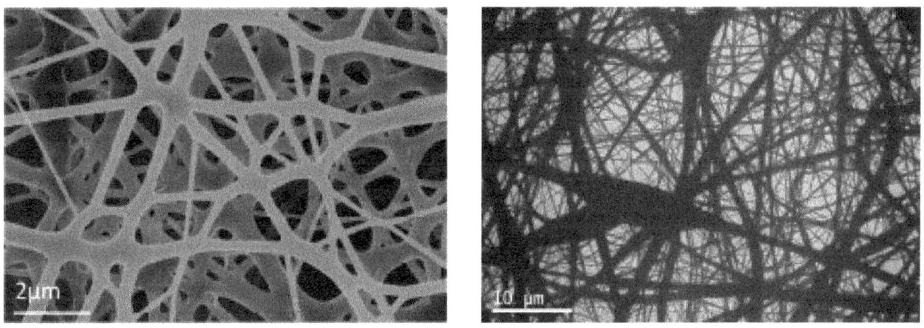

(a) (b)

Figure 2: (a) SEM and (b) TEM images of Pebax nanofibers.

In order to confer antibacterial ability to Pebax, $AgNO_3$ was electrospun with Pebax simultaneously and the as-prepared samples were treated with UV light to decompose $AgNO_3$ to metallic Ag. As shown in Figure 1(c), the Ag/Pebax composite was pale yellow in color which was obviously different from that of pure Pebax (Figure1(a)) and $AgNO_3$/Pebax composite (Figure 1(b)). The color change should come from the presence of metallic Ag in Ag/Pebax composite, so we further studied the composite by TEM. As shown in Figures 3(b)–3(f), the metallic Ag was in near spherical morphology and Ag nanoparticles were dispersed homogeneously within the fiber matrix. Although the number density of Ag nanoparticles in the Pebax matrix was increased with the $AgNO_3$ content, no aggregation was observed. In addition, the Ag particle diameter distribution of Ag/Pebax-2 to Ag/Pebax-5 is shown in Figures 3(g) to 3(j), respectively, which was obtained by measuring about 200 Ag nanoparticles for each sample. Both of the particle diameters of the two samples ranged between 13.5 nm and 16.5 nm, with average particle size at 14.8, 14.9, 14.7, and 14.8 nm for Ag/Pebax-2 to Ag/Pebax-5, respectively, on the assumption of Gaussian distribution. Since the density of Ag particle in Ag/Pebax-1 was very low, its average particle size was calculated to be 14.8 nm by using the arithmetic mean value of all Ag particles. The above results indicated that the mean particle size of Ag nanoparticles nearly did not change with the increase of Ag content, which may be due to the fact that the Ag content was very low and $AgNO_3$ was completely dissolved in the solution; thus the $AgNO_3$ crystals after electrospinning were distributed uniformly in the Pebax matrix, resulting in the fact that the Ag nanoparticles reduced were also monodisperse and the mean particle size only fluctuated slightly at different Ag loadings. Additionally, the EDS spectrum of Ag/Pebax-5 in Figure 3(k) confirmed the ultralow content of Ag content in the composite. The crystalline information of Ag/Pebax-5 was analyzed by XRD. As shown in Figure 3(l), the diffraction peak at 20° was ascribed to pure Pebax 2533 [25]. According to JCPDS 04–0783 (Ag), the crystalline Ag should be revealed by diffraction peaks at 2θ values of 38.12°, 44.28°, and 64.43°. However, no distinctive diffraction of silver was observed in the Ag/Pebax composite even for Ag/Pebax-5 with the highest concentration of $AgNO_3$ (0.25%) (Figure 3(l)). It was probably because of the fact that the concentration of Ag included in the composite was below the detection limit of the XRD method [26].

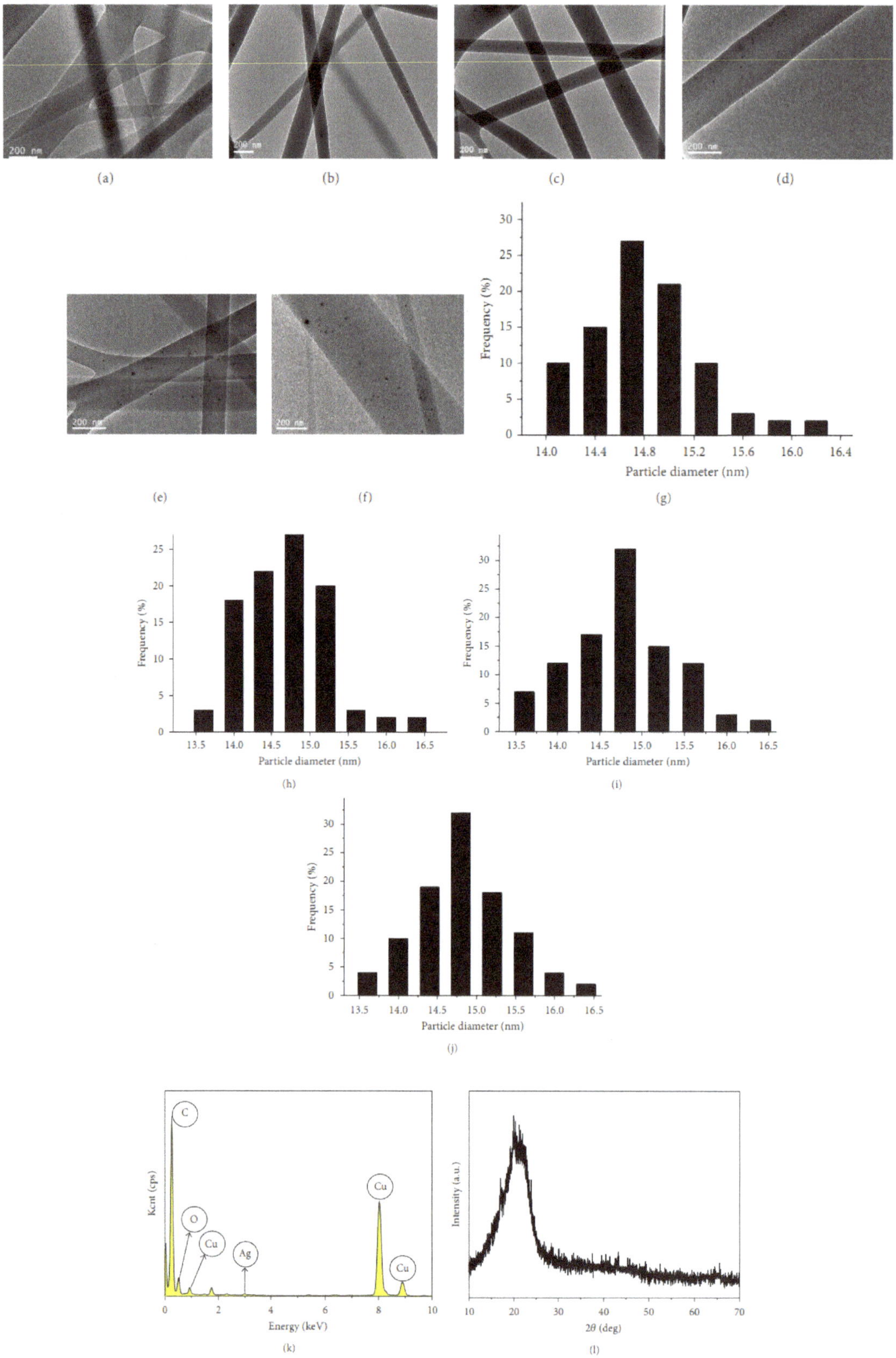

Figure 3: ((a)–(f)) TEM images of (a) Pebax, (b) Ag/Pebax-1, (c) Ag/Pebax-2, (d) Ag/Pebax-3, (e) Ag/Pebax-4, and (f) Ag/Pebax-5. ((g)–(j)) The particle size distribution of Ag nanoparticles of (g) Ag/Pebax-2, (h) Ag/Pebax-3, (i) Ag/Pebax-4, and (j) Ag/Pebax-5. (k) The EDS spectrum of Ag/Pebax-5. (l) The XRD patterns of Ag/Pebax-5.

To investigate the effect of silver on the crystallinity of Pebax, we performed DSC measurements in a N_2 atmosphere. Formula for calculating the degree of crystallization (X_c) was $X_c = \Delta H/\Delta H_o$, where ΔH is the heat of fusion of the sample in this experiment and ΔH_o is the heat of fusion of pure Pebax with 100% crystallinity which is a constant here. Figure 4(a) shows the DSC thermograms for Pebax and the Ag/Pebax composite between −40° C and 80° C. It was found that the ΔH of Ag/Pebax composites was lower than that of pure Pebax, indicating that the X_c of Pebax polymer was reduced in the presence of Ag nanoparticles. Furthermore, the reduction in crystallinity was enhanced with the increase of the $AgNO_3$ content (Figure 4(b)). In addition, the glass transition temperature (T_g) for Pebax, Ag/Pebax- 1, Ag/Pebax-2, Ag/Pebax-3, Ag/Pebax-4, Ag/Pebax-5, and Ag/Pebax-6 is −6.80∘ C, −7.34° C, −7.52° C, −7.95° C, −8.13° C, −8.25° C, and −8.36° C, respectively, indicating that the T_g of Pebax polymer is gradually reduced with the increase of the AgNO3 content. Generally, the regular structure of ether segments is in favor of the crystallization [27]. In this work, with the increase of Ag content, the regularity of the internal structure of Pebax was reduced, whereas the flexibility of the polymer chain was improved [28], which may result in the decrease of the X_c and T_g.

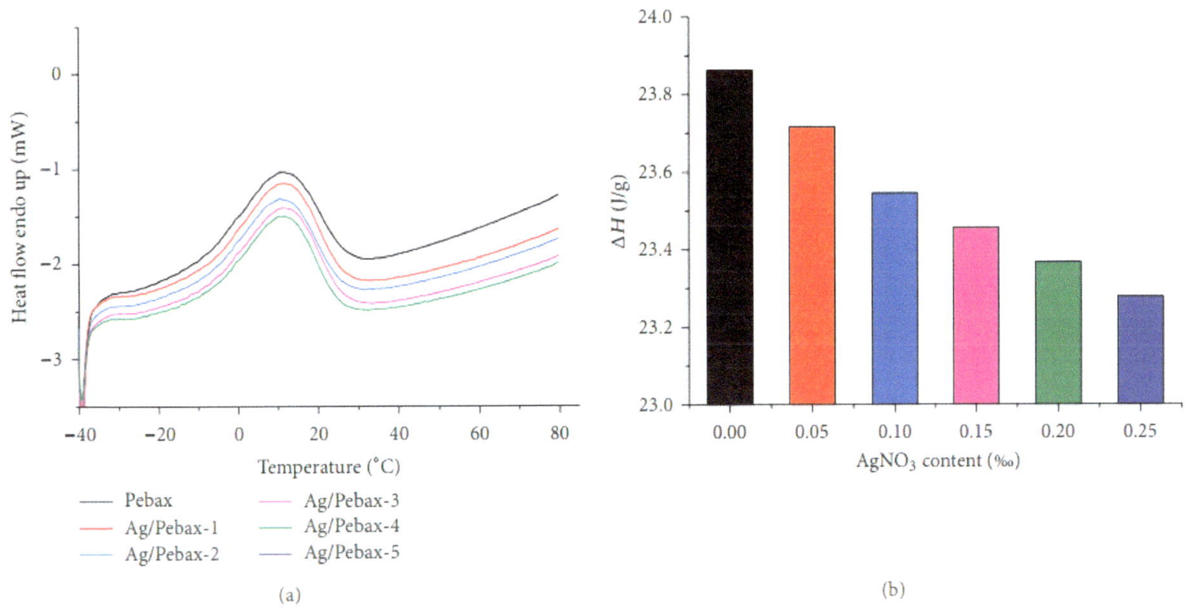

Figure 4: (a) DSC diagrams and (b) the heat of fusion for pure Pebax nanofibers and Ag/Pebax composites with different $AgNO_3$ contents.

The thermal degradation behaviors of Ag/Pebax nanofibers were studied with a TGA under nitrogen atmosphere. With the increase of Ag content, the onset decomposition temperature of the Ag/Pebax composite gradually decreased. As shown in Figure 5, the onset decomposition temperature of Pebax, Ag/Pebax-1, Ag/Pebax-2, Ag/Pebax-3, Ag/Pebax-4, and Ag/Pebax-5 is 350.14°C, 330.62°C, 310.77°C, 291.30°C, 273.10°C, and 250.21°C, respectively. Because the Ag nanoparticles dispersed uniformly in the nanofibers and their particle size nearly did not change at different Ag contents, the increase of Ag content in nanofibers will increase the number of Ag nanoparticles, which can weaken the tightness of the Pebax molecule arrangement and expand the volume of noncrystalline region in the nanofiber, resulting in the decrease of thermal stability of the whole composite, that is, a drop of the onset decomposing temperature. Additionally, Ag nanoparticles may also play a catalytic role in the decomposition during DSC measurement.

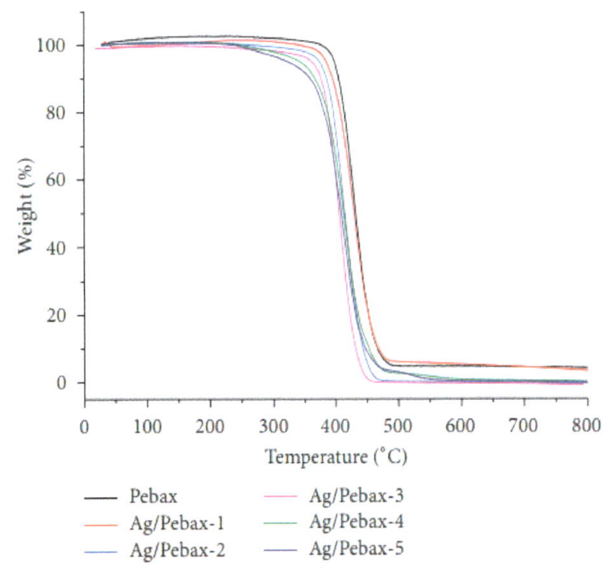

Figure 5: TGA data of pure Pebax nanofibers and Ag/Pebax composite with different $AgNO_3$ contents.

The chemical state and near surface composition of the Ag/Pebax composite were studied by XPS. With a survey range of 0–1200 eV, four elements can be detected: carbon, silver, nitrogen, and oxygen (Figure 6). Figure 7 shows the high-resolution XPS spectra of Ag/Pebax-3. From the analysis of C 1s spectrum (Figure7(a)), we can see carbon atoms in the C–C bond, C–O bond, C=O bond, and N–C=O at 284.5 eV, 285.6 eV, 286.5 eV, and 288.1 eV, respectively. The peak at 532.1 eV in the O 1s spectrum (Figure 7(b)) is attributed to the oxygen in the C–O bond and the peaks at 531.1 eV and 533.7 eV are assigned to the oxygen of C=O and O–C=O bonds, respectively [29]. The N 1s spectrum presents a peak at 399.4 eV which comes from the nitrogen in the –CONH– group. In the XPS spectrum of Ag 3d (Figure 7(d)), two peaks were observed at 368.2 and 374.2 eV, which can be ascribed to Ag ($3d_{5/2}$) and Ag ($3d_{3/2}$) peaks, respectively. Moreover, the splitting of the 3d doublet of Ag was 6.0 eV, which indicates that these Ag/Pebax composites only contain metallic Ag (Ag^0) [30–32], indicating the complete reduction of $AgNO_3$ by UV.

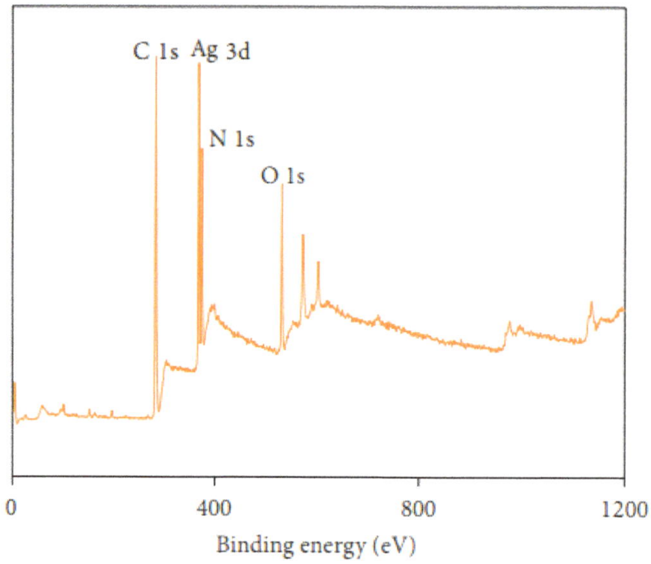

Figure 6: XPS spectrum of Ag/Pebax-3.

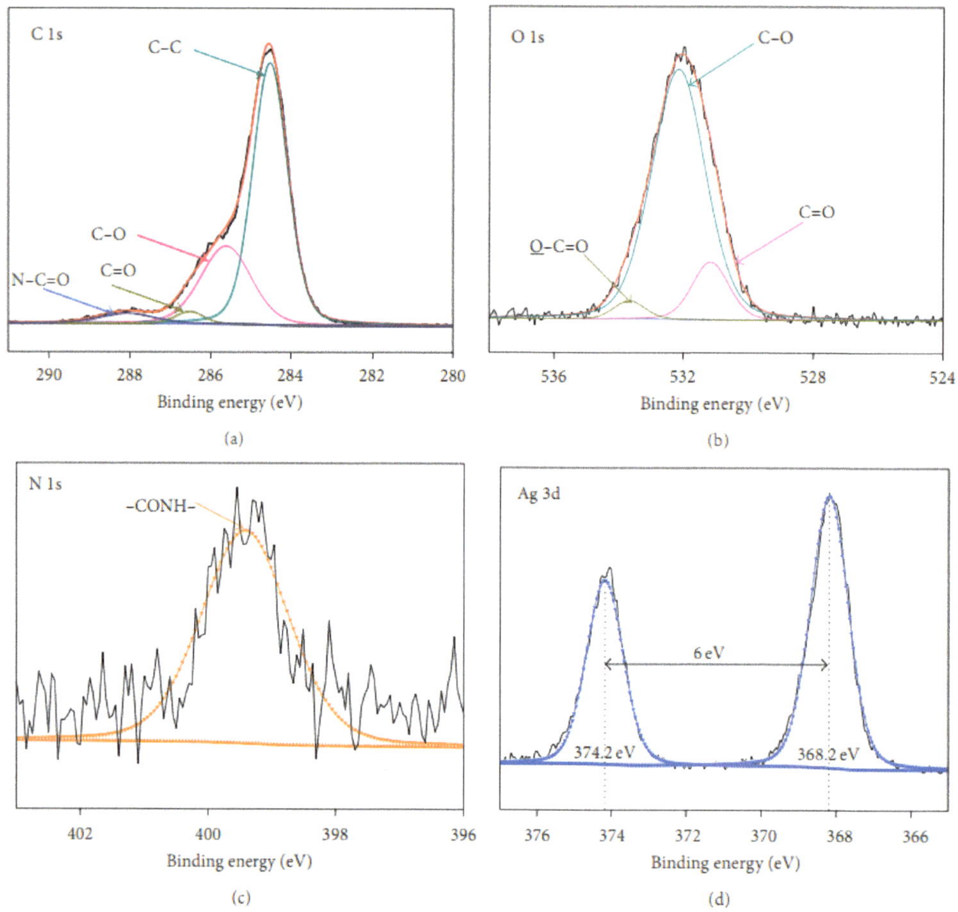

Figure 7: XPS deconvolution spectrum of (a) C 1s, (b) O 1s, (c) N 1s, and (d) Ag 3d core levels of Ag/Pebax-3.

The potential use of Ag/Pebax composite as antibacterial material was assessed by observing their antibacterial activity against two kinds of common bacteria, E. coli and S. aureus. The electrospun Pebax without Ag was evaluated in a control experiment. The bacteria were incubated in a growth medium containing various nanofibers and the antibacterial properties of samples were tested by membrane adhering method, following the antibacterial testing standard (JISZ 2801:2010). The changes in the content of Ag had a significant effect on the antibacterial properties of Ag/Pebax composite. The number density of bacterial colony for pure Pebax was high (Figure 8(a)), and both of the antibacterial rate (1) and antimicrobial activity (2) were nearly zero (Figure9); that is, there was no antibacterial activity observed for Ag-free pure Pebax. In contrast, the samples added with small amount of $AgNO_3$ in the Pebax precursor solution showed an inhibitory effect on E. coli and S. aureus. Although the content of $AgNO_3$ was only 0.05% (Figures 8(b) and 9(b)), the antibacterial rate againstE. coli and S. aureus can reach 88.2% and 76.1%, respectively. Furthermore, the antibacterial properties of Ag/Pebax composite were enhanced gradually with the increase of $AgNO_3$ concentration (Figures 8(b)–8(d), 9(b)–9(d) and 10). When the $AgNO_3$ content reached 0.15%, no bacterial colony can be observed (Figures 8(d) and 9(d)) and the inhibition rates all larger than 99.9%, and the value of antimicrobial activity against E. coli and S. aureus was 5.8 and 5.6, respectively (Figure 10). Further increasing the content of Ag failed to enhance the antibacterial properties of Ag/Pebax any more (Figures 8(e)–8(f), 9(e)–9(f) and 10). Because the Pebax did not show antibacterial ability, Ag nanoparticles were responsible for the antibacterial ability of Ag/Pebax composites. It was reported that the antimicrobial property of material containing silver is mainly determined by the total release rate of silver ions [32]. In this work, owing to the stable particle size, the increase of Ag content will lead to the enlargement of the number and the total surface area of Ag nanoparticles; thus the release rate of silver ions was accelerated, giving rise to the improvement of antibacterial ability of Ag/Pebax composites (Figure 10). According to the antimicrobial testing standard JISZ 2801:2010, the sample whose antimicrobial activity is greater than 2 (R>2) passes the antibacterial test of the strain R>2 (means the antibacterial efficiency can reach 99% or more). According to the experimental results, it can be concluded that the Ag/Pebax composite with $AgNO_3$ content larger than 0.15% presented excellent antibacterial properties, and this Ag content was much lower than that of Ag/polymer composites reported in the literature (usually larger than 1 wt.%) [1, 7, 8, 11, 33–35], making Ag/Pebax a promising antibacterial material with potential application in a number of fields, such as biomedical materials, sports apparatus, and laminating films.

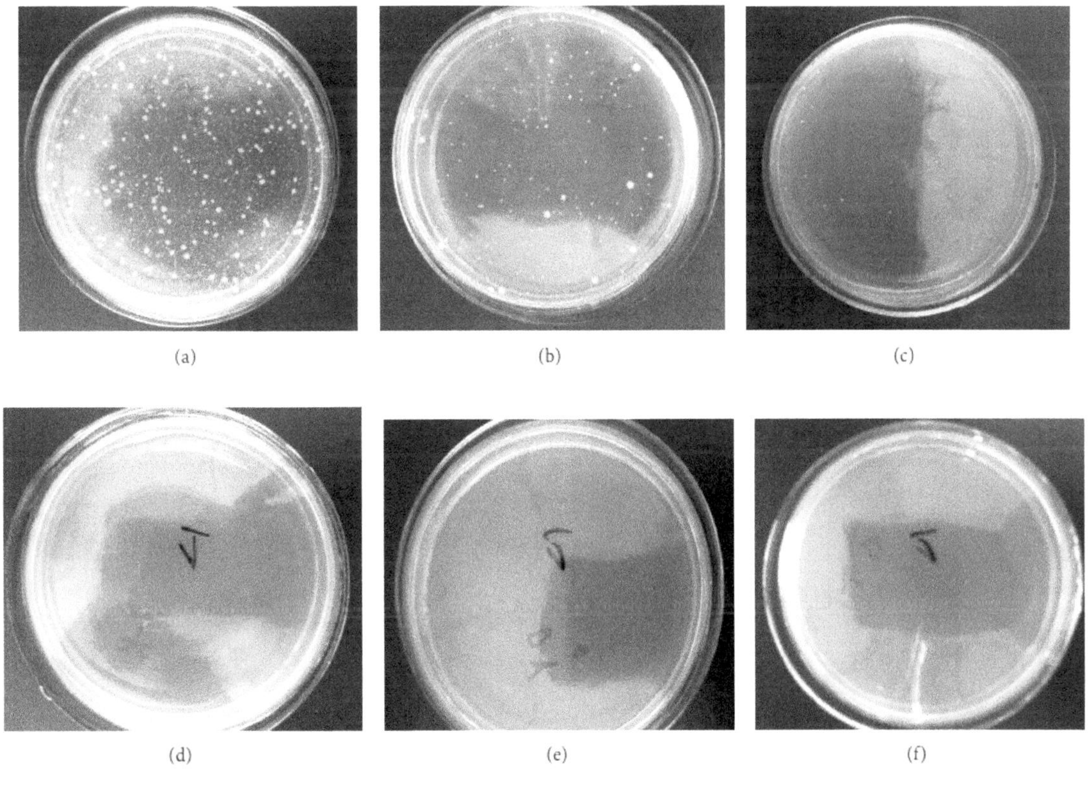

Figure 8: Comparison of the antibacterial ability of (a) Pebax, (b) Ag/Pebax-1, (c) Ag/Pebax-2, (d) Ag/Pebax-3, (e) Ag/Pebax-4, and (f) Ag/Pebax-5 against E. coli.

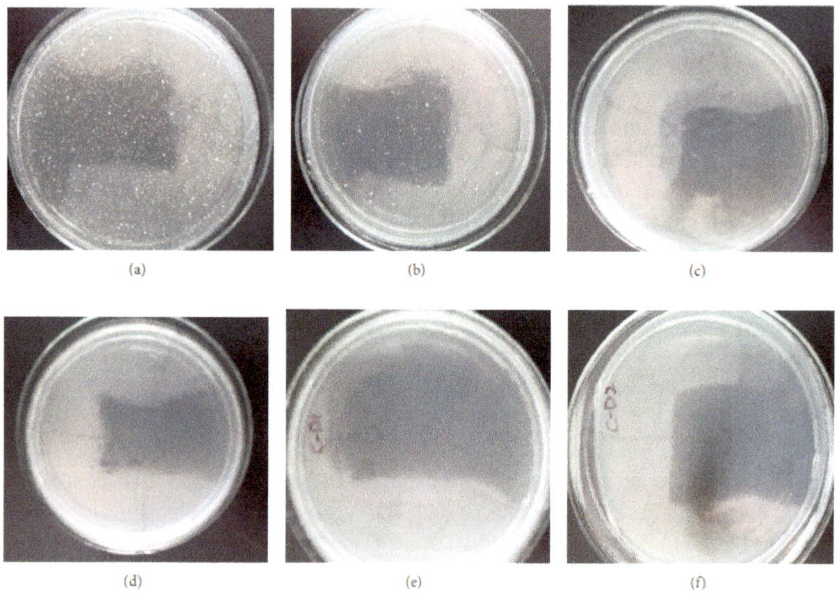

Figure 9: Comparison of the antibacterial ability of (a) Pebax, (b) Ag/Pebax-1, (c) Ag/Pebax-2, (d) Ag/Pebax-3, (e) Ag/Pebax-5, and (f) Ag/Pebax-6 against S. aureus.

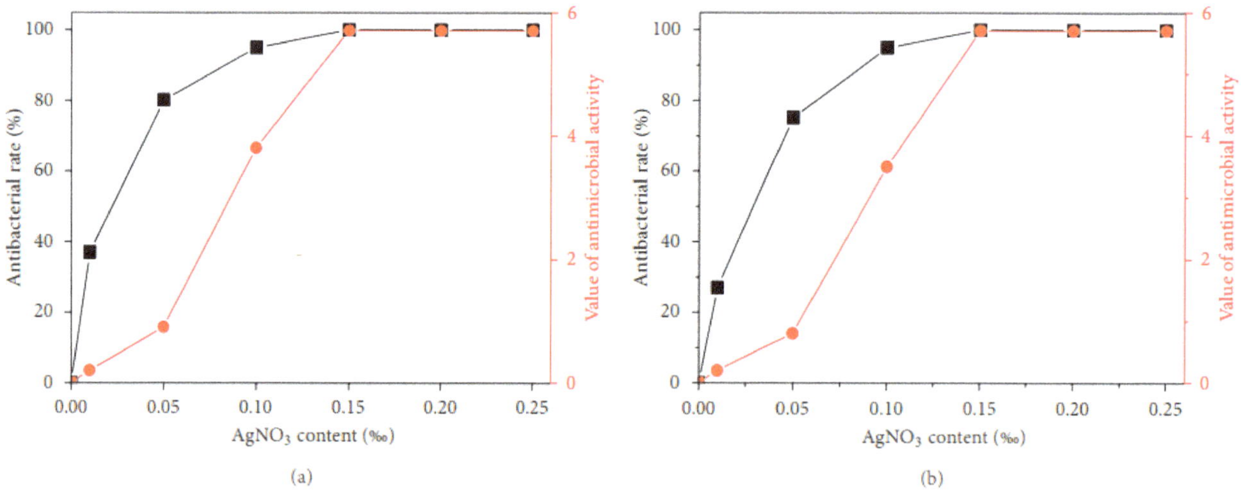

Figure 10: The antibacterial rate and antimicrobial activity of Ag/Pebax against E. coli andS. aureus as a function of $AgNO_3$ content in the precursor.

CONCLUSIONS

This study has described a simple route to prepare well-dispersed and size-uniform Ag nanoparticles in Pebax nanofibers via an electrospinning process. Ag nanoparticles with an average diameter of ca. 15 nm were dispersed homogeneously in Pebax nanofibers. The antimicrobial activities of Ag/Pebax composite against E. coli and S. aureus were evaluated by membrane adhering method. Study results indicated that the Ag/Pebax composites presented outstanding antimicrobial properties at ultralow Ag content; for example, the inhibition rate can reach >99.9% and antimicrobial activity against E. coli and S. aureus was 5.8 and 5.6 at the critical value of $AgNO_3$ content (0.15), respectively, which suggested that the Ag/Pebax composite can be used as a promising antimicrobial material.

ACKNOWLEDGMENTS

This work was financially supported by the National Natural Science Foundation of China under Grant (no. 21373034); the Specially Hired Professorship-Funding of Jiangsu province under Grant (no. SCZ1211400001); the start-up funds from Changzhou University Jiangsu province; Jiangsu Key Laboratory of Advanced Catalytic Material and Technology; and Key Laboratory of Fine Petrochemical Engineering and PAPD of Jiangsu Higher Education Institutions. This trademark belongs to Arkema France and is related to polyether block amides.

REFERENCES

1. M. S. Islam and J. H. Yeum, "Electrospun pullulan/poly(vinyl alcohol)/silver hybrid nanofibers: preparation and property characterization for antibacterial

activity," Colloids and Surfaces A: Physicochemical and Engineering Aspects, vol. 436, no. 5, pp. 279–286, 2013.

2. G. Jin, M. P. Prabhakaran, B. P. Nadappuram, G. Singh, D. Kai, and S. Ramakrishna, "Electrospun poly(L-lactic acid)-co-poly(ϵ-caprolactone) nanofibres containing silver nanoparticles for skin-tissue engineering," Journal of Biomaterials Science, Polymer Edition, vol. 23, no. 18, pp. 2337–2352, 2012.
3. T. Hamouda, A. Myc, B. Donovan, A. Y. Shih, J. D. Reuter, and J. R. Baker Jr., "A novel surfactant nanoemulsion with a unique non-irritant topical antimicrobial activity against bacteria, enveloped viruses and fungi," Microbiological Research, vol. 156, no. 1, pp. 1–7, 2001.
4. T. Yuranova, A. G. Rincon, A. Bozzi et al., "Antibacterial textiles prepared by RF-plasma and vacuum-UV mediated deposition of silver," Journal of Photochemistry and Photobiology A: Chemistry, vol. 161, no. 1, pp. 27–34, 2003.
5. Q. L. Feng, J. Wu, G. Q. Chen, F. Z. Cui, T. N. Kim, and J. O. Kim, "A mechanistic study of the antibacterial effect of silver ions on Escherichia coli and Staphylococcus aureus," Journal of Biomedical Materials Research, vol. 52, no. 4, pp. 662–668, 2000.
6. I. Sondi and B. Salopek-Sondi, "Silver nanoparticles as antimicrobial agent: a case study on E. coli as a model for Gram-negative bacteria," Journal of Colloid and Interface Science, vol. 275, no. 1, pp. 177–182, 2004.
7. Y. Matsumura, K. Yoshikata, S. Kunisaki, and T. Tsuchido, "Mode of bactericidal action of silver zeolite and its comparison with that of silver nitrate," Applied and Environmental Microbiology, vol. 69, no. 7, pp. 4278–4281, 2003.
8. F. A. Sheikh, N. A. M. Barakat, M. A. Kanjwal et al., "Electrospun antimicrobial polyurethane nanofibers containing silver nanoparticles for biotechnological applications," Macromolecular Research, vol. 17, no. 9, pp. 688–696, 2009.
9. H. H. Chae, B. Kim, K. S. Yang, and J. I. Rhee, "Synthesis and antibacterial performance of size-tunable silver nanoparticles with electrospun nanofiber composites," Synthetic Metals, vol. 161, no. 19-20, pp. 2124–2128, 2011.
10. W. K. Son, J. H. Youk, and W. H. Park, "Antimicrobial cellulose acetate nanofibers containing silver nanoparticles," Carbohydrate Polymers, vol. 65, no. 4, pp. 430–434, 2006.
11. C. D. Saquing, J. L. Manasco, and S. A. Khan, "Electrospun nanoparticle-nanofiber composites via a one-step synthesis," Small, vol. 5, no. 8, pp. 944–951, 2009.
12. K. A. Khalil, H. Fouad, T. Elsarnagawy, and F. N. Almajhdi, "Preparation and characterization of electrospun PLGA/silver composite nanofibers for biomedical applications," International Journal of Electrochemical Science, vol. 8, no. 3, pp. 3483–3493, 2013.
13. H. J. Jin, M. O. Hwang, S. Y. Jin, H. L. Kwang, I. Chin, and M. Kim, "Preparation and characterization of electrospun poly(L-lactic acid-co-succinic acid-co-1,4-butane diol) fibrous membranes," Macromolecular Research, vol. 13, no. 1, pp. 73–79, 2005.
14. R. Sridhar, S. Sundarrajan, J. R. Venugopal, R. Ravichandran, and S. Ramakrishna, "Electrospun inorganic and polymer composite nanofibers for biomedical applications," Journal of Biomaterials Science, Polymer Edition, vol. 24, no. 3, pp. 365–385, 2013.
15. T. H. Gilman, "Methods of applying a hydrophilic coating to a substrate, and substrates having a hydrophilic coating," U.S. Patent. no. 0015192 A1, 2012.
16. D. J. Buckley, D. G. Wildes, W. Lee, and W. B. Griffin, "Materials and processes for bonding acoustically neutral structures for use in ultrasound catheters," U.S. Patent 0256502 A1, 2010.
17. C. L. Restuccia and C. LoFaro, "Structural composite material with improved acoustic and vibrational damping properties," U.S. Patent. No. 0170746 A1. 2010.
18. N. L. Le, Y. Wang, and T. S. Chung, "Pebax/POSS mixed matrix membranes for ethanol recovery from aqueous solutions via pervaporation," Journal of Membrane Science, vol. 379, no. 1-2, pp. 174–183, 2011. ·
19. E. V. Konyukhova, A. I. Buzin, and Y. K. Godovsky, "Melting of polyether block amide (Pebax): the effect of stretching," Thermochimica Acta, vol. 391, no. 1-2, pp. 271–277, 2002.
20. C. Cheng, R. Tang, and F. Xi, "Preparation and aggregation of polypseudorotaxane from dendronized poly(methacrylate)-poly(ethylene oxide) diblock copolymer and α-cyclodextrin," Macromolecular Rapid Communications, vol. 26, no. 9, pp. 744–749, 2005.
21. Q. T. Nguyen, J. Sublet, D. Langevin et al., "CO_2 permeation with Pebax -based membranes for global warming reduction," Membrane Gas Separation, vol. 9, no. 1, pp. 255–277, 2010.
22. Q. Shi, N. Vitchuli, J. Nowak et al., "Durable antibacterial Ag/polyacrylonitrile (Ag/PAN) hybrid nanofibers prepared by atmospheric plasma treatment and electrospinning," European Polymer Journal, vol. 47, no. 7, pp. 1402–1409, 2011.
23. "Microbiology of food and animal feeding stuffs-horizontal method for the enumeration of microorganisms–colony-count technique at 30°C," ISO 4833.
24. ISO 22196, "Plastics—measurement of antimicrobial activity on plastics surfaces," 2007.
25. V. Nafisi and M. B. Hägg, "Development of dual layer of ZIF-8/Pebax-2533 mixed matrix membrane for CO_2 capture," Journal of Membrane Science, vol. 459, no. 1, pp. 244–255, 2014.
26. J. W. Rhim, L. F. Wang, and S. I. Hong, "Preparation and characterization of agar/silver nanoparticles composite films with antimicrobial activity," Food Hydrocolloids, vol. 33, no. 2, pp. 327–335, 2013.
27. C. Chen, L. Wang, and Y. Huang, "Electrospinning of thermo-regulating ultrafine fibers based on polyethylene glycol/cellulose acetate composite," Polymer, vol. 48, no. 18, pp. 5202–5207, 2007.

28. B. C. Hancock, S. L. Shamblin, and G. Zografi, "Molecular mobility of amorphous pharmaceutical solids below their glass transition temperatures," Pharmaceutical Research, vol. 12, no. 6, pp. 799–806, 1995. ·

29. Q. Fu, A. Halim, J. Kim et al., "Highly permeable membrane materials for CO_2 capture," Journal of the American Chemical Society, vol. 132, no. 1, pp. 13769–13778, 2013.

30. O. Akhavan, "Lasting antibacterial activities of Ag-TiO_2/Ag/a-TiO_2 nanocomposite thin film photocatalysts under solar light irradiation," Journal of Colloid and Interface Science, vol. 336, no. 1, pp. 117–124, 2009.

31. O. Akhavan and E. Ghaderi, "Bactericidal effects of Ag nanoparticles immobilized on surface of SiO_2thin film with high concentration," Current Applied Physics, vol. 9, no. 6, pp. 1381–1385, 2009.

32. S. Zhang, F. Peng, H. Wang, H. Yu, J. Yang, and H. Zhao, "Electrodeposition preparation of Ag loaded N-doped TiO_2 nanotube arrays with enhanced visible light photocatalytic performance," Catalysis Communications, vol. 12, no. 8, pp. 689–693, 2011.

33. D. P. Dowling, K. Donnelly, M. L. McConnell, R. Eloy, and M. N. Arnaud, "Deposition of anti-bacterial silver coatings on polymeric substrates," Thin Solid Films, vol. 398-399, pp. 602–606, 2001.

34. Q. Shi, N. Vitchuli, J. Nowak et al., "One-step synthesis of silver nanoparticle-filled nylon 6 nanofibers and their antibacterial properties," Journal of Materials Chemistry, vol. 21, no. 28, pp. 10330–10335, 2011. ·

35. A. Celebioglu, Z. Aytac, O. C. Umu, A. Dana, T. Tekinay, and T. Uyar, "One-step synthesis of size-tunable Ag nanoparticles incorporated in electrospun PVA/cyclodextrin nanofibers," Carbohydrate Polymers, vol. 99, no. 2, pp. 808–816, 2014.

CHAPTER 16

FUNCTIONAL THERMOPLASTIC MATERIALS FROM DERIVATIVES OF CELLULOSE AND RELATED STRUCTURAL POLYSACCHARIDES

Yoshikuni Teramoto

Course of Applied Life Science, Faculty of Applied Biological Sciences, Gifu University, 1-1 Yanagido, Gifu 501-1193, Japan

ABSTRACT

This review surveys advances in the development of various material functionalities based on thermoplastic cellulose and related structural polysaccharide derivatives. First, the dependence of thermal (phase) transition behavior on the molecular composition of simple derivatives is rationalized. Next, approaches enabling effective thermoplasticization and further incorporation of material functionalities into structural polysaccharides are discussed. These approaches include: (a) single-substituent derivatization, (b) derivatization with multi-substituents, (c) blending of simple derivatives with synthetic polymers, and (d) graft copolymerization. Some examples addressing the control of supramolecular structures and the regulation of molecular and segmental orientations for functional materials fabrication, which have especially progressed over the past decade, are also addressed. Attractive material functions include improved mechanical performance, controlled biodegradability, cytocompatiblity, and optical functions.

INTRODUCTION

Cellulose derivatives are a class of compounds that originated with nitrate (celluloid), the first man-made thermoplastic to be industrialized in the 19th century, and still retain their industrial importance because of their spinnability, film formability and transparency, strength and tenacity, sorption performance, and other useful properties. Representative commercially available cellulose derivatives and their formulae are listed in Scheme 1. Currently, the conversion of cellulose to advanced materials has attracted considerable attention in connection with green and sustainable industrial development [1,2,3,4,5].

In general, strong intermolecular and intramolecular hydrogen bonding in unmodified cellulose and other structural polysaccharide molecules gives rise to a theoretical glass transition temperature (T_g) exceeding their thermal decomposition point, limiting thermal processability. Derivatization enables the modification of original bulk material properties such as thermal behavior through relatively simple reactions that effectively exploit side-group reactivity. However, the relationship between molecular structure and physical properties and the principles governing the derivatization need to be elucidated in detail.

Citation: Teramoto, Y. Functional Thermoplastic Materials from Derivatives of Cellulose and Related Structural Polysaccharides. Molecules 2015, 20, 5487-5527, doi:10.3390/molecules20045487

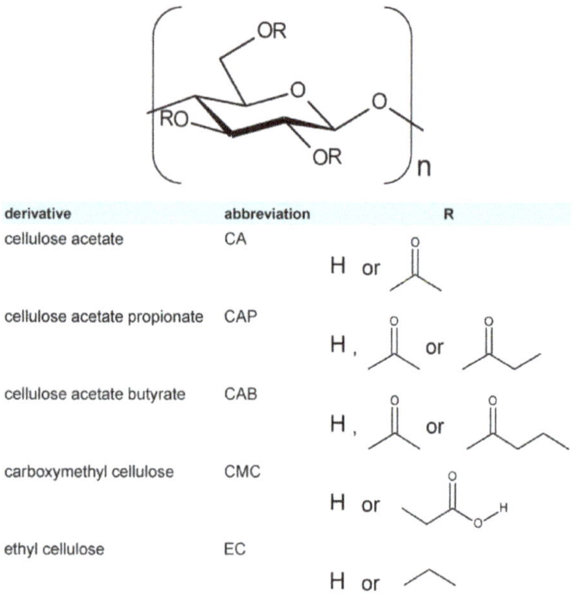

Scheme 1: Representative commercially available cellulose derivatives.

Numerous investigations have addressed cellulose chemical modifications such as ester synthesis [1,6], graft copolymerization [7,8], and nanocellulose formation [9]. While intentionally avoiding studies on the synthetic aspects of cellulose and related structural polysaccharide derivatives, the present review covers the development of material functionalities for systems containing thermoplastic derivatives of cellulose and related polysaccharides because few reviews have dealt with this subject. Therefore, a systematic interpretation of fundamental thermal properties of simple derivatives is discussed first (Section 2.1), and various material functionalization approaches developed over the last decade to derivatize cellulose and related polysaccharides as thermoplastic substances are subsequently reviewed. For convenience, these modifications, which facilitate thermoplasticization and further material functionalization of structural polysaccharides, are grouped into four classes: (a) single-substituent derivatization (Section 2), (b) derivatization with multiple substituents (Section 3), (c) blending of simple derivatives with synthetic polymers (Section 4), and (d) graft copolymerization (Section 5). Highlighted material functionalities consist of improved mechanical performance, controlled biodegradability, cytocompatiblity, and optical functions. Specific progress in the fabrication of functionalized bulk materials has also rapidly occurred over the past decade as a result of the control of supramolecular structures and the regulation of molecular and segmental orientations. This topic will be reviewed separately in Section 6.

SINGLE SUBSTITUENT DERIVATIVES

Fundamental Aspects of Thermal Properties and Phase Behavior

For polymer solutions, intrinsic viscosity [η] and molecular weight M are related by the Mark-Houwink-Sakurada equation as follows:

$$[\eta] = K_m M^a \quad (1)$$

where K_m and a are parameters characteristic of the polymer/solvent combination at a given temperature. Cellulose derivatives generally show exponent a values ranging from 0.8 to 1.0, indicative of semi-rigid polymer chains. Their properties are dominated by a balance between the semi-rigidity of the cellulosic trunk and the nature of side chains. Thermal properties of polymer materials remarkably depend on the backbone flexibility, steric hindrance, and chain-chain polar interaction. The semi-rigid cellulosic main chain is essentially internally plasticized by the introduction of bulky side-chain groups leading to intermolecular hydrogen bond scission.

In general, for cellulose derivatives, thermal and dissolution properties are determined by the degree of substitution (DS) defined as the average number (up to three) of substituted hydroxyl groups per anhydroglucose unit. As demonstrated by differential scanning calorimetry (DSC) thermograms (Figure 1), T_g decreases with increasing DS for cellulose acetate (CA). However, triacetylated cellulose (DS = 2.95) tends to crystallize easily because of its highly symmetric structure. Irrespective of side-chain moieties, cellulose derivatives presenting a low DS (~1.0) often show water solubility, and solubility in hydrophobic organic solvents generally increases with increasing DS.

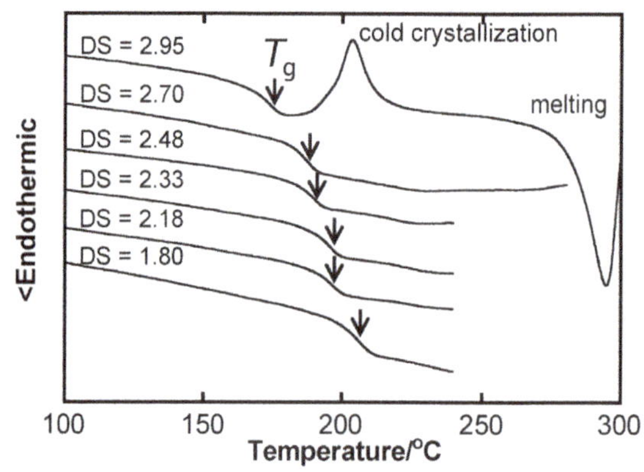

Figure 1: DSC thermograms of CA displaying different DS values obtained during the second heating scan. Arrows indicate T_g positions taken as midpoints of heat flow discontinuity.

Cellulose derivatives presenting a DS value of approximately 3 exhibit a complicated thermal transition behavior depending on the side chain size. Thermal (melt) processability is achievable by incorporating bulky ester substituents into the cellulose backbone [1,10,11,12,13,14]. "Melt processing," however, implies the application of softening behavior including phenomena of glass transition, liquid crystalline transition, and crystallite melting, if any. It is therefore safe to put "melt" into "thermal" processing in order to avoid confusion with melting of crystallites in a thermodynamic sense.

The thermally processable temperature of cellulose triesters decreases rapidly when the ester substituent becomes bulkier. Figure 2 shows typical DSC thermograms acquired during the second heating scan at 20 °C/min of cellulose triesters bearing linear side chains as a function of side chain length m, where m is defined as the number of the carbon atoms forming the side chain skeleton, just after quick quenching (cooling at ~80 °C/min) from the isotropic state (200 °C). Interpreting the thermal events observed by DSC analysis often causes confusion. The dramatic lowering of baseline shift temperatures may be considered as a decrease in T_g for substituents ranging between acetate (m = 2) and hexanoate (m = 6). T_g remains more or less constant (or even rises slightly) for substituents displaying longer chains. The drop in T_g may be modeled as an internal plasticization process of all cellulosic molecules containing methylene groups, almost conforming with the Fox relationship [15]. The failure of waxy triesters ($m \geq 10$, Figure 2) to follow this rule may originate from the formation of highly-ordered domains (pseudo-crystallites) mainly comprising longer side chains and the development of a liquid crystalline (LC) phase [15,16]. Eventually, the glass transition appears more latent, and (side-chain pseudo-) crystallite melting (T_m) and molecular thermotropic LC isotropization (T_i) become more easily detectable (Figure 2). The LC phase of the cellulose esters, characterized by a low birefringence and high viscosity compared to cholesteric phases, is classified as a columnar structure, in which molecules adopt a two-dimensional hexagonal packing order in the plane normal to the chain axis with two chains being included in a unit lattice [16].

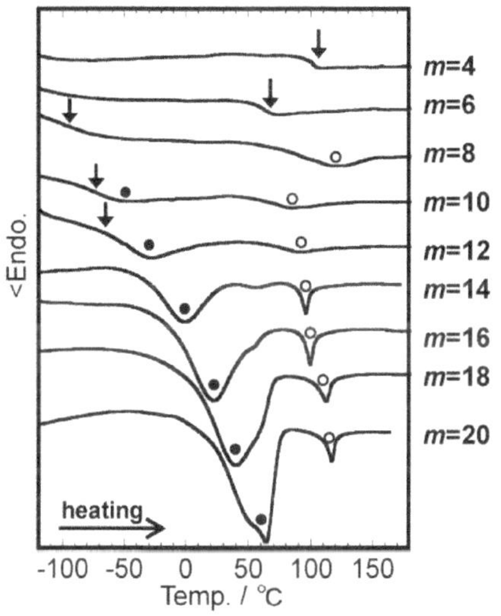

Figure 2: DSC thermograms of cellulose triesters presenting different m values obtained during the second heating scan. Arrows mark T_g positions while the filled and open circles represent T_m and T_i values, respectively.

Fukuda *et al.* have elegantly summarized the complicated thermal transition behavior of tri-substituted cellulose derivatives in terms of thermotropic LC formation [16]. Figure 3 shows the thermal transition temperatures of cellulose triesters as well as cellulose triethers and tri-*O*-carboxymethyl cellulose (CMC) triesters as a function of m. For example, m = 7 represents heptyl cellulose (Cell-O-C_7H_{15}), cellulose heptanoate (Cell-O-CO-C_6H_{13}), and carboxymethyl cellulose pentyl ester (Cell-O-CH_2-$COOC_5H_{11}$). In all cases, the melting temperature T_m decreases similarly with increasing m. Alkyl ethers organize into a cholesteric phase (vertically hatched region) while alkyl esters give a columnar phase (horizontally hatched region). However, CMC derivatives hardly exhibit any liquid crystallinity regardless of m. According to the Flory theory, cholesteric phases are believed to originate essentially from "hard interactions," which represent orientation-dependent excluded-volume effects between "hard" rods or Kuhn segments exclusively interacting through short-range repulsions that preclude the intrusion of one rod or segment into the space occupied by another [17]. In the columnar phase, it is important to account for energetic contributions of "soft interactions," that is the effects of intermolecular attractive forces, such as van der Waals forces between alkyl side chains, similar to the ones between solvent-solute pairs, in addition to hard interactions. In this regard, CMC derivatives may be considered to miss these mesophase-stabilizing hard and soft interactions.

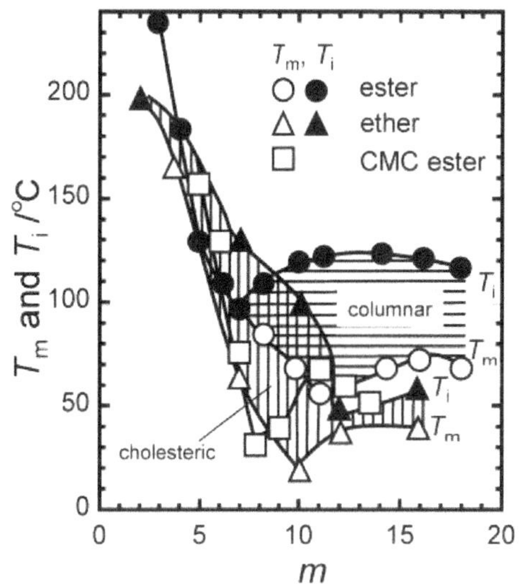

Figure 3: Melting temperature T_m and isotropization temperature T_i of normal alkyl ester and ether derivatives of cellulose. Ester derivatives of tri-*O*-carboxymethyl cellulose (CMC) hardly form a mesophase regardless of m. Reproduced with permission from ref. [16].

Systematic Studies of Related Polysaccharides

Associated with the extensive studies of cellulose derivatives described above, derivatives of other carbohydrates have prompted similar interest for the formulation of thermal properties. Provided in abundance by marine crustaceans, chitin is a representative animal-derived structural polysaccharide. Structurally similar to cellulose, it may be regarded as a cellulose molecule in which the hydroxyl group at the C2 position is replaced by an acetylamino group. Chitosan is a collective name for chitin deacetylated to different degrees. Fully deacetylated chitin is soluble in dilute aqueous acid solutions, whereas its ~50% deacetylated counterpart readily dissolves in neutral water. Chitinous compounds differ from cellulosics by the presence of (acetyl) amino groups in their structure, resulting in distinct biological functions, such as bio-assimilability and antibacterial activity. Chitin and chitosan are therefore expected to find wide application in medical, pharmaceutical, food, and textile industries.

Teramoto *et al.* synthesized a series of single-substituent chitin diester derivatives bearing normal acyl groups (Chitin-O-CO-$C_{m-1}H_{2m-1}$) using pyridine, *p*-toluenesulfonyl chloride, and normal alkanoic acid in a *N,N*-dimethylacetamide (DMAc)–LiCl homogeneous system [18]. The products (DS = 1.7–1.9) showed an m-dependent thermal transition behavior. Specifically, no evident transition was observed for m = 4–10. On the other hand, a glass transition was detected for m = 12 and 14, and a pseudo-first-order phase transition was recorded for m = 16–20. These transitions usually occurred below room temperature by DSC. Wide-angle X-ray diffractometry (WAXD) patterns at 20 °C displayed a sharp diffraction peak (2θ = 2°–7°) and diffuse halo (2θ ≈ 20°) for the chitin diesters (Figure 4). The*d*-spacing (1.5–3.6 nm) increased with increasing *m* to yield two stages of mutually different increasing rates, consistent with the systematic *m*-dependence of the period of the main chain layered structure. The molecular assembly of the chitin diesters exhibited "dual mesomorphy": nematic ordering for the semi-rigid carbohydrate trunk and smectic ordering for the flexible side chains. In relation to this depiction, the typical phase diagram of the already-described cellulose ester consisted of crystalline solid, mesomorphic, and isotropic regions. In contrast, chitin diester bulk samples were continually birefringent at any temperature during heating and eventually underwent thermal degradation without exhibiting any evident flow. In addition, chitin diesters showed poor solubility in common organic solvents, such as acetone, tetrahydroflan, dimethyl sulfoxide, DMAc, toluene, and chloroform. These properties of the chitin diesters are in stark contrast to those of the cellulose ester series [12,15,16] and aromatic liquid-crystalline polyesters [19], in which the acylation was conducted primarily to plasticize the respective rigid main-chain polymers. Therefore, the acylation hardly produced an "internal plasticization" of the rigid chitinous polymer. A quasi-complete diesterification may induce a suitable conformational symmetry around the chitin backbone axis and growth in the lateral dimension via amide groups, which are prone to form a strong intermolecular hydrogen bonding, thereby resulting in the nematic alignment of the main chains as rather stiffer trunks through side chain interlocking.

Defined as heteroglycans consisting of xylose, mannose, arabinose, glucose, galactose, and 4-*O*-methyl-D-glucuronic acid, hemicelluloses constitute about 25%–35% of plants. They exhibit high structural diversity by incorporating different types of side chains and glycosidic linkages in their backbone. Together with cellulose, they form a complex structure in cell walls, providing structure and rigidity to plants. A predominant type of hemicellulose called xylan mainly consists of β-1,4-linked D-xylopyranose units and displays various chemical structures, which may even be branched, depending on its source. Iwata *et al.* have synthesized a series of high-molecular weight xylan esters comprising different alkyl side chain length (*m* = 2–12) by homogeneous reactions [20]. The esterification improved the solubility of xylan in chloroform, but thermal transitions, such as melting and glass transitions, were not detected by DSC between −20 °C and 250 °C. WAXD profiles showed two main diffraction peaks, similarly to those of the chitin diesters (Figure 4). Mechanical properties of the xylan esters were dependent on the alkyl chain length. Specifically, the tensile strength of solution-cast films decreased with increasing *m*, and xylan butyrate (*m* = 4) exhibited the highest tensile strength (29 MPa). On the other hand, the elongation at break increased with increasing *m* and amounted to 19%, 46%, and 44% for xylan butyrate (*m* = 4), decanoate (*m* = 10), and laurate (*m*= 12), respectively. Subsequently, a similar derivatization was applied to konjac glucomannan (GM), a hemicellulose comprising β-1,4-linked D-glucose and D-mannose residues in its main chain with branching through the C3 carbon of glucosyl or mannosyl residues [21], and curdlan (CD), a linear β-1,3-D-glucan [22]. Similarly to the xylan esters, the tensile strength and Young's modulus tended to decrease while the elongation at break increased with increasing *m*for both esterified hemicellulose series. Thermally pressed GM propionate (*m* = 3) and butyrate films (*m* = 4) presented the highest tensile strength (*ca.* ~53 MPa) while the GM laurate film (*m* = 12) showed the highest elongation at break (426%). As displayed in Figure 5, WAXD profiles revealed that low symmetry and branched GM derivatives exhibited completely amorphous structures. In contrast, CD derivatives bearing shorter acyl side chains (*m* ≤ 6) rapidly crystallized, in agreement with DSC measurements. No LC formation was observed for these hemicellulosic systems.

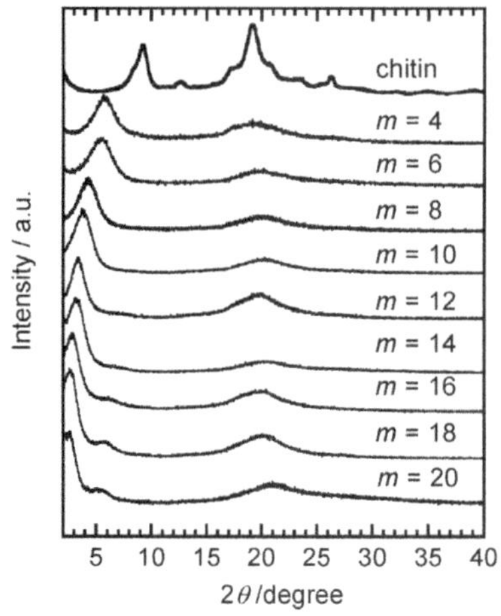

Figure 4: WAXD profiles of chitin and its diester derivatives (*m* = 4–20) at 20 °C. Reprinted with permission from ref. [18].

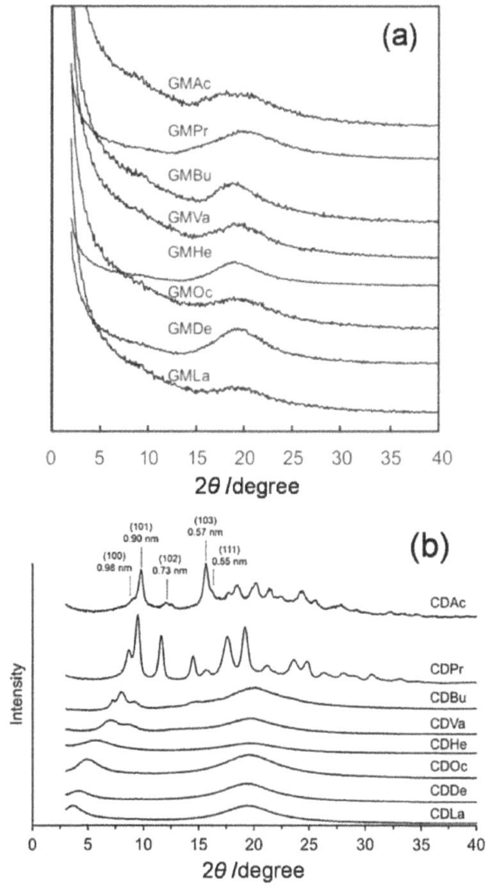

Figure 5. WAXD profiles of (**a**) konjac glucomannan (GM) and (**b**) curdlan (CD) ester derivatives. Abbreviations: Ac, acetate (*m* = 2); Pr, propionate (*m* = 3); Bu, butyrate (*m* = 4); Va, valerate (*m* = 5); He,

hexanoate ($m = 6$); Oc, octanoate ($m = 8$); De, decanoate ($m = 10$); La, laurate ($m = 12$). Reproduced with permission from refs. [21] and [22] for (a) and (b), respectively.

DERIVATIZATION WITH MULTIPLE SUBSTITUENTS

Ductility and toughness are generally incompatible in cellulose-based materials. Single bulky substituents not only promote the expression of thermoplasticity but also tend to soften cellulose derivatives [23], as may be deduced from their thermal behavior and the mechanical data of hemicellulose derivatives (see above, Section 2.2). This suggests that size and DS control of single substituents are not sufficient to balance the mechanical performance (strength) and thermal moldability of cellulose derivatives. The thermotropic nature may also lead to difficulties in their application as structural bulk materials. Although mixed cellulose acetate propionates (CAPs) and butyrates (CABs) have been synthesized commercially (Scheme 1), these materials are virtually unprocessable by thermal methods without the addition of large amounts of external low-molecular-weight plasticizers. However, the introduction of small amounts of ester substituents larger than C4 into CA results in a rapid decline in the thermally processable temperature [24]. This incorporation of multiple substituents has been widely conducted to improve both the mechanical performance and thermal processability and to add specific material functionalities.

Industries have recently deployed tremendous efforts to develop new cellulose-based thermoplastic polymers combining excellent thermal processability and tunable mechanical performance by multi-substitution. Aranishi *et al.*(Toray Industries, Inc., Tokyo, Japan) developed the world's first melt-spun cellulosic fiber Foresse® [25] while the production of conventional cellulosic filaments, such as viscose rayon and cellulose acetate, requires harmful organic reagents and solvents. Substituent and DS optimization gave rise to cellulose mixed esters displaying excellent thermal flowability and fiber mechanical properties. This melt spinning process readily leads to a wide variety of fiber cross sections, such as trilobal, hollow, and sea-islands-type conjugated fibers (Figure 6). The fabric obtained from the cellulosic fiber exhibits soft touch, hygroscopicity, good color, and luster, making it suitable for textiles, such as apparel. Iji *et al.* (NEC Corporation, Tokyo, Japan) produced cellulose-based bioplastics by derivatization of cellulose diacetate (CDA) using 3-pentadecylphenoxy acetic acid (PAA) chloride (Figure 7) called a modified cardanol [26]. Derived from cashew nut shells, cardanol is a phenol derivative bearing a linear unsaturated hydrocarbon side chain (carbon number: 15) [26]. Esterification of CDA with PAA chloride produced thermoplastic PAA-bonded CDA exhibiting high tenacity (elongation at break exceeding 10% while keeping maximum bending strength (>80 MPa)), heat resistance, and water resistance.

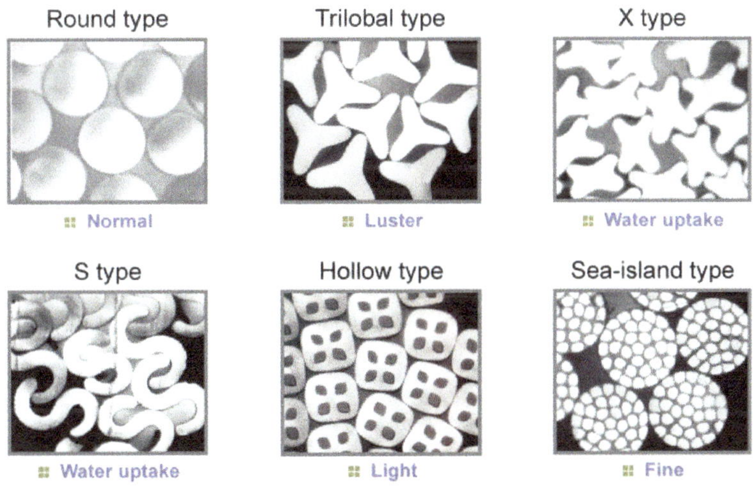

Figure 6: Various fiber cross-sections obtained for the first melt-spun cellulose fiber "Foresse®." Reproduced with permission from ref. [25].

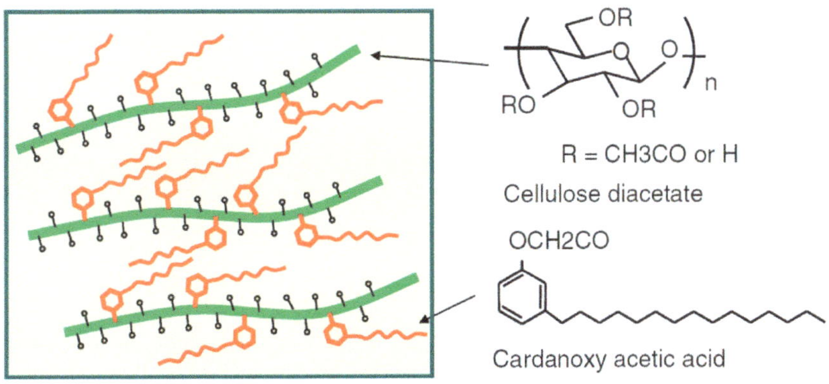

Figure 7: Hypothetical structure of cardanol-bonded CDA. Reproduced with permission from ref. [26]. Copyright 2011 The Society of Polymer Science, Japan (SPSJ).

Sawai *et al.* (FUJIFILM Corporation, Tokyo, Japan) precisely demonstrated that the size and number of introduced multi-substituents, such as combinations of methyl/2-ethylhexanoyl (2-Eh) and acetyl/2-Eh groups, significantly enhanced cellulose derivative plasticity [23]. An association of small and large substituents also dramatically improved the balance between mechanical properties and plasticity (Figure 8). Furthermore, the addition of ether/ester groups together boosted the impact strength of the derivatives, in good correlation with the degree of prevalence of soft segments estimated by dynamic mechanical analysis (DMA) and solid-state NMR. These results may stem from three factors related to the volume and motion of the cellulosic molecular chains. (i) A constant number of molecular motion elements can absorb energy in high-speed destruction tests, such as impact tests at room temperature. (ii) The increase in the degree of freedom in space promotes molecular motions within gaps in response to impact. (iii) The weakened hydrogen bonds led to a symmetrical uniaxial cylindrical structure comprising cellulosic molecules. The materials were processed by injection molding and showed good rheological and processing characteristics. Building on these results, Yao *et al.* (Fuji Xerox Co., Ltd., Tokyo, Japan) proposed multisubstituted cellulose derivatives bearing some residual hydroxyl groups to develop highly thermally processable materials without reduction in mechanical strength and heat resistance [27]. An exploration of appropriate halogen-free flame retardants as additives highlighted compatible composite systems combining the derivatives with acrylonitrile butadiene styrene (ABS) resin at adequate molar ratio, molecular weight, and melt viscosity. The composites were eventually mounted as components for business machines. Similarly, a good thermoplasticity–modulus balance was obtained by mixed esterification for paramylon, a storage β-1,3-glucan polysaccharide for *Euglena gracilis* [28].

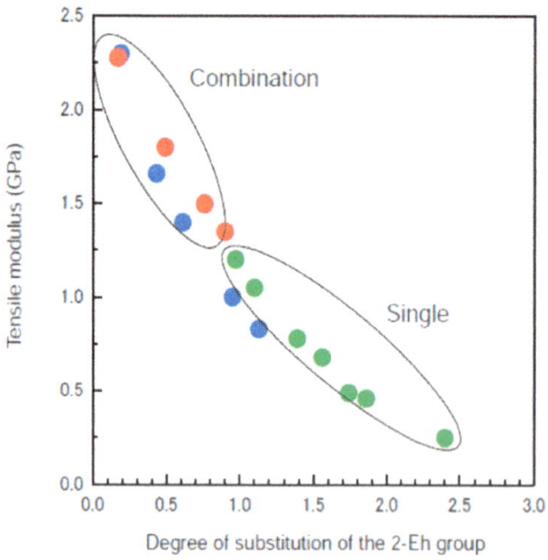

Figure 8: Tensile strength as a function of substitution using methyl/2-ethylhexanoyl (2-Eh) (blue), acetyl/2-Eh (red), and 2-Eh groups (green). Reprinted with permission from ref. [23].

Furthermore, practical and natural polymer materials have inspired researchers to develop derivatives displaying interesting functions as bulk materials. Chang *et al.* fabricated high performance films using cellulose butyral synthesized from native cellulose [29,30] based on the commercial production of poly(vinyl butyral) (PVB) from poly(vinyl alcohol) [31]. Bulk PVB acts as an interlayer in laminated glass because the film adheres well to glass surfaces and provides excellent mechanical resistance to the laminate against breakage [32]. Acetalization generates 1,3-dioxane rings, which may contribute to the impact resistance and toughness of PVB films. Inspired by the PVA-based commodity, the researchers attempted to create these interlayers from cellulose. As summarized in Figure 9, a two-step synthesis was adopted to produce the cellulose butyral derivative. Specifically, etherification of cellulose with glycidol in NaOH/urea aqueous solution [33] gave *O*-(2,3-dihydroxypropyl) cellulose (DHPC), which underwent butyralization to generate the desired derivative [29]. The product displayed at least two annular structures, such as five- and eight-membered rings, on the side chains and was easily molded into thin films by hot pressing. In addition to being amorphous (T_g < 25 °C) and highly transparent, the cellulose butyral films conveniently provided a slightly lower refractive index than conventional glass (~1.51). Despite their insolubility in water, they showed a good adhesive property to glass plates because of their amphiphilic character. DMA results revealed that the original semi-rigid cellulosic molecule was highly plasticized by a surrounding layer of dihydroxypropyl sequences. The flexible polyether side-chains were primarily responsible for the ductile behavior observed in DHPC and cellulose butyral films in tensile tests. The butyral group improved the toughness of the DHPC films during stretching because chain entanglement was more effective in the ring structures. These cellulose butyral films presented comparable optical and mechanical performances to commercial PVB films, suggesting their potential applicability as interlayers for laminated glass.

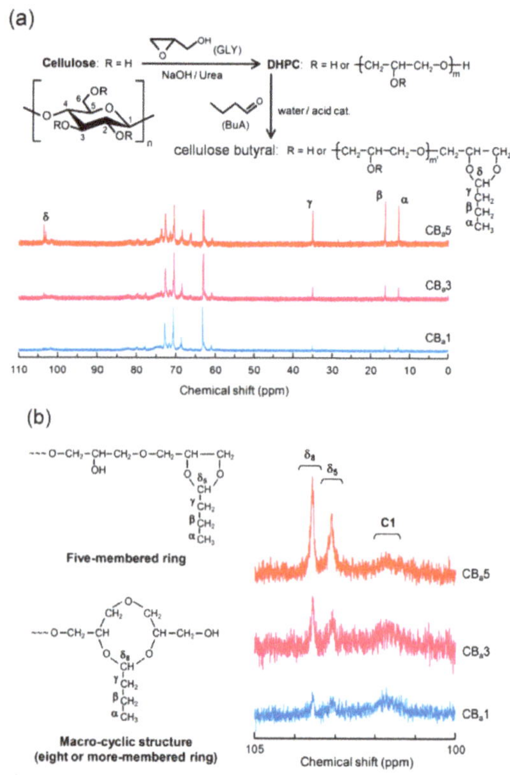

Figure 9: (**a**) Derivatization reactions and ^{13}C-NMR spectra of cellulose butyral products in pyridine-d_5 showing the butyral carbon assignment; (**b**) enlarged ^{13}C-NMR spectra between 100 and 105 ppm and peak assignments. The DHPC precursor comprised 81.1 wt % dihydroxypropyl unit. Butyral molar substitutions of the products are 0.20, 0.94, and 3.60 for CB_a1, CB_a3, and CB_a5, respectively. Reprinted with permission from ref. [29].

Cellulose derivatization does not always involve simple organic moieties. Aoki *et al.* have prepared cellulose derivatives containing a reversible cross-linkable mercapto group by esterification of CA with mercaptoacetic acid (MA) to generate permanent wave products for hair [34]. The CA–MA products showed a shape memory–recovery behavior in film (Figure 10) and fiber forms through adequate redox treatments because of the keratin fiber-like reversible association–dissociation of the mercapto cross-links. Dimethyl sulfoxide (DMSO) acted as an oxidant while 2-mercaptoethanol or ammonium mercaptoacetic acid served as reducing reagents. The progress of the redox reactions was monitored by using a confocal depth scanning technique in Raman spectroscopy. The cross-linking effect on CA-MA film thermal and viscoelastic properties was also estimated by DSC and DMA. The loss tan δ peak alternately declined and recovered in the T_g region according to the repeated redox treatments, consistent with a reversible variation in viscoelasticity resulting from S–S cross-linking and cleavage on oxidation and reduction, respectively. These functional CA–MA derivatives are expected to find application as shape memory-recovery materials.

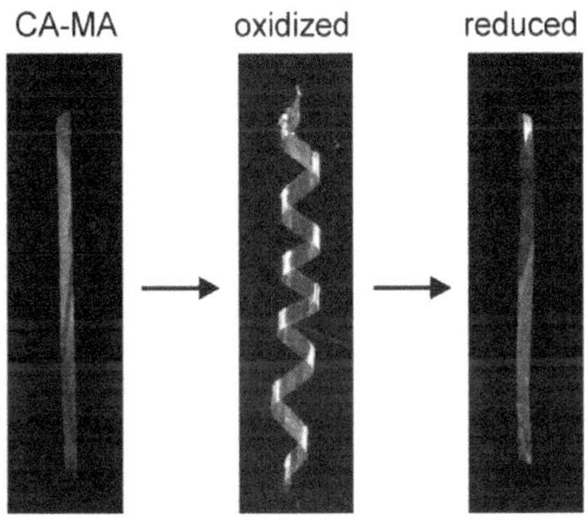

Figure 10: Shape memory–recovery behavior of a *ca.* 6 μm-thick CA-MA film. Reprinted with permission from ref. [34].

BLENDS OF SIMPLE DERIVATIVES WITH SYNTHETIC POLYMERS

As discussed above, side chains introduced into cellulosic molecules by derivatization generally play an effective role as "internal" plasticizers while "external" plasticizers usually facilitate the thermal molding of commodity-type cellulose derivatives. However, the more convenient "external" plasticization using low-molecular-weight substances typically leads to fume generation problems during thermal molding and plasticizer bleed-out over long-term use. The mixing of flexible higher-molecular-weight polymers as plasticizers offers an effective alternative against these problems. However, one cannot assume that any two polymers necessarily form a homogeneous blend [35]. Indeed, immiscibility occurs so frequently that miscibility/compatibility should be treated as an exception to the rule and is only observed under a precisely defined set of conditions, such as the existence of specific intermolecular interactions. This phenomenon can be rationalized by an equation developed from the thermodynamic description first proposed independently by Huggins and Flory in 1941 for a binary polymer mixture:

$$\Delta G_{mix} = RT\left(\frac{V_{mix}}{V_0}\right)\left(\frac{\phi_A}{n_A}\ln\phi_A + \frac{\phi_B}{n_B}\ln\phi_B + \chi_{AB}\phi_A\phi_B\right) \tag{2}$$

where ΔG_{mix} is the change in Gibbs free energy upon mixing, V_{mix} is the volume of the mixture, V_0 is the molar volume of the component, ϕ is the volume fraction, n is the degree of polymerization, R is the gas constant, and T is the temperature. Indices A and B represent the two separate polymers in the mixture. Polymer-polymer miscibility is therefore clearly determined by a delicate balance of enthalpic and entropic forces, where the entropic contribution is significantly smaller than in small molecule liquid mixtures (solutions). Indeed, the combinatorial entropy term, which is highly responsible for miscibility in solvents and polymer solutions, is particularly small in high-molecular-weight polymer blends because of a large degree of polymerization n. Therefore, polymer–polymer miscibility mainly depends on the negative heat of mixing, which results from electrostatic interactions, hydrogen bonding, dipole-dipole interactions, and dispersion forces between macromolecules. Polymer-polymer miscibility is commonly estimated by DSC determination of T_g for the blends. If a binary polymer system exhibits a single glass transition between the T_g values of individual components and its T_g shows a composition-dependent shift, the system is considered highly miscible on the T_g-detection scale, which is usually assumed to be smaller than a couple of tens of nanometers [30].

Blend with N-Vinyl Pyrrolidone-Based Polymers

Despite its growing use as an industrial product, CA shows a high T_g, limiting its thermal processability. Poly(*N*-vinyl pyrrolidone) (PVP) and its copolymers form miscible blends with CA and other cellulose ester (CE) derivatives by hydrogen bonding between the CE residual hydroxyl groups and the *N*-vinyl pyrrolidone (VP) carbonyl groups [36,37,38,39,40]. Ohno and Nishio characterized the miscibility of CA blends involving a random copolymer comprising VP and methyl methacrylate (MMA) units (P(VP-*co*-MMA)) in detail by DSC determination of T_g [37]. Poly(methyl methacrylate) (PMMA) plays an important role in optical and medical materials development because of its unique performance and biocompatibility. Figure 11 displays a miscibility map for the CA/P(VP-*co*-MMA) system as a function of the DS of CA and VP composition in P(VP-*co*-MMA). CA form fully homogeneous blends with P(VP-*co*-MMA) on a scale of a few nanometers when the DS of CA is below 2.75 and the VP fraction exceeds 30 mol % in the copolymer.

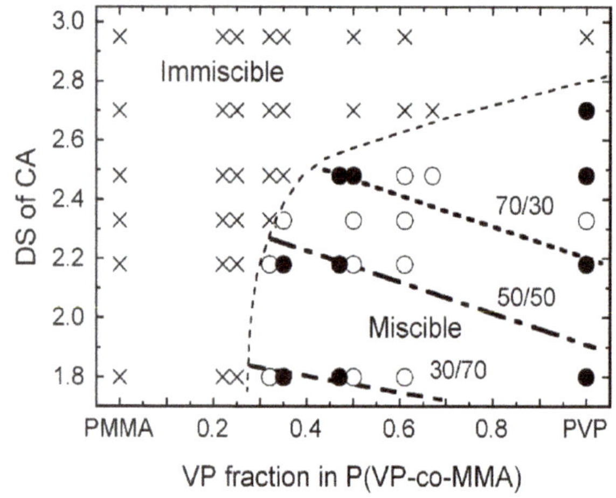

Figure 11: Miscibility map of the CA/P(VP-*co*-MMA) blend. Symbols indicate whether a given CA/P(VP-*co*-MMA) blend is miscible (○ and ●) or immiscible (×). Closed circles represent the blend series used in the subsequent molecular orientation study. Three lines, designated as 70/30, 50/50, and 30/70, represent critical blend compositions. Miscible polymer combinations fitting the respective lines produce a birefringence-free material at the indicated critical composition. Reprinted with permission from ref. [40].

In general, the miscibility of CEs with VP-containing vinyl (co)polymer largely depends on a small difference in acyl substituent side-chain length (carbon number, *m*) in the employed CE, as suggested by a comparative study of blend miscibility involving cellulose propionate (CP) and butyrate (CB) in addition to CA [41,42]. The critical DS values required to obtain miscible blends of CP (2.65) and CB (2.50) with PVP were appreciably lower than for CA because the frequency of hydrogen-bonding interactions was reduced by steric hindrance from the bulkier propionyl or butyryl substituents. Unlike the situation in the CA series, high-substituted CPs (DS > 2.65) and CBs (DS > 2.5) make a miscible pair with some of the copolymers (not rich in VP) despite no contribution of the hydrogen-bonding interaction. This unique miscibility behavior, *i.e.*, advent of a miscibility window, was attributed to CE-copolymer attraction driven by repulsion between monomer units in the vinyl copolymer.

Blends with Aliphatic Polyesters

A derivatization process, such as acylation with straight-chain fatty acids, can produce miscible blends of originally hydrophilic polysaccharides and synthetic hydrophobic macromolecules. A strong correlation has been reported between the primary structure of CE derivatives bearing various normal acyl side groups comprising *m* carbons and their miscibility with the biodegradable aliphatic polyester poly(ε-caprolactone) (PCL) [43,44]. The thermal behavior of the binary polymer blends observed by DSC clearly explained miscibility in terms of *m* and DS. Any CE derivatives showed good miscibility with PCL for *m* = 3–5 and DS ≥ 2.2 (Figure 12). In particular, butyrate (*m* = 4) exhibited a comparatively low DS of 1.9. However, most commodity-type CA (*m* = 2) was immiscible with PCL, even in a highly acetylated state. Cellulose caproate (*m* = 6) and enanthate (*m* = 7), which presented moderately long side-chains, displayed a relatively low degree of miscibility with PCL.

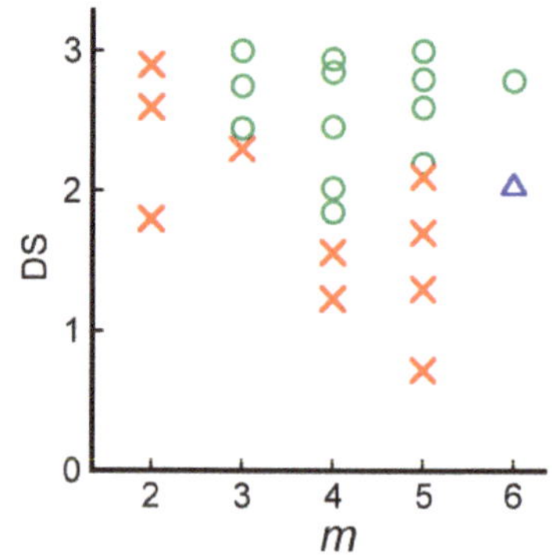

Figure 12: Miscibility map of different CE/PCL blends as a function of side-chain carbon number (*m*) and acyl DS. Symbols indicate whether a given CE/PCL blend is miscible (circle), immiscible (cross), or partially miscible (triangle). Reproduced with permission from ref. [45].

Teramoto *et al.* have conducted a similar investigation with chitin [45]. They synthesized acyl chitin (Acyl-Ch) derivatives containing a normal acyl group (*m* = 2–6) at different DS through a homogeneous reaction between crab-shell chitin and acyl chlorides in DMAc–LiCl solution. NMR analysis quantitatively demonstrated that the acylations occurred at C3/C6 hydroxy protons and C2 amino proton(s), which differs from the chitin diester synthesis mentioned in Section 2.2. It was therefore possible to determine the separate amide-DS and ester-DS values as well as their sum (total-DS). Figure 13 shows the estimated miscibility of Acyl-Ch/PCL blends as a function of *m* and their respective DS value. The critical total-DS (total-DS_{cr}) value for achieving miscibility decreased with increasing *m* but generally surpassed the critical ester-DS for CE/PCL systems (Figure 12) for the same *m* value. In addition, miscibility increased with increasing ester-DS (Figure 13b) but lower correlation was observed between miscibility and amide-DS (Figure 13c).

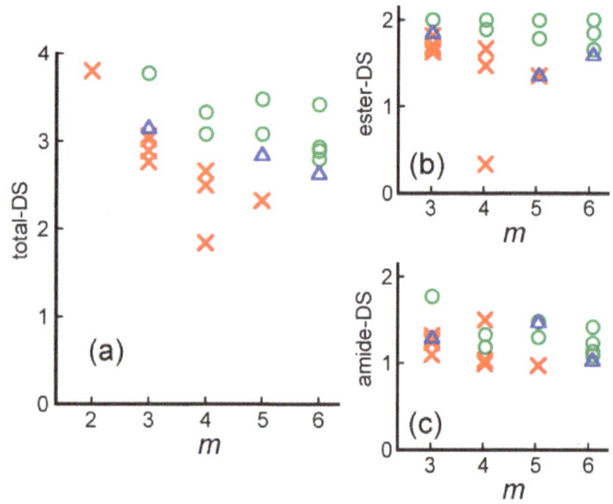

Figure 13. Miscibility maps of different Acyl-Ch/PCL blends as a function of side-chain carbon number *m* and substitution parameters: (**a**) total-DS; (**b**) ester-DS; and (**c**) amide-DS. Symbols indicate whether a given Acyl-Ch/PCL blend is miscible (circle), immiscible (cross), or partially miscible (triangle). Reproduced with permission from ref. [45].

Polymer T_g is directly related to their cooperative segmental motion [35], which involves 15 to 30 statistical segments. Such a cooperative motion requires sufficient thermal energy to ensure suitable ease of rotation about main chain bonds and overcome local bonding. The use of T_g to determine polymer-polymer miscibility is based on the idea that a singular T_g value indicates a domain size smaller than a few tens of nanometers. However, if one component is crystalline, such as PCL, its crystallinity is likely to regress upon addition of a second component (e.g., CE or Acyl-Ch).

^{13}C cross-polarization magic angle-spinning (CP/MAS) NMR provides information on interactions between constituent polymers at the molecular level (≤1 nm). If two different polymers mutually interact at the molecular level in a binary blend, the electron density around carbon atoms bearing the interacting groups becomes perturbed. This significantly changes the ^{13}C-NMR chemical shift and/or signal shape of the blend compared with those of the individual polymers. However, these spectroscopic measurements do not provide any evidence regarding specific intermolecular interactions between polysaccharide derivatives and PCL molecules in the blends.

To determine whether miscibility was achieved in a series of polymer blends, the researchers inferred a possible contribution of dipole-dipole interactions between PCL carbonyl groups and CE and Acyl-Ch side groups. Structural affinity between acyl side groups and PCL repeating units was also considered as a potentially crucial factor to miscibility. For cellulose (CB) and chitin butyrates (ChB) (Figure 14), the butyryl side group is structurally identical with the PCL repeating unit if the glucopyranose carbon atoms are taken into account. On the other hand, in the Acyl-Ch series, the *N*-acyl substitution at the C2 position cannot improve this affinity contrary to esterification at C3/C6 positions. Therefore, higher total-DS_{cr} values are required to ensure miscibility. Actually, the miscibility increased when ester-DS increased (Figure 13b) but the correlation between miscibility and amide-DS was lower (Figure 13c). The presented information strongly suggests a direct correlation between the primary structure of the polysaccharide derivatives and their miscibility with the hydrophobic polymer.

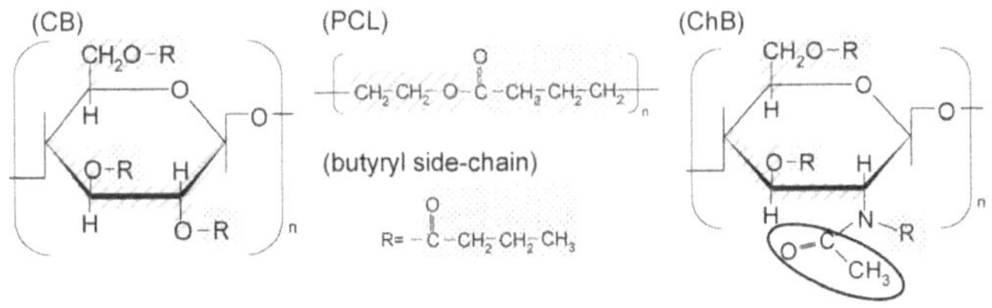

Figure 14: Schematic representation showing the structural similarity between the PCL repeating unit and the cellulose (CB) and chitin butyrate (ChB) butyryl side groups. Reproduced with permission from ref. [45].

The degree of miscibility significantly affects material performances, such as mechanical strength and cytocompatibility [46]. Teramoto *et al.* have fabricated thermoplastic transparent films from ChB/PCL blends [46]. Even though the mechanical properties of the immiscible (IM) blends did not improve, the ductility of ChB increased in miscible (M) and partially miscible systems (PM), making these blends highly processable (Figure 15). DSC analysis showed that PM blends slightly shifted the T_g of the ChB component but enhanced that of PCL. These blends exhibited extremely high ductility (tensile strength ≈15 MPa; elongation at break >200%), which may result from their ideal balance between PCL microcrystallite size and dispersion as well as the moderate domain size (20–50 nm) of the amorphous blend components. These blend films were subsequently subjected to alkaline hydrolysis (2 M NaOH/37 °C/48 h) to enhance their surface hydrophilicity and cell accessibility. Attenuated total reflection (ATR)-FTIR spectroscopy results revealed that the PCL component and ChB ester side chains were selectively removed from the hydrolyzed PM and IM film surface domains, causing a localization in chitin concentration in these surface domains. L929 fibroblast cells adhered well and prolifically distributed on PM film surfaces (Figure 16). Therefore, these materials exhibit great potential for use in tissue engineering as thermoplastic cell scaffolds. These scaffolds adopt various three-dimensional forms under degrees of miscibility selected according to adequate DS and post treatment variation.

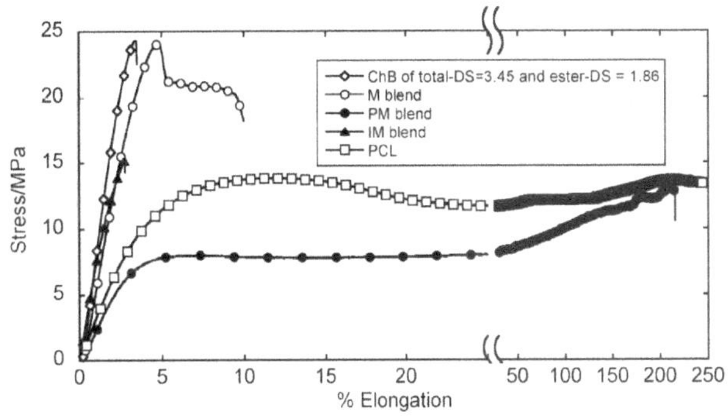

Figure 15: Stress–strain curves of ChB (total-DS = 3.45; ester-DS = 1.86) and PCL and their IM, PM, and M blends (ChB/PCL = 50/50 (*w/w*)). Reproduced with permission from ref. [46].

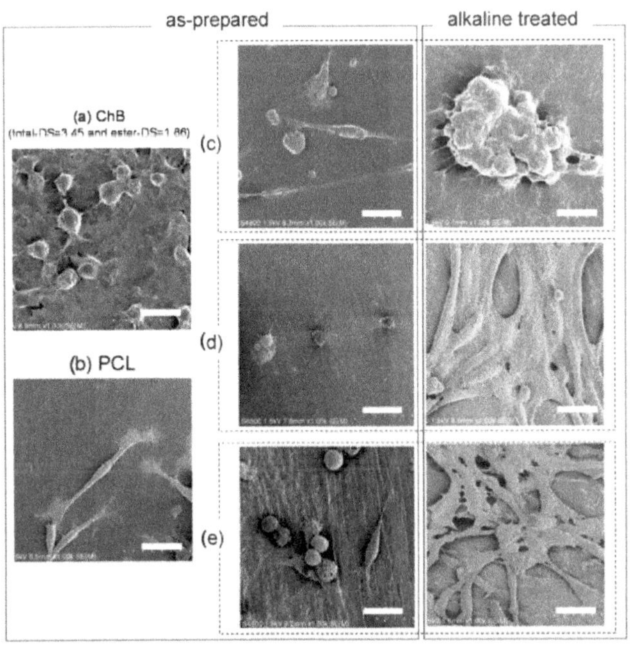

Figure 16: Field emission scanning electron microscope (FE-SEM) images of cell growth on films of (**a**) ChB (total-DS = 3.45; ester-DS = 1.86); (**b**) PCL; and (**c**) M; (**d**) PM, and (**e**) IM blends (original ChB/PCL weight ratio = 50/50). Images (**c**)–(**e**) include those of alkaline treated films. Scale bars denote 20 μm. Reproduced with permission from ref. [46].

GRAFT COPOLYMERS

Graft copolymerization offers a practical way to modify cellulosic molecules and alter woody materials surface properties. It may also prove useful to improve some original properties of polysaccharides and promote novel bulk functions in the copolymer products. In contrast to extensive structural and physical studies for block copolymer and polymer blends, systematic investigations on graft copolymer morphology, thermal transition behavior, and mechanical properties had been restricted to few systems including synthetic/synthetic polymer pairs in the past century [47]. Considerable efforts have been deployed to establish the relationship between copolymer compositions and properties in detail and functionalize cellulosic graft copolymers.

In contrast to unmodified cellulose, cellulose derivatives exhibit a relatively good but DS-dependent solubility in some organic solvents, making reaction possible in an appropriate homogeneous system. For DS < 2.5, CA is usually biodegradable [48,49] and its functionalization may benefit many chemical industries as well as agroindustrial, sanitary, and bio-related fields. Graft copolymerization provides an alternative to overcome above-mentioned problems related to low-molecular-weight "external" plasticizers. Aliphatic hydroxy acids or cyclic esters are attractive monomers for grafting CA for environmental conformity.

Higher-molecular-weight poly(hydroxy alkanoate) s (PHAs) are biodegradable polymers synthesized by polycondensation of hydroxy acids [50] and ring-opening polymerization of cyclic esters, such as lactides and lactones. The ring-opening polymerization is initiated efficiently using hydroxo-initiators [51] involving CAs [52,53,54] and CEs except the esters substituted completely such as cellulose triacetate.

Molar substitution (MS) and oxyalkanoyl DS, defined as the average number of introduced oxyalkanoyl units and that of hydroxyl groups substituted for oxyalkanoyl units per CA anhydroglucose residue, respectively, are important parameters for CE-*graft*-PHAs graft copolymer characterization. They are typically determined by ^{1}H NMR, as illustrated in Figure 17 for CA-*graft*-poly(lactic acid) (CA-*g*-PLA), giving the average degree of polymerization (DPs) of the PHA side chain as MS/(oxyalkanoyl DS). Equations used to calculate MS, oxyalkanoyl (lactyl) DS, and DPs are shown in Figure 17. Here, CE_x-*g*-PHA_y corresponds to CE-*g*-PHA copolymer for which ester-DS = x and oxyalkanoyl MS = y.

Synthesis of Biodegradable Cellulosic Graft Copolymers and Formulation of Basic Physical Properties

Teramoto and Nishio prepared $CA_{2.15}$-*g*-PLAs with a wide range of composition by three grafting methods, including (1) copolycondensation of lactic acid, (2) ring-opening copolymerization of L-lactide in DMSO, and (3) bulk ring-opening copolymerization similar to method (2) but without DMSO [55]. DSC measurements revealed that all copolymer products exhibited a single T_g (Figure 18). When lactyl MS increased to 8, T_g decreased sharply from 202 °C, corresponding to that of original CA, to *ca.* 60 °C, approximating that of plain PLA (Figure 19). When MS exceeded 14, the as-prepared graft copolymers formed a PLA side-chain crystalline phase (Figure 20). Tensile measurements of melt-quenched $CA_{2.15}$-*g*-PLA film sheets conducted at 80 °C–100 °C showed that their stretchability increased drastically with increasing PLA content and, for MS ≥ 14, the elongation at rupture reached a maximum of*ca.* 2000%.

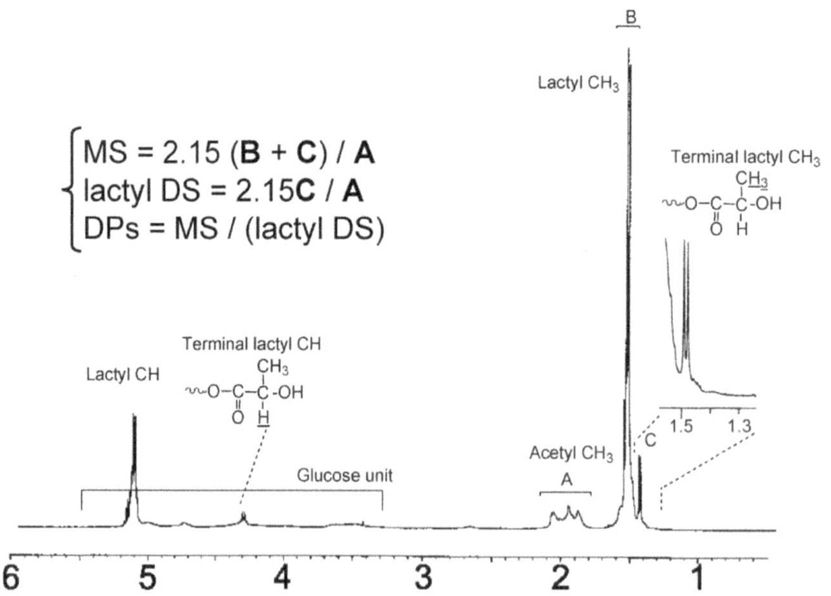

Figure 17: ^{1}H-NMR spectrum of a CA-*g*-PLA for which acetyl-DS = 2.15, MS = 7.86, DPs = 8.93, and lactyl DS = 0.88. Reproduced with permission from ref. [55].

Beside $CA_{2.15}$-*g*-PLA, CA-*g*-PHAs were subsequently synthesized by applying earlier chemical polymerization examples involving (R,S)-β-butyrolactone (BL), δ-valerolactone (VL), and ε-caprolactone (CL) [56]. To enhance the diversity of molecular architectures using this graft series, acetyl-DS was varied for the starting polymer CA to control the intramolecular density of grafts. The resulting CA-*g*-PHAs were characterized to establish a relationship between molecular architecture and thermal transition behavior in terms of semi-empirical equations proposed for polymer blends and comb-like polymers. Fox and Gordon–Taylor equations were assumed unsuitable for depicting the composition dependence of observed T_g for this graft copolymer series. However, if the molecular weight dependence of the side-chain component T_g is taken into account, the composition dependence of T_g obeys a simple mixing rule between $CA_{2.15}$ and PLA ingredients for $CA_{2.15}$-*g*-PLAs although the CA-*g*-PHA copolymers consist of originally immiscible components in this case. On the other hand, the T_g expression proposed for the comb-like polymer model [57] may provide an alternative depiction of the composition dependence of T_g for CA graft copolymers exhibiting acetyl-DS ranging from 1.75 to 2.15 despite some limits in its application to the assessment of multiple data-dependent parameters. However, the observed T_g clearly deviated from the calculated curve for the CA graft series at acetyl-DS values of 2.45 and 2.98, suggesting a possible phase separation of the two copolymer components. This comb-like polymer approximation would not be applicable to cases where the molecular weight of the backbone segments between two adjacent grafting positions is actually too high.

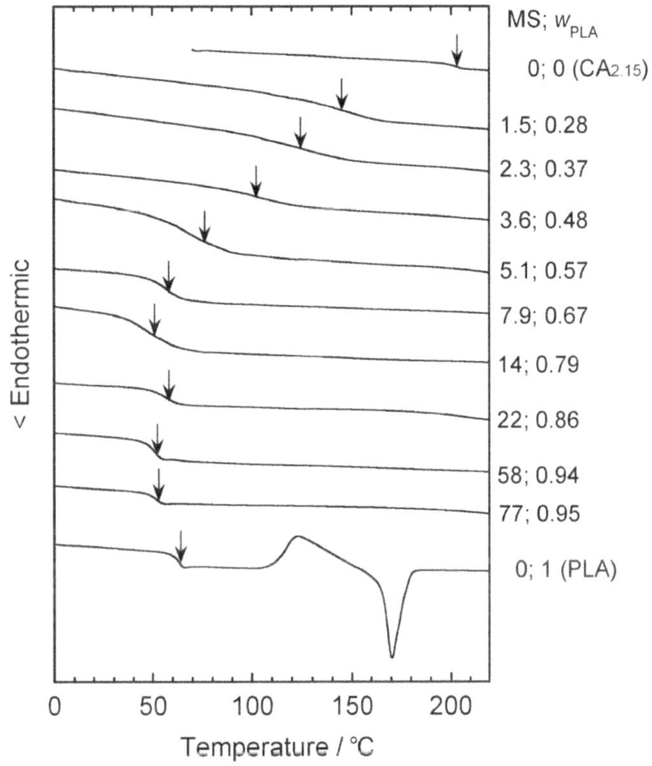

Figure 18: DSC thermograms of $CA_{2.15}$, PLA, and $CA_{2.15}$-*g*-PLAs obtained at various MS values during in the second heating scan. Arrows mark the T_g positions taken as the midpoints of the heat flow discontinuity. Reproduced with permission from ref. [55].

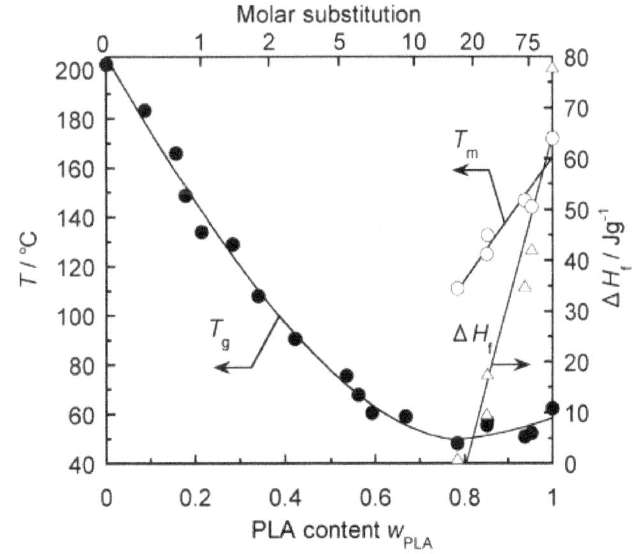

Figure 19: Composition dependence of the thermal transition parameters T_g, T_m, and H_f, estimated by DSC for $CA_{2.15}$-*g*-PLA samples. T_m and H_f were obtained in the first run while T_g was evaluated in the second heating scan. Reproduced with permission from ref. [55].

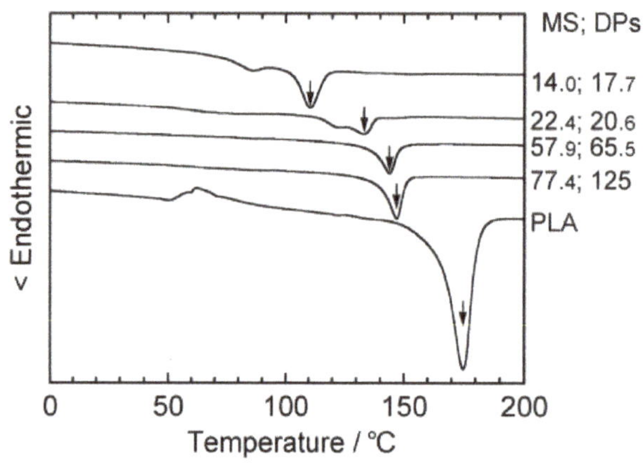

Figure 20: DSC thermograms of $CA_{2.15}$-*g*-PLA samples at different MS values. Arrows indicate T_m positions taken as endothermic peak maxima. Reproduced with permission from ref. [55].

Thermal Treatment Effect on the Development of Supramolecular Structures

The development of supramolecular structures of $CA_{2.15}$-*g*-poly(L-lactide) (PLLA) was investigated via physical aging and crystallization experiments under isothermal conditions [58]. For copolymers presenting MS values of 4.7 and 22 aged at different temperatures below or close to their T_gs (77 °C and 55 °C, respectively) for a definite time period, an optimum temperature of 50 °C was obtained to detect the largest enthalpy relaxation ΔH by DSC. The time evolution of ΔH at 50 °C was analyzed satisfactorily using the Kohlrausch–Williams–Watts stretched exponential function (Figure 21). Overall relaxation times (τ) were somewhat longer for the graft copolymers than for pristine PLLA. The regression analysis provided an exponent β characterizing the relaxation time distribution in addition to fairly large values for the aged copolymers compared with the relaxation behavior of conventional amorphous polymers and PLLA homopolymer. The anchoring of PLLA chains, which mainly contribute to the observed relaxation, onto the semirigid $CA_{2.15}$ backbone may significantly restrict their motions, resulting in a narrower distribution of the relaxation enthalpy.

An isothermal crystallization experiment conducted using a polarized optical microscope (POM) showed that $CA_{2.15}$-*g*-PLLAs underwent spherulite growth for MS values ranging from 22 to 77 (Figure 22) [53]. The optical texture comprised extinction rings, forming a band pattern of concentric circles, which had never been reported for plain PLLA and other graft copolymers. When MS increased, the texture became distorted and lost contrast while the growth rate dropped dramatically. Therefore, PLLA-rich graft copolymers shared common features with some crystalline polymer blends containing low concentrations of compatible polymer diluents regarding spherulite texture and growth rate. The slower growth kinetics caused by the anchoring effect was analyzed quantitatively to estimate the surface free energy σ_e of the folded PLLA lamellar crystals constituting the respective spherulites. This analysis was performed using a folded-chain crystallization formula expanded for a binary system composed of a miscible crystalline/amorphous polymer pair. Despite little MS-related systematic changes, the dependence of σ_e on the copolymer composition was interpretable through a convincing discussion on the stability parameter ϕ estimated from melting point measurements.

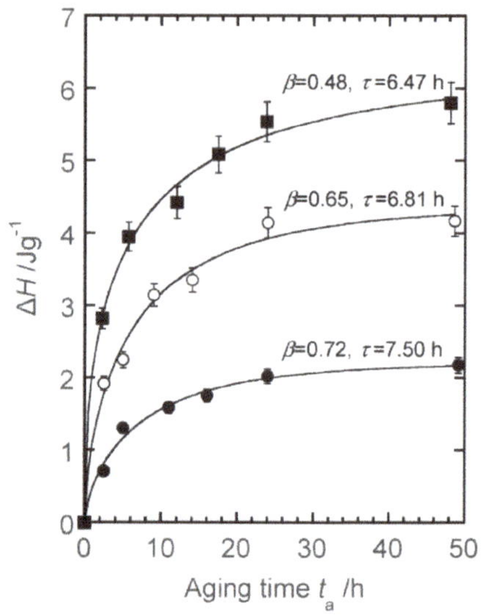

Figure 21: Time evolution of relaxation enthalpy ΔH estimated for PLLA (■) and $CA_{2.15}$-*g*-PLLAs at MS values of 4.7 (●) and 22 (○). The aging temperature amounted to 50 °C. Reproduced with permission from ref. [58].

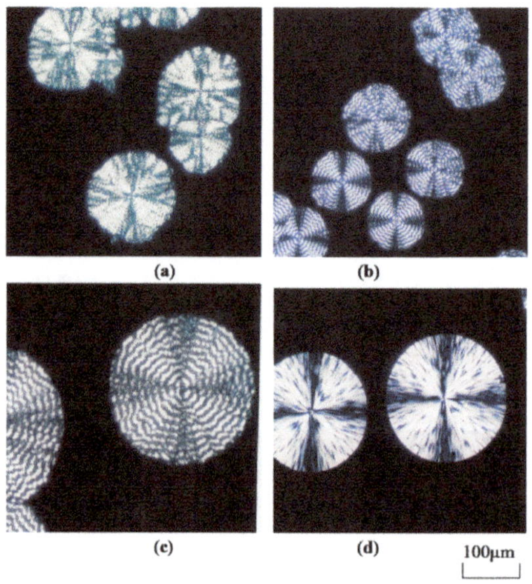

Figure 22: Polarized optical micrographs of typical spherulites observed for $CA_{2.15}$-*g*-PLLAs and plain PLLAs: (**a**) MS = 22 at T_{ic} = 107 °C and *t* = 4000 min; (**b**) MS = 58 at T_{ic} = 119 °C and *t* = 365 min; (**c**) MS = 77 at T_{ic} = 110 °C and *t* = 300 min; (**d**) PLLA at T_{ic} = 130 °C and *t* = 30 min. T_{ic} is the isothermal crystallization temperature and *t* is the elapsed time. Reproduced with permission from ref. [58].

Enzymatic Hydrolysis and Surface Morphological Characterization

Two compositions of the $CA_{2.15}$-*g*-PLLA series (MS = 4.7 (L) and 22 (H)) were subjected to enzymatic hydrolysis using Proteinase K [59]. A q-series of tested film specimens were only quenched from the molten state, and others were further annealed above and below their T_gs. Weight loss data recorded as a function of elapsed time revealed that the hydrolysis rate decreased as a result of the graft modification itself and the low PLLA content unless PLLA crystallinity developed. The adjacent hydrophobic CA2.15 backbone may seriously hinder the enzymatic attack of the PLLA side chains. Heat treatments, followed by physical aging or partial crystallization of the originally amorphous materials, also reduced the degree of enzymatic hydrolysis. Thus, the enzymatic degradation of $CA_{2.15}$-*g*-PLLAs was temporally controlled through variations of the molecular compositional factor and supramolecular rearrangement possibly occurring during isothermal treatments, such as tighter PLLA graft packing and free volume reduction.

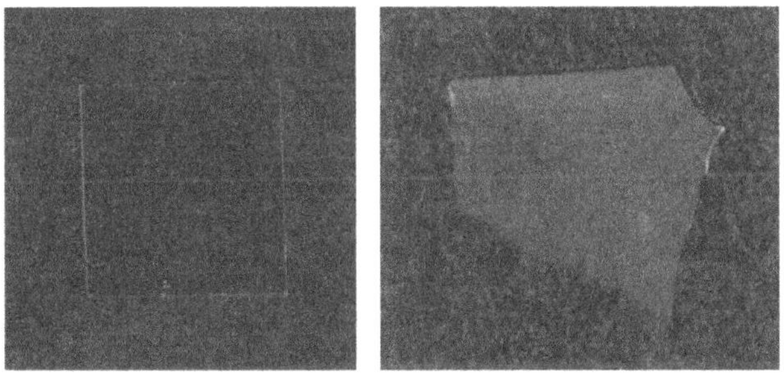

Figure 23: Photographs of an H-q sample (**a**) before and (**b**) after a 20-day enzymatic hydrolysis. Reproduced with permission from ref. [59].

To determine the spatial development of the enzymatic hydrolysis, surfaces of q-series degraded copolymer films were characterized by atomic force microscopy (AFM) and ATR-FTIR. The AFM study showed that the enzymatic hydrolysis produced a more undulated surface exhibiting several protuberances measuring several-hundred nanometers in height and a few micrometers in width. The hydrolysis generated finer and uniform protuberances on the L-q film surface while much deeper erosions, but somewhat irregular in amplitude, prevailed on the degraded H-q film surface at a higher PLLA content. ATR-FTIR measurements revealed that the hydrolyzed film surfaces selectively released lactyl units and absorption intensities were consistent with the AFM results. The degraded film specimens often displayed an iridescent color (Figure 23). This may be attributed to the interference of visible light wavelengths resulting from the diffuse reflection between the protuberances formed on film surfaces, which is virtually interpretable as a thin-layered gradation of refractive indices near the film surface. The complementary observations of after-effects of the enzymatic hydrolysis for the graft copolymers embody a conception of "spatiotemporally controlled degradation" for design of novel polymeric materials endowed with multi-functionalities, which involves degradation rate regulation, surface modification, and ensuing physical property improvement.

Molecular and Segmental Dynamics Characterized by Relaxation Analysis

Molecular and segmental dynamics of CA-*g*-PCL and CB-*g*-PCL components were investigated by dynamic mechanical, dielectric, and nuclear magnetic relaxation measurements in addition to their mixing scale [60].

Different relaxation processes originating from the molecular motions of the relevant structural units were detected by DMA for CA-*g*-PCL and CB-*g*-PCL series. Molecular mobility as a whole was enhanced when the introduction rate of the flexible PCL side chains increased for the copolymer series. Both CA-*g*-PCL polymer constituents tended to undergo phase separation on a *ca.*15 nm scale because of the poor miscibility between CA trunk and PCL side chains. In contrast, CB trunk chains mixed more intimately with aliphatic polyester side chains in CB-*g*-PCLs, reflecting the high miscibility of the CB/PCL pair as shown in Figure 12. However, CB-based copolymer components tended to adopt a heterogeneous mixing state at lower MS (0.5–0.6) because the anchoring of PCL chains onto the semirigid CB trunk affects chain mobility.

Dielectric relaxation spectroscopic analysis revealed that the introduction of CL units hardly affected the chain segmental velocity of the CA trunk. The constant relaxation time distribution for the α relaxation of the CA component suggested that there was almost no correlation in chain segmental dynamics between CA and PCL components. In contrast, the CB-*g*-PCL series exhibited a higher cooperation in segmental dynamics between trunk and side chain polymers, which may stem from the original miscibility of the CB/PCL pair. This enhanced miscibility, which originates from structural affinity and dipole-dipole interactions between butyryl groups and PCL repeating units, appeared to give rise to a dynamical interaction even on a local structural scale.

The ^1H spin-lattice relaxation time in the rotating frame ($T_{1\rho}{}^H$) determined by solid-state NMR (Figure 24) for specific carbons in a multicomponent polymer system provides an estimate of mixing homogeneity at the ^1H spin-diffusion length scale (2–4 nm). In general, $T_{1\rho}{}^H$ values are obtained by fitting the decaying carbon resonance intensity with the following single-exponential equation:

$$M(t) = M(0)\exp(-t/T_{1\rho}^H) \tag{3}$$

where *M*(*t*) is the magnetization intensity observed as a function of the spin-locking time *t*. In practice, TH1ρ is determined from the slope of the plot of ln[*M*(*t*)/*M*(0)] against *t*. Data confirmed that molecular mobility as a whole was promoted by the introduction of flexible PCL side chains for MS < 4. The promoted mobility is an alternative expression of the internal plasticization effect. However, for graft copolymers (MS > 7) containing a distinct crystalline phase expected to show a slower magnetization decay, the logarithmic *M*(*t*) data hardly fitted a single straight line. In this case, the normalized *M*(*t*) was simulated using a bi-exponential function involving two relaxation times as [61]:

$$M(t)/M(0) = x_f\exp(-t/T_{1\rho,fast}^H)x_s\exp(-t/T_{1\rho,slow}^H) \tag{4}$$

where TH1ρ,fast and TH1ρ,slow represent TH1ρ values for flexible (faster decay) and rigid (slower decay) components, respectively, and x_f and x_s are the corresponding fractions.

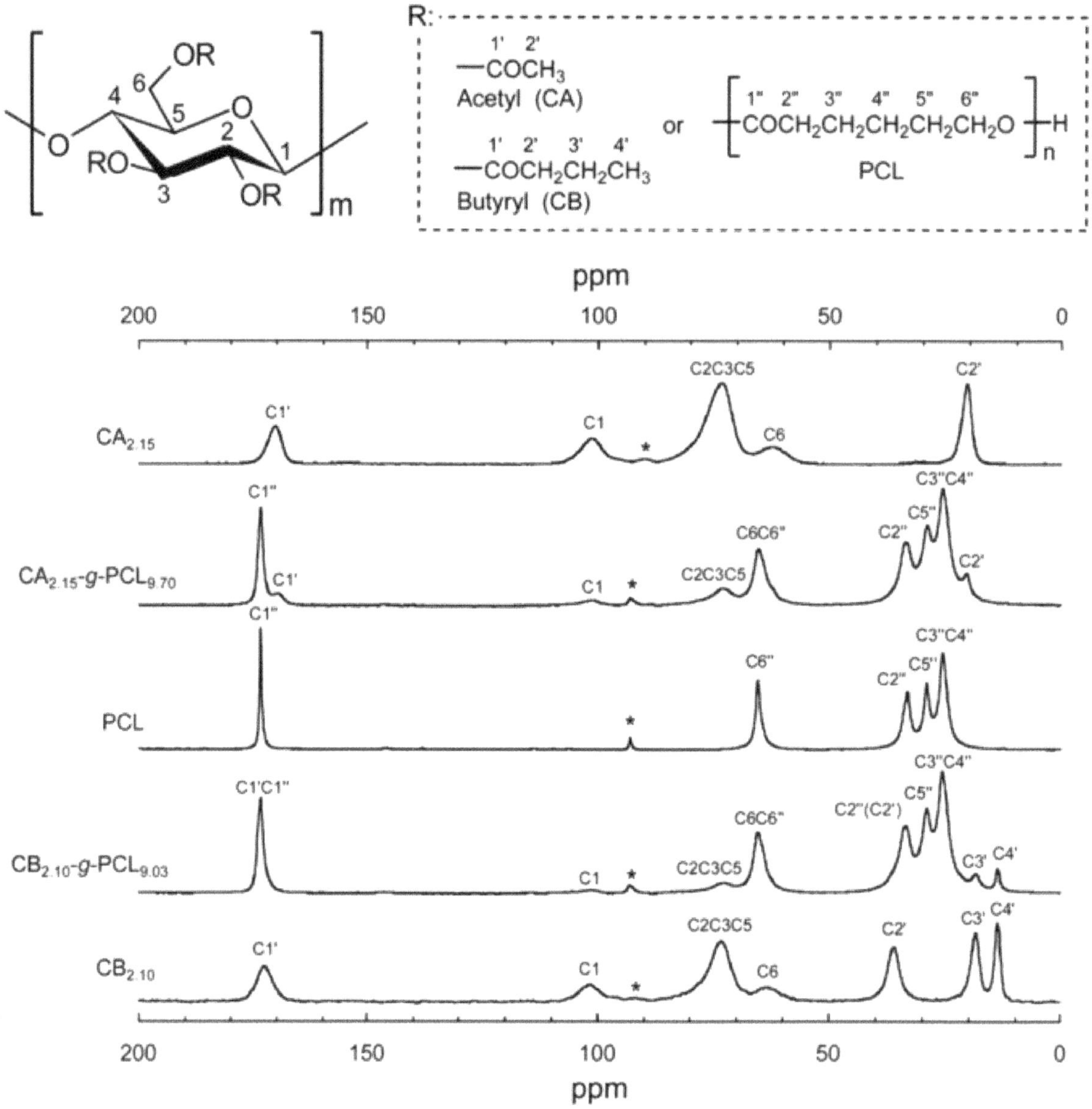

Figure 24: Solid-state ^{13}C CP/MAS NMR spectra of $CA_{2.15}$ (DS = 2.15), $CA_{2.15}$-*g*-$PCL_{9.70}$, plain PCL, $CB_{2.10}$-*g*-$PCL_{9.03}$, and $CB_{2.10}$, and their peak assignments. Asterisks denote a spinning sideband overlapping with the C4 pyranose carbon resonance signal. The C2" peak for $CB_{2.10}$-*g*-$PCL_{9.03}$ overlaps with the low intensity peak of the butyryl substituent C2›. Reproduced with permission from ref. [60].

When CA acted as the trunk, the magnetization decay of the acetyl C2› resonance yielded a TH1ρ value that almost coincided with that of another TH1ρ obtained by monitoring the pyranose C2/C3/C5 signal, as demonstrated for $CA_{2.15}$-*g*-$PCL_{9.70}$ in Figure 25a. TH1ρ data for CEs, PCL, and CE-*g*-PCLs are listed in Table 1. These values corresponding to the CA component were intermediate between TH1ρ,fast and TH1ρ,slow values for the crystallizable PCL. These observations suggest that the acetyl group is firmly restrained to the cellulose backbone and the molecular mobility of the unified CA trunk is somewhat restricted by the contiguous PCL crystalline domains in PCL-rich CA-based copolymers. On the other hand, the decay of the butyryl C4› signal was characterized by two different TH1ρvalues while that of the skeletal C2/C3/C5 signal provided a single TH1ρ for CB-based copolymers exhibiting MS values exceeding 7. The representative behavior of $CB_{2.10}$-*g*-$PCL_{9.03}$ is shown in Figure 25b. Shorter and longer TH1ρvalues obtained for the butyryl C4› signal decay were associated with amorphous and ordered phases, respectively, by comparison with other TH1ρ data for CB-*g*-PCL samples displaying the same butyryl DS. This observation indicates that the butyryl substituent may be free from restraint to the cellulose backbone and partly intrude into the PCL lamellar crystal surface, in agreement with crystallization kinetic studies [62].

Table 1: TH1ρ data for CA, CB, PCL, and their graft copolymers. Reproduced with permission from ref. [55].

Samples	$T_{1\rho}^{H}$/ms			
	CA or CB Component		**PCL Component**	
	Pyranose C2C3C5	**Acetyl C2' or Butyryl C4'**	**C3''C4''**	**C5''**
$CA_{2.15}$	13.7	13.9	–	–
$CA_{2.15}$-*g*-$PCL_{0.27}$	12.9	13.3	12.8	12.3
$CA_{2.15}$-*g*-$PCL_{0.87}$	8.46	7.87	8.02	8.30
$CA_{2.15}$-*g*-$PCL_{1.30}$	6.99	7.11	4.72	3.72
$CA_{2.15}$-*g*-$PCL_{2.50}$	2.79	2.74	2.20	1.71
$CA_{2.15}$-*g*-$PCL_{9.70}$	4.05	3.79	3.01[a]/24.3[b]	3.63[a]/24.6[b]
$CA_{2.45}$	16.6	15.9	–	–
$CA_{2.45}$-*g*-$PCL_{0.11}$	14.0	14.4	13.6	13.2
$CA_{2.45}$-*g*-$PCL_{0.22}$	11.4	11.7	9.03	7.23
$CA_{2.45}$-*g*-$PCL_{1.20}$	6.21	5.64	4.02	4.17
$CA_{2.45}$-*g*-$PCL_{2.50}$	3.02	2.68	2.05	2.39
$CA_{2.45}$-*g*-$PCL_{9.30}$	7.49	8.31	4.10[a]/22.5[b]	4.83[a]/23.3[b]
$CA_{2.98}$	15.7	15.4	–	–
$CA_{2.98}$-*g*-$PCL_{0.22}$	13.7	13.9	8.95	9.03
$CA_{2.98}$-*g*-$PCL_{0.55}$	12.3	14.0	7.62	6.39
$CA_{2.98}$-*g*-$PCL_{2.07}$	6.34	5.64	3.49	4.52
$CA_{2.98}$-*g*-$PCL_{9.20}$	7.09	7.87	2.15[a]/22.0[b]	1.75[a]/21.6[b]
$CB_{2.10}$	9.20	8.64	–	–
$CB_{2.10}$-*g*-$PCL_{0.16}$	7.45	6.73	8.58	6.42
$CB_{2.10}$-*g*-$PCL_{0.60}$	6.58	6.23	6.47	5.23
$CB_{2.10}$-*g*-$PCL_{2.33}$	3.24	3.11	2.48	2.69
$CB_{2.10}$-*g*-$PCL_{9.03}$	3.94	2.02[a]/18.0[b]	3.86[a]/27.3[b]	3.00[a]/31.1[b]
$CB_{2.50}$	7.60	7.90	–	–
$CB_{2.50}$-*g*-$PCL_{0.26}$	7.09	7.26	n.d.[c]	n.d.
$CB_{2.50}$-*g*-$PCL_{1.37}$	3.89	3.23	3.45	n.d.
$CB_{2.50}$-*g*-$PCL_{3.49}$	3.26	3.50	3.96	3.55
$CB_{2.50}$-*g*-$PCL_{7.42}$	4.82	3.20[a]/18.3[b]	3.68[a]/29.4[b]	4.83[a]/31.6[b]
$CB_{2.93}$	8.26	7.60	–	–
$CB_{2.93}$-*g*-$PCL_{0.23}$	7.73	7.67	n.d.	n.d.
$CB_{2.93}$-*g*-$PCL_{0.50}$	5.81	5.36	n.d.	n.d.
$CB_{2.93}$-*g*-$PCL_{3.58}$	3.25	2.80	3.64	3.00
$CB_{2.93}$ *g* $PCL_{12.6}$	n.d.[c]	3.12[a]/18.7[b]	7.20[a]/35.0[b]	7.29[a]/38.0[b]
PCL	–	–	6.29[a]/60.2[b]	7.17[a]/61.8[b]

[a] $T_{1\rho,fast}^{H}$; [b] $T_{1\rho,slow}^{H}$; [c] undetectable.

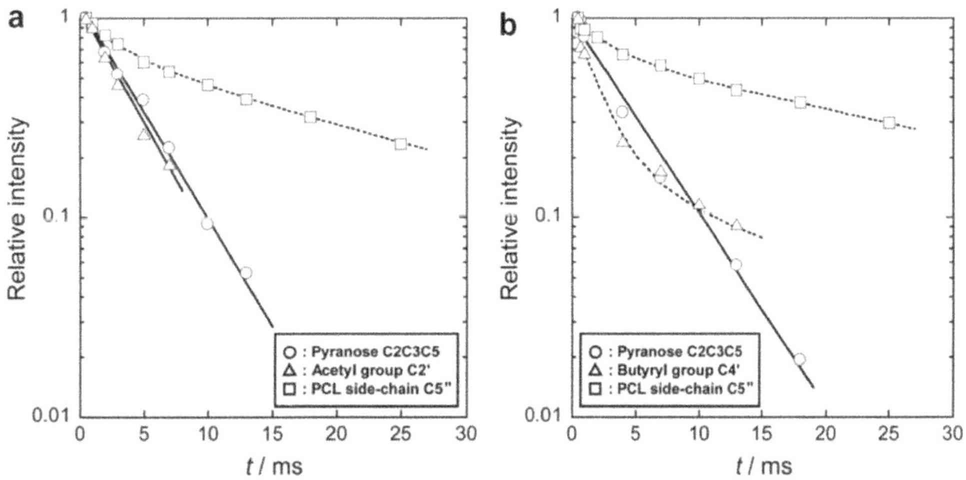

Figure 25: Semilogarithmic plots of ^{13}C resonance intensity decay as a function of spin-locking time *t* for (**a**) $CA_{2.15}$-*g*-$PCL_{9.70}$ and (**b**) $CB_{2.10}$-*g*-$PCL_{9.03}$ films. Straight lines represent fits to a single-exponential function (see Equation (3)) for the acetyl C2› in $CA_{2.15}$-*g*-$PCL_{9.70}$ and for the pyranose C2/C3/C5 in $CA_{2.15}$-*g*-$PCL_{9.70}$ and $CB_{2.10}$-*g*-$PCL_{9.03}$. Dashed-line curves correspond to fits to a double exponential function (see Equation (4)), where $x_f = 0.34$ and $x_s = 0.66$ for the PCL side-chain C5» in $CA_{2.15}$-*g*-$PCL_{9.70}$, $x_f = 0.82$ and $x_s = 0.18$ for the butyryl C4› in $CB_{2.10}$-*g*-$PCL_{9.03}$, and $x_f = 0.32$ and $x_s = 0.68$ for the PCL side-chain C5» in $CB_{2.10}$-*g*-$PCL_{9.03}$. Reproduced with permission from ref. [60].

Extension of Reaction Systems for Graft Copolymerization

The effective use of polysaccharide hydroxyl groups as initiators for the ring-opening polymerization of lactides and lactones to fabricate graft copolymers is expected to witness growth. Kitaoka *et al.* prepared partially substituted cellulose graft copolymers by regioselective derivatization of cellulose with L-lactide and CL in homogeneous DMAc–LiCl solution using native cellulose fiber as a starting material [63]. The resulting copolymers exhibited good solubility in common organic solvents. Enzymatic degradation of the copolymers occurred immediately upon cellulase treatment, even under mild conditions that do not hydrolyze commercial CDA. Vidéki *et al.* investigated the effect of external and internal (grafting) plasticization of CA using PCL and found that the external plasticizer was less efficient at reducing the stiffness of CA than grafted PCL chains [64]. Yuan *et al.* introduced PCL and PLLA block copolymers as grafted side chains in ethyl cellulose (EC) [65]. The EC-*g*-PCL spherulites formed sharply defined extinction rings. These researchers also fabricated thermoplastic cellulose-*g*-PLLA graft copolymers by ring-opening polymerization of L-lactide in homogeneous 1-allyl-3-methylimidazolium chloride ionic liquid solution using 4-dimethylaminopyridine as an organic catalyst and unmodified cellulose as a starting material [66]. Lönnberg *et al.* performed the ring-opening polymerization of CL and L-lactide from unmodified filter paper and paper preactivated with xyloglucan-bis(methylol)-2-methylpropanamide (XG-bis-MPA) and 2,2-bis(methylol)propionic acid (bis-MPA), respectively [67]. The bis-MPA activated paper exhibited the highest grafting efficiency, giving the most promising results regarding compatibility improvement between components and thus improving the mechanical properties. AFM observations showed that grafting produced a less-defined fibrillar structure and much smoother surface, indicating that the polymer covered the paper surface. Habibi *et al.* reported that PCL was covalently grafted onto the surface of cellulose nanocrystals suspended in dry toluene [68]. The obtained PCL-grafted cellulose nanocrystals exhibited significantly enhanced mechanical performances compared to those dispersed in PCL by solution casting. Enomoto-Rogers and Iwata synthesized di-*O*-(6-azidohexanoyl)-xylan-*graft*-PLLA coplymers by grafting propargyl-terminated PLLA onto di-*O*-(6-azidohexanoyl)-xylan ($XylC6N_3$) via click chemistry [69]. DSC measurements revealed that grafted PLLA side chains played an effective role as internal plasticizers for $XylC6N_3$ and native xylan.

ORIENTATION CONTROL

Processing methods such as stretching, spinning, and rolling are applicable to polymeric materials to produce polymer films and fibers exhibiting direction-dependent physical properties. This anisotropic behavior is at least partly attributable to the preferred orientation of molecular chains in non-crystalline regions, crystallite orientation, and, to some extent, the orientation of the supramolecular structure itself. Because a variety of thermal processing methods can be potentially employed for the thermoplasticized cellulosics mentioned above, it is expected to introduce some specific material functions based on anisotropy into the systems. It is therefore important to clarify the relationship between molecular orientation behavior and physical properties.

Controlling the optical anisotropy of CE-based materials has recently attracted increasing interest [40,70]. Specifically, CA is expected to provide a film base for optical devices used to regulate the polarization of light. Triacetyl cellulose (TAC) has been applied to liquid crystal display (LCD) panels as a protective layer for their poly(vinyl alcohol) polarizing films. However, the resulting birefringence creates imperfect optical compensation, leading to a color shift at oblique incidence and/or a reduction in black/white contrast [71]. Therefore, the selection, regulation, or compensation of linearly polarized light in these materials rests on the development of methods that control their optical properties.

The macromolecule orientation distribution is typically characterized by (i) WAXD, which only applies to crystalline polymers, (ii) dichroic infra-red or UV-visible spectroscopy and birefringence measurements, which provide secondary orientation functions, as well as (iii) fluorescence polarization measurements and laser Raman spectroscopy, which simultaneously give the statistic second and fourth moments. Interestingly, a combination of methods (ii) and (iii) enables an evaluation of the degree of orientation and the orientation distribution. This review focuses on examples that combine fluorescence polarization and birefringence (*n*) measurements. In fluorescence polarization measurements, the fluorescent 4,4›-bis(2-benzoxazolyl)stilbene (BBS, Figure 26a; length: *ca.* 2.5 nm) is mixed with polymer films and utilized as a probe to estimate their overall molecular orientation. Optical birefringence determines the orientation of polymer chains that display inherent anisotropy to their polarizability. Therefore, the difference between the dimensions of BBS and the polarizability of anisotropic units (statistical segment length) provides a precise depiction of molecular orientation (Figure 26b).

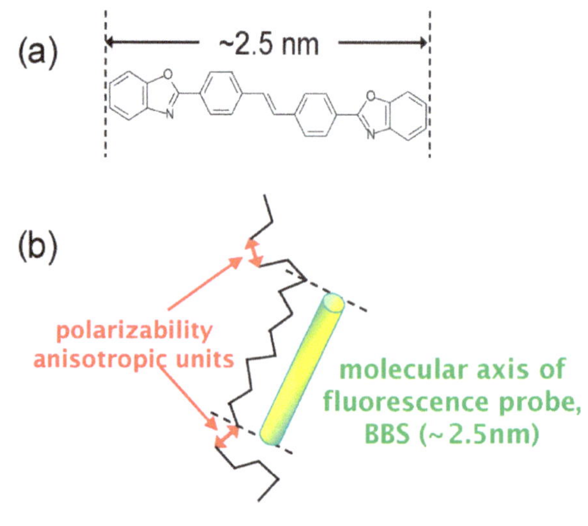

Figure 26: (**a**) Molecular structure of BBS and (**b**) schematic representation of the dimensional and directional differences between the polarizability anisotropic units and BBS.

Ohno and Nishio evaluated the molecular orientation and optical anisotropy resulting from the stretching of the already-described miscible blend films of CA and VP-based vinyl polymers [40]. P(VP-*co*-MMA) copolymers showed negative birefringence (Figure 27). Birefringence development in the blends was widely controllable in degree and polarity by altering the DS of CA, the VP/MMA ratio in P(VP-*co*-MMA), and the mixed polymer proportions because the miscible polymer components adopted a cooperative orientation during the uniaxial drawing process (Figure 27). At a specific blend composition (Figure 11), the drawn film behaved like an optically isotropic medium even though it was expected to be mechanically anisotropic after deformation.

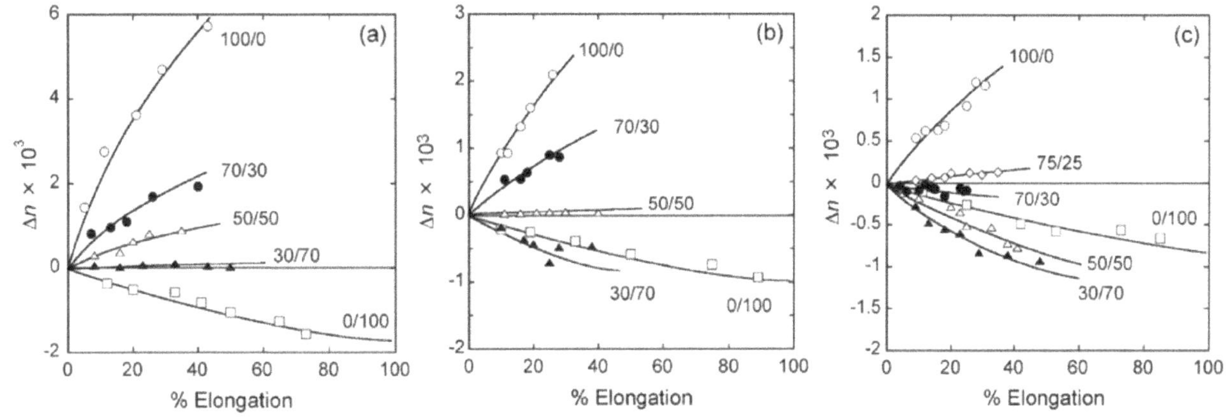

Figure 27: Birefringence of CA/P(VP-*co*-MMA) drawn films prepared at different DS and VP/MMA ratio as a function of elongation: (**a**) DS = 1.80 and VP/MMA = 35:65; (**b**) DS = 2.18 and VP/MMA = 47:53; (**c**) DS = 2.48 and VP/MMA = 50:50. Reprinted with permission from ref. [40].

On the other hand, practical properties and functions of cellulosic graft copolymers as thermoplastic solids may change dramatically according to their polymer chain orientation, which results from their manufacturing into geometrically defined products (e.g., films and filaments). Unohara *et al.* examined the stretching-induced molecular orientation and optical anisotropy in $CA_{2.15}$-*g*-PLLA films in connection with their grafted side chain length [72]. To address the rigidity of the trunk polymers, the flexible poly(vinyl acetate-*co*-vinyl alcohol) (PVAVAc; VA/VAc molar ratio = 0.358:0.642) was chosen as an additional trunk polymer instead of the semi-rigid $CA_{2.15}$. The overall orientation was estimated using the statistical second ($\langle\cos^2\omega\rangle$) and fourth ($\langle\cos^4\omega\rangle$) moments obtained by fluorescence polarization and demonstrated that all stretched films exhibited a positive orientation function (*i.e.*, $f = ((3\langle\cos^2\omega\rangle - 1)/2) > 0$) that increased with increasing deformation. The degree of molecular orientation was higher in the $CA_{2.15}$ graft series containing a semi-rigid trunk but decreased monotonically with increasing PLLA side chain content in both film types (Figure 28).

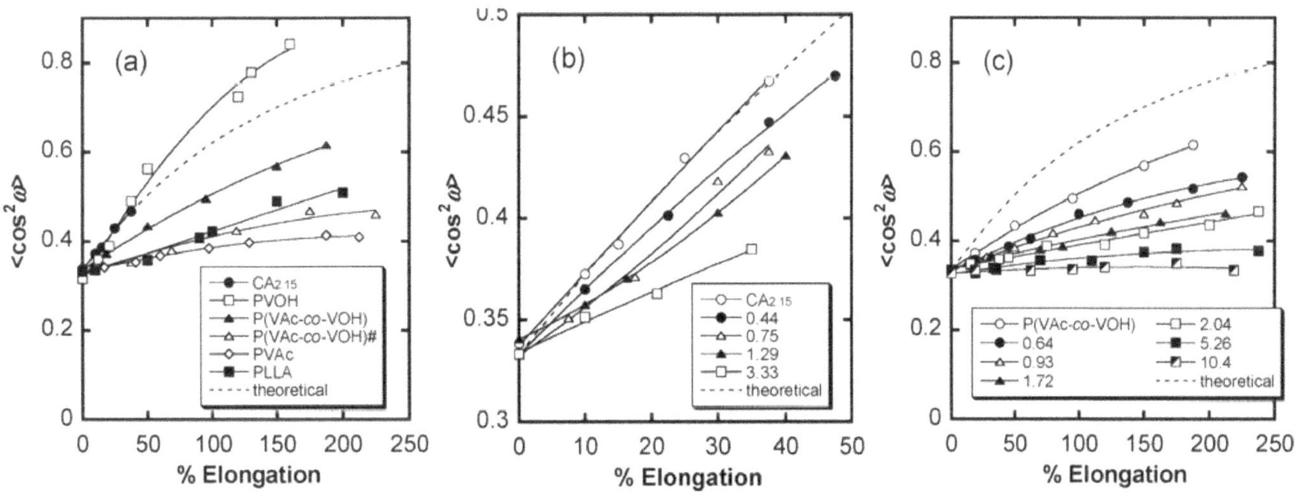

Figure 28: $\langle\cos^2\omega\rangle$ plots of films comprising $CA_{2.15}$-*g*-PLLA (**a**); PVAVAc-*g*-PLLA (**b**); and their component polymers and related ones (**c**) as a function of elongation. Broken lines represent theoretical curves obeying the Kratky-type affine deformation model. Numerals shown in (b) and (c) represent OH-MS values for graft copolymers. The parameter OH-MS was mainly chosen to characterize the grafting degree, defined as the average number of introduced lactyl units per hydroxyl group in the original CA or PVAVAc used as trunk. PVAVAc# is an additional copolymer displaying a VA/VAc molar ratio of 0.164/0.836. Reproduced with permission from ref. [72].

In terms of optical anisotropy (Figure 29), $CA_{2.15}$-*g*-PLLA films constantly exhibited a positive birefringence ($n > 0$) upon stretching whereas PVAVAc-*g*-PLLA drawn films displayed a negative value. This contrast in polarity reflects the difference in intrinsic birefringence between the two trunk polymers. Of particular interest was the discovery of a discontinuous change in n for a given stage of elongation with copolymer composition that is indicated with MS. This may stem from the different localized orientations of the attached PLLA chain segments. Lactyl units positioned close to the graft joint are arranged perpendicular to the trunk chain, which is most closely aligned to the drawing direction, and make a negative contribution to birefringence. On the other hand, lactyl units located far away from the joint are preferentially oriented in the drawing direction and make a positive contribution to the total birefringence. The situation is illustrated schematically in Figure 30.

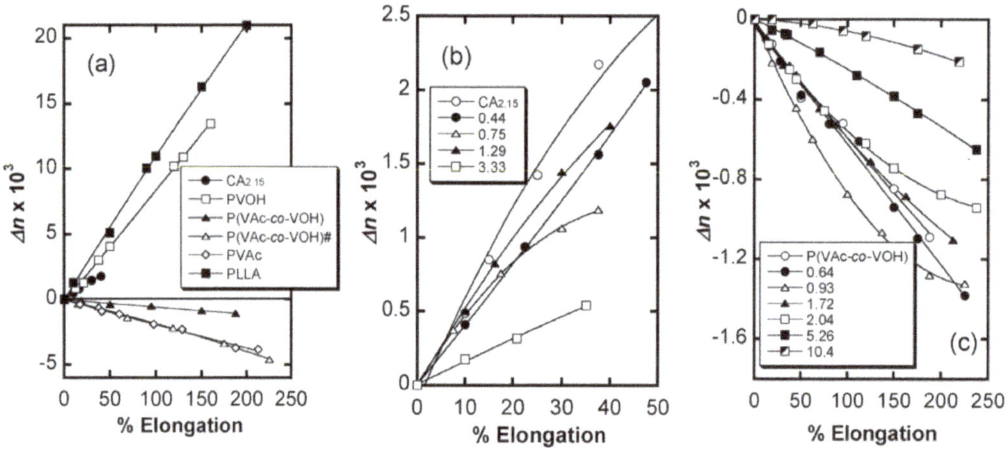

Figure 29: Birefringence plots of films comprising $CA_{2.15}$-*g*-PLLA (**b**), PVAVAc-*g*-PLLA (**c**), and their component polymers and related ones (**a**) as a function of elongation. Numerals in (b) and (c) correspond to OH-MS values for graft copolymers. PVAVAc# is an additional copolymer presenting a VA/VAc molar ratio of 0.164/0.836. Reproduced with permission from ref. [72].

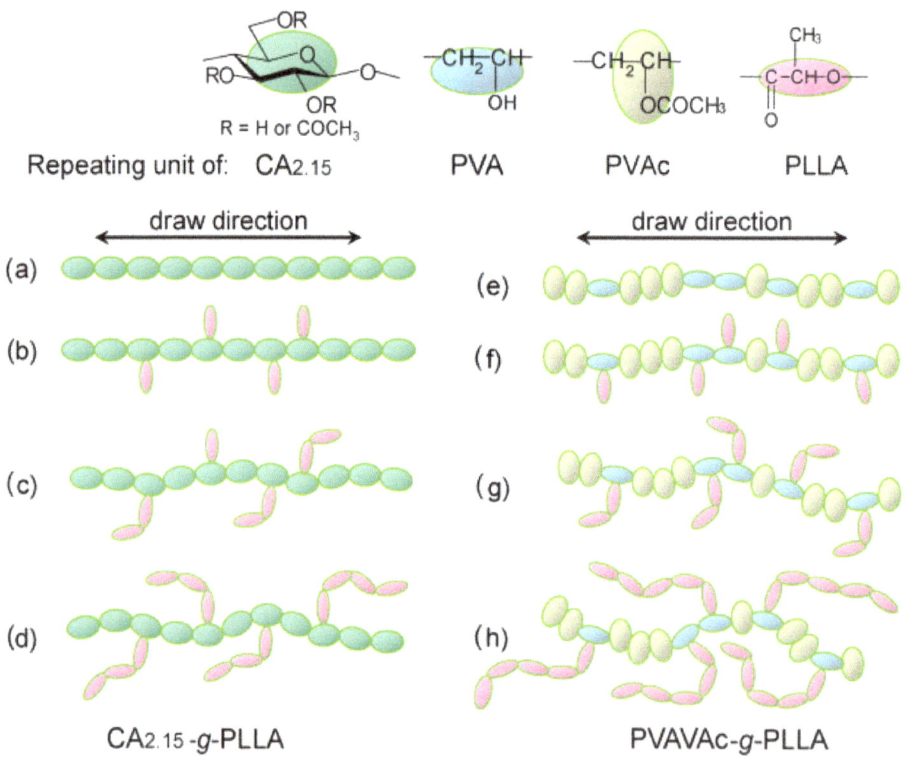

Figure 30: Schematic representation of orientational birefringence in $CA_{2.15}$-*g*-PLLA (**a–d**) and PVAVAc-*g*-PLLA (**e–h**). Polymeric chains appear as polarizability ellipsoid sequences of the constituent monomeric units. Reproduced with permission from ref. [72].

Very recently, Yamanaka *et al.* prepared graft copolymers composed of $CA_{2.15}$ and PMMA at various ratios via atom-transfer radical polymerization (ATRP) using a CuBr/*N,N,N',N,"N"*-pentamethyldiethylenetriamine catalytic system [73]. PMMA was chosen because of its importance in optical applications. These CA-*g*-PMMA graft copolymers were prepared in three steps (Figure 31), including (a) the ATRP initiator (2-bromoisobutyryl groups) introduction onto CA chains to obtain the macroinitiator CAmBBr, (b) ATRP grafting, and (c) the dehalogenation (hydrogenolysis) of the PMMA side-chain terminal. All reactions were conducted in homogeneous solutions and molecular characterization was performed by NMR and GPC analyses before and after deliberate cleavage of the PMMA grafts. These measurements revealed that the polydispersity index remained below 1.2 even if the graft molecular weight increased up to ~3000. Therefore, the graft chains were clearly produced via a well-controlled ATRP mechanism. The *n* value (Figure 32) at any given stage of the copolymer film elongation also decreased rapidly with increasing MS, leading to a transition from the positive *n* of pristine $CA_{2.15}$ to a negative value above 65% PMMA (the highest MS in the graft series used). Figure 33 shows a schematic representation of orientational birefringence in CA-*g*-PMMA. These graft copolymers may therefore find use as a highly functional material whose optical anisotropy can be controlled through birefringence compensation between the oriented trunk and graft chains. This material may play a significant role in ensuring molecular orientation growth and facilitate the design of optical retardant films exhibiting zero-birefringence regardless of their elongation.

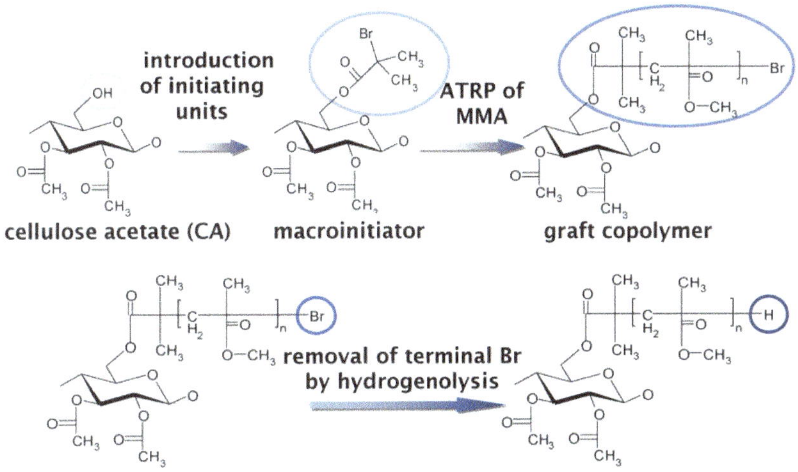

Figure 31: Synthesis of CA-g-PMMA.

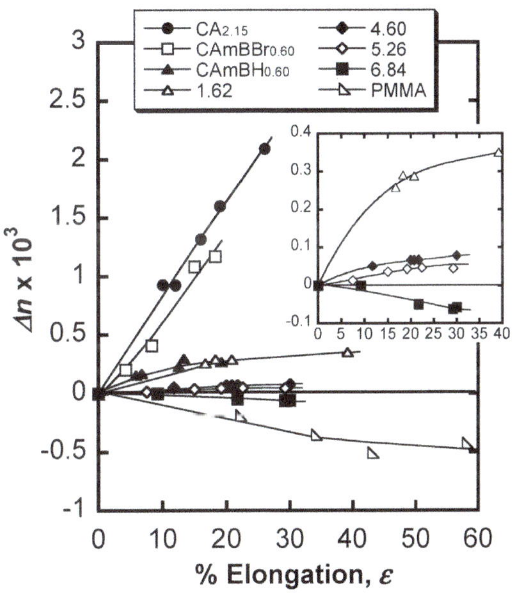

Figure 32: Birefringence plots of films comprising CA-*g*-PMMA copolymers and their component polymers and related ones as a function of elongation. Numerals represent MS values for the CA-*g*-PMMA series. Inset: graft copolymer data on an enlarged scale. Reproduced with permission from ref. [73].

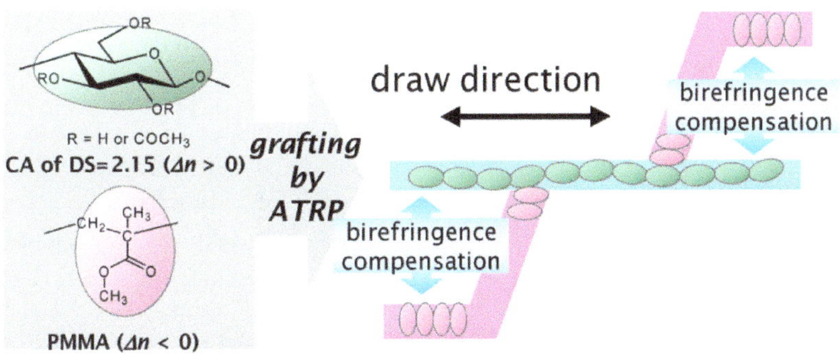

Figure 33: Schematic representation of orientational birefringence in CA-*g*-PMMA. Reproduced with permission from ref. [73].

CONCLUSIONS

This review focuses on the functionalization of cellulose and related polysaccharides as thermoplastic bulk materials. Four major systems, including single-substituent derivatives, derivatives with multiple substituents, polymer blends, and graft copolymers, were addressed. Studies summarized in this article highlighted highly complicated thermal (phase) transition behaviors in these derivatives despite their simple appearance. The accumulation of inclusive data on the relationship between the structure and thermal and mechanical properties is of primary importance for application of structural polysaccharide derivatives as thermoplastic materials. In conjunction with a rising impetus for the development of recyclable and/or renewable materials using biomass, it is a great reassurance that the active utilization of cellulose-based materials is practically conducted by industry, in particular via multiple derivatization. Polysaccharide derivatives utilized as commodity-type products and materials exerting special functions, such as stimulation response, biodegradability, separation, adsorption, and specific mechanical and optical performance, are expected to play crucial roles in numerous fields. Because substituents can be introduced with comparative ease in most cases, the material function added to the polysaccharides is important. This functionalization of polysaccharides will prove fruitful if controlled at the molecular primary and higher-order levels, such as specific phase developments as well as molecular and segmental orientations. Continuing research efforts are expected to expand the industrial use of a wide spectrum of polysaccharide derivative-based materials.

ACKNOWLEDGMENTS

The author is greatly indebted to Yoshiyuki Nishio of Kyoto University for his encouragement and helpful discussion. The author also wishes to thank H. Yano, H. Yamanaka, and H. Hashiwaki for their extensive studies. Studies conducted by the author were partly financed by the JST A-STEP program (No. AS231Z01846) and by a Grant-in-Aid for Scientific Research (23580228, 26620178, and 26712016) from JSPS.

REFERENCES

1. Edgar, K.J.; Buchanan, C.M.; Debenham, J.S.; Rundquist, P.A.; Seiler, B.D.; Shelton, M.C.; Tindall, D. Advances in cellulose ester performance and application. *Prog. Polym. Sci.* 2001, *26*, 1605–1688.
2. Klemm, D.; Heublein, B.; Fink, H.P.; Bohn, A. Cellulose: Fascinating biopolymer and sustainable raw material. *Angew. Chem. Int. Ed.* 2005, *44*, 3358–3393.
3. Nishio, Y. Material functionalization of cellulose and related polysaccharides via diverse microcompositions. *Adv. Polym. Sci.* 2006, *205*, 97–151.
4. Klemm, D.; Kramer, F.; Moritz, S.; Lindstrom, T.; Ankerfors, M.; Gray, D.G.; Dorris, A. Nanocelluloses: A new family of nature-based materials. *Angew. Chem. Int. Ed.* 2011, *50*, 5438–5466.
5. Eichhorn, S.J. Cellulose nanowhiskers: Promising materials for advanced applications. *Soft Matter* 2011, *7*, 303–315.
6. Heinze, T.; Liebert, T. Unconventional methods in cellulose fictionalization. *Prog. Polym. Sci.* 2001, *26*, 1689–1762.
7. Roy, D.; Semsarilar, M.; Guthrie, J.T.; Perrier, S. Cellulose modification by polymer grafting: A review. *Chem. Soc. Rev.* 2009, *38*, 2046–2064. [PubMed]
8. Carlmark, A.; Larsson, E.; Malmström, E. Grafting of cellulose by ring-opening polymerization—A review. *Eur. Polym. J.* 2012, *48*, 1646–1659.
9. Habibi, Y. Key advances in the chemical modification of nanocelluloses. *Chem. Soc. Rev.* 2014, *43*, 1519–1542. [PubMed]
10. Malm, C.; Mench, J.; Kendall, D.; Hiatt, G. Aliphatic acid esters of cellulose: Properties. *Ind. Eng. Chem.* 1951, *43*, 688–691.
11. Morooka, T.; Norimoto, M.; Yamada, T.; Shiraishi, N. Dielectric properties of cellulose acylates. *J. Appl. Polym. Sci.* 1984, *29*, 3981–3990.
12. Sealey, J.; Sarmaranayake, G.; Todd, J.; Glasser, W.G. Novel cellulose derivatives. IV. Preparation and thermal analysis of waxy esters of cellulose. *J. Polym. Sci. Part B Polym. Phys.* 1996, *34*, 1613–1620.
13. Vaca-Garcia, C.; Borredon, M.; Gaseta, A. Determination of the degree of substitution (DS) of mixed cellulose esters by elemental analysis. *Cellulose* 2001, *8*, 225–231.
14. Glasser, W.G. Prospects for future applications of cellulose acetate. *Macromol. Symp.* 2004, *208*, 371–394.
15. Vaca-Garcia, C.; Gozzelino, G.; Glasser, W.G.; Borredon, M. Dynamic mechanical thermal analysis transitions of partially and fully substituted cellulose fatty esters. *J. Polym. Sci. Part B Polym. Phys.* 2003, *41*, 281–288.
16. Fukuda, T.; Sugiura, M.; Takada, A.; Sato, T.; Miyamoto, T. Characteristics of cellulosic thermotropics. *Bull. Inst. Chem. Res. Kyoto Univ.* 1991, *69*, 211–218.
17. Flory, P.J. Molecular theory of liquid crystals. *Adv. Polym. Sci.* 1984, *59*, 1–36.
18. Teramoto, Y.; Miyata, T.; Nishio, Y. Dual Mesomorphic assemblage of chitin normal acylates and rapid enthalpy relaxation of their side chains. *Biomacromolecules* 2006, *7*, 190–198. [PubMed]
19. Plate, N.; Shbaev, V.; Petrukhin, B.; Zubov, Y.; Kargin, V. Structure of crystalline polymers with unbranched long side chains. *J. Polym. Sci. Part A Polym. Chem.* 1971, *9*, 2291–2298.
20. Fundador, N.; Enomoto-Rogers, Y.; Takemura, A.; Iwata, T. Syntheses and characterization of xylan esters. *Polymer* 2012, *53*, 3885–3893.
21. Enomoto-Rogers, Y.; Ohmomo, Y.; Takemura, A.; Iwata, T. Syntheses of glucomannan esters and their thermal andmechanical properties. *Carbohydr. Polym.* 2014, *101*, 592–599. [PubMed]
22. Marubayashi, H.; Yukinaka, K.; Enomoto-Rogers, Y.; Takemura, A.; Iwata, T. Curdlan ester derivatives: Synthesis, structure, and properties. *Carbohydr. Polym.* 2014, *103*, 427–433. [PubMed]
23. Sawai, D.; Nozoe, Y.; Yoshitani, T.; Tsukada, Y. Curdlan ester derivatives: Synthesis, structure, and properties. *FUJIFILM Res. Dev.* 2012, *57*, 55–58.
24. Glasser, W.G.; Sarmaranayake, G.; Durmay, M.; Dave,

V. Novel cellulose derivatives. III. Thermal analysis of mixed esters with butyric and hexanoic acid. *J. Polym. Sci. Part B Polym. Phys.* 1995, *33*, 2045–2054.

25. Kano, H.; Aranishi, Y.; Maeda, Y.; Nishio, Y. Chemical modification of cellulose for melt spinning process. In Proceedings of the International Symposium on Fiber Science and Technology (ISF2014), Tokyo, Japan, 1 October 2014; pp. G1–G36.
26. Iji, M.; Moon, S.; Tanaka, S. Hydrophobic, mechanical and thermal characteristics of thermoplastic cellulose diacetate bonded with cardanol from cashew nutshell. *Polym. J.* 2011, *43*, 738–741.
27. Yao, K.; Okoshi, M.; Kawashima, M.; Yamanoi, K. The development of flame-resistant bio-based plastic from inedible wood material. *Fuji Xerox Tech. Rep.* 2013, *22*, 96–103. (In Japanese).
28. Shibakami, M.; Tsubouchi, G.; Hayashi, M. Thermoplasticization of euglenoid β-1,3-glucans by mixedesterification. *Carbohydr. Polym.* 2014, *105*, 90–96. [PubMed]
29. Chang, C.; Teramoto, Y.; Nishio, Y. Synthesis of *O*-(2,3-dihydroxypropyl) cellulose in NaOH/urea aqueous solution: As a precursor for introducing "necklace-like" structure. *J. Polym. Sci. Part A Polym. Chem.* 2013, *51*, 3590–3597.
30. Chang, C.; Teramoto, Y.; Nishio, Y. High performance films of cellulose butyral derivative having a necklace-like annular structure in the side chains. *Polymer* 2014, *55*, 3944–3950.
31. Culter, R.; Kleinlein, B. Effect of the hydroxyl content and molecular weight of polyvinyl butyral on tape properties. *J. Eur. Ceram. Soc.* 2009, *29*, 3211–3218.
32. Jeong, H.; Rooney, M.; David, D.; MacKnight, W.J.; Karasz, F.E.; Kajiyama, T. Miscibility of polyvinyl butyral/nylon 6 blends. *Polymer* 2000, *41*, 6003–6013.
33. Cai, J.; Zhang, L.; Liu, S.; Liu, Y.; Xu, X.; Chen, X.; Chu, B.; Guo, X.; Xu, J.; Cheng, H.; *et al.* Dynamic self-assembly induced rapid dissolution of cellulose at low temperatures. *Macromolecules* 2008, *41*, 9345–9351.
34. Aoki, D.; Teramoto, Y.; Nishio, Y. SH-containing cellulose acetate derivatives: Preparation and characterization as a shape memory-recovery material. *Biomacromolecules* 2007, *8*, 3749–3757. [PubMed]
35. Utracki, L.A. *Polymer Alloys and Blends*; Hanser Gardner Publications: Munich, Germany, 1990.
36. Miyashita, Y.; Suzuki, T.; Nishio, Y. Miscibility of cellulose acetate with vinyl polymers. *Cellulose* 2002, *9*, 215–223.
37. Ohno, T.; Yoshizawa, S.; Miyashita, Y.; Nishio, Y. Interaction and scale of mixing in cellulose acetate/poly(*N*-vinyl pyrrolidone-*co*-vinyl acetate) blends. *Cellulose* 2005, *12*, 281–291.
38. Ohno, T.; Nishio, Y. Cellulose alkyl ester/vinyl polymer blends: Effects of butyryl substitution and intramolecular copolymer composition on the miscibility. *Cellulose* 2006, *13*, 245–259.
39. Ohno, T.; Nishio, Y. Estimation of miscibility and interaction for cellulose acetate and butyrate blends with *N*-vinylpyrrolidone copolymers. *Macromol. Chem. Phys.* 2007, *208*, 622–634.
40. Ohno, T.; Nishio, Y. Molecular orientation and optical anisotropy in drawn films of miscible blends composed of cellulose acetate and poly(*N*-vinylpyrrolidone-*co*-methyl methacrylate). *Macromolecules* 2007, *40*, 3468–3476.
41. Sugimura, K.; Teramoto, Y.; Nishio, Y. Cellulose propionate/poly(*N*-vinyl pyrrolidone-*co*-vinyl acetate) blends: Dependence of the miscibility on propionyl DS and copolymer composition. *Cellulose* 2013, *20*, 239–252.
42. Sugimura, K.; Teramoto, Y.; Nishio, Y. Blend miscibility of cellulose propionate with poly(*N*-vinyl pyrrolidone-*co*-methyl methacrylate). *Carbohydr. Polym.* 2013, *98*, 532–541. [PubMed]
43. Nishio, Y.; Matsuda, K.; Miyashita, Y.; Kimura, N.; Suzuki, H. Blends of poly(ε-caprolactone) with cellulose alkyl esters: Effect of the alkyl side-chain length and degree of substitution on miscibility. *Cellulose* 1997, *4*, 131–145.
44. Kusumi, R.; Inoue, Y.; Shirakawa, M.; Miyashita, Y.; Nishio, Y. Cellulose alkyl ester/poly(ε-caprolactone) blends: Characterization of miscibility and crystallization behaviour. *Cellulose* 2008, *15*, 1–16.
45. Sugimoto, M.; Kawahara, M.; Teramoto, Y.; Nishio, Y. Synthesis of acyl chitin derivatives and miscibility characterization of their blends with poly(ε-caprolactone). *Carbohydr. Polym.* 2010, *79*, 948–954.
46. Hashiwaki, H.; Teramoto, Y.; Nishio, Y. Fabrication of thermoplastic ductile films of chitin butyrate/poly(ε-caprolactone) blends and their cytocompatibility. *Carbohydr. Polym.* 2014, *114*, 330–338. [PubMed]
47. Takano, A.; Kondo, K.; Ueno, M.; Ito, K.; Kawahara, S.; Isono, Y.; Suzuki, J.; Matsushita, Y. Preparation and morphology of model graft copolymers of the A_3B_2 type with different graft junction points. *Polym. J.* 2001, *33*, 732–740.
48. Buchanan, C.M.; Gardner, R.M.; Komarek, R.J. Aerobic biodegradation of cellulose acetate. *J. Appl. Polym. Sci.* 1993, *47*, 1709–1719.
49. Sakai, K.; Yamauchi, T.; Nakasu, F.; Ohe, T. Biodegradation of cellulose acetate by *Neisseria sicca. Biosci. Biotechnol. Biochem.* 1996, *60*, 1617–1622. [PubMed]
50. Ajioka, M.; Enomoto, K.; Suzuki, K.; Yamaguchi, A. Basic properties of polylactic acid produced by the direct condensation polymerization of lactic acid. *Bull. Chem. Soc. Jpn.* 1995, *68*, 2125–2131.
51. Kricheldorf, H.R.; Kreiser-Saunders, I.; Boettcher, C. Polylactones: 31. Sn(II)octoate-initiated polymerization ofl-lactide: A mechanistic study. *Polymer* 1995, *36*, 1253–1259.
52. Yoshioka, M.; Hagiwara, N.; Shiraishi, N. Thermoplasticization of cellulose acetates by grafting of cyclic esters.*Cellulose* 1999, *6*, 193–212.

53. Hatakeyama, H.; Yoshida, T.; Hatakeyama, T. The effect of side chain association on thermal and viscoelastic properties of celluloses acetate based polycaprolactones. *J. Therm. Anal. Calorim.* 2000, *59*, 157–168.

54. Teramoto, Y.; Yoshioka, M.; Shiraishi, N.; Nishio, Y. Plasticization of cellulose diacetate by graft copolymerization of epsilon-caprolactone and lactic acid. *J. Appl. Polym. Sci.* 2002, *84*, 2621–2628.

55. Teramoto, Y.; Nishio, Y. Cellulose diacetate-*graft*-poly(lactic acid)s: Synthesis of wide-ranging compositions and their thermal and mechanical properties. *Polymer* 2003, *44*, 2701–2709.

56. Teramoto, Y.; Ama, S.; Higeshiro, T.; Nishio, Y. Cellulose acetate-*graft*-poly(hydroxyalkanoate)s: Synthesis and dependence of the thermal properties on copolymer composition. *Macromol. Chem. Phys.* 2004, *205*, 1904–1915.

57. Reimschuessel, H. On the glass transition temperature of comblike polymers: Effects of side chain length and backbone chain structure. *J. Polym. Sci. Part A Polym. Chem.* 1979, *17*, 2447–2457.

58. Teramoto, Y.; Nishio, Y. Biodegradable cellulose diacetate-*graft*-poly(l-lactide)s: Thermal treatment effect on the development of supramolecular structures. *Biomacromolecules* 2004, *5*, 397–406. [PubMed]

59. Teramoto, Y.; Nishio, Y. Biodegradable cellulose diacetate-*graft*-poly(l-lactide)s: Enzymatic hydrolysis behavior and surface morphological characterization. *Biomacromolecules* 2004, *5*, 407–414. [PubMed]

60. Kusumi, R.; Teramoto, Y.; Nishio, Y. Structural characterization of poly(ε-caprolactone)-grafted cellulose acetate and butyrate by solid-state ^{13}C-NMR, dynamic mechanical, and dielectric relaxation analyses. *Polymer*2011, *52*, 5912–5921.

61. Huang, J.M.; Yang, S.J. Studying the miscibility and thermal behavior of polybenzoxazine/poly(ε-caprolactone) blends using DSC, DMA, and solid state ^{13}C-NMR spectroscopy. *Polymer* 2005, *46*, 8068–8078.

62. Kusumi, R.; Teramoto, Y.; Nishio, Y. Crystallization behavior of poly(ε-caprolactone) grafted onto cellulose alkyl esters: Effects of copolymer composition and intercomponent miscibility. *Macromol. Chem. Phys.* 2008, *209*, 2135–2146.

63. Mayumi, A.; Kitaoka, T.; Wariishi, H. Partial substitution of cellulose by ring-opening esterification of cyclic esters in a homogeneous system. *J. Appl. Polym. Sci.* 2006, *102*, 4358–4364.

64. Vidéki, B.; Klébert, S.; Pukánszky, B. External and internal plasticization of cellulose acetate with caprolactone: Structure and properties. *J. Polym. Sci. Part B Polym. Phys.* 2007, *45*, 873–883.

65. Yuan, W.; Yuan, J.; Zhang, F.; Xie, X. Syntheses, characterization, and *in vitro* degradation of ethyl cellulose-*graft*-poly(ε-caprolactone)-*block*-poly(l-lactide) coplymers by sequential ring-opening polymerization.*Biomacromolecules* 2007, *8*, 1101–1108. [PubMed]

66. Yan, C.; Zhang, J.; Lv, Y.; Yu, J.; Wu, J.; Zhang, J.; He, J. Thermoplastic cellulose-*graft*-poly(l-lactide) copolymers homogeneously synthesized in an ionic liquid with 4-dimethylaminopyridine catalyst. *Biomacromolecules* 2009,*10*, 2013–2018. [PubMed]

67. Lönnberg, H.; Zhou, Q.; Brumer, H.; Teeri, T.; Malmström, E.; Hult, A. Grafting of cellulose fibers with poly(ε-caprolactone) and poly(l-lactic acid) via ring-opening polymerization. *Biomacromolecules* 2006, *7*, 2178–2185. [PubMed]

68. Habibi, Y.; Goffin, A.; Schiltz, N.; Duquesne, E.; Dubois, P.; Dufresne, A. Bionanocomposites based on poly(ε-caprolactone)-grafted cellulose nanocrystals by ring-opening polymerization. *J. Mater. Chem.* 2008, *18*, 5002–5010.

69. Enomoto-Rogers, Y.; Iwata, T. Synthesis of xylan-*graft*-poly(l-lactide) copolymers via click chemistry and their thermal properties. *Carbohydr. Polym.* 2012, *87*, 1933–1940.

70. Yamaguchi, M.; Masuzawa, K. Birefringence control for binary blends of cellulose acetate propionate and poly(vinyl acetate). *Eur. Polym. J.* 2007, *43*, 3277–3282.

71. Nakayama, H.; Fukagawa, N.; Nishiura, Y.; Yasuda, T.; Ito, T.; Mihayashi, K. Development of low-retardation TAC film for protection films of LCD's polarizer. *J. Photopolym. Sci. Technol.* 2006, *19*, 169–173.

72. Unohara, T.; Teramoto, Y.; Nishio, Y. Molecular orientation and optical anisotropy in drawn films of cellulose diacetate-*graft*-PLLA: Comparative investigation with poly(vinyl acetate-*co*-vinyl alcohol)-*graft*-PLLA. *Cellulose*2011, *18*, 539–553.

73. Yamanaka, H.; Teramoto, Y.; Nishio, Y. Orientation and birefringence compensation of trunk and graft chains in drawn films of cellulose acetate-*graft*-PMMA synthesized by ATRP. *Macromolecules* 2013, *46*, 3074–3083.

CHAPTER 17

CHARACTERIZATION OF FLEXIBLE COPOLYMER OPTICAL FIBERS FOR FORCE SENSING APPLICATIONS

Marek Krehel[1,2], René M. Rossi[1], Gian-Luca Bona[1,2] and Lukas J. Scherer[1].

[1]Empa, Swiss Federal Laboratories for Materials Science and Technology, Laboratory for Protection and Physiology, Lerchenfeldstrasse 5, St. Gallen 9014, Switzerland

[2]ETH Zurich, Swiss Federal Institute of Technology, Department of Information Technology and Electrical Engineering, Gloriastrasse 35, Zurich 8092, Switzerland

ABSTRACT

In this paper, different polymer optical fibres for applications in force sensing systems in textile fabrics are reported. The proposed method is based on the deflection of the light in fibre waveguides. Applying a force on the fibre changes the geometry and affects the wave guiding properties and hence induces light loss in the optical fibre. Fibres out of three different elastic and transparent copolymer materials were successfully produced and tested. Moreover, the influence of the diameter on the sensing properties was studied. The detectable force ranges from 0.05 N to 40 N (applied on 3 cm of fibre length), which can be regulated with the material and the diameter of the fibre. The detected signal loss varied from 0.6% to 78.3%. The fibres have attenuation parameters between 0.16–0.25 dB/cm at 652 nm. We show that the cross-sensitivies to temperature, strain and bends are low. Moreover, the high yield strength (0.0039–0.0054 GPa) and flexibility make these fibres very attractive candidates for integration into textiles to form wearable sensors, medical textiles or even computing systems.

INTRODUCTION

Optical fibres-based sensors have numerous advantages compared to electrical or chemical devices, e.g., insensitivity to electromagnetic fields, water and corrosion resistance, compact size and light weight [1,2]. Due to these benefits, optical fibre based sensors are used in numerous applications to detect changes in chemical and/or physical parameters [1,3–6]. Various methods to detect pressure using optical fibres are known. Wang *et al.* proposed a pressure sensor that employs the effect of photoelasticity [7]. Urban *et al.* designed a pressure sensor that relies on FBG lateral deformation [8]. Within the Seventh Framework Programme "Ofseth", optical fibres using optical time-domain reflectometry for distributed respiration measurement and optical fibres with long period gratings in microstructured POF were incorporated into elastic fabrics to measure the breathing rate and the breathing volume by measuring the changes in applied pressure [9–11]. The most common method to determine the pressure using optical fibres is by using fibre Bragg gratings [8,12–14]. To implement optical fibre based pressure sensors into textile fabrics, the fibres should be highly flexible, which is not the case for glass-based fibres most often used for fibre Bragg grating based sensors. Moreover, sensors based on fibre Bragg gratings have

Citation: Krehel, M.; Rossi, R.M.; Bona, G.-L.; Scherer, L.J. Characterization of Flexible Copolymer Optical Fibers for Force Sensing Applications. Sensors 2013, 13, 11956-11968, doi:10.3390/s130911956

to be single mode and are dependent on temperature, strain and pressure simultaneously and require compensation of some of the parameters in order to sense one of them [15,16].

In this study, elastically deformable light pipes, so called multimode optical fibres which react to applied pressure by deflecting light in the fibre structure, are reported. Due to their high flexibility and their appropriate tensile strength (from 0.0033 to 0.0056 GPa), they meet the critical criteria to form medical textiles [17,18]. Long-term body monitoring solutions using textile fabrics have been intensively studied in recent years [19,20]. Plastic optical fibres have several advantages compared to textile electronics when wearing them close to the body, e.g., comfort, ease of movements, and reduced movement artefacts [17]. The electronics involved in the proposed solutions often influence negatively the haptic of medical textiles. Flexible and smart fibres allow the separation of the rigid electronics from the measured region and thus from the body. Possible body parameters which could be measured with the proposed solution are, e.g., muscles activity, motion detections or breath monitoring [21]. For technical applications, these pressure sensors can be used for counting occupied seats in airplanes, cars or for monitoring immobile patients in hospitals to prevent decubitus.

EXPERIMENTAL SECTION

Materials and Methods

All the polymers used were purchased from WackerChemie AG (München, Germany). The polymers were all from the Geniomer group and are copolymers containing a soft silicon part and a hard polyurethane part. Geniomer 100-HDS and Geniomer 100 consist of about 70% silicon and 30% polyurethane, while Geniomer 175 consists of only 10% polyurethane and 90% silicon and has a lower processing temperature than the two Geniomer 100 polymers. Geniomer 100-HDS is Geniomer 100 with the addition of highly dispersed silicate particles.

Fibres Extrusion with Melt Flow Index

The transparent light pipes were produced by a melt flow index apparatus (MFI), model 7085.15a, provided by ZWICK (Ulm, Germany) with an external drawing engine. In Table 1 the extrusion parameters at which the fibres were extruded are shown. All tested materials were firstly dried in a vacuum oven for 8 h at 80 °C. 7 g were used for each test. Fibres made of Geniomer 100 were extruded with three different diameters. Since this material showed the best melt flow characteristics of all three polymers, the diameter of the light pipes could be easily varied with this polymer. Three different fibre diameters made of this material were studied: 0.45 mm, 0.75 mm, and 0.85 mm. Consequently, these optical fibres can support thousands of optical modes although the fibre does not feature an outer cladding to protect the surface from scratches or other defects.

Table 1: Summary of parameters used for the fibre extrusions

Material type	Fibre Diameter [mm]	Temperature [°C]	Pre-worming time [s]	Load [kg]	Engine Speed [arbitrary units]
Geniomer 100	0.45	163	240	1.2	5–6
Geniomer 100	0.75	160	240	2.16	10
Geniomer 100	0.85	165	240	3.8	2–3
Geniomer 175	0.5	160	240	2.16	10
Geniomer 100-HDS	0.5	160	240	2.16	10

Light Loss

The optical loss measurements were performed with the cut-back method [22]. The medical laser used therefore had a wavelength of 652 nm with an intensity of 100 mW and was provided by AOL Medical Instruments. The 1 m long fibres were connected to the mode mixer (coupled to the light source) using F-SMA connectors from Thorlabs GmbH (Newton, MA, USA). The following equation was used to compute the light attenuation [23]. The optical loss measurements were repeated five times.

$$A = 10 \cdot log_{10}\left(\frac{I}{I_0}\right) \quad (1)$$

where: I - Input light intensity, I_0- Output light intensity.

The optical loss measurements were repeated five times. Intrinsic losses over a whole visible spectrum range from 300 nm to 750 nm of Geniomer 100 were performed using a Lambda 900 UV-VIS spectrophotometer provided by Perkin Elmer. The spectrometer measures continuously and independently (by means of beam splitting) the intensity of the reference beam and uses the value to compute reflection or transmission coefficients. To determine the intrinsic loss of the copolymers (Figure 1), a polymer cube with the dimension 50 × 10 × 5 mm^3 was formed. First the Fresnel reflection coefficient of the material was measured. This was done by measuring the light transmission through the thinnest part of the sample (5 mm), where the light absorption could be neglected. Afterwards, the light transmission through the 50 mm part was measured. The spectrophotometer was set to give the light transmission values from 300 nm to 750 nm in steps of 1 nm. For the force measurement setup, the extruded fibres were connected via a mode mixer to a Halogen-Deuterium Lamp model L10290 provided by Hamamatsu Photonics K.K. (Hamamatsu City, Japan), with SMA 905 connectors. In order to achieve flat fibre facets for good light coupling efficiency, fibres were cut with unused scalpel blades. Due to the elastic behaviour of the fibre material, polishing was not performed [17]. The ends of the fibres were connected by custom developed connectors to the measurement head RW3701-2 (Gigahertz Optik, Türkenfeld, Germany) and the outcoming light was measured using an Optometer P9710 detector (Gigahertz Optik). The force was applied with a Zwick INSTRON tensile testing machine model 4502 over the length of 3 cm. To study the influence of the temperature on the sensing system, the optical fibres were heated by a hot plate (Heidolph D-91126 type: MR Hei-Standard, Schwabach, Germany) and monitored by a temperature controller (Heidolph type: EHT Hei-Con). As a light source, an LED IFE97 and as a detector a photodiode IFD91 were used, both provided by Industrial Fiber Optics (Tempe, AZ, USA).

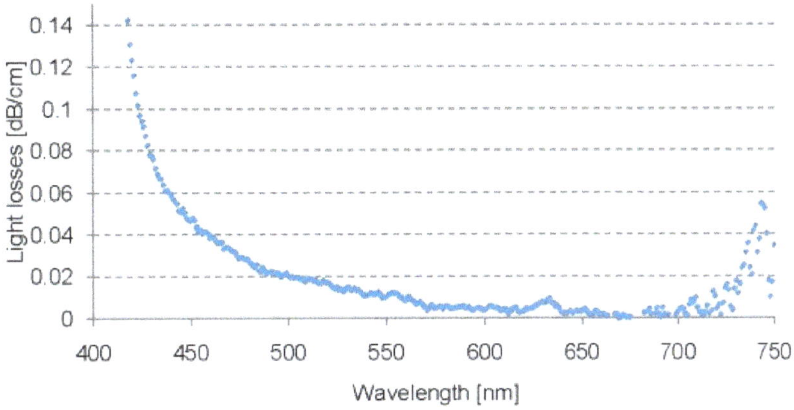

Figure 1: Intrinsic losses of the Geniomer 100-HDS polymer copolymer (measured from a polymer cube)

RESULTS AND DISCUSSION

Optical Properties of the Extruded Optical Fibres

The force sensor is based on the losses in light transmission along the light guides when the fibres are compressed. Thus it was essential to check whether the extruded fibres have reasonable light transmission parameters, *i.e.*, low optical losses. Since the final goal is to incorporate the fibres to medical textiles with a max. length of 1 m, the maximum light attenuation should not exceed 0.2–0.3 dB/cm at 652 nm. Light losses at this level allow the detection of the transmitted light of a 1 m long light guide.

Light Attenuation

In Table 2, a summary of light attenuation measurements at 652 nm is listed. No significant differences between the different materials and different diameters were observed. The optical signal loss varied between 0.16 and 0.25 dB/cm. That means that all fibres produced with the MFI apparatus were suitable for light transmissions over short distances of up to 1 m.

Table 2: Summary of light attenuation measurements of the fibres produced in this study

Type of a Fibre	Light Attenuation [dB/cm]	σ [1] [dB/cm]
Geniomer 100-HDS d = 0.5 mm	0.25	0.09
Geniomer 175 d = 0.5 mm	0.19	0.07
Geniomer 100 d = 0.75	0.22	0.06
Geniomer 100 d = 0.85 mm	0.20	0.09
Geniomer 100 d = 0.45 mm	0.24	0.12

[1] Standard deviation.

Intrinsic Light Absorption Spectrum Measured at 652 nm

Figure 1 shows the intrinsic absorption spectrum of the light guide made of Geniomer 100-HDS as explained in Section 2.2. Since the overall attenuation of the light pipes was much higher (Table 2), the main light loss was due to extrinsic losses caused by the extrusion process (irregular fibre surface, bubbles or other inhomogeneities in the material). The absorption spectrum of Geniomer 100-HDS showed a local absorption band at around 635 nm, which corresponds to the 6th overtone of the C–H vibrations [24]. In the range of 660–700 nm, the absorption became lower than 0.01 dB/cm. Below 450 nm and above 700 nm the material starts to absorb strongly.

Force Sensing

The force sensing setup is presented below in Figure 2. The signal was measured from the middle of the three fibres. The outer fibres were used to support the weight.

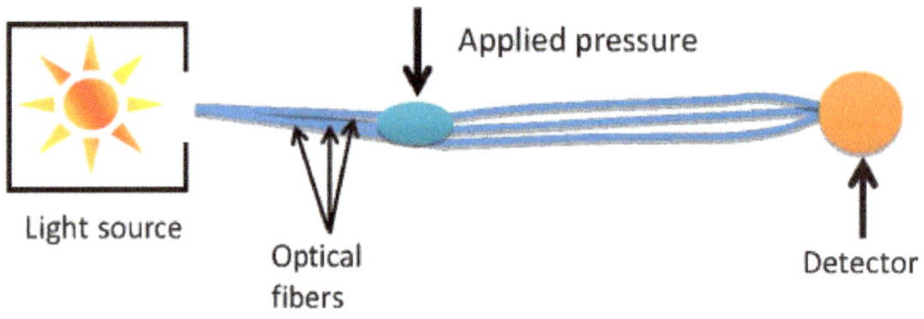

Figure 2: Force sensing setup.

In Figure 3, a scheme representing the pressure sensing principle is shown. The applied force resulted in an elliptical deformation of the fibre cross section. From the side projection a kind of cavity was observed, as schematically illustrated in Figure 3. This deflection increased the out-coupling of the light in the region under pressure due to the geometrical deformation of the fibre. Since the degree of deflection was directly related to the applied force, the force could be quantitatively determined.

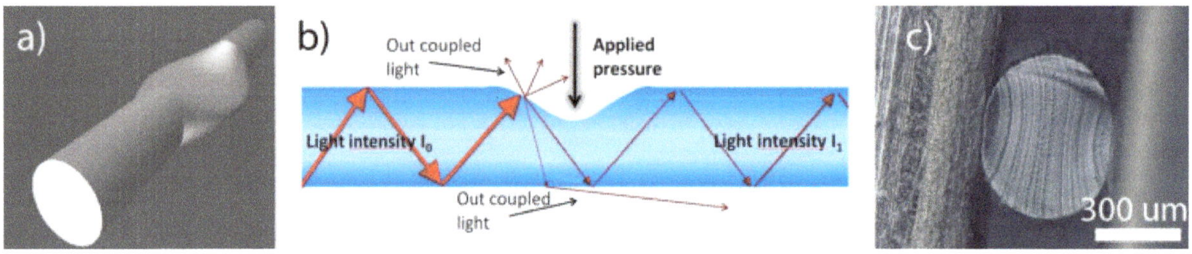

Figure 3: (a,b) Show schematically the influence of applied pressure on the elastic optical fibre. In (**c**) The cross-deformation of the fibre is demonstrated with a micrograph.

From the strain-stress curve (Figure 4) it can be concluded that the fibre made of Geniomer 175 had a smaller elastic modulus than the fibres made of the other two Geniomer materials and thus should have the highest response for applied pressure. The Young moduli of Geniomer 100 and Geniomer 100-HDS are similar and big differences in the light transmission could not be observed while applying pressure. However, the strain-stress curve was measured along the fibres and not perpendicularly (as the force was applied) which can cause a discrepancy between young moduli and the drop in light transmission.

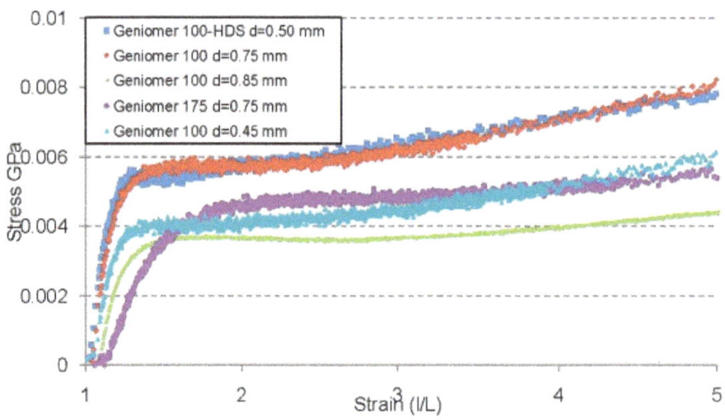

Figure 4: Stress strain curve of the different fibres produced in this study. The fibres made of Geniomer 175 show the lowest young's modulus.

The force sensing experiments were conducted with the three different types of fibre materials. In order to check the repeatability, the measurements for each force were repeated three times. Each time the force was applied, it was completely released afterwards to check if the geometry of the fibre relaxed completely to its initial state. Firstly, the force of 1 N was applied three times to check whether the fibre was sensitive to the force at this level. The forces were then increased with increments of 10 N until the normalized signal output after releasing was below 90% of the original signal. If the signal did not recover 100%, the fibre was deformed not only elastically but also plastically. If the original shape was not recovered it was a sign that plastic deformation occurred [25]. The stress which is needed to start the plastic deformation is defined as the yield strength which was different for the three materials (Table 3).

Table 3: Yield strength of the extruded fibres

Copolymer Type	Yield Strength [GPa]	σ [GPa]
Geniomer 100-HDS d = 0.5 mm	0.0054	0.0002
Geniomer 175 d = 0.5 mm	0.0045	0.0001
Geniomer 100 d = 0.45 mm	0.0039	0.0001

Depending on the copolymer material used for extrusion, the applied forces caused different light transmission responses (Table 4). This was expected due to the different yield strengths shown in Table 3. Geniomer 100-HDS showed full reversibility up to 20 N and was therefore the material which was best suited for large forces. At the level of 30 N the signal went back to 95.8% ± 1.1%. Although the fibre did not relax completely at 30 N and 40 N, the measured signal after applying pressure was completely reproducible (σ = 0.0%) at these forces. Geniomer 100 and Geniomer 175 showed plastic deformation already at 10 N. Geniomer 100 did only reach 91% ± 2.5% of the initial signal after releasing from 10 N and was therefore the material with the lowest applicable forces for the sensing system. The sensitivity of the applied force on the fibre deflection is related to the young›s modulus of the material. As expected from the strain-stress curve, the fibre made of Geniomer 175 had the highest response to applied force, as shown in Table 4. The drop in the signal of 20.8% ± 0.5% for 1 N was twice as much than the fibres produced from Geniomer 100 and Geniomer 100-HDS as presented in Table 4 (Sections 2 and 4). Due to the highest sensitivity of the fibres extruded of Geniomer 175, these fibres were chosen to measure the smallest detectable force. Forces ranging from 0.05 N to 0.5 N were applied three times (over 3 cm fibre length). From Section 1 in Table 4 it can be concluded that all the measured forces were successfully detected and the measurements were fully reproducible (σ = 0.0%). The smallest detectable force was as low as 0.05 N.

Table 4: Summary of force influence on light transmission

	Polymer type	Diameter [mm]	Force [N]	Normalized signal	σ	Normalized signal after releasing force	σ
			0.05	99.4%	0.1%	100.0%	0.0%
			0.1	98.8%	0.1%	100.0%	0.0%
			0.2	97.8%	0.0%	100.0%	0.0%
1	Geniomer 175	0.5	0.5	94.3%	0.2%	100.0%	0.0%
			1	79.2%	0.5%	100.0%	0.0%
			10	42.0%	0.4%	100.0%	0.0%
			20	25.0%	0.7%	91.5%	1.7%
2	Geniomer 100	0.45	1	90.9%	1.0%	100.0%	0.0%
			10	61.6%	1.4%	91.0%	2.5%
			1	90.0%	0.3%	100.0%	0.0%
			10	55.1%	0.6%	100.0%	0.0%
3	Geniomer 100-HDS	0.5	20	40.6%	0.5%	100.0%	0.0%
			30	30.4%	0.0%	95.8%	1.1%
			40	21.7%	0.0%	90.6%	0.9%
			1	94.3%	0.2%	100.0%	0.0%
4	Geniomer 100	0.75	10	71.3%	0.6%	100.0%	0.0%
			20	53.0%	0.5%	90.7%	0.3%
			1	91.7%	1.4%	100.0%	0.0%
			10	71.4%	0.7%	100.0%	0.0%
5	Geniomer 100	0.85	20	54.5%	0.5%	100.0%	0.0%
			30	43.1%	0.8%	95.6%	1.0%
			40	31.9%	1.1%	91.7%	0.9%

Since the light losses caused by the applied forces are associated to the fibre deflection, not only the polymer material but also the diameter of the fibre influenced the measurement range. This was demonstrated with the polymer Geniomer 100. By decreasing the fibre diameter, the fibres became more sensitive to the applied force. The fibre with the diameter of 0.85 mm was not sensitive enough at low forces (1 N) to get accurate and repeatable values as shown in Table 4. However, the thicker fibre showed a larger measurement window; forces of up to 40 N could be measured reversibly, while the thinnest fibre (0.45 mm) could only be used for forces of up to 10 N.

Fibre Deflection against Signal Drop

In Figure 5 the relative fibre deflections in the force direction *vs.* the signal drop of all fibres (different polymers and different diameters) are presented. Only the measurements with strictly elastic deformations were included in the figure. The measurements of the fibre deflection were performed simultaneously with the force measurements using the tensile testing machine. Figure 5 clearly shows that the signal drop is correlated to the fibre deflection, independently of the polymer material and the fibre diameter. This shows that only the deflection in force direction influences the amount of the out-coupled light and that the elastic properties of the material has only a minor influence on the signal change. However, the sensitivity of the deflection towards the applied force is material dependent as shown above.

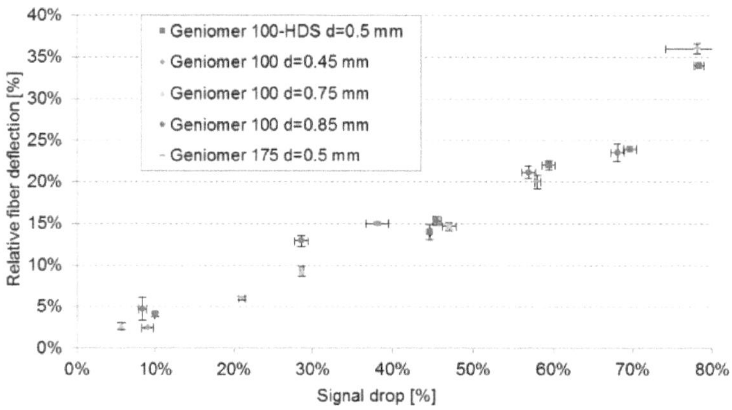

Figure 5: Fibre deflection *vs.* signal drop of fibres made of different materials and different diameters measured at 652 nm. Only the measurements with elastic deflections are shown.

Cross Sensitivity

Since this sensing technique relies on intensity measurements, it is essential to assess the influence of other parameters that can influence the light intensity. Below, we describe the influence of axial strain and bend losses on the fibre. The motion artifacts are not taken into account since they can be filtered out with signal processing.

Axial Strain Influence on the Sensor Signal

In order to determinate whether the sensor is cross sensitive to axial strain, the fibre made of Geniomer 100 (d = 0.75 mm) was stitched into a woven textile (code name FRTT1022; provided by Forster Rohner, St. Gallen, Switzerland) as presented in Figure 6.

According to the technical data sheet of this textile, the elastic deformation is 4%. The light intensity signal was measured simultaneously while the fabric was stretched with increments of 1.33%. In Table 5 the relative textile elongation *vs.* normalized signal is presented. It can be noticed that a loss in light transmission was detected when the fibre was stretched. At the maximum elongation of the fabric that was used, the signal dropped by 97% of the initial signal. This signal drop of 3% was lower than the signal drop of the smallest applied force in this study which was 1 N (signal drop of 5.7%). This leads us to the conclusion that axial strain would only influence the accuracy of the measurements at very low pressure. However, the influence of axial strain at low pressure can be bypassed by using a textile with a lower flexibility.

Bend Losses

To determine the influence of bend losses, the textile presented in Figure 5 was placed on three cylinders with radii 3, 5 and 6.5 cm respectively. Those numbers were chosen as the representative of human body parts: forearm, arm and thigh. In Table 6, the light loss *vs.* bend radius is presented. The signal when bended with radii 5 and 6.5 cm decreased less than 1%, which is most likely caused by applied pressure and contact with the cylinder. When the fabric was stronger bended with a radius of 3 cm, the light loss increased by up to ~3.5%. The reason of the light loss was a combination of loss caused by increased imprinting of the matrix textile to the elastic optical fibre and of optical bend loss. This could be minimized by placing a cladding on the core of this waveguide, which in turn would protect the core from textile imprints. However, even at this level, the light loss was relatively small (detectable forces start at around 5.7% light loss).

Table 6 shows the signal changes when a pressure was applied after the fibre was bent with a defined radius. As can be seen, the light loss was lower when the bending was narrower. The differences between the different radii became more significant when less force was applied.

Temperature Dependency

All the three Geniomer materials used have a Tg of around −130 °C and therefore have no phase change at higher temperature due to the amorphic nature of these polymers. In order to check the temperature dependence of the sensing properties, the experiments were performed under different temperatures. A force of 3 N was always applied to the Geniomer 100-HDS fibre with a diameter of 0.5 mm to show the reversibility at a certain temperature. Since this sensing system is supposed to be integrated into medical textiles in close contact with human beings, the measurements of the light losses with applied forces were performed at temperatures ranging from 25 to 45 °C with intervals of 10 °C (Figure 7). As can be seen in Figure 7, the temperature had no major influence on the light loss of the fibre while a force was applied.

Inaccuracy

Applied strain, bends and temperature change can influence the data. However, the accumulated errors should not exceed 10% of the signal change under the tested conditions. To determine for instance the breathing rate or to monitor the seat occupation, 10% of signal change would be acceptable. When detecting very low forces, as it would be the case for heart rate monitoring, the cross sensitivity becomes more important. Influences on the signal change have to be controlled by using other sensors (e.g., temperature sensor) or by controlling the environment of the fibre (e.g., avoiding of strain on the textile by using a non-elastic textile matrix). Since the optical power of LEDs is temperature-dependent, a control measurement may be necessary. The power fluctuation of the LED can be monitored by installing another LED of the same type for reference measurements into the measurement setup.

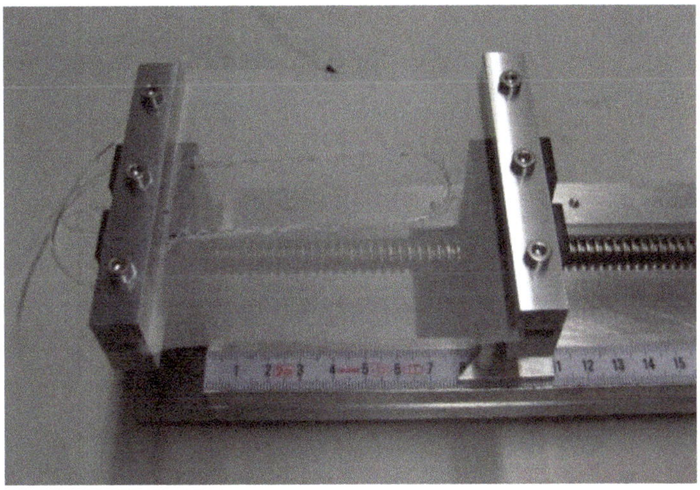

Figure 6: Measurement setup for axial strain influence.

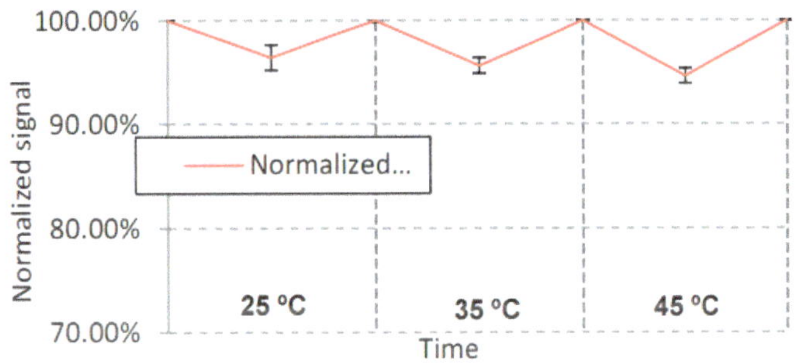

Figure 7: Signal dependency on the temperature. First, the attenuation was measured at a given temperature without applying any force. Then a force of 3 N was applied. The experiments were performed between 25 and 45 °C with 10 °C intervals and at a wavelength of 652 nm.

Table 5: Strain influence on the light transmission trough the sensor

Relative Textile Elongation	Normalized Signal
0.00%	100.0%
1.33%	99.6%
2.67%	99.0%
4.00%	97.0%

Table 6: Influence of fibre bendings on the relative light loss

Bending Radius [cm]	Light Loss by Bending	σ	Light Loss when 1 N Force Applied	σ	Light Loss when 20 N Force Applied	σ
3	3.43%	0.24%	5%	0.4%	43.1%	0.2%
5	0.91%	0.26%	7.9%	0.4%	44.6%	0.3%
6.5	0.89%	0.93%	8.8%	0.4%	45.1%	0.6%

CONCLUSIONS

We demonstrate a simple way to manufacture force sensors based on light pipes in the form of multimode optical fibres made of copolymers. The working principle of the sensor is the use of the deflection of the fibre structure when a force is applied. By using materials with different young's modulus, the sensitivity of the material can be tuned according to the desired application in the range of 0.05–40 N over a fibre length of 3 cm. The fibres produced from Geniomer 175 are best suited for low forces, while the Geniomer 100-HDS fibres are ideal sensors for larger forces. Moreover, due to the flexibility and the high mechanical strength of the material, the proposed sensor can be easily integrated into textiles to form textile-based force sensors with possible applications as seat occupation monitoring in automotive or aeroplanes or as force sensors for medical applications (prevention for decubitus or breathing monitoring). Since the flexible fibres and not the electronics have to be placed on the measuring place, this sensing principle has no negative influence on the flexibility and the haptic of the textile. Only one wavelength is used for the measurement, which simplifies the electronic system consisting of an LED and a photo detector like e.g., a photodiode. With a thin polymer cladding with a lower refractive index than the copolymer waveguide core, the robustness of the sensor could be improved further. Scratches and imperfections at the fibre surface would then not deteriorate or mask the sensing signal.

ACKNOWLEDGMENTS

This work was financially supported by NanoTera (TecInTex). Gratefully acknowledged is the support of Katrin Schöller, Benjamin Leuthold, Benno Wüst, Pierluigi Barbadoro, Alexander Haag, Youna Maussion, and Jörg Gschwend.

REFERENCES

1. Giallorenzi, T.G.; Bucaro, J.A.; Dandridge, A.; Sigel, G.H.; Cole, J.H.; Rashleigh, S.C.; Priest, R.G. Optical fiber sensor technology. *IEEE J. Quantum Elect.* 1982, *18*, 626–665.
2. Liehr, S.; Lenke, P.; Wendt, M.; Krebber, K.; Seeger, M.; Thiele, E.; Metschies, H.; Gebreselassie, B.; Munich, J.C. Polymer optical fiber sensors for distributed strain measurement and application in structural health monitoring. *IEEE Sens. J.* 2009, *9*, 1330–1338.
3. Ding, J.Y.; Shahriari, M.R.; Sigel, G.H. Fiber optic ph sensors prepared by sol-gel immobilization technique. *Electron. Lett.* 1991, *27*, 1560–1562.
4. Rothmaier, M.; Selm, B.; Spichtig, S.; Haensse, D.; Wolf, M. Photonic textiles for pulse oximetry. *Opt. Express*2008, *16*, 12973–12986.
5. Witt, J.; Narbonneau, F.; Schukar, M.; Krebber, K.; de Jonckheere, J.; Jeanne, M.; Kinet, D.; Paquet, B.; Depre, A.; D'Angelo, L.T.; *et al.* Medical textiles with embedded fiber optic sensors for monitoring of respiratory movement. *IEEE Sens. J.* 2012, *12*, 246–254.
6. Rantala, J.; Hannikainen, J.; Vanhala, J. Fiber optic sensors for wearable applications. *Pers. Ubiquit. Comput.*2011, *15*, 85–96.
7. Wang, A.B.; He, S.; Fang, X.J.; Jin, X.D.; Lin, J.X. Optical fiber pressure sensor based on photoelasticity and its application. *J. Lightwave Technol.* 1992, *10*, 1466–1472.
8. Urban, F.; Kadlec, J.; Vlach, R.; Kuchta, R. Design of a pressure sensor based on optical fiber bragg grating lateral deformation. *Sensors* 2010, *10*, 11212–11225.
9. Grillet, A.; Kinet, D.; Witt, J.; Schukar, M.; Krebber, K.; Pirotte, F.; Depre, A. Optical fiber sensors embedded into medical textiles for healthcare monitoring. *IEEE Sens. J.* 2008, *8*, 1215–1222.
10. Witt, J.; Schukar, M.; Krebber, K. Medicinal textiles with integrated polymer-optical fibers for respiration monitoring. *Tm-Tech. Mess.* 2008, *75*, 670–677.
11. De Jonckheere, J.; Jeanne, A.; Grillet, A.; Weber, S.; Chaud, P.; Logier, R.; Weber, J.L. Ofseth: Optical Fibre Embedded into Technical Textile for Healthcare, an Efficient Way to Monitor Patient under Magnetic Resonance Imaging. Proceedings of the 29th Annual International Conference of the IEEE Engineering in Medicine and Biology Society (EMBC'07), Lyon, France, 22–26 August 2007; pp. 3950–3953.
12. Zhang, Z.C.; Deng, G.; Dai, Y.B.; Liu, Y.J.; Leng, J.S. A film pressure sensor based on optical fiber bragg grating. *Smart Sens. Phenom. Technol. Netw. Syst.* 2010, *7648*, 764809:1–764809:7.
13. Tjin, S.C.; Hao, J.Z.; Lam, Y.Z.; Ho, Y.C.; Ng, B.K. A pressure sensor using fiber bragg grating. *Fiber Integr. Opt.*2001, *20*, 59–69.
14. Kaczmarek, T.; Kaczmarek, Z. Modified fiber bragg grating pulse pressure sensor. No. 66081e. *Light. Their Appl. III* 2007, *6608*, 66081E:1–66081E:4.
15. Tanaka, N.; Okabe, Y.; Takeda, N. Temperature-compensated strain measurement using fiber bragg grating sensors embedded in composite laminates. *Smart Mater. Struct.* 2003, *12*, 940–946.
16. Guan, B.O.; Tam, H.Y.; Chan, H.L.W.; Choy, C.L.; Demokan, M.S. Discrimination between strain and temperature with a single fiber bragg grating. *Microw. Opt. Technol. Lett.* 2002, *33*, 200–202.
17. Rothmaier, M.; Luong, M.P.; Clemens, F. Textile pressure sensor made of flexible plastic optical fibers. *Sensors*2008, *8*, 4318–4329.
18. Selm, B.; Gurel, E.A.; Rothmaier, M.; Rossi, R.M.; Scherer, L.J. Polymeric optical fiber fabrics for illumination and sensorial applications in textiles. *J. Intel. Mater. Syst. Struct.* 2010, *21*, 1061–1071.
19. Lymberis, A.; Dittmar, A. Advanced wearable health systems and applications. *IEEE Eng. Med. Biol. Mag.* 2007,*26*, 29–33.
20. Narbonneau, F.; Kinet, D.; Paquet, B.; Depré, A.; Jonckheere, J.D.; Logier, R.; Zinke, J.; Witt, J.; Krebber, K. Smart textile embedding optical fibre sensors for healthcaremonitoring during MRI. *Adv. Sci. Technol.* 2008, *60*, 134–143.
21. Meyer, J.; Lukowicz, P.; Troster, G. Textile pressure sensor for muscle activity and motion detection. Proceedings of 10th IEEE International Symposium on Wearable Computers, Montreux, Switzerland, 11–14 October 2006; pp. 69–72.
22. Ziemann, O.; Krauser, J.; Zamzow, P.E.; Daum, W. *Pof-handbook*; Springer: Berlin, Germany, 2008.
23. Kuzyk, M.G. *Polymer Fiber Optics*; CRC/Taylor & Francis: Boca Raton, FL, USA, 2007; pp. 97–98.
24. Koike, Y.; Koike, K. Progress in low-loss and high-bandwidth plastic optical fibers. *J. Polym. Sci. Pol. Phys.* 2011,*49*, 2–17.
25. Lubliner, J. *Plasticity Theory*; Macmillan Publishing: New York, NY, USA, 2008.

CHAPTER 18 MODIFICATION OF POLY(ETHER ETHER KETONE) POLYMER FOR FUEL CELL APPLICATION

Devesh Shukla,[1] Yuvraj Singh Negi,[1] and Vijai Kumar[2]

[1]Department of Polymer and Process Engineering, Indian Institute of Technology, Roorkee 247001, India
[2]Central Institute of Plastic Engineering and Technology, Lucknow 226008, India

ABSTRACT

Polyelectrolyte membrane (PEM) is an important part of PEM fuel cell. Nafion is a commercially known membrane which gives the satisfactory result in PEM fuel cell operating at low temperature. Present research paper includes functionalization of Poly(ether ether ketone) (PEEK) polymer with phosphonic acid group. The functionalization was done with the help of nickel-based catalyst. Further, the polymer was characterized by the FTIR, EDAX, DSC, TGA, and ^{1}H NMR, and it was found that PEEK polymer was functionalized with phosphonic acid group with good thermal stability in comparison to virgin PEEK. Finally, the thin films of functionalized polymer were prepared by solution casting method, and proton conductivity of film samples was measured by impedance spectra whose value was found satisfactory with good thermal stability in comparison to commercially available Nafion membrane.

INTRODUCTION

PEEK is a semicrystalline aromatic polymer having high melting temperature (335°C) and very low or no solubility in ordinary solvents at room temperature. Sulfuric acid is the common solvent which dissolves and also sulfonates PEEK at room temperature. Dilute solutions at high temperatures can also be obtained in hydrofluoric acid, trifluoromethanesulfonic acid, dichlorotetrafluoroacetone monohydrate, phenol 1,2,4-trichlorobenzene, and benzophenone [1]. Therefore, modification of PEEK is typical to some extent but desired for many applications such as polymer electrolyte membrane fuel cell (PEMFC) where acid functionalized polymer is required. In case of PEEK, functionalization in polymer stage is mostly reported on sulfonation and nitration [2]. Phosphonation of polymer is one of the methods to modify their properties such as proton conductivity, mechanical and thermal stability, and in polymer stage it was reported by various groups. Alexandratos et al. [3] reported the phosphonation of polystyrene using PCl_3 which gave the cross-linked polymer. However, Jakoby et al. [4] reported that the highly substituted phosphonic acid derivative of polyphenylsulfone (PPS) was achieved via the Pd-dibenzylideneacetone (dba) complexed catalyst in solvent diphenyl ether. Increase of reaction temperature from 90°C to 120°C also enhanced the substitution level. Bock et al. [5] reported phosphonation of the polysulfones (PSU) using nickel chloride and silylated phosphonates. The route was found less

Citation: Devesh Shukla, Yuvraj Singh Negi, and Vijai Kumar, "Modification of Poly(ether ether ketone) Polymer for Fuel Cell Application," Journal of Applied Chemistry, vol. 2013, Article ID 386903, 7 pages, 2013. doi:10.1155/2013/386903

expensive and gave the halogen free product in comparison to palladium-based catalytic route. Lafitte and Jannasch [6] reported a phosphonation of polysulfones via the lithiation of polymer and subsequent conversion of lithiated sites in to phosphonate unit. The substitution level in lithiation route was found very low.

The above efforts have been done on polysulfones polymer. However, in present paper we have done phosphonation of PEEK polymer by utilizing the well-known Michaelis-Arbuzov Reaction route [7]. Various characterization techniques such as FTIR, EDAX, and NMR supported the phosphonation PEEK with high thermal stability in comparison to PEEK polymer. The proton conductivity of P-PEEK film was also compared with Nafion at >100°C which showed slightly lower value than Nafion.

EXPERIMENTAL

Materials and Methods

The commercial available GatoneTM PEEK polymer was obtained by Gharda Chemical India; however, reagent grade $NiCl_2$ and Triethyl phosphate were purchased from Sigma-Aldrich, India. Liquid bromine was obtained from MERK India Ltd, India. Dimethyl acetamide (DMAC) and N-Methyl-2-pyrrolidone (NMP) were purchased from HiMedia Laboratories, India.

Synthesis of Brominated PEEK (Br-PEEK)

All glass wares and PEEK polymer were completely dried at 110°C overnight before use. The reaction assembly was established in fume-hood with help of three necks round bottom flask, mechanical stirrer and gas washing bottle attached with nitrogen supply. 15 g (0.052 mol.) of completely dried PEEK powder were charged in flask containing a mixture of 4 mL (0.156 mol.) liquid bromine and 150 mL of concentrated H_2SO_4. The reaction mixture was stirred at ambient temperature for 24 h. The brominated samples were collected at 3 h, 6 h, 9 h, and 24 h time intervals and attributed as 3 h, 6 h, 9 h, and 24 h respectively. The polymer was precipitated into ice cold distilled water and further washed with several portions of distilled water, methanol, respectively. To ensure bromine free sample, the polymer was extracted with methanol for 8 h. A white modified polymer was obtained which was finally dried in vacuum oven. To obtain the powder sample, the polymer was grinded in Retsch MM 400 Mixture Mill. Modified polymer was readily soluble in a common organic solvent such as Dimethyl sulfoxide (DMSO), DMAC, Chloroform, and Tetrahydrofuran (THF).

Aryl Phosphonation of Monobrominated PEEK (Br-PEEK)

(a) *Tetrakis(triethyl phosphite)nickel(0)*: Tetrakis(triethy lphosphite)nickel(0) compound was synthesized by the previously reported method [7]. About 4.0 g of Tetrakis(triethy l phosphite)nickel(0) (melting point 106–109°C) was obtained.

(b) *Phosphonation of Br-PEEK*: Solution of 12 g (0.033 mol) 3 h Br-PEEK in 150 mL DMAC was prepared in a reactor equipped with oil bath, gas inlet and outlet, thermometer, condenser, and dropping funnel. A cold trap was attached to the head of the reflux condenser. The temperature of solution was raised up to 100°C. Subsequently, a solution of freshly prepared tetrakis(triethylphosphite)nickel(0) in DMAC was transferred in to the reaction flask and finally raised the temperature up to 180–200°C. The concentration of catalyst was varied as 5, 10, 15, and 20 mol% and attributed as 5 m, 10 m, 15 m, and 20 m. Application of nitrogen stream provides the inert medium and simultaneously drives off evolved vapors. A homogeneous, brownish solution was obtained. Above the 160°C, 2–4 drops of tricthylphosphite were added in the reaction flask. Further, the batch was kept under slow stirring at 180–200°C for 8–10 h. The reaction mixture was turned black as the reaction proceeded. A solid polymer was obtained by precipitation of reaction mixture into methanol. Polymer content was decanted and subsequently suspended in water. For removal of catalyst residue, the water-suspended polymer was sonicated in Ultrasonicator (UTS-120, Unitech Ultrasonics) for 15 min. Polymer was filtered and dried at 100°C. Hydrolysis of phosphonated polymer (P-PEEK) was carried out by refluxing with concentrated HCl for 50 h. After complete washing with distilled water and drying in vacuum oven, the white solid polymer was obtained.

The characterizations of polymers were done by the following equipment—elemental analysis by EDAX detector equipped in FEI Quanta 200FC SEM system, IR spectra by Thermo-Nicolet Nexus FT-IR spectrometer, ^1H NMR by Bruker AC500 MHz NMR spectrometer, and thermal analysis by EXSTAR TG/DTA 6300. Proton conductivity of P-PEEK film sample was determined by Impedance Gain Phase Analyzer (Solartron SL1260) in which Julabo S25 was equipped for temperature variation.

RESULTS AND DISCUSSION

Characterization of Brominated PEEK (Br-PEEK)

Elemental Analysis of Br-PEEK by EDAX

The wt% of C, O, and Br in mono- and dibrominated PEEK was calculated theoretically (Table 1) and also measured by EDAX system (Figure 1 and Table 2). EDAX system attached with SEM was unable to measure the hydrogen weight. Therefore, during theoretical calculation we have eliminated the hydrogen weight and assumed the total weight as 100%. Br level increased with increasing the duration of reaction (Figure 2). The wt.% of Br in sample 3 h was found equal to the calculated wt% of monobrominated PEEK, while in sample 9 h it was found to be close to the calculated wt.% of dibrominated PEEK that is about 37% (Table 2). The wt% was correlated with degree of substitution (DS) by using [12, 13].

$$DS = \frac{mY}{(100W - W_sY)}, \quad (1)$$

where m is molar mass of repeat unit, W is the formula weight of the substituent group, W_s is the net increase in formula weight due to the introduction of substituent group per unit, and Y is the wt.% of substituent. The value of DS was found to be 0.89 (3 h), 1.60 (6 h), 1.85 (9 h), and 2.65 (24 h).

Table 1: Calculated wt% of elements in Br-PEEK

Element	PEEK	Br-PEEK	$(Br)_2$-PEEK
C	82.61	64.06	52.32
O	17.39	13.48	11.01
Br	0	22.45	36.66

Table 2: Measured wt% of elements in Br-PEEK

Element	wt%				
	PEEK	3 h	6 h	9 h	24 h
C	86.63	66.71	59.48	54.33	48.27
O	13.37	12.86	11.73	10.8	8.31
Br	—	20.43	28.79	34.87	43.42

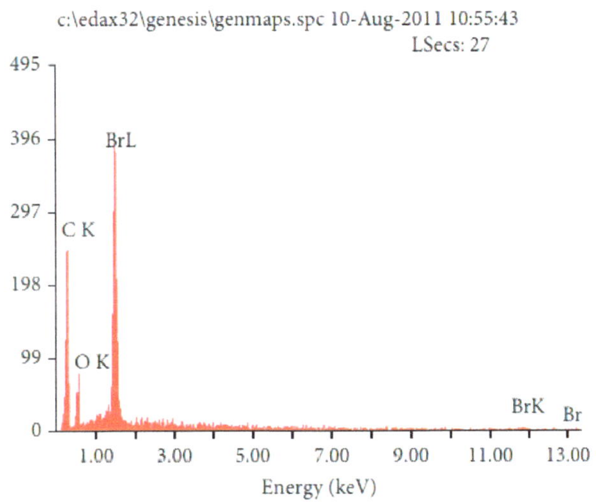

Figure 1: EDAX of 3 h Br-PEEK.

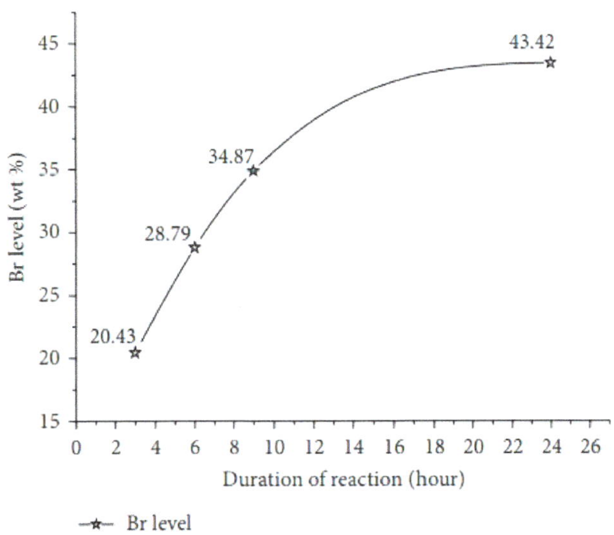

Figure 2: Bromine level in Br-PEEK polymer.

FTIR Analysis of Br-PEEK

FTIR spectrum of pure PEEK polymer was reported by the Nguyen and Ishida [14], Varsanyi [15], and Bellamy [16]. The typical absorption band reported in PEEK polymer was found almost same as Br-PEEK (Figure 3). In general, the C-Br stretching vibration in organic compound was reported in the range of 700–500 cm^{-1} while in [17] it was reported that aliphatic hydrocarbon showed the C-Br stretching in the range of 700–500 cm^{-1}whereas aromatic hydrocarbon showed the C-Br stretching in the region of 1200–1000 cm^{-1} (strong multiple bands). A new doublet band was observed near 1066–1053 cm^{-1} in Br-PEEK which may be due to C-Br stretching vibration. Since pure PEEK has many absorption bands in the region of 700–500 cm^{-1} therefore it is very difficult to assign the specific bands for C-Br stretching in this region.

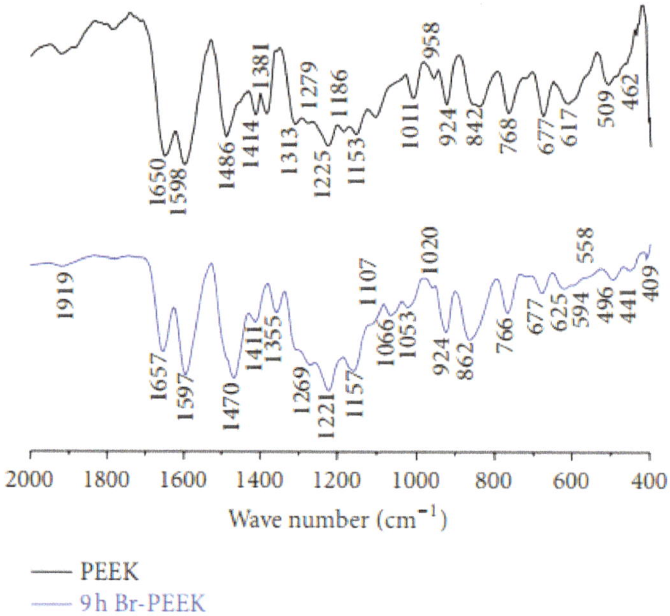

Figure 3: FTIR of PEEK and 9 h Br-PEEK.

^{1}H NMR Spectra of Br-PEEK

Since pure PEEK is insoluble in organic solvents except in strong acids, therefore its ^{1}H-NMR spectrum could not be recorded. The ^{1}H NMR of brominated sample is shown in Figure 4. Br-PEEK (3 h) showed signal between the region 7.78–7.85 (m) ppm, 7.40 (d) ppm, 7.26–7.28 (dd) ppm, 7.17–7.24 (m) ppm, and 7.12–7.14 (dd) ppm. These signals were observed in the aromatic region and associated with Ar-H. The signal 7.40 (d) ppm was found due to substitution of bromine group in the hydroquinone segment.

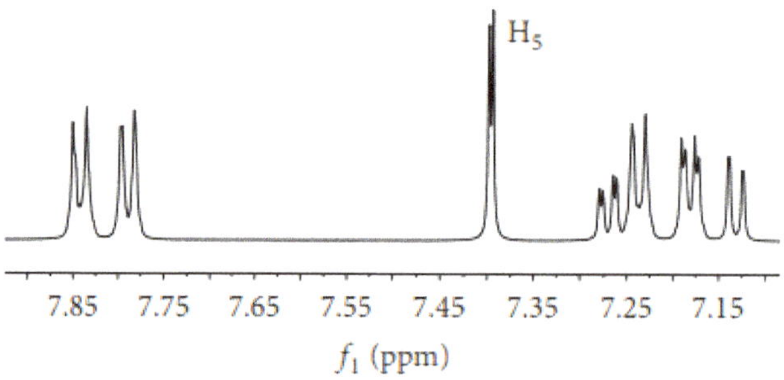

Figure 4: ^{1}H NMR of 3 h Br-PEEK.

Further, a distinct signal at 8.00 (d) ppm and 7.62 (dd) ppm was obtained in 9 h and 24 h samples which may be associated with the neighbouring proton of substituted benzophenone segment of Br-PEEK. The substitution in benzophenone segment of PEEK polymer was also reported by the Conceição et al. [18]. The DS in Br-PEEK was also determined by measuring the ratio of intensity of the distinct signal with integrated intensity of the signals corresponding to all the other aromatic hydrogens [10]. The DS obtained by the NMR was found lower in comparison to the DS obtained by the elemental analysis (EDAX data) (0.74 (3 h), 1.23 (6 h), 1.65 (9 h), 2.38 (24 h)).

Characterization of Phosphonated PEEK (P-PEEK)

FTIR Spectra of P-PEEK

P-PEEK was characterized by the FTIR spectrometer. The main reported absorption bands are as follows. González et al. [19] assigned the P=O stretching in the range of 1240–807 cm^{-1}. Trotta et al. [20] assigned the P=O stretching at 1240 cm^{-1}. According to Stuart [21] the characteristic infrared bands in phosphonated aromatic compounds are as follows: (1) aromatic P=O stretching in between 1350–1300 cm^{-1}, (2) aromatic P–C stretching in between 1450–1430 cm^{-1}, and (3) phosphorus ester P–OH stretching in the region 1040–930 cm^{-1}. The spectrum of pristine PEEK shows many absorption bands in the above mentioned region such as asymmetric stretching of diphenyl ether groups in the region of 1227 cm^{-1} and 1190 cm^{-1}. Therefore, a clear identification of addition of phosphonate ester group was somewhat difficult. From the comparison of spectra of P-PEEK ester and PEEK (Figure 5), we found the new absorption band at 1250–1233 cm^{-1}, 1470 cm^{-1}, and doublet at 1036 cm^{-1} which might be due to the presence of P=O stretching, P–C stretching, and phosphorus ester P–OH stretching, respectively. All other characteristics band of PEEK such as presence of carbonyl group at 1651 cm^{-1} remained the same in P-PEEK derivatives.

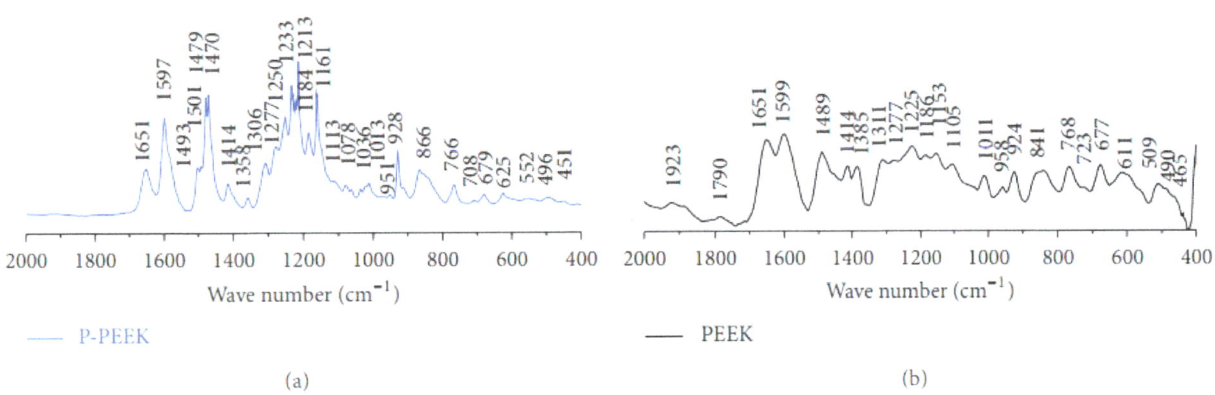

Figure 5: FTIR of P-PEEK ester (20 m).

Elemental Analysis by EDAX

The wt.% of different elements in P-PEEK ester derivatives were determined by EDAX (Figure 6) and also calculated theoretically by ignoring the hydrogen atom as like in Br-PEEK. Table 3 represented the calculated and measured wt% of different elements in monophosphonated P-PEEK ester. The wt% of phosphorous in different P-PEEK ester samples were found lower than the calculated wt% of monophosphonated PEEK (7.69 wt%). It was observed that P-PEEK samples have some bromine. The DS was calculated according to (1) whose values were ~0.17, 0.30, 0.45 and 0.71 in P-PEEK ester 5 m, 10 m, 15 m and 20 m respectively.

Table 3: Element % in P-PEEK ester by EDAX

Element	Calculated wt%	Measured wt% P-PEEK ester			
		5 m	10 m	15 m	20 m
C	68.49	65.94	67.34	70.41	72.01
O	23.82	13.69	15.06	18.46	19.93
P	7.69	1.73	2.98	4.15	6.02
Br	—	18.46	13.86	6.07	0.82
Ni	—	0.18	0.76	0.91	1.22

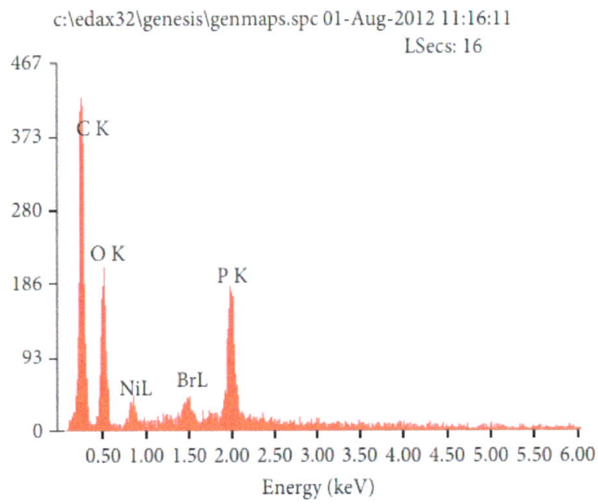

Figure 6: EDAX of P-PEEK ester (20 m).

Solubility of P-PEEK Ester and P-PEEK Acid

The solubility of P-PEEK ester and P-PEEK acid were determined by visual inspection in various solvent at room temperature and also at high temperature >150°C. It was found that both are insoluble in almost all the organic solvents. NMP sparingly dissolved the P-PEEK acid and ester both only at high temperature. Further, addition of 2-3 drops of acid such as HBr or H_2SO_4 fairly dissolved the both form of polymer (Table 4).

Table 4: Solubility of P-PEEK acid and ester in various solvents

Solvent	Phosphonated PEEK	
	Room temp.	Temp. > 150°C
DMSO	C	C
DMAC	C	C
NMP	C	B
DMF	C	B
DMSO⁺	C	B
DMAC⁺	C	B
NMP⁺	C	A
DMF⁺	C	C
4-Chlorophenol⁺	C	C
THF⁺	C	C
Cyclohexanone⁺	C	C

⁺Solvent and 2-3 drops of acid HBr/H_2SO_4, A: soluble, B: Sparingly soluble, C: not soluble.

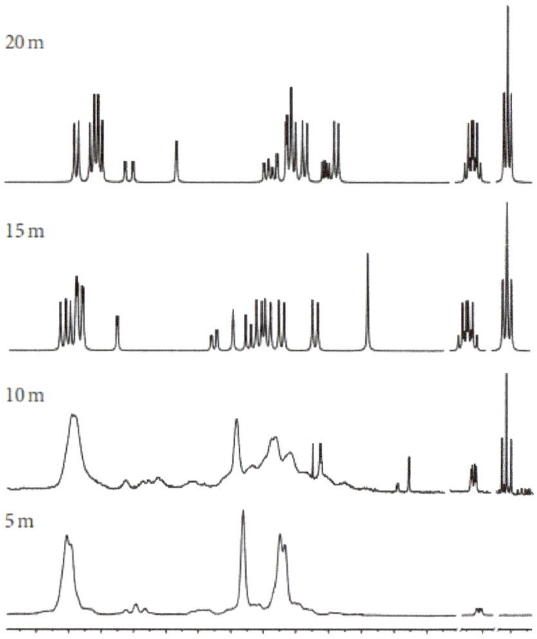

Figure 7: ^{1}H NMR of P-PEEK ester 5 m, 10 m, 15 m, and 20 m.

NMR Characterization of P-PEEK

Solubility of P-PEEK ester was not found in any normal available deuterated solvent. Therefore, characterization of P-PEEK in liquid stage was somewhat difficult. To resolve the problem of solubility, initially a concentrated solution of P-PEEK ester in NMP at higher temperature was prepared which was further utilized in DMSO- d_6 to prepare a solution of P-PEEK ester—NMP/DMSO- d_6 in the ratio of 1/3.

The ^{1}H NMR spectra of NMP and P-PEEK ester—NMP in DMSO- d_6 were determined, respectively. NMP showed the signal at 3.4 ppm, 2.85 ppm, 2.37 ppm, and 2.30 ppm. The signal near 2.5 ppm was associated with DMSO-d_6. These entire signals were also found in P-PEEK ester-NMP sample.

P-PEEK ester gave signal in both aromatic and also in aliphatic regions (Figure 7) due to presence of aromatic protons and protons of ethyl ester at 4.46 ppm and 1.19 ppm. Aryl phosphonation effected a downfield shift of proton neighbouring to phosphonic ester group as well as a broadening and splitting of signals into a doublet of doublets [5]. The broadening, splitting of signal, and downfield shifting were also found in case of P-PEEK samples as the degree of phosphonation increases (Figure 7).

The degree of phosphonation can be calculated from the signal ratio of the ethyl ester protons and the protons of aromatic region [4] which was observed 0.14, 0.25, 0.42, and 0.68 in P-PEEK ester 5 m, 10 m, 15 m, and 20 m, respectively. Phosphonation in PEEK polymer was also confirmed by the solid state ^{31}P NMR spectrum of P-PEEK (Figure 8) which showed only a single signal at 12.52 ppm. Single peak was also reported by Bock et al. [5] in case of ^{31}P NMR spectra of P-PSU acid.

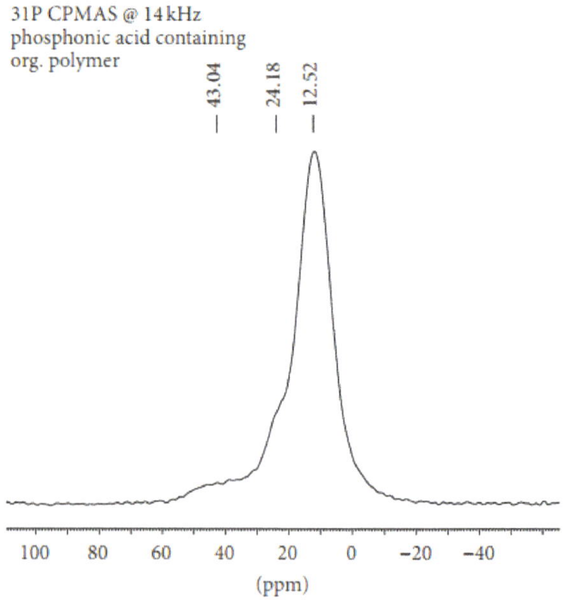

Figure 8: Solid state ^{31}P NMR of P-PEEK acid.

Thermal Analysis of P-PEEK

P-PEEK (20 m) showed T_g 220°C and Tm ~ 381°C which was about 44°C higher than the PEEK polymer. It shows that P-PEEK exhibited higher thermal stability than PEEK polymer

(T_g 152°C, Tm 337°C) (Figure 9). A significant wt. loss was found at 313°C which may be possible due to the conversion of phosphoric acid segment in to anhydride (Figure 10). P-PEEK also showed the higher residual mass, which may be possible due to presence of incombustible phosphorus oxides.

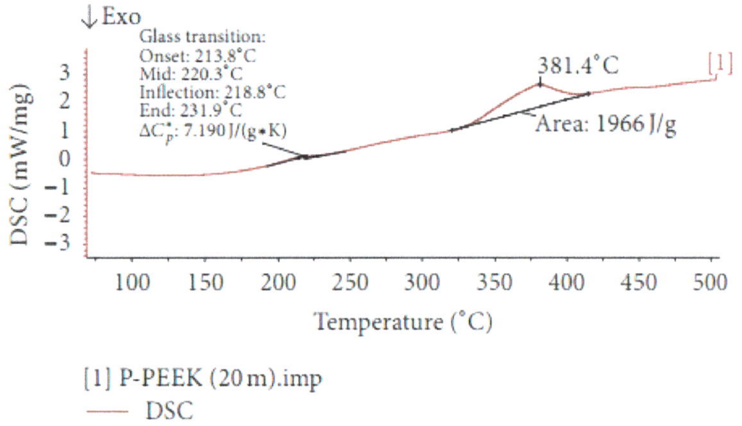

Figure 9: DSC of P-PEEK (20 m).

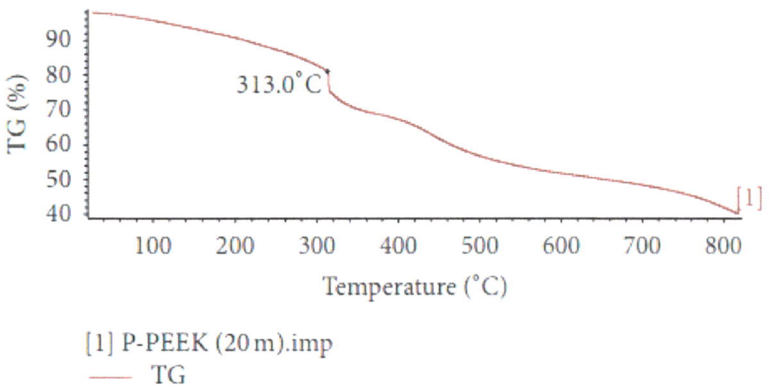

Figure 10: TGA of P-PEEK (20 m).

Proton Conductivity P-PEEK

Proton conductivity of P-PEEK was determined by casting the film on Indium-tin oxide (ITO) coated glass sheet using the solvent NMP. Coated glass sheet was further dipped in water for 24 h. The thickness of polymer film was determined by taking the difference in glass sheet thickness before and after the coating. The polymer coated sheet was sandwiched with ITO coated glass sheet, and impedance was measured by varying the frequency 100 Hz to 32 MHz, AC voltage 50–500 mV in Impedance Gain Phase Analyzer (Solartron SL1260). Julabo S 25 was used for variation in temperature during the impedance measurement. Proton conductivity was calculated by the following equation $\sigma = L/RA$, where L, R, *and* A are the distance between the two electrodes, membrane resistance, and cross-sectional area of the membrane, respectively [22–24]. The value of R was determined by plotting the graph between the impedance of real and imaginary plot (Figure 12). The point of resistance was that where the imaginary part was found to be zero. Conductivity of the P-PEEK samples increases with temperature and degree of phosphonation. The proton conductivity P-PEEK-20 m was found 0.0069 S cm^{-1} at 25°C, 0.055 at 65°C, 0.083 S cm^{-1} at 90°C, and 0.114 cm^{-1} at 120°C (Figure 11 and Table 5). The Nafion membrane has been reported best ionic conductor at low temperature and high humidity [8, 25]. S-PEEK membrane showed comparable proton conductivity with Nafion at high degree of sulfonation and humidity. Both Nafion and S-PEEK membrane exhibited loss of conductivity and physical failure at temperature >100°C. The loss of conductivity was found greater in S-PEEK. Although conductivity of P-PEEK was found to be slightly lower than Nafion at >100°C but its thermal stability was found higher, and it is expected that P-PEEK will show better durability with acceptable proton conductivity in comparison to both Nafion and S-PEEK.

Table 5: Proton conductivity of polymer membranes

Membrane	Proton conductivity (S cm^{-1})		
Nafion	0.06–0.100 (25–30°C) [8]	0.19 (90°C) [8]	0.210 (120°C) [9]
SPEEK	0.008 (25°C) (DS 0.80) [10]	0.076 (100°C) (DS 0.80) [10]	0.033 (120°C) (DS 0.83) [11]
P-PEEK (20 m)	0.0069 (25°C)	0.0550 (90°C)	0.114 (120°C)

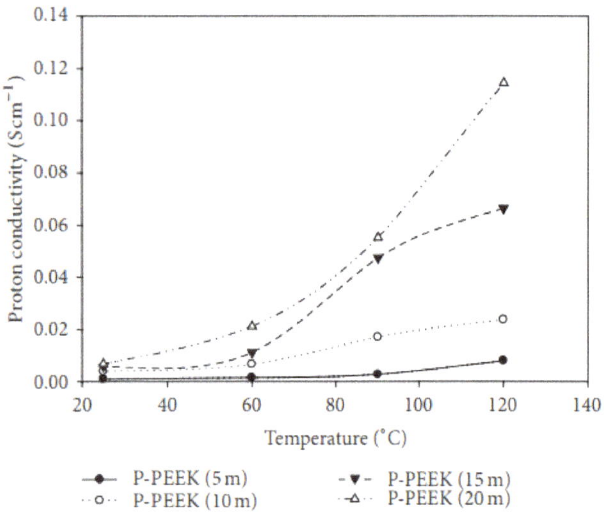

Figure 11: Proton conductivity of P-PEEK samples with temperature.

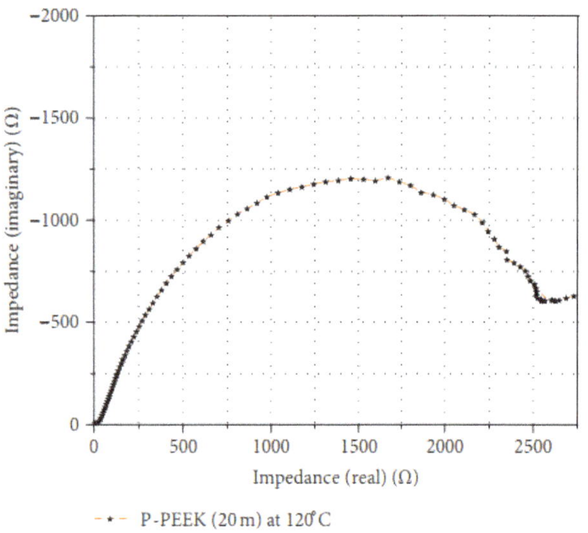

Figure 12: Nyquist plot of P-PEEK (20 m).

CONCLUSIONS

Commercial PEEK polymer can be brominated with only a small excess of bromine in H_2SO_4. However, no catalyst was used during the reaction. FTIR, EDAX, and ^{1}H-NMR indicated that PEEK can be brominated readily. Further, conversion of Br-PEEK in to P-PEEK was carried out in presence of nickel-based catalyst. P-PEEK was characterized by various techniques, and it was found that P-PEEK showed high thermal stability in relation with PEEK polymer and solubility in NMP at higher temperature. The proton conductivity of P-PEEK was compared with Nafion at >100°C which showed slightly low value but thermal stability of P-PEEK was found good. Thus, it is expected that P-PEEK will show better durability with acceptable proton conductivity in comparison to both Nafion and S-PEEK at high temperature.

ACKNOWLEDGMENTS

One of the authors is thankful for financial support from the Ministry of Human Resource Development (MHRD), India, and also for NCL Pune, India, for carrying out solid state NMR.

REFERENCES

1. J. E. Mark, Polymer Data Handbook, Oxford University Press, 1999.
2. D. Shukla, Y. S. Negi, J. S. Uppadhyaya, and V. Kumar, "Synthesis and modification of poly (ether ether ketone) and their properties: a review," Polymer Reviews, vol. 52, no. 2, pp. 189–228, 2012.
3. S. D. Alexandratos, M. A. Strand, D. R. Quillen, and A. J. Walder, "Synthesis and characterization of bifunctional phosphinic acid resins," Macromolecules, vol. 18, no. 5, pp. 829–835, 1985.
4. K. Jakoby, K. V. Peinemann, and S. P. Nunes, "Palladium-catalyzed phosphonation of polyphenylsulfone,"Macro molecular Chemistry and Physics, vol. 204, no. 1, pp. 61–67, 2003.
5. T. Bock, R. Mülhaupt, and H. Möhwald, "Halogen-free polyarylphosphonates and polyelectrolyte membranes for PEMFC by Nickel-Catalyzed phosphonylation with silylated phosphates,"Macromolecular Rapid Communications, vol. 27, no. 24, pp. 2065–2071, 2006.
6. B. Lafitte and P. Jannasch, "Phosphonation of polysulfones via lithiation and reaction with chlorophosphonic acid esters," Journal of Polymer Science A, vol. 43, no. 2, pp. 273–286, 2005.
7. T. M. Balthazor and R. C. Grabiak, "Nickel-catalyzed Arbuzov reaction: mechanistic observations,"Journal of Organic Chemistry, vol. 45, no. 26, pp. 5425–5426, 1980.
8. S. Slade, S. A. Campbell, T. R. Ralph, and F. C. Walsh, "Ionic conductivity of an extruded Nafion 1100 EW series of membranes," Journal of the Electrochemical Society, vol. 149, no. 12, pp. A1556–A1564, 2002.
9. N. Miyake, J. S. Wainright, and R. F. Savinell, "Evaluation of a sol-gel derived nafion/silica hybrid membrane for proton electrolyte membrane fuel cell applications: I. Proton conductivity and water content," Journal of the Electrochemical Society, vol. 148, no. 8, pp. A898–A904, 2001.·
10. S. M. J. Zaidi, S. D. Mikhailenko, G. P. Robertson, M. D. Guiver, and S. Kaliaguine, "Proton conducting composite membranes from polyether ether ketone and heteropolyacids for fuel cell applications,"Journal of Membrane Science, vol. 173, no. 1, pp. 17–34, 2000.·
11. B. Lakshmanan, W. Huang, D. Olmeijer, and J. W.

Weidner, "Poly(ether ether ketone) membranes for elevated temperature PEMFCs," Electrochemical and Solid-State Letters, vol. 6, no. 12, pp. A282–A285, 2003.

12. A. Philippides, P. M. Budd, C. Price, and A. V. Cuncliffe, "The nitration of polystyrene," Polymer, vol. 34, no. 16, pp. 3509–3513, 1993.

13. "Degree of Substitution," in Encyclopedia of Polymer Science and Technology, John Wiley & Sons, 2002.

14. H. X. Nguyen and H. Ishida, "Molecular analysis of the crystallization behavior of poly(aryl-ether-ether-ketone)," Journal of Polymer Science B, vol. 24, no. 5, pp. 1079–1091, 1986.

15. G. Varsanyi, Assignments for Vibrational Spectra of Seven Hundred Benzene Derivatives, Adam Hilger, London, UK, 1974.

16. J. L. Bellamy, The Infrared Spectra of Complex Molecule, Chapman and Hall, London, UK, 3rd edition, 1975.

17. Sadtler Handbook of Infrared Spectra, Bio-Rad Laboratories, Informatics Division.

18. T. F. Conceição, J. R. Bertolino, G. M. O. Barra, S. L. Mireski, A. C. Joussef, and A. T. N. Pires, "Preparation and characterization of poly (ether ether ketone) derivatives," Journal of the Brazilian Chemical Society, vol. 19, no. 1, pp. 111–116, 2008.

19. L. González, O. Mó, M. Yáñez, and J. Elguero, "Very strong hydrogen bonds in neutral molecules: the phosphinic acid dimers," Journal of Chemical Physics, vol. 109, no. 7, pp. 2685–2693, 1998.

20. F. Trotta, S. C. Mio, K. Martina, E. Drioli, and P. Finocchiaro, "New poly ether ether ketones containing phosphorus for membrane preparation," Asia-Pacific Journal of Chemical Engineering, vol. 5, no. 1, pp. 249–255, 2010.

21. B. H. Stuart, "Organic molecules," in Infrared Spectroscopy, Fundamentals and Applications, pp. 71–93, John Wiley & Sons, 2005.

22. C. Zhao, H. Lin, K. Shao et al., "Block sulfonated poly (ether ether ketone)s (SPEEK) ionomers with high ion-exchange capacities for proton exchange membranes," Journal of Power Sources, vol. 162, no. 2, pp. 1003–1009, 2006.

23. P. Xing, G. P. Robertson, M. D. Guiver, S. D. Mikhailenko, K. Wang, and S. Kaliaguine, "Synthesis and characterization of sulfonated poly (ether ether ketone) for proton exchange membranes," Journal of Membrane Science, vol. 229, no. 1-2, pp. 95–106, 2004.

24. N. Zhang, G. Zhang, D. Xu et al., "Cross-linked membranes based on sulfonated poly (ether ether ketone) (SPEEK)/Nafion for direct methanol fuel cells (DMFCs)," International Journal of Hydrogen Energy, vol. 36, no. 17, pp. 11025–11033, 2011.

25. G. Alberti, M. Casciola, L. Massinelli, and B. Bauer, "Polymeric proton conducting membranes for medium temperature fuel cells (110–160°C)," Journal of Membrane Science, vol. 185, no. 1, pp. 73–81, 2001.

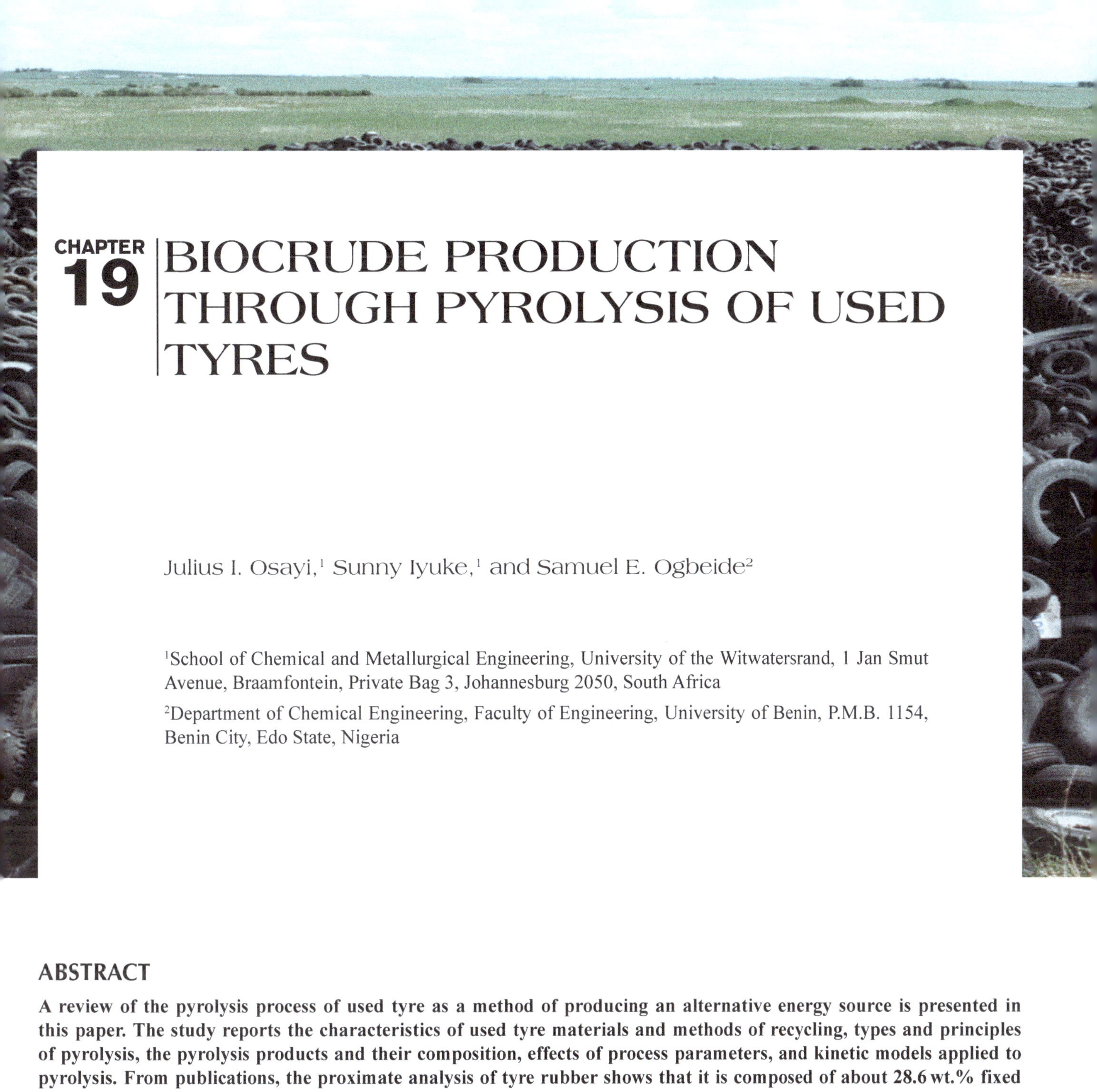

CHAPTER 19

BIOCRUDE PRODUCTION THROUGH PYROLYSIS OF USED TYRES

Julius I. Osayi,[1] Sunny Iyuke,[1] and Samuel E. Ogbeide[2]

[1]School of Chemical and Metallurgical Engineering, University of the Witwatersrand, 1 Jan Smut Avenue, Braamfontein, Private Bag 3, Johannesburg 2050, South Africa

[2]Department of Chemical Engineering, Faculty of Engineering, University of Benin, P.M.B. 1154, Benin City, Edo State, Nigeria

ABSTRACT

A review of the pyrolysis process of used tyre as a method of producing an alternative energy source is presented in this paper. The study reports the characteristics of used tyre materials and methods of recycling, types and principles of pyrolysis, the pyrolysis products and their composition, effects of process parameters, and kinetic models applied to pyrolysis. From publications, the proximate analysis of tyre rubber shows that it is composed of about 28.6 wt.% fixed carbon, 62 wt.% volatile material, 8.5 wt.% ash, and 0.9 wt.% moisture. Elemental analysis reveals that tyre rubber has an estimated value of 82 wt.% of C, 8 wt.% of H, 0.4 wt.% of N, 1.3 wt.% of S, 2.4 wt.% of O, and 5.9 wt.% of ash. Thermogravimetry analysis confirms that the pyrolysis of used tyre at atmospheric pressure commences at 250°C and completes at 550°C. The three primary products obtained from used tyre pyrolysis are solid residue (around 36 wt.%), liquid fraction or biocrude (around 55 wt.%), and gas fraction (around 9 wt.%). Although there is variation in the value of kinetic parameters obtained by different authors from the kinetic modeling of used tyre, the process is generally accepted as a first order reaction based on Arrhenius theory.

INTRODUCTION

One of the main challenges of modern society is the rising rate of solid waste generated by man's activities which has poised a major environmental concern [1–3]. The disposal of used tyres and other polyisoprene based products is a large fraction of such problems as 1.3 billion tyres are estimated to reach their end of life cycle annually worldwide [4–9]. This is because of their excellent properties which have made them useful in all areas of human life [10]. However, these excellent properties also place them at a disadvantage [3, 11, 12] as they are not biologically degradable leading to problems with their disposal [2, 7, 11, 13–16]. Over the years landfill and open dumping (stock piling) were the common ways in handling the problem of used tyres. However, landfills take up valuable land space due to the bulky nature of tyres which cannot be compacted neither does it degrade easily [7, 11, 17–19]. Dumped used tyres in massive stockpiles do not only occupy a large land space but also serve as a potential health and environmental hazard due to the possibility of a fire outbreak with high emissions of toxic gases and as a breeding ground for disease carrying vectors

Citation: Julius I. Osayi, Sunny Iyuke, and Samuel E. Ogbeide, "Biocrude Production through Pyrolysis of Used Tyres," Journal of Catalysts, vol. 2014, Article ID 386371, 9 pages, 2014. doi:10.1155/2014/386371

[7, 15, 16, 20,21]. Despite other tyre recycling options such as reclaiming, grinding, incineration, retreading, and so forth being used, these processes have their setback and do not fully utilize used tyres as an excellent material for energy recovery [8, 22]. For this reason, pyrolysis has received renewed attention as the process conditions can be optimized to produce high valued products [4–6, 11, 23–27]. Pyrolysis mainly involve the decomposition of polyisoprene materials to low molecular weight component at high temperature (250°C–900°C) in an inert atmosphere [3, 7, 22, 28]. The process being an excellent energy recovery route can be performed under atmospheric or reduced pressure. It is an environmentally friendly technique to thermally decompose a wide variety of wastes, including used tyres [2, 4, 6, 11, 15, 16, 20, 27, 29–44]. The three basic products of pyrolysis are solid residue (char), liquid (biocrude), and gases [6]. Solid residue may be used in the manufacture of activated carbon, reinforcement in rubber industry, or as solid fuel. The pyrolytic liquid can serve as replacements for conventional liquid fuel due to its high gross calorific value of about 41–44 MJ/kg. It can also be used as petroleum refinery feedstock or a source of chemicals for wide industrial applications. Gaseous fraction can be used as fuel or as a source of energy for the pyrolysis process [22, 28, 29, 45–49].

This paper presents a literature review of the pyrolysis process with a focus on the characteristics of used tyres materials and methods of recycling, types and principles of pyrolysis, properties of the pyrolysis products, effects of process parameters, and kinetic models applied to pyrolysis.

USED TYRES

Used tyres also referred to as waste tyres can be defined as tyres that have expired as a result of exceeding their production life span or are no longer safe for usage due to defects, such as degradation of its physical composition/structure from use and cannot be retreaded. It is one of the most challenging hazardous solid wastes facing modern society, particularly in developing countries [3, 10, 11]. It is estimated that the EU, Japan, and USA generate 6×10^6 tons of waste tyres annually [6, 22, 23, 46, 64]. According to a report, 160,000 tons of waste tyres are generated in South Africa annually and up to 28 million used tyres are dumped unlawfully or burnt [70]. This figure is projected to increase by 9.3 million yearly [70]. The nonbiodegradable nature of this polymer material makes its disposal difficult. Disposal by massive stockpiling and landfilling is one of the common ways of handling waste tyre, but this requires a large space as the volume of tyres cannot be compacted. These also poise the danger of the possibility of fire outbreak with the emission of harmful gases [1,7, 19, 21, 22, 64, 71–75].

Methods of Recycling of Used Tyres

The various methods in which used tyres are recycled are discussed below.

Retreading Method

Retread also referred to as "remold" involves buffing away of the remaining tread in a spent tyre and replacing it with a new tread rubber strip by vulcanization. It is carried out only on used tyres casing that have been inspected and repaired. Although it is economically advantageous, product quality confidence is a major challenge of the process.

Mechanical or Cryogenic Method

Both mechanical and cryogenic (grinding of waste tyres at temperatures of −80°C to −100°C using liquid nitrogen) recycling of used tyres involve the milling of tyres to produce ground rubber of different particle sizes. The fine grounded rubber can be used in varieties of application such as an additive in road asphalting, sports and children playground surfacing, carpets, and other rubber products. However, the limited market for the product and a high cost of running the process are a disadvantage.

Reclaiming Rubber Raw Materials Methods

Different methods and processes for reclaiming rubber have been developed. The most relevant of them are thermomechanical reclaiming [48], microwave reclaiming [49], mechanical shearing process [69], reclaiming by biotechnological process, ultrasonic reclaiming [76], reclaiming by renewable resource materials, reclaiming by use of different chemical agents [9], and pyrolysis of waste tyre [77]. These methods actually help in transforming used tyres from a three dimensionally interlinked thermoset polymer state to a two-dimensional polymer exhibiting the properties of a virgin rubber. The high cost of the process, quality of products, and lack of acceptance of reclaim rubber by industries as a raw material are a setback on the process.

Although used tyres can be used as a source of fuel in cement kilns by combustion method, it is not economically wise and environmentally friendly [1, 4, 7].

Characteristics and Composition of Used Tyres

Tyres are made up of different types of rubber elastomers (natural or synthetic rubber) with varying composition, carbon black, stabilizers, antioxidants, sulphur, hydrocarbon oils, zinc oxides, textile or steel cords [4, 6, 11, 69], and so forth. Proximate and ultimate analyses of used tyres reported by various authors are shown in Tables 1 and 2. The difference in values can be attributed to the manufacturers' formulation which is closely guarded and determines to a large extent the weight percent of the various components of tyres.

Table 1: Proximate analysis of used tyre rubber

Author	Moisture (wt.%)	Ash (wt.%)	Volatile (wt.%)	Fixed carbon (wt.%)	HHV (MJ/kg)
This study	0.5	16.0	56.4	27.1	31.2
Donatelli et al. [31]	0.8	4.4	61.3	33.5	37.1
Kar [32]	1.72	19.13	59.69	19.45	27.37
Chang [33]	1.31	10.21	62.32	26.26	33.24
Trongkaew et al. [14]	1.22	8.73	62.4	27.6	35.7
González et al. [19]	0.7	8.0	61.9	29.5	36.2
Su and Zhao [34]	0.8	7.41	64.64	27.15	39.56
Juma et al. [9]	1.72	14.01	61.61	22.66	—
de Marco et al. [35]	58.8	27.7	1.72	—	—
Williams and Bottrill [36]	0.8	2.4	66.5	30.3	—

Table 2: Ultimate analysis of used tyre rubber

Author	C (wt.%)	H (wt.%)	N (wt.%)	S (wt.%)	O (wt.%)	Ashes (inorganic)
Donatelli et al. [31]	85.2	7.3	0.4	2.3	0.4	4.4
Kar [32]	67.08	6.12	0.17	2.05	24.58	—
Chang [33]	74.41	6.94	0.21	1.60	5.02	—
Trongkaew et al. [14]	80.10	7.48	0.42	1.54	10.50	—
González et al. [19]	86.7	8.10	0.40	1.40	1.30	2.1
Su and Zhao [34]	81.84	6.06	1.77	1.64	0.48	—
Juma et al. [9]	81.24	7.36	0.49	1.99	8.92	—
de Marco et al. [35]	74.20	5.80	0.3	1.50	4.70	13.5
Williams and Bottrill [36]	85.8	8.0	0.4	1.0	2.3	2.4

Study of Thermal Degradation of Used Tyre

Thermogravimetry analysis is used to study the general behavior of thermal degradation of used tyres [13]. Both the thermogravimetry (TG) and differential thermogravimetry (typical curve is shown in Figures 1 and 2) are standard techniques used in the study of mass change of tyre rubber sample as a function of temperature and time [73, 74, 78]. With the results data (shown in Table 3) obtained in this study and those of other authors in the literature, it can be understood that there is more than one degradation temperature zone during tyre rubber pyrolysis. This also gives an idea of the commencement and end point temperature for the different rubber components in the waste tyre [3, 10]. For example, NR (natural rubber), SBR (styrenebutadiene rubber), and BR (butadiene rubber) decompose at a temperature range of 375°C to 500°C.

Table 3: Commencement and end temperatures of pyrolysis of used tyre

Author	Pyrolysis commencement temperature (°C)	Pyrolysis end temperature (°C)	Heating rate (°C/min)
This study	210	520	20
Juma et al. [9]	250	550	5
Conesa et al. [38]	250	500	5
Su and Zhao [34]	200	500	10
Senneca et al. [39]	200	450	5

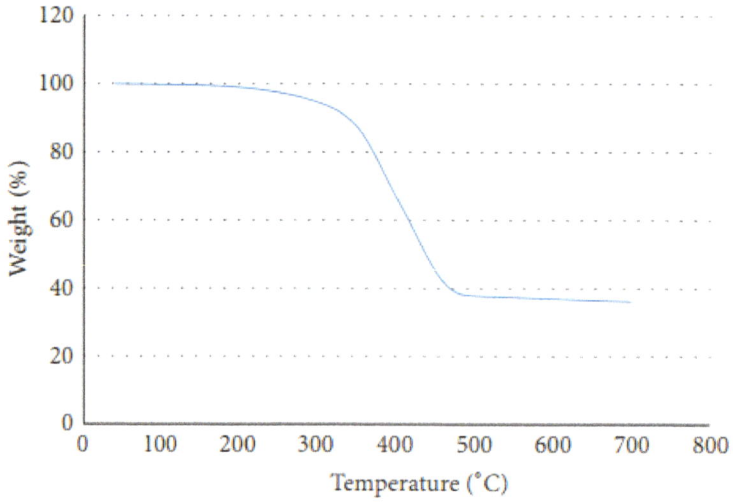

Figure 1: Typical TGA curve of used tyre sample.

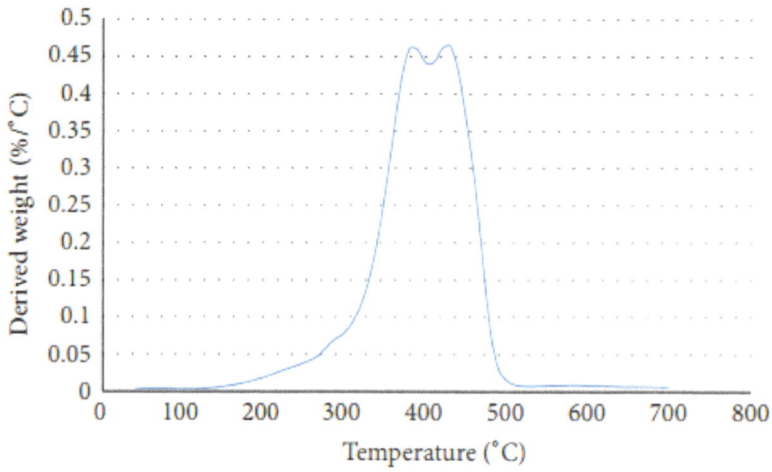

Figure 2: Typical DTG curve of used tyre sample.

PYROLYSIS, PROCESS CONDITIONS, AND REACTORS

Pyrolysis mainly involves the thermal degradation of tyre rubber at high temperatures (250–900°C) in an oxygen absent environment. It can be performed under vacuum or atmospheric pressure [23, 29]. Amongst other methods such as combustion and gasification used to extract energy from biomass, pyrolysis has received more attention in the area of research because the process conditions can be optimized to produce high energy density liquids, char, and gas. Also, the condensable fraction (biocrude) can be stored and easily transported to where it can be most proficiently utilized. Tyre pyrolysis can be said to be made up of three stages, that is, the release of volatile and moisture at lower temperature succeeded by the thermal decomposition of natural rubber (NR) and the decomposition of polybutadiene (BR) and polybutadiene-styrene rubber at higher temperature, respectively [79, 80].

The process conditions that can influence the percentage weight of each fraction of pyrolysis include pressure, heating rate, temperature, feed particle sizes, catalysis, flow rate of inert carrier gas, configuration of reactor used [2], and so forth.

Pyrolysis Reactors

The reactor is a very important part of the pyrolysis process. Over the years, the innovation and technological advancement of pyrolysis has advanced considerably. Researchers have developed and studied different reactors and processes to the point where pyrolysis is now an acceptable technique to the deriving of gaseous, liquid, solid fuels, and chemicals. Table 4 presents the chronology of pyrolysis process from the 1970s to date. Reactors can greatly influence the pyrolysis performance, product quality, and yield [2, 46]. However, each reactors type has its benefits and limitations. Table 5 shows different types of reactors and their heating methods.

Table 4: Chronology of processes used for pyrolysis of used tyres

Researcher(s)	Reactor used	Process type	Year
Gotshall [50]	n.r	Destructive distillation	1970
Alpert [51]	n.r	Hydroconversion	1972
Grannen and Robinson [52]	Microwave	Microwave pyrolysis	1974
Crane and Kay [53]	n.r	Pyrolization	1976
Herbold [54]	n.r	Subatmospheric pyrolysis	1978
Kaminsky and Sinn [55]	Fluidized bed	Pyrolysis	1980
Ito et al. [56]	Fluidized bed and incineration fluidized bed chamber	Pyrolysis and incineration	1982
Engman et al. [57]	Quartz tube	Pyrolysis	1984
Bouvier and Gelus [58]	Heated cylindrical glass flask	Tyre copyrolysis with oil	1986
Chou et al. [59]	Quartz tube	Flash pyrolysis	1988
Roy et al. [60]	Batch reactor	Vacuum pyrolysis	1990
Roy et al. [61]	Batch reactor	Vacuum pyrolysis	1992
Pakdel and Roy [62]	Continuous feed reactor	Vacuum pyrolysis	1994
Conesa et al. [63]	Fluidized bed reactor	Pyrolysis	1996
Leung and Wang [64]	Platinum pan	Pyrolysis and combustion	1998
Mastral et al. [30]	Swept fixed bed	Pyrolysis and hydropyrolysis	2000
Galvagno et al. [65]	Rotary kiln reactor	Pyrolysis	2002
Laresgoiti et al. [66]	Autoclave reactor	Pyrolysis	2004
Murillo et al. [67]	Fixed bed reactor	Pyrolysis	2006
Zhang et al. [12]	Fixed bed reactor	Vacuum pyrolysis	2008
Aylón et al. [68]	Moving bed	Pyrolysis	2010
Aydin and Ilkilic [15]	Fixed bed reactor	Pyrolysis	2012
Undri et al. [4]	Microwave oven	Microwave assisted pyrolysis	2014
This study	Plasma reactor	Catalytic copyrolysis	2014

n.r.: not reported.

Table 5: Reactors and their heating methods [23, 40–42]

Reactor type	Heating method
Fixed bed	Heated wall surface
Bubbling fluidized bed	Heated recycle gas
Circulating fluidized bed	Wall and sand heating
Rotating cone	Gasification of char to heat
Auger	Fire tube
Fluidized bed/quartz	Solar
Plasma	Radio frequency
Vacuum	Direct contact with hot surface
Ablative	Wall heating

Types of Pyrolysis

Pyrolysis process can be performed under different operating conditions which can be used in classifying it. They are differentiated by residence time of the pyrolysed material in the reactor, process temperature, feed particle size, heating rate [81–83], and so forth. These include the following.

Slow Pyrolysis

The solid residence time(s) in the reactor is 450–550, heating rate (°C) is 0.1–1, and feed particle size (mm) is 5–50 with temperature (°C) of 550–950. This process enhances char production and is unlikely to be unsuitable for high quality bio-oil production. Also, due to high residence time secondary reaction is favorable as cracking of primary product occurs which could adversely affect bio-oil yield and quality [84] (Table 6).

Table 6: Physical and chemical properties of used tyre oil and diesel fuel [15]

Fuel/property	Diesel fuel	Used tyre oil
Density (kg/m^3) @ 20°C	845	945
Kinematic viscosity @ 40°C in mm^2/sec	3.4	3.8
Flash point	60°C	50°C
Diesel index	55.2	41.9
Heat value (MJ/kg)	46	43.34
Sulphur amount (%)	0.1	0.906
Cetane number	53	44

Fast Pyrolysis

Fast pyrolysis involves the rapid heating of the feed material to a high temperature in the absence of oxygen with a short residence time of the condensable vapor in the reactor. Its operating parameters are solid residence time between 0.5 and 10s, heating rate of 10–200°C, feed particle size less than 5 mm, and reaction temperature of 550–1200°C. The technology has received much popularity in producing liquid fuels and a range of specialty and commodity chemicals. Typically on weight basis, fast pyrolysis yields 60%–75% pyro oil with 15%–25% compared to other processes; it has reasonably low investment costs and high energy efficiencies particularly on a small scale [85].

Flash Pyrolysis

This process is characterized by residence time of less than 0.5 s, high heating rate of more than 200°C, particle size of less than 0.2 mm, and high reaction temperature of more than 1000°C. However, the major technological challenge of the process is poor thermal stability, solids in the oil, and production of pyrolytic water [86, 87].

Catalytic Pyrolysis

Catalytic pyrolysis is a pyrolysis process that includes the use of a catalyst. The catalyst helps enhance the pyrolysis reaction kinetics by cracking down higher molecular weight hydrocarbon compounds to lighter hydrocarbon products. It has been reported that the use of catalyst in tyre pyrolysis systems can greatly influence the composition, quality, and yield of products [6, 88]. Examples of catalysts used in tyre pyrolysis include Na_2CO_3, NaOH, MgO, $CaCO_3$, aluminium-based catalyst, perlite, CaC_2, $Cu(NO_3)_2$ zeolite-based catalyst, and so forth [6, 7, 89]. Operating conditions can determine different product distribution for different catalysts. Pyrolysis catalyst can be grouped based on their method of application. The first group is when the catalyst is added to the feedstock before being fed into the reactor. The second group is when catalyst is added after the feed is already heated up in the reactor allowing it to have immediate contact with vapors, solid, and char. The third group is when the catalyst is placed in another reactor located downstream from the pyrolysis reactor [46, 90].

Chemical Reaction of the Catalytic Pyrolysis

The chemical reaction of catalytic pyrolysis is an endothermic process. Three stages involved include dehydration, fragmentation, and product formation which occur at different temperatures during the process. It is carried out in an inert environment in order to avoid combustion. The mechanism of the reaction is shown below:

$$\underset{\text{polyisoprene}}{[CH_2CH{=}C(CH_3)CH_2]_n} \xrightarrow[+\Delta H]{\text{catalyst, No } O_2} \text{char, gas, oil} \tag{1}$$

where ΔH = heat.

Pyrolysis Products and Their Composition

The three primary products obtainable from tyre rubber pyrolysis are listed and discussed below.

(1). Solid residue or char (34–38 wt.% of feedstock) is made up of carbon black and other nonvolatile materials such as tyre rubber additives like zinc, sulphur, silica, clays, and so forth, initially present in the tyre. This char can be used as solid fuel, carbon black, or as precursor for activated carbon manufacturing [22]. From the literature, elemental analysis shows that the solid residue contains 71 wt.% of C, 13.3 wt.% of O, 5.4 wt.% of Fe, 2.8 wt.% of S, 2.3 wt.% of Zn, 1.3 wt.% of Ca, and 0.3 wt.% of Al.

(2). Liquid product (48–58 wt.% of feedstock) also known as pyrolysed tyre oil or biocrude is the most significant product of the process. It is gotten from the condensation of vapor of a pyrolysis reaction. Several publications on the study of the properties of the pyrolytic oil abound [18, 45]. It is reported that the calorific value of the oil is 42 MJ/kg, percentage content of sulphur ranges between 1.2 and 1.36, and the oil is a mixture of C6–C24 organic compounds (53.4–74.8%) specified to be paraffins, olefins, and aromatic compounds with some nitrogenated (2.47–3.75%) and some oxygenated compounds (2.29–4.85%) [17, 18, 28]. Such pyrolytic oil can be used directly as low-sulphur-emission fuels or blend with petroleum products such as gasoline, feedstock for petroleum refinery, and as an important source for chemicals in chemical industries due to high concentration of benzene, toluene, xylene, and limonene [18]. Elemental analysis of the tyre pyrolysis oil shows that it contains 86.871 wt.% of C, 10.07 wt.% of H, 1.169 wt.% of N, 0.906 wt.% of S, and 1.169 wt.% of O.

(3). Gas fraction (9–14 wt.% of feedstock) from the pyrolysis process is mainly composed of H_2, CO, CH_4, CO_2, C_2H_4, H_2S, and other light hydrocarbons with SO_2, NO_x, CO, and PAHs reported as the main pollutant gases. It has a high heating value of about 37 MJ/m^3, an amount of energy enough for the pyrolysis process.

It is worthy to note that the composition of the various fractions is influenced by the pyrolysis conditions used and the composition of the tyre [2, 3, 7, 10, 17].

PYROLYSIS KINETICS

Kinetic models for used tyre pyrolysis have been presented by various authors [91–97]. Generally, the degradation of the tyre rubber is split into two or more steps, while thermogravimetry data and Arrhenius equation form the basis of the models. In the kinetic analysis of the data for the tyre decomposition, the Arrhenius theory based on first order reaction is mostly assumed by researchers:

$$k = A\exp\left(-\frac{E}{RT}\right), \tag{2}$$

Where k is the rate constant, A is the preexponential factor (min^{-1}), E is the activation energy (J/mol), and R and T are the universal gas constant ($8.314\,Jmol^{-1}\,k^{-1}$) and temperature of reaction (K), respectively. Equation (3) can be used to represent the rate of tyre decomposition:

$$\frac{d\alpha}{dt} = k(1-\alpha)^n, \tag{3}$$

Where α is the fraction of reactant decomposed at time t and n is the order of reaction. Application of the above given equations for tyre pyrolysis under different conditions along with experimental conditions and obtained kinetic parameters published in the literature by some authors is given below.

Leung and Wang [64]. Consider the following.

Model:

$$\frac{d\alpha_T}{dt} = \sum_{i=1}^{3}\frac{d\alpha_i}{dt} = \sum_{i=1}^{3} A_i \exp\left(-\frac{E_i}{RT}\right)(1-\alpha), \tag{4}$$

Where α_T (min^{-1}) is normalized mass loss rate, A is preexponential factor, E is activation energy, R is ideal gas constant, T is absolute temperature, t is time, and index (i = 1, 2, 3) is reaction.

Experimental Conditions: Temperature: 20–600°C, particle size: 0.355–0.425 mm, heating rate: 10–60°C/min, and used mobile gas: N_2. The kinetic parameters obtained by Leung and Wang are given in Table 7.

Table 7: Kinetic parameters obtained by Leung and Wang [64]

Heating rate	Activation energy, E (KJ/mol)		Preexponential factor, A (min^{-1})		Reaction order, n	
	Lower temperature	Higher temperature	Lower temperature	Higher temperature	Lower temperature	Higher temperature
10	164.5	136.1	6.29×10^{13}	2.31×10^{9}	1	1
30	180.9	133.6	1.32×10^{14}	2.09×10^{9}	1	1
45	203.4	107.0	7.58×10^{15}	3.34×10^{7}	1	1
60	218.7	99.1	1.13×10^{17}	1.02×10^{7}	1	1

Williams and Besler [69]. Consider the following.

Model:

$$\frac{dw}{dt} = -k\left(W - W_f\right), \tag{5}$$

Where W is the weight of sample at time t, W_f is the weight of residue at the end of the reaction, and represents the rate constant defined by Arrhenius equation.

Experimental Conditions: Temperature: 720°C, sample sizes: <1 mm, heating rate: 5°C/min, and used mobile gas: N_2. Table 8 shows the kinetic parameters obtained.

Table 8: Kinetic parameters obtained by Williams and Besler [69]

Heating rate	Activation energy, E (kJ/mol)		Preexponential factor, A (min^{-1})		Reaction order, n	
	Sample A	Sample B	Sample A	Sample B	Sample A	Sample B
5	142.7	120.8, 145.0	2.1×10^{8}	$9.3 \times 10^{7}, 1.1 \times 10^{8}$	—	—
20	90.8	128.3, 137.7	2.6×10^{4}	$3.3 \times 10^{5}, 2.1 \times 10^{8}$	—	—
40	70.4	66.1, 136.2	1.3×10^{3}	$5.9 \times 10^{2}, 6.3 \times 10^{7}$	—	—
80	66.4	55.6	1.1×10^{3}	1.3×10^{2}	—	—

CONCLUSION

Biocrude production from used tyres and natural rubber is a viable means of an alternative renewable energy source that can help caution the fast depletion of crude oil reserves, its fast rising cost due to high demand, along with its adverse negative environmental impact. This review reports the present stage of research in used tyre pyrolysis. From the literature, the proximate analyses of tyre rubber show that it is composed of about 28.6 wt.% fixed carbon, 62 wt.% volatile material, 8.5 wt.% ash, and 0.9 wt.% moisture. Elemental analyses reveal that tyre rubber has an estimated value of 82 wt.% of C, 8 wt.% of H, 0.4 wt.% of N, 1.3 wt.% of S, 2.4 wt.% of O, and 5.9 wt.% of ash. Both the proximate and ultimate analysis value of the used tyres is dependent on the tyre formulation by the tyre manufacturers. Thermogravimetry analysis confirms that the pyrolysis of used tyre at atmospheric pressure commences at about 250°C and completes at around a temperature of

550°C. Usually, during tyre pyrolysis, there exist different degradation temperatures. This was also confirmed by the thermogravimetric analysis performed in our laboratory. It was also observed that pyrolysis products yields and their characteristics are greatly affected by the composition of the feed, operating conditions, and the specific properties of the system used.

Thus, results vary from different researchers making it difficult to compare results. Nevertheless, the three primary products obtained from used tyres pyrolysis are solid residue or char (around 36 wt.%), liquid fraction or biocrude (around 55 wt.%), and gas fraction (around 9 wt.%) with average higher heating values (HHV) of 28 MJ kg^{-1}, 42 MJ kg^{-1}, and 36 MJ N^{-1} m^{-3}, respectively.

Studies have also shown that the main components of the liquid fractions are aliphatic and aromatic hydrocarbons and hydroxyl compounds while the gases that make up the gas fractions are H_2, CO, CH_4, CO_2, C_2H_4, H_2S, and other light hydrocarbons with SO_2, NO_x, CO, and polycyclic aromatic hydrocarbons (PAHs) reported as the main pollutant gases. The char contains high fixed-carbon content and inorganic matter.

Although there is a wide difference in results of kinetic modeling of used tyres by different authors in the value of kinetic parameters obtained, generally it is accepted that the process is a first order reaction based on Arrhenius theory.

Therefore, there is the need to investigate the effect of copyrolysis of natural rubber (biomass obtained from rubber tree) and used tyre on the pyrolysis product yields in relation to the following.(i)Composition and quantity of gas yield. (ii)Enhancing the possibility of more valuable chemical products from the liquid fraction such as limonene which is closely related to polyisoprene (NR) present in used tyre.(iii) Analyzing the amount of sulphur content in the biocrude (liquid fraction).(iv)The quality of char produced in comparison with commercial carbon black will be the area of our research.

ACKNOWLEDGMENTS

Financial support from the National Research Foundation (NRF) under the South Africa NRF Focus Area and student bursaries made available by the University of the Witwatersrand are deeply appreciated.

REFERENCES

1. M. Bianchi, G. Bortolani, M. Cavazzoni et al., "Preliminary design andnumerical analysis of a scrap tyre pyrolysis system," Energy Procedia, vol. 45, pp. 111–120, 2014.
2. J. D. Martinez, A. Veses, A. M. Mastral et al., "Co-pyrolysis of biomass withwaste tyres: upgrading of liquid to bio-fuel," Fuel Processing Technology, vol. 119, pp. 263–271, 2014.
3. S. Frigo, M. Seggiani, M. Puccini, and S. Vitolo, "Liquid fuel production from waste tyre pyrolysis and its utilization in a diesel engine," Fuel, vol. 116, pp. 399–408, 2014.
4. A. Undri, L. Rosi, M. Frediani, and P. Frediani, "Upgraded fuel from microwave assisted pyrolysis of waste tyre," Fuel, vol. 115, pp. 600–608, 2014.
5. H. Hu, Y. Fang, H. Liu et al., "The fate of sulfur during rapid pyrolysis of scarp tyres," Chemosphere, vol. 97, pp. 102–107, 2014.
6. P. T. Williams, "Pyrolysis of waste tyres: a review," Wastemanagement, vol. 33, pp. 1714–1728, 2013.
7. A. Quek and R. Balasubramanian, "Liquefaction of waste tyres by pyrolysis for oil and chemicals—a review," Journal of Analytical and Applied Pyrolysis, vol. 101, pp. 1–16, 2013.
8. S. Karthikeyan, C. Sathiskumar, and S. R. Moorthy, "Effect of process parameters on tire pyrolysis: a review," Journal of Scientific and Industrial Research, vol. 71, no. 5, pp. 309–315, 2012.
9. M. Juma, Z. Korenova, J. Markos, J. Annus, and L. Jelemenšky, "Pyrolysis and combustion of Scrap tyre,"Petroleum & Coal, vol. 48, no. 1, pp. 15–26, 2006.
10. J. Haydary, Z. Korenova, L. Jelemensky, and J. Markos, "Thermal decomposition of waste polymers," inThermoPhysics, pp. 62–68, 2008.
11. M. R. Islam, M. S. H. K. Tushar, and H. Haniu, "Production of liquid fuels and chemicals from pyrolysis of Bangladeshi bicycle/rickshaw tire wastes," Journal of Analytical and Applied Pyrolysis, vol. 82, no. 1, pp. 96–109, 2008.
12. X. Zhang, T. Wang, L. Ma, and J. Chang, "Vacuum pyrolysis of waste tires with basic additives," Waste Management, vol. 28, no. 11, pp. 2301–2310, 2008.
13. D. Pradhan and R. K. Singh, "Thermal pyrolysis of bicycle waste tyre using batch reactor," International Journal of Chemical Engineering and Application, vol. 2, no. 5, pp. 332–336, 2011.
14. P. Trongkaew, T. Utistham, P. Reubroycharoen, and N. Hinchiranan, "Photocatalytic desulfurization of waste tire pyrolysis oil," Energies, vol. 4, no. 11, pp. 1880–1896, 2011.
15. H. Aydin and C. Ilkilic, "Optimization of fuel production from waste vehicle tyres by pyrolysis and resembling to diesel fuel by various desulfurization methods," Fuel, vol. 102, pp. 605–612, 2012.
16. H. Schnecko, "Rubber recycling," Macromolecular Symposia, vol. 135, pp. 327–343, 1998.
17. R. Aguado, M. Olazar, D. Vélez, M. Arabiourrutia, and J. Bilbao, "Kinetics of scrap tyre pyrolysis under fast heating conditions," Journal of Analytical and Applied Pyrolysis, vol. 73, no. 2, pp. 290–298, 2005. ·
18. A. M. Cunliffe and P. T. Williams, "Influence of process conditions on the rate of activation of chars derived from pyrolysis of used tires," Energy and Fuels, vol. 13, no. 1, pp. 166–175, 1999.
19. J. F. González, J. M. Encinar, J. L. Canito, and J. J. Rodríguez, "Pyrolysis of automobile tyre waste. Influence of operating variables and kinetics study," Journal of Analytical and Applied Pyrolysis, vol. 58-59, pp. 667–683, 2001.
20. M. Bajus and N. Olahova, "Thermal conversion of scrap tyres," Petroleum & Coal, vol. 53, pp. 98–105, 2011.
21. C. Roy, H. Darmstadt, B. Benallal, and C. Amen-Chen,

"Characterization of naphtha and carbon black obtained by vacuum pyrolysis of polyisoprene rubber," Fuel Processing Technology, vol. 50, no. 1, pp. 87–103, 1997.

22. S. Boxiong, W. Chunfei, L. Cai, G. Binbin, and W. Rui, "Pyrolysis of waste tyres: the influence of USY catalyst/tyre ratio on products," Journal of Analytical and Applied Pyrolysis, vol. 78, no. 2, pp. 243–249, 2007.
23. M. I. Jahirul, M. G. Rasul, A. A. Chowdhury, and N. Ashwath, "Biofuels production through biomass pyrolysis—a technical review," Energies, vol. 5, pp. 4952–5001, 2012.
24. G. Harrison and A. B. Ross, "Use of tyre pyrolysis oil for solvent augmentation in two-stage coal liquefaction," Fuel, vol. 75, no. 8, pp. 1009–1013, 1996.
25. A. Hershaft, "Solid waste: treatment technology," Environmental Science and Technology, vol. 6, no. 5, pp. 412–421, 1972.
26. S. R. Fix, "Microwave devulcanizationof rubber," Elastomerics, vol. 112, no. 6, pp. 38–40, 1980.
27. A. A. Phadke, A. K. Bhattacharya, S. K. Chakraborty, and S. K. De, "Studies of vulcanization of reclaimed rubber," Rubber Chemistry and Technology, vol. 56, no. 4, pp. 726–736, 1983.
28. A. I. Isayev, S. P. Yushanov, and J. Chen, "Ultrasonic devulcanization of rubber vulcanizates. I: process model," Journal of Applied Polymer Science, vol. 59, no. 5, pp. 803–813, 1996.
29. M. Beecham, Global Market Review of Automotive Tyres—Forecasts to 2014, Pub ID JA1867566, Aroq Limited, Bromsgrove, UK, 2008.
30. A. M. Mastral, R. Murillo, M. S. Callen, and T. Garcia, "Optimisation of scrap automotive tyres recycling into valuable liquid fuels," Resources, Conservation and Recycling, vol. 29, no. 4, pp. 263–272, 2000.
31. A. Donatelli, P. Iovane, and A. Molino, "High energy syngas production by waste tyres steam gasification in a rotary kiln pilot plant. Experimental and numerical investigations," Fuel, vol. 89, no. 10, pp. 2721–2728, 2010.
32. Y. Kar, "Catalytic pyrolysis of car tire waste using expanded perlite," Waste Management, vol. 31, no. 8, pp. 1772–1782, 2011.
33. Y. M. Chang, "On pyrolysis of waste tire: degradation rate and product yields," Resources, Conservation and Recycling, vol. 17, no. 2, pp. 125–139, 1996.
34. Y. Su and B. Zhao, "Pyrolysis of waste tire and its model," in Proceedings of the 4th International Conference on Bioinformatics and Biomedical Engineering (iCBBE '10), June 2010.·
35. I. R. de Marco, M. F. Laresgoiti, M. A. Cabrero, A. Torres, M. J. Chomón, and B. Caballero, "Pyrolysis of scrap tyres," Fuel Processing Technology, vol. 72, no. 1, pp. 9–22, 2001.
36. P. T. Williams and R. P. Bottrill, "Sulfur-polycyclic aromatic hydrocarbons in tyre pyrolysis oil," Fuel, vol. 74, no. 5, pp. 736–742, 1995.
37. J. Yang, P. A. Tanguy, and C. Roy, "Heat transfer, mass transfer and kinetics study of the vacuum pyrolysis of a large used tire particle," Chemical Engineering Science, vol. 50, no. 12, pp. 1909–1922, 1995.
38. J. A. Conesa, R. Font, A. Fullana, and J. A. Caballero, "Kinetic model for the combustion of tyre wastes,"Fuel, vol. 77, no. 13, pp. 1469–1475, 1998.
39. O. Senneca, P. Salatino, and R. Chirone, "Fast heating-rate thermogravimetric study of the pyrolysis of scrap tyres," Fuel, vol. 78, no. 13, pp. 1575–1581, 1999.
40. L. Tang and H. Huang, "Plasma pyrolysis of biomass for production of syngas and carbon adsorbent,"Energy and Fuels, vol. 19, no. 3, pp. 1174–1178, 2005.
41. A. Domínguez, Y. Fernández, B. Fidalgo, J. J. Pis, and J. A. Menéndez, "Bio-syngas production with low concentrations of CO_2 and CH_4 from microwave-induced pyrolysis of wet and dried sewage sludge,"Chemosphere, vol. 70, no. 3, pp. 397–403, 2008.
42. X. Zhao, Z. Song, H. Liu, Z. Li, L. Li, and C. Ma, "Microwave pyrolysis of corn stalk bale: a promising method for direct utilization of large-sized biomass and syngas production," Journal of Analytical and Applied Pyrolysis, vol. 89, no. 1, pp. 87–94, 2010.
43. A. Quek and R. Balasubramanian, "Mathematical modeling of rubber tire pyrolysis," Journal of Analytical and Applied Pyrolysis, vol. 95, pp. 1–13, 2012.
44. M. Balat, M. Balat, E. Kirtay, and H. Balat, "Main routes for the thermo-conversion of biomass into fuels and chemicals. Part 1: pyrolysis systems," Energy Conversion and Management, vol. 50, no. 12, pp. 3147–3157, 2009.
45. T. Bridgwater, Pyrolysis of Biomass, IEA Bioenergy Task 34, Bioenergy Research Group, Aston University, Birmingham, UK, 2007.
46. A. V. Bridgwater, S. Czernik, and J. Piskorz, "An overview of fast pyrolysis," in Progress in Thermochemical Biomass Conversion, vol. 2, pp. 977–997, 2001.
47. A. Demirbaş and G. Arin, "An overview of biomass pyrolysis," Energy Sources, vol. 24, no. 5, pp. 471–482, 2002.
48. R. Aguado, M. Olazar, B. Gaisán, R. Prieto, and J. Bilbao, "Kinetic study of polyolefin pyrolysis in a conical spouted bed reactor," Industrial and Engineering Chemistry Research, vol. 41, no. 18, pp. 4559–4566, 2002.
49. T. Cornelissen, J. Yperman, G. Reggers, S. Schreurs, and R. Carleer, "Flash co-pyrolysis of biomass with polylactic acid. Part 1: influence on bio-oil yield and heating value," Fuel, vol. 87, no. 7, pp. 1031–1041, 2008.
50. W. W. Gotshall, "Reinforcing agent from scrap rubber char," US Patent 3644131 A, February 1972.
51. S. B. Alpert, "Hydroconversion of waste natural rubber and synthetic rubbers," US Patent 3704108, November 1972.
52. E. A. Grannen and L. Robinson, "Microwave pyrolysis of wastes," US Patent 3843457, October 1974.
53. G. Crane and E. L. Kay, "Pyrolization," US Patent 3966487, June 1976.
54. O. Herbold, "Method and apparatus for the pyrolysis of

waste products," US Patent 4084521, April 1978.

55. W. Kaminsky and H. Sinn, "Pyrolysis of plastic waste and scrap tyres using a fluidized process," inThermal Conversion of Solid Wastes and Biomass, J. L. Jones and S. B. Radding, Eds., vol. 130 of American Chemical Society Symposium Series, pp. 423–439, 1980.

56. K. Ito, Y. Hirayama, Y. Ishii, and N. Ando, "Pyrolyzing apparatus," US Patent 4324620, April 1982.

57. H. Engman, H. T. Mayfield, T. Mar, and W. Bertsch, "Classification of bacteria by pyrolysis-capillary column gas chromatography-mass spectrometry and pattern recognition," Journal of Analytical and Applied Pyrolysis, vol. 6, no. 2, pp. 137–156, 1984.

58. J. M. Bouvier and M. Gelus, "Pyrolysis of rubber wastes in heavy oils and use of the products," Resources and Conservation, vol. 12, no. 2, pp. 77–93, 1986.

59. M. I. M. Chou, M. A. Lake, and R. A. Griffin, "Flash pyrolysis of coal, coal maceral, and coal-derived pyrite with on-line characterization of volatile sulfur compounds," Journal of Analytical and Applied Pyrolysis, vol. 13, no. 3, pp. 199–207, 1988.

60. C. Roy, B. Labrecque, and B. de Caumia, "Recycling of scrap tires to oil and carbon black by vacuum pyrolysis," Resources, Conservation and Recycling, vol. 4, no. 3, pp. 203–213, 1990.

61. V. Roy, B. de Caumia, and C. Roy, "Development of a gas-cleaning system for a scrap-tire vacuum-pyrolysis plant," Gas Separation and Purification, vol. 6, no. 2, pp. 83–87, 1992.

62. H. Pakdel and C. Roy, "Simultaneous gas chromatographic—Fourier transform infrared spectroscopic—mass spectrometric analysis of synthetic fuel derived from used tire vacuum pyrolysis oil, naphtha fraction," Journal of Chromatography A, vol. 683, no. 1, pp. 203–214, 1994.

63. J. A. Conesa, R. Font, and A. Marcilla, "Gas from the pyrolysis of scrap tires in a fluidized bed reactor,"Energy and Fuels, vol. 10, no. 1, pp. 134–140, 1996.

64. D. Y. C. Leung and C. L. Wang, "Kinetic study of scrap tyre pyrolysis and combustion," Journal of Analytical and Applied Pyrolysis, vol. 45, no. 2, pp. 153–169, 1998.

65. S. Galvagno, S. Casu, T. Casabianca, A. Calabrese, and G. Cornacchia, "Pyrolysis process for the treatment of scrap tyres: preliminary experimental results," Waste Management, vol. 22, no. 8, pp. 917–923, 2002.

66. M. F. Laresgoiti, B. M. Caballero, I. de Marco, A. Torres, M. A. Cabrero, and M. J. Chomón, "Characterization of the liquid products obtained in tyre pyrolysis," Journal of Analytical and Applied Pyrolysis, vol. 71, no. 2, pp. 917–934, 2004.

67. R. Murillo, E. Aylón, M. V. Navarro, M. S. Callén, A. Aranda, and A. M. Mastral, "The application of thermal processes to valorise waste tyre," Fuel Processing Technology, vol. 87, no. 2, pp. 143–147, 2006. ·

68. E. Aylón, A. Fernández-Colino, R. Murillo, M. V. Navarro, T. García, and A. M. Mastral, "Valorisation of waste tyre by pyrolysis in a moving bed reactor," Waste Management, vol. 30, no. 7, pp. 1220–1224, 2010. ·

69. P. T. Williams and S. Besler, "Pyrolysis-thermogravimetric analysis of tyres and tyre components," Fuel, vol. 74, no. 9, pp. 1277–1283, 1995.

70. J. Bi, X. Guo, M. Liu, and X. Wang, "High effective dehydration of bio-ethanol into ethylene over nanoscale HZSM-5 zeolite catalysts," Catalysis Today, vol. 149, no. 1-2, pp. 143–147, 2010.

71. J. A. Conesa and A. Marcilla, "Kinetic study of the thermogravimetric behavior of different rubbers,"Journal of Analytical and Applied Pyrolysis, vol. 37, no. 1, pp. 95–110, 1996.

72. L. M. Mahlangu, Waste tyre management problems in South Africa and the possible Opportunities that can be created through the recycling therefore [M.S. thesis], 2009.

73. J. H. Chen, K. S. Chen, and L. Y. Tong, "On the pyrolysis kinetics of scrap automotive tires," Journal of Hazardous Materials, vol. 84, no. 1, pp. 43–55, 2001.

74. M. Liompart, L. Sanchez-Prado, J. P. Lamas, C. Garcia-Jares, E. Roca, and T. Dagnac, "Hazardous organic chemicals in rubber recycled tyre playgrounds and pavers," Chemosphere, vol. 90, pp. 423–431, 2013.

75. J. P. Lin, C. Y. Chang, and C. H. Wu, "Pyrolysis kinetics of rubber mixtures," Journal of Hazardous Materials, vol. 58, no. 1–3, pp. 227–236, 1998.

76. H. Cui, J. Yang, and Z. Liu, "Pyrolysis of tires and tire components by TG/DTA analyzer," Journal of Chemical Industry and Engineering, vol. 50, no. 6, pp. 826–833, 1999.

77. J. G. Brammer and A. V. Bridgwater, "Drying technologies for an integrated gasification bio-energy plant," Renewable & Sustainable Energy Reviews, vol. 3, no. 4, pp. 243–289, 1999.

78. M. Olazar, G. Lopez, M. Arabiourrutia, G. Elordi, R. Aguado, and J. Bilbao, "Kinetic modelling of tyre pyrolysis in a conical spouted bed reactor," Journal of Analytical and Applied Pyrolysis, vol. 81, no. 1, pp. 127–132, 2008.

79. California Integrated waste management Board (CIWMB), "Effects of waste tires, waste tire facilities, and waste tire projects on the environment," CIWMB Report 432-96-029, 1996.

80. S. Ucar, S. Karagoz, A. R. Ozkan, and J. Yanik, "Evaluation of two different scrap tires as hydrocarbon source by pyrolysis," Fuel, vol. 84, no. 14-15, pp. 1884–1892, 2005.

81. E. L. K. Mui, W. H. Cheung, and G. McKay, "Tyre char preparation from waste tyre rubber for dye removal from effluents," Journal of Hazardous Materials, vol. 175, no. 1–3, pp. 151–158, 2010.

82. O. Doğan, M. B. Elik, and B. Özdalyan, "The effect of tire derived fuel/diesel fuel blends utilization on diesel engine performance and emissions," Fuel, vol. 95, pp. 340–346, 2012.

83. S. Murugan, M. C. Ramaswamy, and G. Nagarajan, "The use of tyre pyrolysis oil in diesel engines," Waste Management, vol. 28, no. 12, pp. 2743–2749, 2008.

84. S. Uçar, S. Karagöz, J. Yanik, M. Saglam, and M.

Yuksel, "Copyrolysis of scrap tires with waste lubricant oil," Fuel Processing Technology, vol. 87, no. 1, pp. 53–58, 2005.

85. A. M. Fernández, C. Barriocanal, and R. Alvarez, "Pyrolysis of a waste from the grinding of scrap tyres,"Journal of Hazardous Materials, vol. 203-204, pp. 236–243, 2012.

86. K. Unapumnuk, T. C. Keener, M. Lu, and F. Liang, "Investigation into the removal of sulfur from tire derived fuel by pyrolysis," Fuel, vol. 87, no. 6, pp. 951–956, 2008.

87. S. Murugan, M. C. Ramaswamy, and G. Nagarajan, "Performance, emission and combustion studies of a DI diesel engine using distilled Tyre pyrolysis oil-diesel blends," Fuel Processing Technology, vol. 89, no. 2, pp. 152–159, 2008.

88. S. Murugan, M. C. Ramaswamy, and G. Nagarajan, "Assessment of pyrolysis oil as an energy source for diesel engines," Fuel Processing Technology, vol. 90, no. 1, pp. 67–74, 2009.

89. D. Bunthid, P. Prasassarakich, and N. Hinchiranan, "Oxidative desulfurization of tire pyrolysis naphtha in formic acid/H_2O_2/pyrolysis char system," Fuel, vol. 89, no. 9, pp. 2617–2622, 2010.·

90. M. Miranda, F. Pinto, I. Gulyurtlu, I. Cabrita, C. A. Nogueira, and A. Matos, "Response surface methodology optimization applied to rubber tyre and plastic wastes thermal conversion," Fuel, vol. 89, no. 9, pp. 2217–2229, 2010.

91. O. Doğan, M. B. Elik, and B. Özdalyan, "The effect of tire derived fuel/diesel fuel blends utilization on diesel engine performance and emissions," Fuel, vol. 95, pp. 340–346, 2012.

92. A. Napoli, Y. Soudais, D. Lecomte, and S. Castillo, "Scrap tyre pyrolysis: are the effluents valuable products?" Journal of Analytical and Applied Pyrolysis, vol. 40-41, pp. 373–382, 1997.

93. C. Ilkiliç and H. Aydin, "Fuel production from waste vehicle tires by catalytic pyrolysis and its application in a diesel engine," Fuel Processing Technology, vol. 92, no. 5, pp. 1129–1135, 2011.

94. S. Hariharan, S. Murugan, and G. Nagarajan, "Effect of diethyl ether on tyre pyrolysis oil fueled diesel engines," Fuel, vol. 104, pp. 109–115, 2013.

95. P. Behera and S. Murughan, "Combustion, performance and emission parameters of used transformer oil and its diesel blends in a DI diesel engine," Fuel, vol. 104, pp. 147–154, 2013.

96. A. V. Bridgwater, "Review of fast pyrolysis of biomass and product upgrading," Biomass and Bioenergy, vol. 38, pp. 68–94, 2012.

97. Y. Yang, J. G. Brammer, M. Ouadi et al., "Characterisation of waste derived intermediate pyrolysis oils for use as diesel engine fuels," Fuel, vol. 103, pp. 247–257, 2013.

CHAPTER 20

FABRICATION AND PACKAGING OF FLEXIBLE POLYMERIC MICROANTENNAE FOR IN VIVOMAGNETIC RESONANCE IMAGING

Magdalèna Couty[1,2,3], Marion Woytasik[1,2], Jean-Christophe Ginefri[2,4], Anne Rubin[1,2,5], Emile Martincic[1,2], Marie Poirier-Quinot[2,4], Luc Darrasse[2,4], Fawzi Boumezbeur[5], Franck Lethimonnier[5], Michael Tatoulian[3] and Elisabeth Dufour-Gergam[1,2]

[1]Univ. Paris Sud, Laboratoire IEF (Institut d'Electronique Fondamentale), UMR-8622, Bat. 220, F91405 Orsay, France;

[2]CNRS (Centre National de Recherche Scientifique), F-91405 Orsay, France;

[3]LGPPTS (Laboratoire du Génie des Procédés Plasmas et Traitements de Surface), Chimie ParisTech/UPMC EA3492, 11 rue Pierre et Marie Curie, F-75005 Paris, France;

[4]Univ. Paris Sud, Laboratoire IR4M (Imagerie par Résonance Magnétique Médicale et MultiModalités), UMR-8081, Bat. 220, F-91405, Orsay, France

[5]I2BM (Institut d'Imagerie BioMédicale), Neurospin, CEA Saclay, Bat. 145, F-91191 Gif-sur-Yvette, France;

ABSTRACT

In this paper, we detail how microantennae dedicated to Magnetic Resonance Imaging (MRI) can benefit from the advantages offered by polymer substrates, especially flexibility and dielectric properties. We present a monolithic and wireless design based on the transmission lines between conductor windings on both sides of a dielectric substrate and its fabrication process. This last one requires specific plasma treatments to improve polymer/metal adhesion. We have led a comparative study on the effects of the ageing time on the wettability and the metal adhesion to Kapton and Teflon surfaces. Correlation between wettability (water contact angle) and adhesion (tensile strength) has been established. Then, the use of PolyDiMethylSiloxane (PDMS) as biocompatible packaging material and the optimization of its thickness allows us to conserve suitable f_0 and Q values in a conducting environment such as the biological tissues. These studies allow us to perform 7 Tesla *in vivo* MRI of the rat brain with a high spatial resolution of 100 x 100 x 200 μm^3 and a Signal to Noise Ratio of 80.

INTRODUCTION

The use of polymers in the microsystem field has spread these last years. Indeed, they are low-cost while allowing batch fabrication by microtechnologies. Contrary to the substrates traditionally used in the field, such as silicon or silicone oxide, they offer high flexibility and a wide range of dielectric properties for numerous applications, e.g., stretchable electronics and implantable medical devices. In the case of implantable microdevices, flexibility is especially desirable to reduce the damage inflicted during surgery and for the comfort of the patient during the implantation [1]. Moreover, polymers also have interesting interactions with the biological tissues: the biocompatible ones as packaging materials and the biodegradable ones for drug delivery.

We develop microantennae on polymer substrates for Magnetic Resonance Imaging (MRI) on the microscopic scale so we have considered how this kind of devices can benefit from the advantages described above. Our approach consists in reducing the size of the antenna to increase its sensitivity

Citation: Couty, M.; Woytasik, M.; Ginefri, J.-C.; Rubin, A.; Martincic, E.; Poirier-Quinot, M.; Darrasse, L.; Boumezbeur, F.; Lethimonnier, F.; Tatoulian, M.; Dufour-Gergam, E. Fabrication and Packaging of Flexible Polymeric Microantennae for in Vivo Magnetic Resonance Imaging. Polymers 2012, 4, 656-673, doi:10.3390/polym4010656

and therefore to obtain higher Signal to Noise Ratio (SNR). This approach has been already validated for the classic configuration of macroscopic antennas [2,3]. Miniaturization of macroscopic MRI antennas is limited to the centimeter scale by the use of discrete components such as tuning and matching capacitors [3,4]. To overcome this limitation, a solution is to use a monolithic design based on transmission lines. We have developed a monolithic design based on the transmission lines between conductor windings on both sides of a dielectric substrate, called Multi-turn Transmission Lines Resonator (MTLR) [5]. Coils with similar designs (also based on transmission lines) have been fabricated on various rigid substrates such as silicon dioxide [6] and glass [7], then on flexible polymeric substrates such as polyimide [7,8,9] and Teflon [10,11]. This last substrate has a low loss tangent (less than 0.0003 at 1 MHz) which allows to reduce the parasitic capacitance and maximize the quality factor Q of the antennas [11].

In the case of surface imaging, flexible antennas conform very close to curved surfaces and provide improved images, compared with planar antennas [12]. A concern about flexibility could be the performances of the antennas when they are bent. For surface imaging, the bending is relatively low and no significant variation of the resonant frequency f_0 and the quality factor Q is observed. However, other applications need a stronger bending such as catheter tracking by MRI. Ahmad *et al.* reported that wrapping flexible antennas fabricated on polyimide and SU-8 substrates round the catheter alters their inductances thus shifts f_0 and subsequent tuning and matching are necessary [13]. Ellersiek *et al.* studied the influence of the bending radius on the inductance of the antenna and determined that it decreases to 20% when this radius is very low [14]. Microfabricated catheter-mounted coils are now reasonably developed, and can be bent without significantly degrading their performance [15]. Moreover, this specific application remains the most extreme case of bending conditions for MRI antennas.

When imaging tissues deeply inside the body is required, as microantennae have a limited field of view, they must be implanted [3]. In this case, MRI becomes invasive but offers the possibility of an easy long-term monitoring. Besides flexibility, additional specifications for the design and the packaging have to be defined to reduce as much as possible the invasive aspect of implantation. The antenna should be very small (*i.e.*, about a few millimeters) and without wire connection. Our MTLR design is totally in agreement with these specifications. Furthermore, packaging the antenna with an adequate biocompatible polymer minimizes its impact on the surrounding environment. Reciprocally, the antenna must be protected from the undesirable effects of the biological tissues on its characteristics. Indeed, they have a high permittivity (about 80) and a high conductivity (about 0.7 S/m) thus induce by dielectric coupling a shift-down of f_0 and degradation of Q. This phenomenon has been observed by Yung *et al.* and Bilgen [3,4]. Dielectric and biocompatibility properties of polymers make them also appropriate as packaging materials for implanted MRI microantennae. Bilgen chose to apply a thick biocompatible elastomer coating to increase the distance between the antenna and the tissues and so to reduce the coupling effect [4]. We chose to work with PolyDiMethylSiloxane (PDMS), the most common silicone elastomer, which is widely considered as biocompatible for short-term (days) implantation [16] and has a low permittivity (2.65 at 100 kHz) and a low loss tangent (1×10^{-3} at 100 kHz). Its preparation and deposition need no specific expensive equipment and an accurate control of thickness can be achieved through spin coating parameters (rotation speed, time) and curing conditions (temperature, time). Moreover, thickness of several hundred micrometers may be deposited, which is not possible with other biocompatible coating materials such as parylene.

Despite the numerous advantages of polymers explained above, their main drawback is the poor adhesion of metal to their surface. Surface modifications by plasma treatments are the most common solution to overcome this drawback. Each polymer requires a specific plasma treatment depending on its chemistry. The efficiency of oxygen plasma exposure for polyimide is well-known [17]. In a previous work [12], we used this method to fabricate flexible microantennae (with MTLR design) on Kapton and showed how they improve the *in vivo* surface imaging at 2.35 Tesla of a subcutaneous tumor in mouse by form-fitting to the skin.

However, fluorinated polymers are very hydrophobic [18] and a reliable adhesion of metal to their surfaces is difficult to obtain. Zhang *et al.* fabricated large-sized antennas (about 10 × 10 cm² squares) with copper tape on Teflon films [10], whereas Eroglu *et al.* directly used commercial metalized films [11]. Studies were necessary to find an efficient plasma treatment to improve the metal adhesion to the Teflon surface. These studies led to a low pressure N_2/H_2 plasma [19] and have allowed us to fabricate flexible microantennae (still with MTLR design) on Teflon substrates. The project aims to use MRI as a diagnosis and monitoring tool for tumors and neurodegenerative diseases, *i.e.*, Alzheimer's one, in clinical research on small rodents, and has evolved to 7 Tesla *in vivo* imaging of the rat brain using implanted microantennae. In this paper, after description of the MTLR design and its fabrication process, we present a comparative study on the improvement of polymer/metal adhesion by specific plasma treatments on Kapton and Teflon films, the optimization of the PDMS packaging to conserve the desired electrical characteristics in a conducting environment and finally *in vivo* imaging of the rat brain.

EXPERIMENTAL SECTION

Fabrication Process

The substrates are polyimide 125 μm-thick Kapton HN films from DuPont Nemours® and FEP Teflon films from MICEL®. The antennas are fabricated on these flexible polymer substrates using the copper micromoulding process detailed in [7,20]. Figure 1 summarizes the different steps of the process.

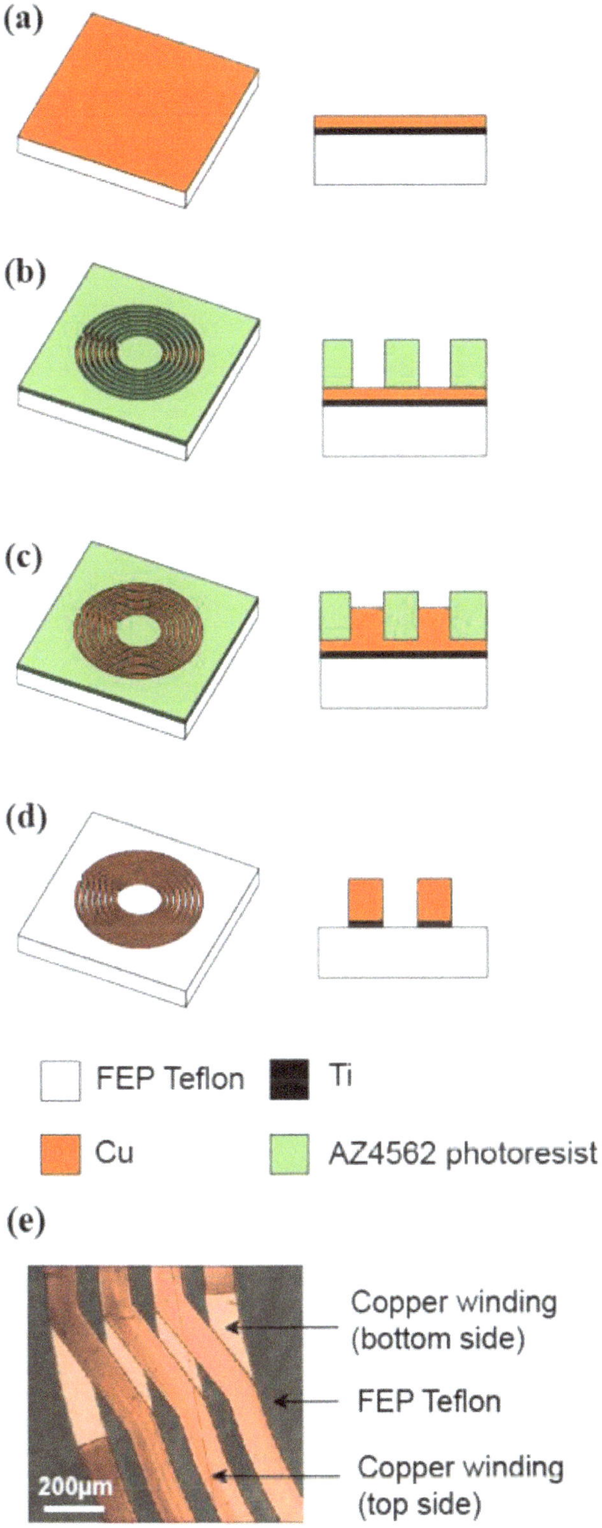

Figure 1: Steps of copper micromoulding process: (**a**) Ti/Cu:10/100 nm sputtering, (**b**) patterning 20 µm AZ4562 photoresist mould by photolithography, (**c**) 10 µm copper electrodeposition, (**d**) photoresist removal and wet etching of seed layer; (**e**) view of the alignment between the copper windings on the top and the bottom sides by transparency through the Teflon substrate.

First, a metallic seed layer Ti/Cu:10/100 nm is deposited by sputtering on the polymer film (a). The metalized film is glued with photoresist on a rigid glass support to ensure a flat surface and the rigidity of the sample for the subsequent steps. The 20 µm-thick AZ4562 mould is patterned by photolithography (b). The patterned mould is filled with 10 µm copper by electrodeposition (c). This thickness allows avoidance of the increase of the series resistance due to the skin effect—skin depth is 3.77 µm at 300 MHz for copper. Finally, the photoresist is removed by acetone and the seed layer is removed by wet etching (d). These steps are repeated for the second side of the substrate, with an accurate alignment of the windings by transparency through glass using a EVG®620 double-side lithography equipment (e).

Plasma Treatments

The Kapton films are treated with an O_2 plasma in a plasma etcher Pico UHP from Diener electronic. The O_2pressure is adjusted to 0.6 mbar and the plasma is produced using a 40 kHz RF generator with a power of 120 W, during a treatment time of 90 s.

The treatment of the FEP Teflon films is performed in a low pressure home-made plasma reactor with an asymmetric electrode configuration described in [21]. The rotary cylindrical grounded electrode and the high-voltage knife electrode are spaced of 8 mm. The cylinder—length 22 cm, diameter 7 cm—is also a carrier for the samples and is protected by a polyethylene film. A base pressure of 3×10^{-2} mbar is reached using a turbo molecular pumping system and a chemical pump. Then, the gaseous mixture N_2/H_2:25/75% (v/v) is introduced and the pressure is adjusted to 0.600 mbar. The plasma is produced between the two electrodes. A Corona discharge is provided by a 70 kHz RF generator with a power of 2 W for an effective treatment time of 5 s.

Tensile Strength Measurements by Pull-Off System

The tensile strength is measured for both untreated and plasma treated Kapton and Teflon samples about 1 cm². A Ti/Cu seed layer has been deposited by sputtering on all these samples (few hours or days after the plasma exposure for treated samples). Measurements of tensile strength are performed using a homemade pull-off set-up similar to described in [22,23]. The samples are prepared using the following method: the bottom side of the metalized polymer film is glued with epoxy on a brass carrier. A metal stud, 4 mm in diameter, is glued on the center of the metallic surface. After one hour in oven at 60 °C to ensure a reproducible curing of the epoxy glue, and a slow cooling to room temperature, the sample is mechanically held in the pull-off system. The pulling force is progressively applied by pumping. The voltage detected by the force sensor is amplified and measured with a numerical multimeter. The calibration of the system, for each amplification range, allows us to deduce the tensile strength from the maximal variation of the voltage (V_{max}) measured at the breaking point. Figure 2 shows the typical acquisition curve using this system.

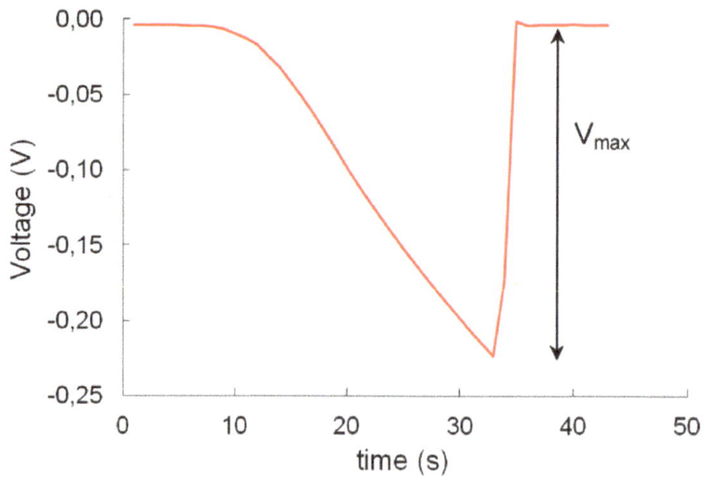

Figure 2: Typical acquisition curve of the voltage with time, from the pull-off system.

Electromagnetic Characterization of the Antennas

To limit the cross effect between neighboring antennas, the batch substrate is cut to obtain single antenna which will be individually characterized. The resonant frequency f_0 and the quality factor Q are measured using the single-loop probe method [24] and a E5061A network analyzer from Agilent Technologies. The magnetic coupling between the antenna and the single loop must be sufficiently weak so that the performances of the antenna are not affected by the presence of the probe and are neither dependent on the input nor on the distance between the probe and the antenna. This condition is achieved when the reflection coefficient at the single loop terminal is less than −40 dB [25].

Saline agar phantoms are prepared with 5% agarose gel in NaCl 0,45% Cooper® solution. Permittivity is about 80 and electrical conductivity about 0.7 S/m, close to the electrical properties of biological tissues in the rat brain. Measurements are performed in air and in agar phantom to evaluate the influence of a conducting environment on the performances of the antenna.

PDMS Packaging

PDMS is prepared using the commercial kit Sylgard184® provided by Dow Corning. The silicone polymer and the curing are mixed in a 10:1 weight ratio and then placed under vacuum for one hour to remove all bubbles. To improve the adhesion of PDMS on the antenna, its surface is treated with a He/O_2:50/50% (v/v) plasma [26,27] in the low pressure reactor described in Section 2.2. The power and pressure conditions are similar to the ones used for the N_2/H_2 treatment. In this case, the surface modifications are more stable because helium species participate to crosslink the upper-most surface layer, minimizing side chain reorganization [28]. Then, the PDMS mixture is deposited by spincoating to ensure accurate control of the thickness through the parameters (speed rotation, time). The PDMS is cured at 95 °C for 30 min in an oven. Under these conditions, the resulting thickness is about 30 µm. This process is repeated for each side of the antenna. To observe the deposition profile of PDMS on the copper windings, an antenna is fractured after cryopreparation. Figure 3 shows a SEM view of the cross section. The gap between the windings is completely filled with PDMS and the surface of the deposit is flat. Note that the difference in the thermal expansion coefficient of the materials may provoke small localized gaps at the copper/PDMS interface.

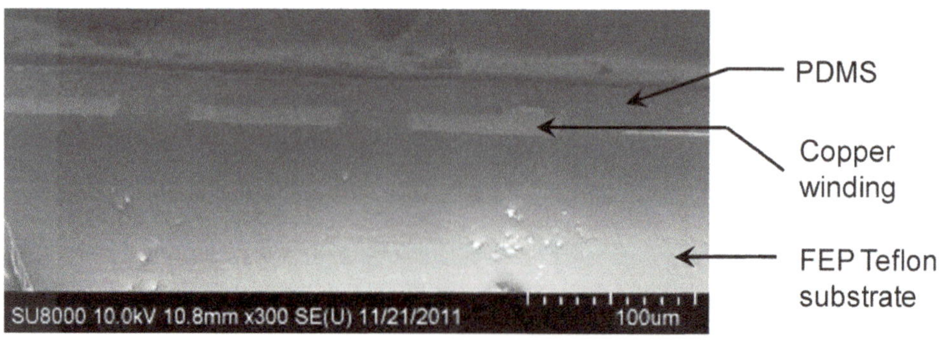

Figure 3: SEM view of the cross section of the PolyDiMethylSiloxane (PDMS) layer deposited on a Teflon antenna.

To investigate the impact of the PDMS thickness on the characteristics of the antenna, the thickness is progressively increased, with a path of 20 µm, and measurements of f_0 and Q factor are systematically made after each increment. The study is led on 3 groups of 2 or 3 exactly identical antennas. Good reproducibility is observed in each group. The size of the antenna (less than 1 cm²) is too small to keep uniform deposition with successive PDMS layers deposited by spincoating. Therefore, 20 µm-thick PDMS films are separately prepared on a FEP Teflon substrate which has not received any plasma treatment and thus has a low surface energy. Adhesion of these films to the FEP surface is very poor and so they may be easily transferred to the surface of the antenna. A surface activation by O_2 plasma (pressure 0.4 mbar, power 240 W, time 90 s in the etcher Pico UHP) between two increments is required.

Magnetic Resonance Imaging

MRI experiments are performed in Neurospin platform in CEA Saclay on a 7 Tesla horizontal bore magnet (Pharmascan, Brüker Biospin, Ettlingen, Germany) driven by a Paravision 5.1 console (Brüker, Ettlingen Germany) and equipped with a 720 mT/m actively shielded gradient device (100 µs rise time). The NMR signal is transferred from the implanted coil to the receiver channel of the MRI unit by inductive coupling via an external pick-up loop (Figure 4).

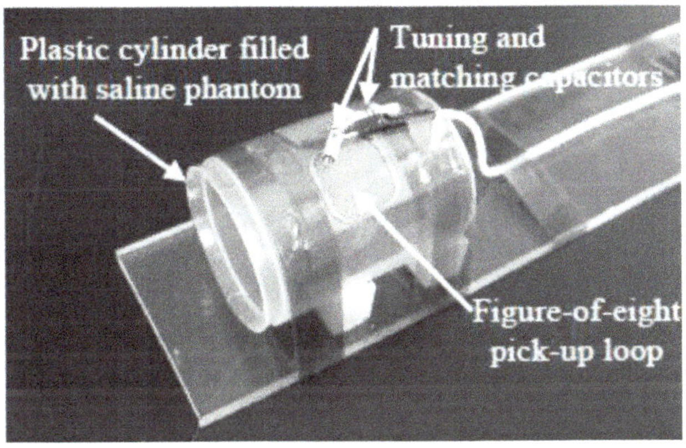

Figure 4: Experimental set-up of the figure-of-eight pick-up loop and the gel phantom for MRI.

The pick-up loop is based on the figure-of-eight design. It consists of two loops connected in series with opposite current flow direction, thus producing a magnetic field perpendicular to the pick-up loop plane. The external pick up loop is tuned using a non magnetic trimmer capacitor (0.8–8 pF) connected in at the pick-up loop terminal. Impedance matching to the 50 Ω input of the receiver channel is achieved by adjusting a non magnetic trimmer capacitor (0.8–8 pF) connected in series in between the tuning capacitor and the copper core of the coaxial cable.

The equivalent electrical circuit of the implanted coil inductively coupled to the external pick-up loop is shown in Figure 5. In practice, the mutual inductance between the implanted coil and the external pick-up loop is fixed. Tuning and matching are performed by iterative adjustments of the capacitors values and typical matching level of −30 dB are achieved. For phantom imaging, the pick-up loop is fixed on a plastic cylinder which contains the agar gel.

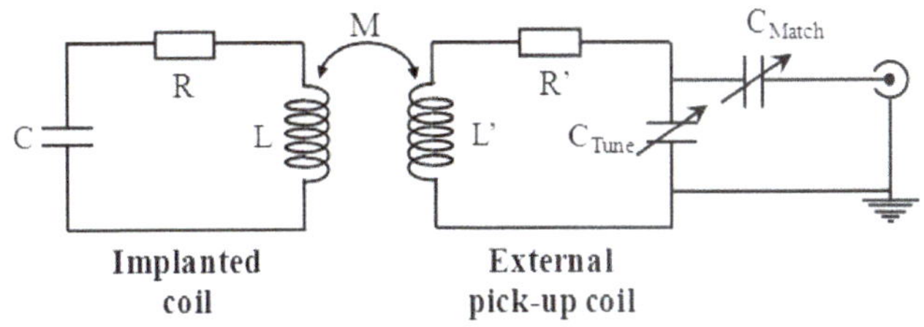

Figure 5: Equivalent electrical circuit of the implanted coil inductively coupled to the external pick-up coil. C_{Tune} and C_{Match} are the tuning capacitor and the matching capacitor, respectively, and consisted of (0.8-8 pF) non magnetic trimmer capacitor.

All *in vivo* experiments are conducted according to EEC guidelines and directives (86/09/EEC). An adult male Sprague-Dawley rat (300 g) is deeply anaesthetized with ketamine/xylazine (50/7.5 mg kg^{-1}, i.p.). The anaesthetized rat is immobilized in a stereotaxic apparatus. After sterilization, the microantenna is surgically implanted at the level of the interhemispheric cleft close to the cingulate cortices. The scalp is sutured before placing the anaesthetized animal inside a cradle apparatus. Animal respiration and temperature are monitored. Upon completion of the *in vivo*MRI experiments, the animal is euthanized and the antenna is retrieved.

RESULTS AND DISCUSSION

Multi-Turn Transmission Line Resonator Design

Multi-turn Transmission Line Resonator (MTLR) design [5] consists in conductor (copper) lines—concentric split-rings connected in series—deposited on both sides of a dielectric substrate (polymer film) as shown in Figure 6. The MTLR structures are self-resonant and achieve both the RF field emission for nuclear magnetic excitation with amplification, and then the amplification and detection of the resulting NMR signal without the need of any wire connection.

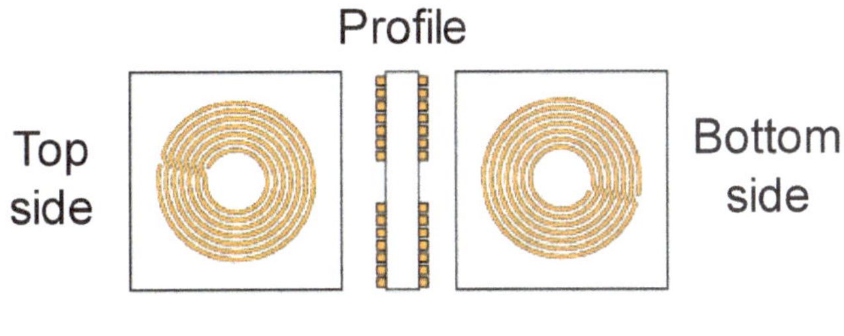

Figure 6: Multi-turn transmission line resonator.

MRI requires that the f_0 of the antenna match the Larmor frequency ν_0, given in Equation (1):

$$\nu_0=\frac{\gamma}{2\pi}B_0 \quad (1)$$

where B_0 is the static magnetic field and γ is the gyromagnetic ratio of the nucleus. The nucleus is generally the proton ^1H

for its abundance in the tissues (mainly composed of water) and its high sensitivity (γ of 42.576 MHz/Tesla). The Larmor frequency for ^{1}H with B_0 = 7 Tesla is 299 MHz. An analytical model has been developed for the MTLR design to define the geometrics parameters of the antennas, depending on the dielectric properties of the substrate and its thickness, to target a given f_0. The full description of the multi-turn transmission line design, including the behavioral analysis of the resonance condition and the formulae used for equivalent transmission line parameters determination can be found in [29]. We expect a decrease of f_0 value with the packaging, according to the permittivity value of PDMS, thus we target a wide f_0 range from 300 to 400 MHz. The geometric parameters are given in Table 1 for this frequency range and Fluorinated Ethylene Propylene (FEP), a variety of Teflon, as substrate. These micrometric dimensions imply to use microfabrication technologies in a clean room.

Table 1: Geometric parameters of the antennas defined for 7 Tesla MRI

Diameter (mm)	Number of turns	Width of the winding (μm)	Gap between two windings (μm)
5.4 to 6.0	6	88	40

Improvement of Polymer/Metal Adhesion by Plasma Treatment

The microantennae are fabricated using the copper micromoulding process described in the Experimental Section. The first step of this process is the deposition of a metallic seed layer by sputtering on the polymer film. As previously detailed, because of the low adhesion of metal to polymer surfaces, specific plasma treatments before metallization are needed. The efficiency of the plasma treatments have been assessed by measuring the Water Contact Angle (WCA) and adhesion properties with tensile strength measurements, for both untreated and treated samples. The study is led on FEP Teflon, our substrate of interest, and on Kapton which is widely used and well-known in our team. Kapton films are treated using O_2 plasma exposure in a reactor placed in the same clean room than the sputtering equipment, allowing the immediate subsequent metallization. The N_2/H_2 specific plasma treatment for Teflon is carried out in another laboratory that implies an incompressible ageing time of 3 h for the surface modifications before metallization. During the transportation, the samples are kept under vacuum to limit the evolution of surface modifications, *i.e.*, adsorption of atmospheric contaminants and short range reorganization of side chains. For this reason, we have studied the influence of the ageing time on the wettability of the surface and on the polymer/metal adhesion for Kapton and Teflon films.

Hydrophilic surfaces are considered to be more suitable to obtain a good adhesion. Figure 7 and Figure 8 show the WCA before and immediately after the plasma exposure for Kapton and Teflon films. In both cases, the WCA is decreased by the plasma treatment. For Kapton, it decreases from 72° to 10°, resulting in a highly hydrophilic surface which should provide a stronger adhesion of metal. For Teflon, the surface is very hydrophobic before the treatment (WCA of 107°). After treatment, the WCA is decreased to 63° which makes the film more likely to ensure a reliable adhesion of metal. The variation induced by the plasma exposure is due to the creation of polar groups on the surface which increases its wettability. In the case of Kapton in oxygen atmosphere, the amount of oxygenated moieties increases, while for Teflon in N_2/H_2 atmosphere fluorine atoms are removed by H radicals and replaced by amino moieties. Detailed characterization of the surface modifications by X-Ray Photoelectron Spectroscopy (for chemical aspects) and Atomic Force Microscopy (for topography aspects) are currently in progress.

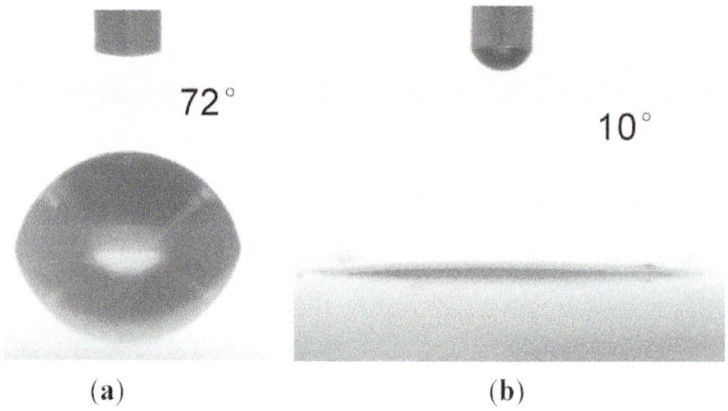

Figure 7: Water Contact Angle (WCA) before (**a**) and after (**b**) plasma treatment for a Kapton film.

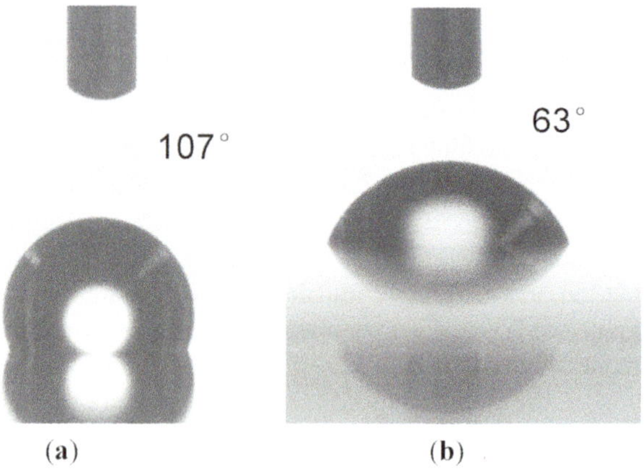

Figure 8: WCA before (**a**) and after (**b**) plasma treatment for a FEP Teflon film.

The evolution of these surface modifications is evaluated by WCA measurements for several days after the treatment (Figure 9). Measurements are also performed on untreated samples as a reference. Both Kapton and Teflon untreated samples show only a negligible variation around the initial value. However, the WCA of treated samples increases with time, especially in the first hours after the plasma exposure. After the first hours, the increase is slower, probably corresponding to the stabilization of the surface modifications. It reaches a value of 27° for Kapton after 7 days while the value for Teflon after 5 days is 76°. The WCA variation may be explained by rearrangement within the modified layer, that is through short-range reorientation of side chains [28] (Figure 10). In all cases, even several days after the treatment, the WCA for treated sample is still lower than that obtained for the untreated sample. The WCA is also still lower for Kapton than Teflon samples, both treated and untreated.

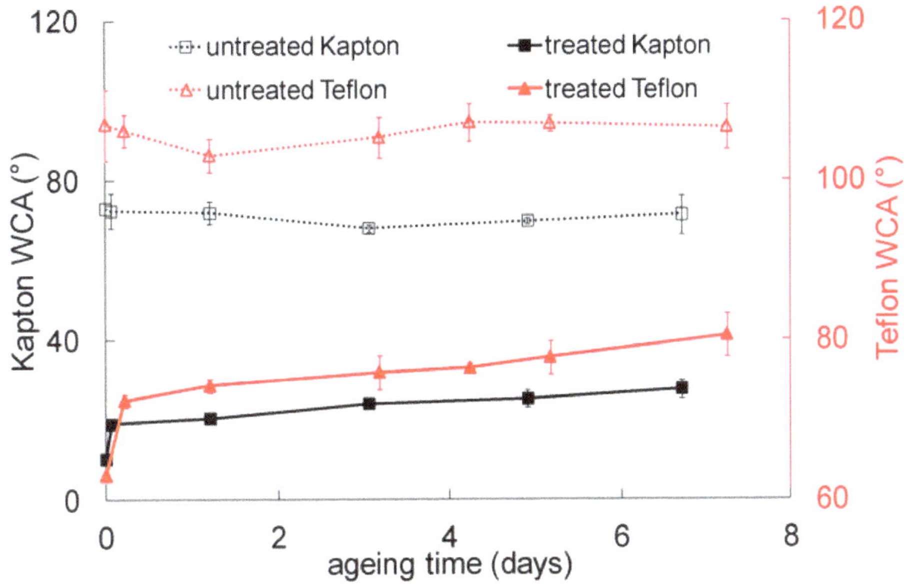

Figure 9: Evolution of WCA with time for untreated and treated Kapton and FEP Teflon films.

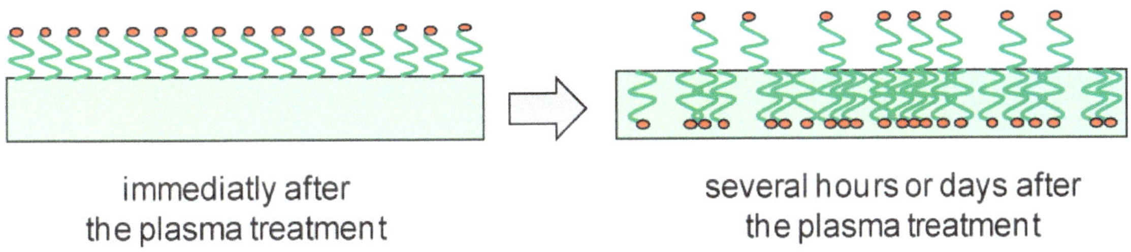

Figure 10: Scheme explaining ageing phenomenon due to short range reorientation of side chains.

To establish a correlation between the WCA and the polymer/metal adhesion, a thin metallic layer is deposited on the treated films after different ageing times: the shortest, *i.e.*, 3 h for Teflon and quasi immediately for Kapton, then 1 day and 5 days. Figure 11 shows the measured tensile strength compared with the ones measured for untreated samples. The tensile strength measured for untreated Teflon is lower than the one for untreated Kapton, in accordance with their WCA values. For immediate metallization of Kapton, the adhesion of the seed layer is increased by factor 3 by the plasma treatment, as it could be expected from the observed WCA decrease. When the ageing time is about few days, the tensile strength is slightly lower but still high compared to the untreated sample. The oxygen plasma exposure of Kapton provides quite stable surface modifications (if conserved under vacuum) which allows ageing times of several days without a critical loss of adhesion. For metallization of Teflon after an ageing time of 3hours, the adhesion of the seed layer is increased by factor 2 compared with the untreated film. Note that the tensile strength in this case is comparable with the one measured for the untreated Kapton sample. These observations are coherent with the WCA measurements. When the metallization is done 1 day after the treatment, the tensile strength shows only a small decrease. However, after 5 days, it becomes comparable to the value without any treatment. The plasma exposure of Teflon allows a significant improvement of the polymer/metal adhesion only for short ageing times about several hours (eventually a day). As the tensile strength and thus the adhesion is lower for Teflon (even with the plasma treatment) than Kapton substrates, it becomes crucial for a reliable fabrication process to limit the evolution of the surface modifications by keeping as short as possible the ageing time between the treatment and the metallization.

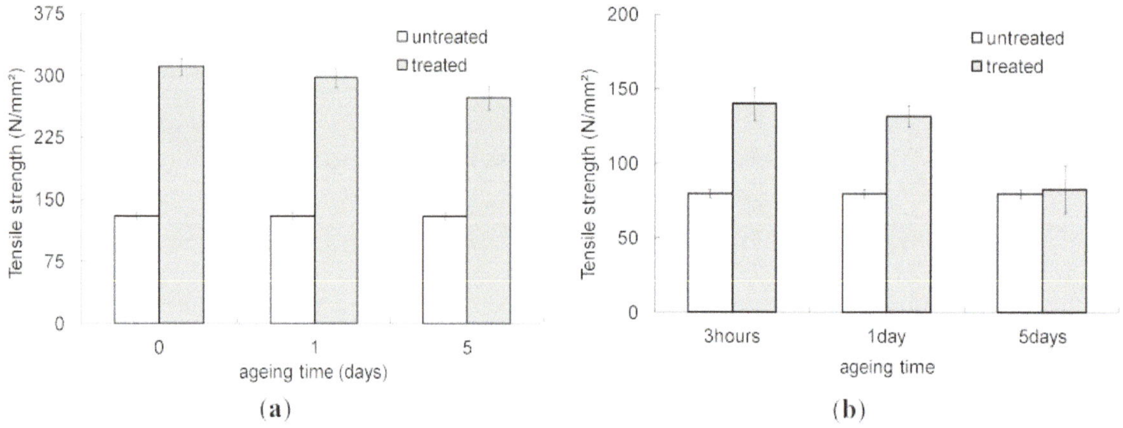

Figure 11: Evolution of tensile strength of the metallic seed layer deposited on untreated and treated Kapton (**a**) and FEP Teflon (**b**) films, with time.

A good correlation can be obtained between the measured tensile strength and the WCA values for Kapton and Teflon surfaces as shown in Figure 12.

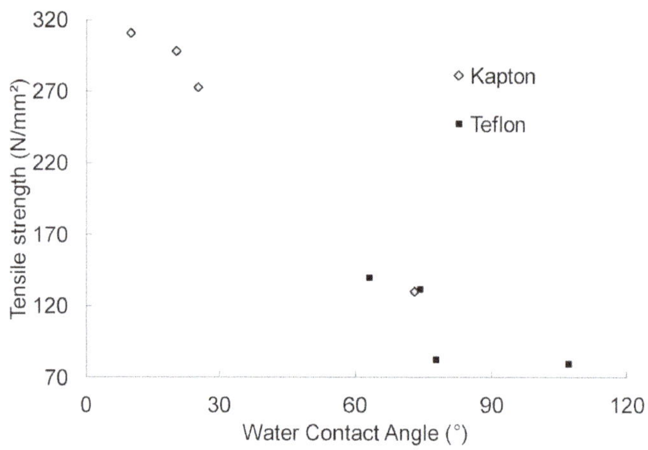

Figure 12: Correlation between the measured tensile strength and WCA values for Kapton and Teflon substrates.

Fabrication and Characterization of Microantennae on Polymer Substrates

The flexible microantennae fabricated on plasma treated FEP Teflon substrates using copper micromoulding process are shown in Figure 13. The characteristics f_0 and Q of the antennas are measured after fabrication without any packaging. The measured f_0 values are in good agreement with the predicted ones using the analytical model described in [29] (Figure 14), with a standard deviation of 3%, from the model. The measured Q values range from 100 to 120, which are excellent, as expected from the low loss tangent of the Teflon substrate.

Figure 13: Flexible microantennae fabricated on plasma treated FEP Teflon substrates.

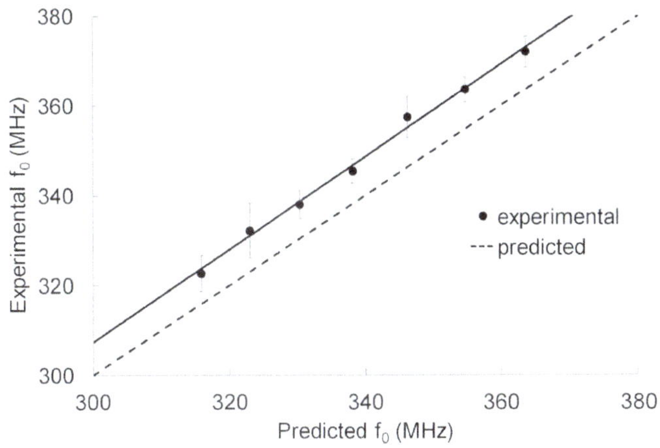

Figure 14: Experimental resonant frequencies *versus* predicted ones.

Optimization of the PDMS Packaging

To ensure biocompatibility of the antennas and to keep the same characteristics independently from the environment, they are coated with PDMS. To investigate the impact of the thickness of this coating both in air and in a conducting environment (saline agar gel phantom simulating the tissues), the PDMS thickness is progressively increased with a path of 20 µm, with thickness ranging from 30 to 400 µm. For each increment, measurements of $f_0$and Q are systematically performed in air and in the gel phantom. Preliminary results have shown that the f_0 value in air decreases with the thickness coating, as expected from the PDMS permittivity value (Figure 15). This decrease of f_0 is not linear and more important after deposition of 30 µm of PDMS, probably because a thin coating is sufficient to confine the most part of the electric field lines. When the electric field lines are totally confined (*i.e.*, thickness about 270 µm), the variation of f_0 in air becomes negligible. In the phantom, with the minimal PDMS thickness (30 µm), f_0 is only half of its value in air. Indeed, a dielectric coupling with the conducting material above the antenna occurs and shifts down the f_0 value. The dielectric coupling decreases with the distance between the top windings and the surface of the antenna, *i.e.*, the packaging thickness. This observation is in agreement with the observation of Bilgen *et al.* [4]. As for measurements in air, the first PDMS layers have a higher impact on the f_0 variation. Finally, the two curves converge to the same value, *i.e.*, 300 MHz, for the same thickness (about 270 µm). For higher PDMS thickness values, f_0 becomes independent from the packaging thickness and from the environment conductivity. Similar behavior was observed by Volland *et al.* with thicknesses of several millimeters on macroscopic antennas [30]. Through the knowledge of the maximal f_0 shift with coating, it is possible to predict the required value just after fabrication to reach the same f_0 value in air and in a conducting environment. Our results clearly indicate that a PDMS thickness higher than 270 µm is suitable to ensure the achievement of the frequency specification for 7 Tesla MRI.

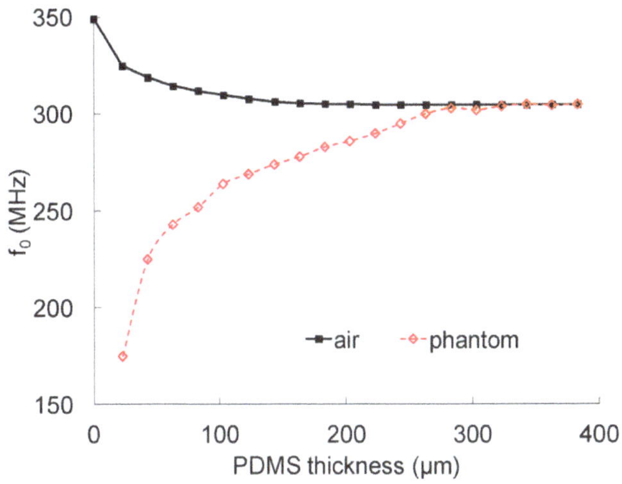

Figure 15: Variation of f_0 with PDMS thickness in air and in the gel phantom.

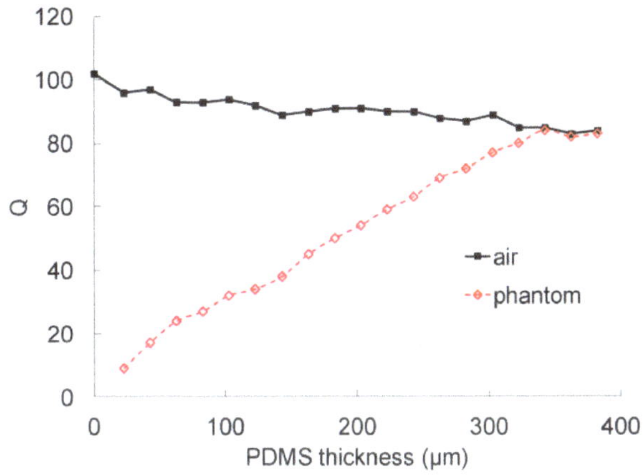

Figure 16: Variation of Q with PDMS thickness in air and in the gel phantom.

The Q value in air decreases with the PDMS thickness, in agreement with the variation of f_0 (Figure 16). In the phantom, when the coating is very thin (30 µm), the Q value is reduced by a factor of 10 compared with the one in air. A similar observation was made by Yung *et al.* [3]. It may be explained by the same reasons as for f_0. The Q value in the phantom quickly increases with the PDMS thickness until the same final value (about 80) as in air, for thicknesses higher than 350 µm. In these conditions, the Q value is totally suitable to obtain high quality images.

In Vivo *Magnetic Resonance Imaging*

After the optimization of the PDMS packaging and successful tests in the gel phantom, we decided to perform *in vivo* imaging of the rat brain. A microantenna with a thick PDMS packaging (about of 300 µm on each side) is implanted at the level of the interhemispheric cleft. An *in vivo* image of the rat brain obtained at 7 Tesla with this implanted microantenna is displayed Figure 17. A 3D FLASH acquisition sequence is used with an in-plane field of view of 2.56 × 1.92 cm^2 and a matrix of 192 × 92. The corresponding in-plane resolution is 100 x 100 μm^2. The entire data set is composed of 64 images (slice thickness of 200 μm) corresponding to an explored volume of 2.56 × 1.92 × 1.28 cm^3 that covers the entire rat brain. The echo time and repetition time are 11 ms and 50 ms, respectively. The total acquisition time is about 40 min.

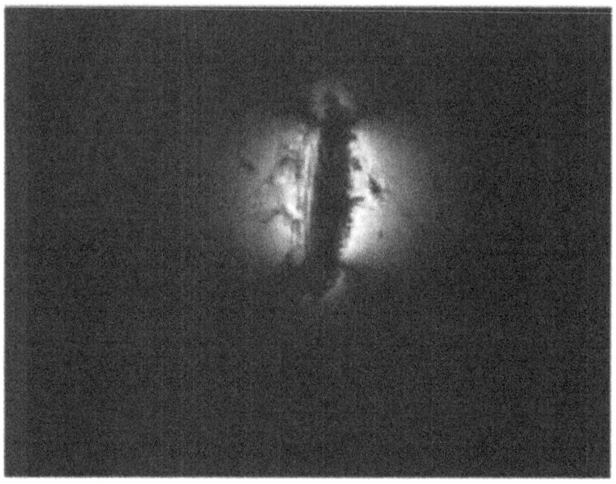

Figure 17: *In vivo* image of the rat brain acquired at 7 Tesla from an implanted microantenna with an optimized (300 μm-thick) PDMS packaging.

The brain image reveals a quite large (about 6 mm in diameter) observable volume with sufficiently high SNR. An average SNR of 80 is measured close to the microantenna. The high spatial resolution achieved in all directions allows the visualization of very fine cerebral structures. The realization of this image with a good SNR value constitutes an experimental validation of our miniaturization and PDMS packaging approaches.

CONCLUSIONS

Polymer thin films have interesting mechanical (flexibility), dielectric (permittivity and loss tangent) and interfacial (biocompatibility) properties which make them suitable both as substrate and as packaging material for implantable MRI microantennae. Dielectric properties of polymer substrates provide a solution to overcome the miniaturization limit, by the use of a monolithic and wireless design based on transmission lines between two conductor windings on both sides of a dielectric material (MTLR design). To benefit from this advantage, specific plasma treatments must be applied to improve the polymer/metal adhesion. We have evidenced the efficiency of O_2 and N_2/H_2 plasma exposure (respectively for Kapton and Teflon films) for this purpose. In the case of Teflon, the ageing time between the surface treatment and the metal deposition is crucial to ensure a reliable adhesion. Good correlation between the wettability (WCA) of the surface and the adhesion (tensile strength) has been established. Additional characterization of the treated surfaces by XPS and AFM techniques should provide a more complete understanding of the induced surface modifications and the observed correlation.

Dielectric properties of polymers offer a solution to the f_0 shift down and the Q degradation induced by the coupling effect in a conducting environment. With PDMS as packaging material, thickness has been optimized in air and in gel phantoms modeling the tissues to obtain characteristics independent from the electrical properties of the surrounding environment. The minimal thickness to achieve this condition is about 300 μm. *In vivo* MR images of the rat brain have been easily obtained at 7 Tesla from an implanted antenna with this packaging thickness. The spatial resolution is 100 × 100 × 200 μm^3 and the SNR is about 80, demonstrating the capability of the system for this application.

We aim to continue to reduce as much as possible the invasiveness of the implantation. The geometric parameters of the antenna are very large compared with the extreme resolution of microfabrication technologies and thus can be still reduced without process concerns. Furthermore, the packaging thickness can be also reduced by the use of other biocompatible materials with higher permittivity than the PDMS one. Surface modifications by plasma treatments could be applied to the packaging material in order to graft appropriate molecules on its surface and so to increase biocompatibility for long-term monitoring. Moreover, a rounded shape for the substrate around the coil that is both adequate to implantation and ergonomic for the surgeon is currently in development.

ACKNOWLEDGMENTS

This work is supported by the European project ISEULT, in the Neurospin platform of CEA Saclay. The authors would like to thank Sylvain Massey for contributiing in the plasma treatments during his post doctoral position in the Laboratoire du Génie des Procédés Plasmas et Traitements de Surface (LGPPTS, Chimie ParisTech) and Boucif Djemaï from the Institute of Biomedical Imaging (I2BM, Neurospin, CEA Saclay) for the surgical implantation of the antennas into the rat brain. A special thanks is also addressed to the staff of the University Technology Center (CTU) for technical expertise and cleanroom facilities.

REFERENCES

1. Polikov, V.S.; Tresco, P.A.; Reichert, W.M. Response of brain tissue to chronically implanted neural electrodes. *J. Neurosci. Methods* 2005, *148*, 1–18.
2. Martin, P.J.; Plewes, D.B.; Henkelman, R.M. MR imaging of blood vessels with an intravascular coil. *J. Mag. Res. Imag.* 1992, *2*, 421–429.
3. Yung, A.C.; Kolowski, P. Signal-to-noise ratio comparison of phased-array *vs.* implantable coil for rat spinal cord MRI. *Magn. Reson. Imaging* 2007, *25*, 1215–1221.
4. Bilgen, M. Magnetic resonance microscopy of spinal cord injury in mouse using a miniaturized implantable RF coil. *J. Neurosci. Methods* 2007, *159*, 93–97.
5. Serfaty, S.; Haziza, N.; Darrasse, L.; Kan, S. Multi-turn split-conductor transmission-line resonators. *Magnet. Reson. Med.* 1997, *38*, 687–689.
6. Renaud, L.; Armenean, M.; Berry, L.; Kleimann, P.; Morin, P.; Pitaval, M.; O'Brien, J.; Brunet, M.; Saint-Jalmes, H. Implantable planar RF microcoils for NMR microspectroscopy. *Sens. Actuat. A* 2002, *99*, 244–248.
7. Coutrot, A.-L.; Dufour-Gergam, E.; Quemper, J.-M.;

Martincic, E.; Gilles, J.-P.; Grandchamp, J.P.; Matlosz, M.; Sanchez, A.; Darrasse, L.; Ginefri, J.-C. Copper micromolding process for NMR microinductors realization.*Sens. Actuat. A* 2002, *99*, 49–54.

8. Uelzen, T.; Fandray, S.; Müller, J. Mechanical and electrical properties of electroplated copper for MR-imaging coils. *Microsyst. Technol.* 2006, *12*, 343–351.
9. Ellersiek, D.; Harms, S.; Casanova, F.; Blümich, B.; Mokwa, W.; Schnakenberg, U. Flexible RF microcoils with integrated capacitor for NMR applications. In Proceedings of MicroMechanics Europe (MME'05), Göteborg, Sweden, 4–6 September 2005; pp. 256–259.
10. Zhang, X.; Ugurbil, K.; Chen, W. Microstrip RF surface coil design for extremely high-field MRI and spectroscopy. *Magnet. Reson. Med.* 2001, *46*, 443–450.
11. Eroglu, S.; Gimi, B.; Roman, B.; Friedman, G.; Magin, R.L. NMR spiral surface microcoils: design, fabrication and imaging. *Concept Magnetic Res.* 2003, *17B*, 1–10.
12. Woytasik, M.; Ginefri, J.-C.; Raynaud, J.-S.; Poirier-Quinot, M.; Dufour-Gergam, E.; Grandchamp, J.-P.; Robert, P.; Gilles, J.-P.; Martincic, E.; Girard, O.; Darrasse, L. Characterization of flexible RF microcoils dedicated to local MRI. *Microsyst. Technol.* 2007, *13*, 1575–1580.
13. Ahmad, M.M.; Syms, R.R.A.; Young, I.R.; Mathew, B.; Casperz, W.; Taylor-Robinson, S.D.; Wadsworth, C.A.; Gedroyc, W.M.W. Catheter-based flexible microcoil RF detectors for internal magnetic resonance imaging. *J. Micromech. Microeng.* 2009, *20*, 074011.
14. Ellersiek, D.; Fassbender, H.; Bruners, P.; Pfeiffer, J.G.; Penzkofer, T.; Mahnken, A.H.; Schmitz-Rode, T.; Mokwa, W.; Schnakenberg, U. A monolithically fabricated flexible resonant circuit for catheter tracking in magnetic resonance imaging. *Sens. Actuat. B* 2009, *144*, 432–436.
15. Syms, R.R.A.; Young, I.R.; Ahmad, M.M.; Rea, M.; Wadsworth, C.A.; Taylor-Robinson, S.-D. Thin-film detector system for internal magnetic resonance imaging. *Sens.Actuat. A* 2010, *163*, 15–24.
16. Mata, A.; Fleischman, A.J.; Roy, S. Characterization of Polydimethylsiloxane (PDMS) properties for biomedical micro/nanosystems. *Biomed. Microdev.* 2005, *7*, 281–293.
17. Egitto, F.D.; Matienzo, L.J. Plasma modification of polymer surfaces for adhesion improvement. *IBM J. Res. Develop.* 1994, *38*, 423–439.
18. Brewis, D.M.; Dahm, R.H. *Adhesion to Fluoropolymers*; Rapra Technology Limited: Shrewsbury, UK, 2006; Volume 16, pp. 17–23.
19. Sarra-Bournet, C.; Ayotte, G.; Turgeon, S.; Massines, F.; Laroche, G. Effects of chemical composition and the addition of H2 in a N2 atmospheric pressure dielectric barrier discharge on polymer surface functionalization. *Langmuir* 2009, *25*, 9432–9440.
20. Woytasik, M.; Grandchamp, J.-P.; Dufour-Gerfam, E.; Gilles, J.-P.; Megherbi, S.; Martincic, E.; Mathias, H.; Crozat, P. Two-dimensional microcoil fabrication process for three-axis magnetic sensors on flexible substrates.*Sens. Actuat. A* 2006, *132*, 2–7.
21. Tatoulian, M.; Shahidzadeh, N.; Arefi-Khonsari, F.; Amouroux, J. Comparison of the efficiency of N2 and NH3 plasma treatments to improve the adhesion of PP films to in-situ deposited Aluminium coatings. Study of ageing phenomena in terms of acid-base properties. *Int. J. Adhesion Adhesives* 1995, *15*, 177–184.
22. Neubauer, E.; Korb, G.; Eisenmenger-Sittner, C.; Bangert, H.; Chotikaprakahn, S.; Dietzel, D.; Mansanares, A.M.; Bein, B.K. The influence of mechanical adhesion of copper coatings on carbon surfaces on the interfacial thermal contact resistance. *Thin Solid Films* 2003, *433*, 160–165.
23. Jeon, B.J.; Lee, S.; Lee, J.K. Adhesion characteristics of copper thin films deposited on PET substrate by electron cyclotron resonance-metal organic chemical vapor deposition. *Surf. Coat. Tech.* 2008, *202*, 1839–1846.
24. Ginefri, J.-C.; Durand, E.; Darrasse, L. Quick measurement of nuclear magnetic resonance coil sensitivity with a single-loop probe. *Rev. Sci. Instrum.* 1999, *70*, 4730–4731.
25. Darrasse, L.; Kassab, G. Quick measurement of NMR-coil sensitivity with a dual-loop probe. *Rev. Sci. Instrum.*1993, *64*, 1841–1844.
26. Chen, Q. Investigation of corona charge stability mechanisms in polytetrafluoroethylene (PTFE) teflon films after plasma treatment. *J. Electrostat.* 2003, *59*, 3–13.
27. Tatoulian, M.; Arefi-Khonsari, F.; Mabille-Rouger, I.; Amouroux, J.; Gheorgiu, M.; Bouchier, D. Role of helium plasma pre-treatment in the stability of the wettability, adhesion and mechanical properties of ammonia plasma-treated polymers: Application to the Al-polypropylene system. *J. Adhesion Sci. Technol.* 1995, *9*, 923–934.
28. Morra, M.; Occhiello, E.; Garbassi, F. Dynamics of Plasma Treatred Polymer Surfaces: Mechanisms and Effects. In Proceedings of the 1st International Conference of Polymer Solid Interfaces, Namur, Belgium, 2–6 September 1991; pp. 407–428.
29. Megherbi, S.; Ginefri, J.-C.; Darrasse, L.; Raynaud, G.; Pône, J.-F. Behavioral Vhdl-Ams model and experimental validation of a Nuclear Magnetic Resonance sensor. *Microsyst. Technol.* 2005, *12*, 38–43.
30. Volland, N.A.; Mareci, T.H.; Constantinidis, I.; Simpson, N.E. Development of an inductively coupled MR coil system for imaging and spectroscopic analysis of an implantable bioartificial construct at 11.1T. *Magnet. Reson. Med.* 2010, *63*, 998–1006.

CHAPTER 21

A MAGNETORESISTIVE TACTILE SENSOR FOR HARSH ENVIRONMENT APPLICATIONS

Ahmed Alfadhel[1], Mohammed Asadullah Khan[1], Susana Cardoso[2,3], Diana Leitao[2,3] and Jürgen Kosel[1]

[1]Computer, Electrical and Mathematical Sciences and Engineering Division (CEMSE), King Abdullah University of Science and Technology (KAUST), Thuwal 23955-6900, Saudi Arabia

[2]INESC-Microsystems and Nanotechnologies (INESC-MN), Rua Alves Redol, 9, Lisbon 1000-029, Portugal

[3]Instituto Superior Técnico IST, Physics Department, Universidade de Lisboa, Lisbon 1049-001, Portugal

ABSTRACT

A magnetoresistive tactile sensor is reported, which is capable of working in high temperatures up to 140 °C. Hair-like bioinspired structures, known as cilia, made out of permanent magnetic nanocomposite material on top of spin-valve giant magnetoresistive (GMR) sensors are used for tactile sensing at high temperatures. The magnetic nanocomposite, consisting of iron nanowires incorporated into the polymer polydimethylsiloxane (PDMS), is very flexible, biocompatible, has high remanence, and is also resilient to antagonistic sensing ambient. When the cilia come in contact with a surface, they deflect in compliance with the surface topology. This yields a change of the GMR sensor signal, enabling the detection of extremely fine features. The spin-valve is covered with a passivation layer, which enables adequate performance in spite of harsh environmental conditions, as demonstrated in this paper for high temperature.

INTRODUCTION

Advances in computing and artificial intelligence have resulted in increasingly capable robotic systems that have the ability of delivering revolutionary advances in the fields of healthcare, remote search and rescue operations, deep sea exploration, mining, *etc*. However, in order to enable these applications, the robots need a plethora of precise sensors, among which tactical sensors with excellent spatial and pressure resolution are of particular importance [1,2,3,4]. In addition to robots, tactical sensors can enhance, for example, prosthetic limbs, artificial skin, and surgical instruments [5,6].

Various tactile pressure sensors have been presented in literature that use different principles for transduction. For example, [7] presents a self-powered tactile sensor. This sensor consists of a polymer polydimethylsiloxane (PDMS)/ Indium Tin Oxide (ITO) textured micro-pyramid sheet and utilizes triboelectric charging to generate a voltage signal that is proportional to the applied stress. This sensor is capable of 400 Pa resolution over a range of 0 to 7.3 kPa. In [8], the deformation of a pyramidal textured PDMS membrane deformed by tactile input is used to scatter light in an acrylic waveguide, which is then captured by a camera and used to determine the stress on the PDMS membrane. This sensor is capable of registering touches as light as 1 kPa over a range of −60 kPa to 60 kPa. A resistive tactile sensor, in which the change in resistance of a Molybdenum Sulfide layer deposited

Citation: Alfadhel, A.; Khan, M.A.; Cardoso, S.; Leitao, D.; Kosel, J. A Magnetoresistive Tactile Sensor for Harsh Environment Applications. Sensors 2016, 16, 650, doi:10.3390/s16050650

on a SU-8 substrate serves as the transduction mechanism, is reported in [9]. This sensor operates linearly between 0 to 100 kPa with a minimum detection threshold of 1.24 kPa.

Polymeric artificial cilia on a magnetic sensor have been presented previously [10]. Polypyrole cilia have been fabricated on top of giant magnetoresistive (GMR) sensors and were tagged with a 300 nm $Co_{50}Fe_{50}$ by sputter deposition. The sensor was tested by agitation at different frequencies and measuring the response due to the cilia bending. The proposed design suffers from several issues such as the easy corrosion of the magnetic tagging, the toxicity of Co, the low stray field expected from the magnetic tagging, and the non-uniform tagging that lead to having cilia with different magnetic and mechanical properties.

However, as capable as the above sensors are, they have not been tested in harsh environmental conditions. Thus, their efficacy, for instance, in high temperature environments is unproven, which is quite important because some tasks, such as fire rescue operations or industrial work, need tactile sensors that can work in adverse conditions. There are conventional pressure sensors that have been tested at high temperatures, however, they are not suitable for tactile applications. For example, [11] uses a low-temperature cofireable ceramic (LTCC) process to make a wireless passive pressure sensor that can work at temperatures up to 400 °C with linear sensitivity between 0 and 700 kPa. Being made of rigid ceramic it is not conformal and its resolution of 2.4 kPa is more suited for high pressure applications than for sensing touch. A high temperature pressure sensor, which uses a SiC membrane and capacitive transduction to achieve a resolution of 5.2 kPa in the range 147–235 kPa, is demonstrated in [12]. While capable of operating at temperatures up to 400 °C, its sensing range and resolution are not good enough to meet the requirement for typical tactile applications. A piezoresistive pressure sensor made on silicon on insulator (SoI) substrate that operates at temperatures up to 500 °C with an ability to resolve pressures up to 20 kPa is reported in [13]. It is not suited for large area applications like artificial skin due to the high cost of the SoI substrate and also the resolution is not suited for sensing gentle touches and textures.

Polymer composites have gained attention as potential candidates for sensing and energy harvesting as they offer various benefits such as ease of manufacture, tunable properties, mechanical flexibility, low cost, low toxicity, and conformality. Depending upon the doping material and the polymer itself, the composite can be conductive, piezoelectric, triboelectric, magnetic, *etc.* [14,15,16,17,18,19,20,21,22,23]. Composites have been made using super paramagnetic nanobeads for various applications. However, since superparamagnetic particles have very low remanent magnetization, they require an externally generated magnetic field for operation, which greatly reduces power efficiency and portability.

For our tactical sensor we use a magnetic nanocomposite comprising of the polymer PDMS and iron nanowires (NWs). PDMS has very low stiffness, is chemically resistant, and is biocompatible. Due to the large shape anisotropy, iron NWs exhibit high remanent magnetization. Thus, this magnetic nanocomposite exhibits large remanent magnetization and flexibility. Iron NWs are also not cytotoxic [24] and thus the composite is suitable for medical applications. Typical electrodeposited iron NWs are polycrystalline which can be oxidized if exposed to the environment, reducing the magnetization, and hence, degrading the sensor's performance. This has been studied previously via corroding polycrystalline iron NWs by oxidizing them through annealing at 150 °C [24]. The formation of a magnetite shell around an iron core was observed, which happened by oxygen diffusion between the grain boundaries of iron NWs, starting from the surface of the NWs. The iron core's diameter decreased as the oxide shell thickness increased until the NWs were fully oxidized, yielding a polycrystalline magnetite structure. As a result, only 40% of the magnetization was maintained by the magnetite NWs [24]. In this work, the issue of corrosion is addressed by embedding the polycrystalline iron NWs inside of a polymer, which provides a high level of protection.

The reported sensor consists of a 3 × 3 array of 1 mm long, 200 μm wide cilia made up of the magnetic nanocomposite mounted on top of a 37 × 16 array of spin-valve elements connected in series (total of 592 elements). It occupies an area of 1 mm × 1 mm. Due to the low Young's modulus and large aspect ratio, the cilia are sensitive to both in-plane as well as out of plane mechanical forces. This causes them to respond with significant deflection to even the lightest of touch. This sensor is also capable of vibration and fluid flow detection. As the electrical connections to the spin-valve elements are not affected by the mechanical forces applied, the sensor is robust. Unlike conventional capacitive sensors, this sensor can be used both in dry and wet conditions. It is also capable of studying both dynamic and static forces.

This paper builds upon a previously reported tactile sensor [17,18] and flow sensor [20], which uses a similar transduction mechanism but with a giant magnetoimpedance (GMI) sensor to detect the stray field of the cilia. The GMI sensor needs a high frequency signal for operation, which can prove to be an impediment for certain use case scenarios. In this work, spin-valve GMR sensors are utilized as they can be operated using direct current (DC) signals. This in turn simplifies the circuitry and instrumentation needed for the measurement. The effect of temperature on the performance of the spin-valve elements and the nanocomposite is evaluated and the overall ability of the tactile sensor to withstand high temperature conditions is gauged.

MATERIALS AND METHODS

Tactile Sensing Concept

The magnetic tactile sensing concept utilizes permanent magnetic and highly elastic nanocomposite cilia integrated on a magnetic sensing element. A magnetic sensor detects the change of the magnetic stray field, created by the permanent magnetic cilia, when deflected by an external force (Figure 1). A spin-valve GMR sensor, which is simple to fabricate and can be applied onto flexible substrates, is utilized to measure the change in the magnetic field [25,26,27]. When the cilia are standing in the rest position, their stray field biases the GMR sensor. In the presence of an external force the cilia bend, resulting in a change of the stray field; hence, changing the sensor's resistance.

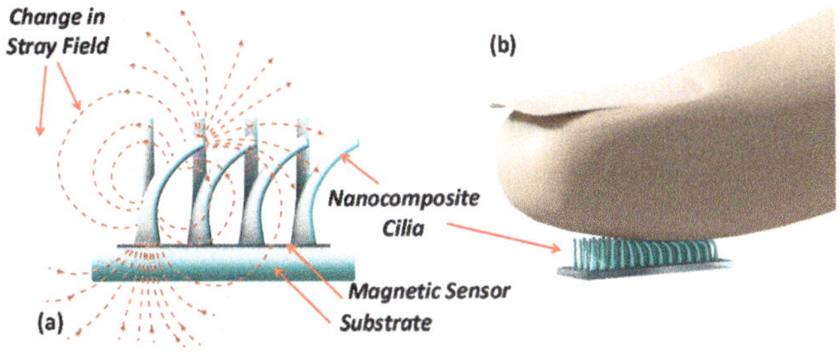

Figure 1: (**a**) Illustration of the tactile sensor working principle. In the standing position, the cilia's stray magnetic field affects the spin-valve giant magnetoresistive (GMR) sensor and biases it to an initial resistance value. In the presence of an external force, the cilia bend and the stray field measured with the spin-valve changes, resulting in a change in the resistance; (**b**) Example of the cilia deflection mechanism when touched with a finger.

The tactile sensor consists of 3 × 3 cilia applied on top of a spin-valve array connected in series forming an active area of 1 × 1 mm². The effective pressure is quantified by calculating the ratio of the applied forces and the active area. Each spin-valve is 60 μm long and 3 μm wide with sensitivity along the width direction.

The artificial cilia deflection due to an external force can be described by the Euler–Bernoulli beam theory (classical beam theory) [20,28]. Using the displacement-force relationship of a cylindrical beam, the displacement δ can be quantified when applying a force F at the free end of a beam with length l, Young's modulus E, and diameter D, such as:

$$\delta = \vec{F}\frac{64l^3}{3\pi ED^4} \tag{1}$$

Fabrication

Iron NWs were fabricated by the electrodeposition into a nanoporous aluminum oxide membrane, a well-known technology to fabricate dense array with high aspect ratio NWs (Figure 2a). The aluminum oxide membrane with dense hexagonally patterned nanopores was prepared by a two-step anodization process, which consisted of an electrochemical oxidation of aluminum substrate using oxalic acid and with controlled temperature and electrical potential [29]. After electrodepositing NWs into the membranes, they were released and dispersed in ethanol. Six micrometer long iron NWs with a diameter of 35 nm were fabricated to act as nanoscale permanent magnets. The dimensions and morphology of released iron NWs were investigated using scanning electron microscopy (Figure 2b). X-ray Diffraction (XRD) experiments were carried out, using a Bruker D8 Discover high-resolution X-ray diffractometer system that provides X-ray radiation with wavelength of 1.5406 Å, to determine the NWs composition and crystal structure. The XRD spectrum in Figure 2c reveals that the NWs are polycrystalline iron (Fe) and magnetite (Fe_3O_4), which is growing as a thin shell around the NWs. Using various characterizations, the same type of NWs has been studied before confirming the iron/magnetite core-shell structure [24].

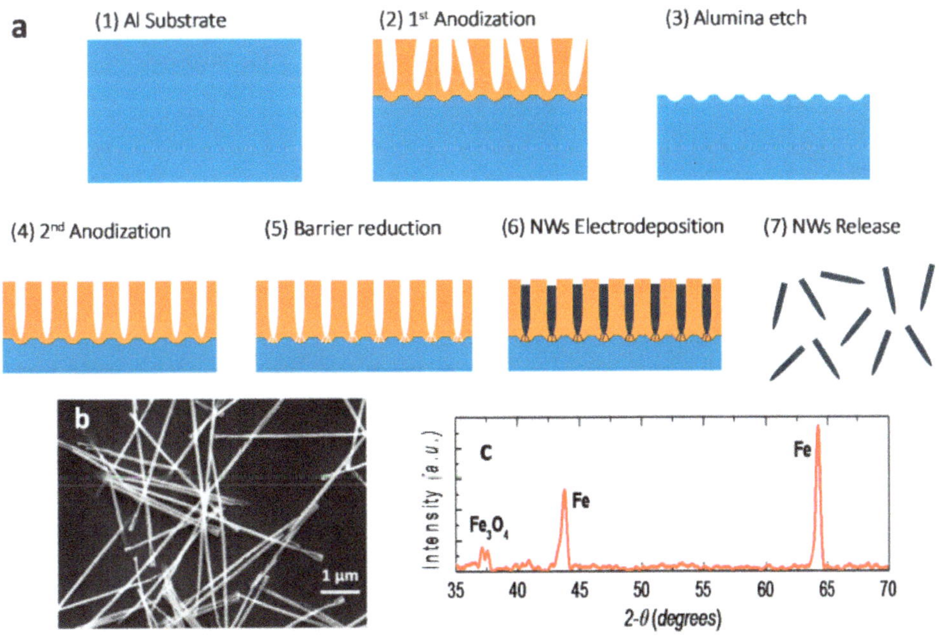

Figure 2: (**a**) Illustration of the electrochemical fabrication process of iron nanowires (NWs). (1) A nanoporous template is formed using an aluminum substrate. (2–4) The aluminum is anodized to create nanoporous alumina in two steps, with an alumina etching step in between to obtain uniform nanopores of ~35 nm in diameter. (5) A barrier reduction step establishes an electrical contact with the bottom electrode. (6) Iron is electrodeposited into the nanoporous membrane. (7) The NWs are released by etching the alumina membrane and dispersed in ethanol; (**b**) Scanning electron microscopy image of released NWs; (**c**) X-ray Diffraction (XRD) result of iron NWs.

The magnetic nanocomposite was synthesized by washing dispersed iron NWs with Sodium dodecyl sulfate (SDS) surfactant, and then mixing them with PDMS (Sylgard 184 Silicone Elastomer, Dow Corning Corporation, Midland, MI, USA). The nanocomposite was then cured, whereby the NWs concentration can be adjusted easily. A NWs to PDMS volume ratio of 14% was chosen to reach a high magnetization of the nanocomposite while avoiding adverse effects on the polymerization process of the PDMS or the elasticity of the cilia.

The magnetic sensing element is a spin-valve sensor, microfabricated on a 150 mm diameter Si (100) wafer passivated with 60 nm of Al_2O_3 deposited by RF magnetron sputtering. Top pinned spin valve sensors were fabricated by ion beam deposition (Nordiko3600 system, from Nordiko Tech. Services, Hampshire, England) with the following structure (thickness in nm): substrate/Ta(1.0)/$Ni_{80}Fe_{20}$(2.8)/$Co_{90}Fe_{10}$(2.5)/Cu(2.1)/$Co_{90}Fe_{10}$(2.3)/$Mn_{78}Ir_{22}$(7.0)/Ta(10.0) [30] (compositions in at%). A 5 mT magnetic field was applied during deposition to induce the free layer and pinned layer easy axis with parallel anisotropies. The sensors were defined by direct write laser lithography (DWL) and by ion milling (Nordiko 3000 system). The electrical leads were patterned by DWL in a 2-contact geometry, and consist of 300 nm sputtered Al ($Al_{98.5}Si_1Cu_{0.5}$ in at%) and 15 nm of TiW(N_2) films defined by liftoff. The sensors consist of arrays of 592 rectangular spin-valve elements (each with a sensing area of 3 μm × 60 μm) connected in series and occupying a 1 × 1 mm^2 total area. The geometry was optimized to minimize the sensor signal-to-noise level [31], because of the total volume of magnetic material, while maintaining a well-defined linearity in the transfer curve [32]. The sensor was passivated with sputtered Al_2O_3 (200 nm) and SiO_2 (200 nm) deposited by magnetron sputtering (final passivation layer) to protect against corrosion and to enable the chemical bonding with the PDMS [33]. The fabrication details are illustrated in Figure 3.

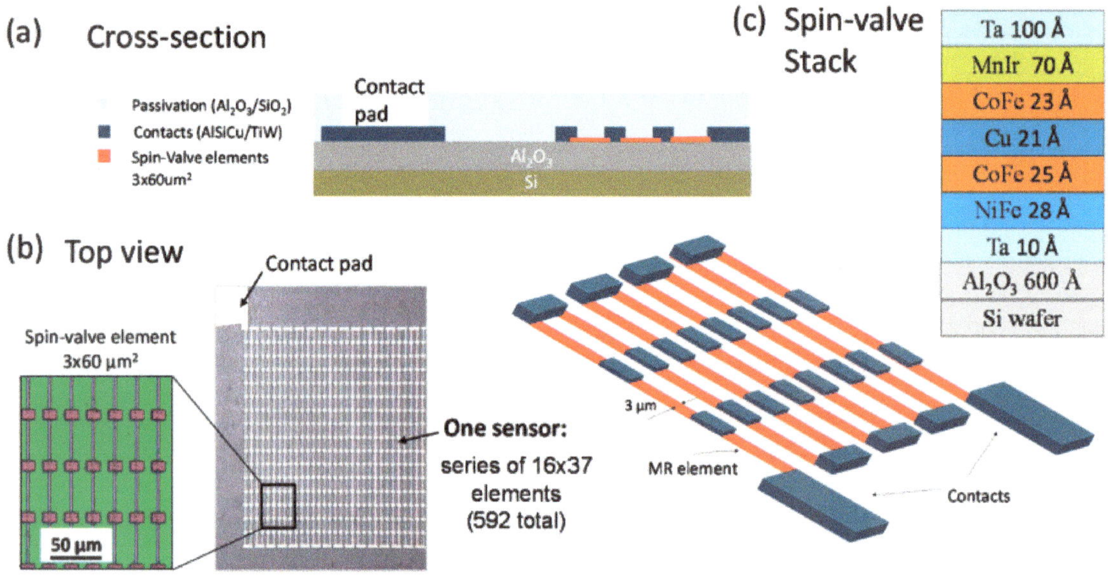

Figure 3: GMR sensor layout and geometry: (**a**) Cross section of the spin-valve sensor array connected in series; and (**b**) Top view of the samples microfabricated; and (**c**) the spin-valve thin films stack.

A 5 μm PDMS layer was spun on top of the magnetic sensor to provide electrical isolation and, at the same time, enhance the adhesion of the cilia to the spin-valve. A master mold technique was utilized to pattern the cilia. The master mold is a 1 mm thick poly(methylmethcrylate) (PMMA) with 200 μm in diameter 3 × 3 array of holes patterned using a CO_2 laser cutter (Universal PLS6.75). The nanocomposite is casted onto the surface of the thin PDMS/spin-valve stack, and the master mold is mounted on top. A uniform magnetic field along the cilia is applied using a C-shaped magnet. This structure is then placed in a desiccator for 30 min to remove any trapped air bubbles and assist in filling the pores. Next, the composite is cured at 90 degrees Celsius for one hour, forming the cilia with aligned NWs on top of the spin-valve. After releasing the cured cilia, they are annealed at 150 °C to increase the Young's modulus and to withstand high temperature operation without significant variation in the material's elasticity, and hence, the operating range. The cilia are then fully magnetized by applying a magnetic field of 10 kOe. The fabrication process is illustrated in Figure 4.

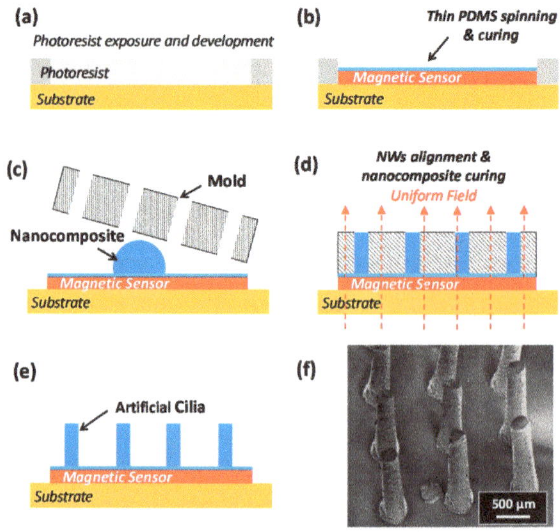

Figure 4: Illustration of the tactile sensor fabrication process. (**a**,**b**) Spin-valves were fabricated using standard lithography process, ion beam deposition and ion milling on a silicon substrate. A thin

polydimethylsiloxane (PDMS) layer was then spun and cured; (**b–e**) A poly(methylmethcrylate) (PMMA) mold technique was used for the fabrication of nanocomposite cilia. A homogeneous magnetic field was applied during the curing process to align the NWs. The mold is peeled off after curing the nanocomposite resulting in cilia mounted on the magnetic sensor; (**f**) Scanning Electron Microscopy image of the cilia array, where each cilium is 1 mm in length and 200 μm in diameter.

Characterization

The magnetic properties of the nanocomposite cilia were studied to confirm the permanent magnetic behavior. The magnetization loops along the cilia in the parallel and perpendicular directions were obtained using a vibrating sample magnetometer (VSM) (PMC MicroMag 3900, Westerville, OH, USA, and superconducting quantum interference device (SQUID) VSM 1500-100 by Quantum design Inc., San Diego, CA, USA).

In order to characterize the spin-valves array, a Helmholtz coil was used to apply a homogeneous magnetic field along the sensitive and non-sensitive directions, and the resistance was measured directly with a source meter (Keithley 2400) at a DC current of 100 μA.

The high elasticity of the nanocomposite is crucial to obtain a high sensitivity with the tactile sensor. Therefore, a study was conducted to investigate if adding NWs to the PDMS affects the elasticity. The Young's modulus of the nanocomposite cilia was obtained by fixing them between the holder plate and the moveable head of an extensometer tensile/compressive force testing system (INSTRON 5966). The system can precisely determine the exerted force and the vertical displacement. By using a 10 N load cell and three cilia that are 1 mm long and 200 μm in diameter, resolutions of 5 mN and 16 μm were obtained for the tensile force and vertical displacement, respectively. Tensile forces were applied to the cilia until achieving a strain of 1.3 (130% stretching). The Young's modulus was calculated as the ratio between stress and strain (slope) in the elastic region.

The performance of the tactile sensor was studied by applying vertical forces at different temperatures while measuring the sensor response. The forces were applied using the computer controlled force testing system, which can precisely determine the amount of force applied to the sensors. The system comes with an oven that allows setting the ambient temperature during the experiment. The resistance variation upon the deflection of the cilia was obtained directly using a source meter (Keithley 2400) that can apply a fixed current value throughout the measurement. The experimental setup is illustrated in Figure 5.

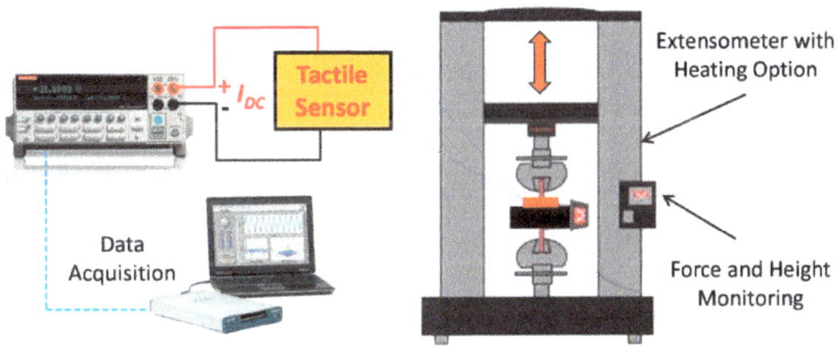

Figure 5: Experimental setup for the tactile sensor testing.

RESULTS AND DISCUSSION

Nanocomposite Characterization

The magnetic properties of the nanocomposite cilia were obtained by measuring the magnetizations loops shown in Figure 6. The cilia with NWs aligned along the parallel direction had a magnetic anisotropy with a saturation magnetization (*Ms*) of 11.7 memu, and remanence to saturation magnetization (*Mr/Ms*) of 93% and 19% along the parallel and perpendicular direction, respectively. The coercivity along the parallel direction was found to be 2.1 kOe, and coercivity along the perpendicular direction was 1.6 kOe. These results confirm both the permanent magnetic behavior of the nanocomposite and the alignment of the NWs.

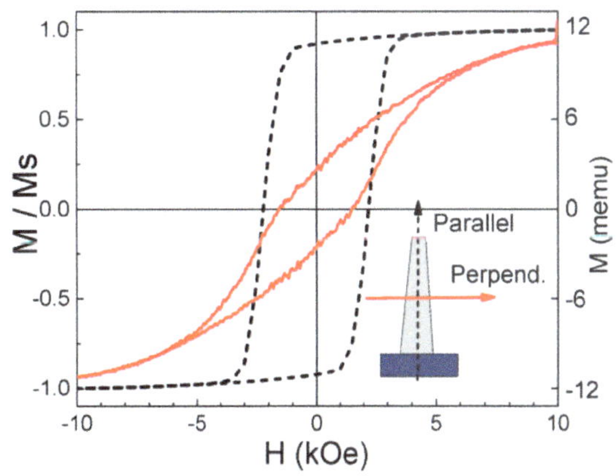

Figure 6: Nanocomposite magnetization loops obtained for parallel and perpendicular direction of a cilia array. The NWs in the cilia are aligned in the parallel direction.

NWs exhibit very promising magnetic properties, suggesting that they could be used as permanent magnets that can operate at high temperatures. Through their strong shape anisotropy, NWs exhibit a large temperature independent coercivity when operating below the Curie temperature, which is 770 °C for iron [34]. Hence, the high magnetization of iron can be utilized while remaining below the Curie temperature. On the other hand, we have shown that the polycrystalline NWs are prone to corrosion [24]. In order to elucidate the thermal stability of the nanocomposite, where the NWs are protected by the polymer, we conducted a temperature dependent study of the magnetization. The hysteresis loops of a 1 mm long and 200 μm in diameter cilia were obtained at a temperature range between 27 °C and 290 °C using SQUID VSM. The cilia were heated in a vacuum oven during the magnetization measurement, and the hysteresis loops were obtained for each temperature value. Within this temperature range, the permanent magnetic behavior of the nanocomposite was maintained, as shown in Figure 7, where only a slight magnetization reduction was observed as a result of Curie's law. The nanocomposite was then cooled down to 27 °C and the magnetization loop was measured again revealing a reduction in the remanence magnetization *Mr* of 16% (compared to the initial remanence magnetization, Mr_0, at 27 °C) with an increase in coercivity compared to the initial state. The reduction in remanence can be caused by trapped water molecules in the PDMS that resulted in slightly oxidizing the NWs in the nanocomposite with the heat assisted oxygen diffusion process, forming a thin shell of magnetite around the NWs. The magnetite shell reduces the diameter of the iron core in the NWs, increasing its coercivity, due to the increased aspect ratio.

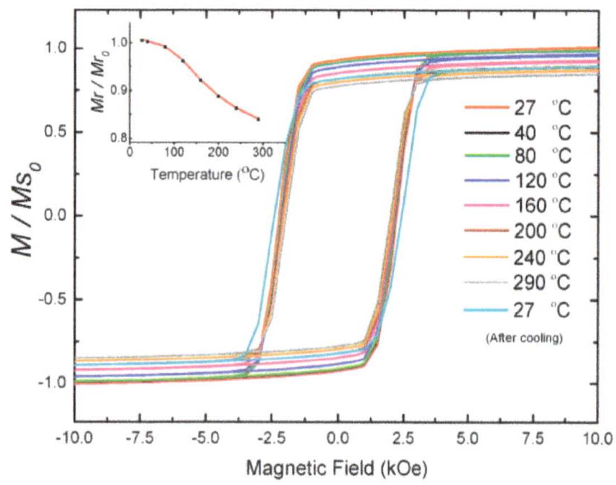

Figure 7: Magnetization loops of nanocomposite cilia at different temperatures from 27 °C to 290 °C, showing the magnetization *M* normalized to the initial saturation magnetization (Ms_0) at 27 °C. Inset: remanence magnetization (*Mr*) as function of temperature normalized to initial remanence magnetization (Mr_0) at 27 °C.

As can be seen from Equation (1), the Young's Modulus affects the sensor signal, since a change in the Modulus changes the cilia deflection. Hence, we investigated the effect of NWs on the nanocomposite's stiffness. The results of the tensile/compressive force tests show that nanocomposite cilia have a Young's modulus of 1.1 MPa, compared to the pure PDMS cilia that have a Young's modulus of 1.04 MPa. Thus, incorporating NWs in the nanocomposite does not have a significant effect on the rigidity of the cilia, and the advantage of the high elasticity of the PDMS is maintained.

GMR Sensor Response and Thermal Stability

The spin-valve GMR sensor array showed a linear response to a magnetic field between 0 and 23 Oe with sensitivities of 57 Ω/Oe and 4.8 Ω/Oe in the sensitive and non-sensitive directions, respectively (Figure 8). The average magnetoresistive (MR) sensitivity was 0.08%/Oe and 0.01%/Oe in the sensitive (along the width of the GMR elements) and non-sensitive directions (along the length of the GMR elements), respectively, when considering the sensor's initial resistance at room temperature, which is 62 kΩ. This anisotropic behavior equips the cilia sensor array with a directional sensitivity.

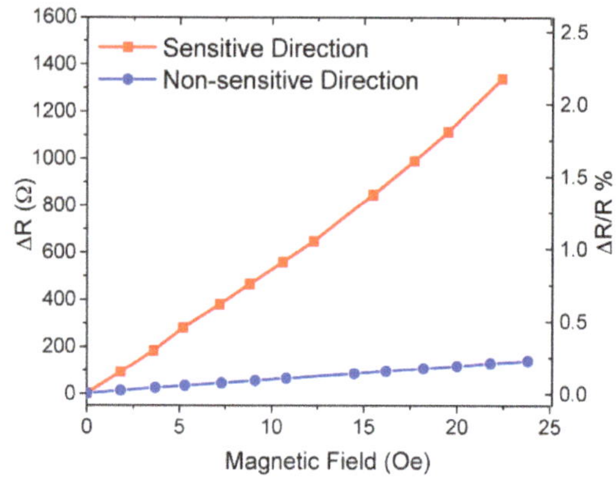

Figure 8: Resistance change ΔR (left axis) and magnetoresistive ratio (right) of the GMR sensor array with an external magnetic field applied to the sensitive and non-sensitive directions.

The thermal stability of the GMR sensor was evaluated under two situations: (i) after magnetic annealing, measuring the MR response after 30 min of consecutive annealing steps up to 350 °C, with a 3 kOe applied field during cooling; and (ii) after heating 4 min on a hot plate without a magnetic field applied. Results are plotted inFigure 9a. When heated and cooled under a magnetic field, the MR signal is stable upon annealing up to 280 °C, which is the onset for the irreversible diffusion among the metallic layers, and is consistent with previous results obtained with CoFe/MnIr-based magnetoresistive stacks [35]. On the contrary, if the sample is heated without a magnetic field above 200 °C (close to the blocking temperature of the MnIr antiferromagnet [36]), the reference layer and pinned layer will rotate coherently and the MR signal is rapidly reduced. This, however, is a reversible process, and the sample can recover the full MR level upon annealing under a magnetic field to set the exchange coupling between the CoFe/MnIr interfaces. Figure 9b shows the resistance change (ΔR) of spin-valve sensors normalized to the initial resistance (R_0) (at room temperature). Three different sensors were tested including: an un-patterned spin-valve, a single microfabricated spin-valve, and the full sensor array used in this work. ($\Delta R/R_0$) is a linear function of temperature change ΔT with a slope of ~0.1%/°C. The signal does not depend on the sensors overall resistance, shape or size.

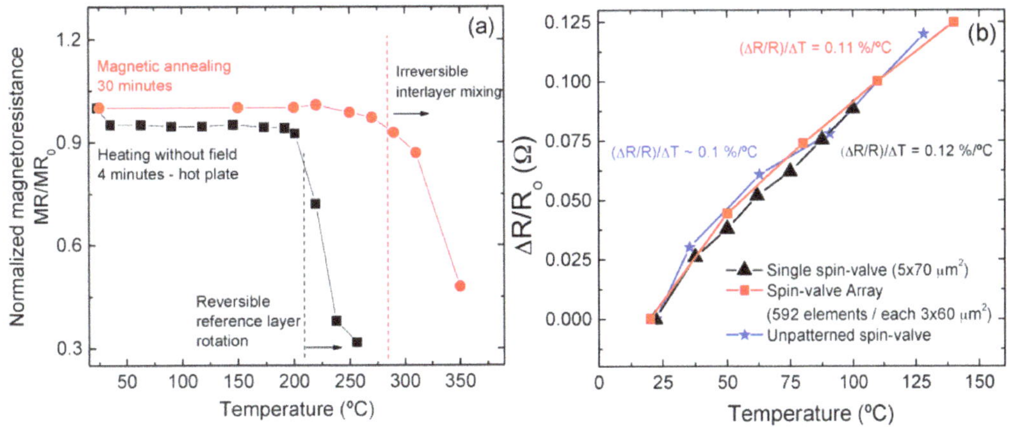

Figure 9: (**a**) Magnetoresistance dependence on temperature when a magnetic field is applied (magnetic annealing) and upon heating for 4 min on a hot plate without field. The data is normalized by the MR value at room temperature, measured before the thermal treatments; (**b**) Resistance dependence on temperature, tested for un-patterned spin-valve, a single microfabricated spin-valve, and the full sensor array, showing a similar dependency with variation of about 0.1%/°C.

Tactile Sensor Response

The tactile sensor was tested by applying vertical forces up to 32 mN (32 kPa) at different temperatures between 20 and 140 °C, and the variation in resistance was recorded. The cilia sensor response to external forces is shown inFigure 10. The results at room temperature show the sensor's ability to detect forces up to 15 mN (15 kPa) with a high sensitivity of 46 Ω/mN (46 Ω/kPa) and with noise of ±30 Ω. Considering this noise level, the resolution is 1.3 mN. Beyond this high sensitivity range, another linear region is found where the signal starts saturating, thereby providing a lower sensitivity. This region corresponds to the nearly full deflection of the cilia. The results for the tactile sensor at the high sensitivity region when operated at elevated temperatures are shown in Table 1. As the temperature increases, the signal exhibits lower resistance changes, which is attributed to the reduction in the nanocomposite's magnetization. It is observed that the nanocomposite is mechanically stable, but undergoes a small reduction in elasticity which slightly hardens the nanocomposite, increases the Young's modulus, and slightly increases the operating range. Previous studies have shown that the mechanical properties are maintained at low heating temperatures. Operating at temperatures greater than 200 °C will reduce the mechanical strength due to the thermal decomposition of PDMS that starts at about 200 °C and peaks at around 310 °C [37].

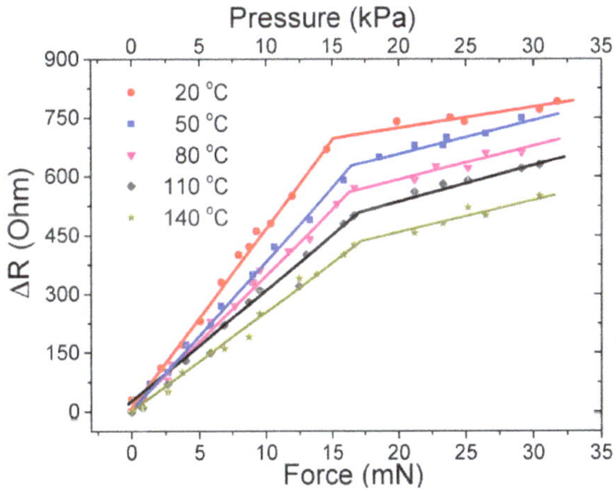

Figure 10: Tactile sensor response between 20 and 140 °C for different vertical forces/pressures. Pressures are calculated by considering the forces acting on the sensor's active area.

Table 1: Tactile sensor results at different temperatures. Pressures are calculated by considering the forces acting on the sensor's active area.

Temperature (°C)	Operating Range (mN) or (kPa)	Sensitivity (Ω/mN) or (Ω/kPa)	Maximum ΔR @ 15 mN (Ω)	Resolution (mN) or (kPa)	Noise Level ± (Ω)
20	0–15.0	46	700	1.31	30
50	0–16.2	39	581	1.69	33
80	0–16.5	35	531	2.17	38
110	0–17.1	29.4	462	2.85	42
140	0–17.4	25	385	3.91	49

The magnetization is reduced from both Curie's law and the slight corrosion of the iron NWs due to the heat assisted oxygen diffusion in the NWs. The latter is attributed to trapped water molecules in the PDMS, leading to corrosion of the NWs [20]. This is in accordance with a previous study [20] in which bare NWs were exposed to air, a nanocomposite sample was exposed to air, and another nanocomposite sample was kept in water. It was found that PDMS offers a good level of protection, and the magnetization of NWs inside of it slightly drops over time. After 10 days, the remanent magnetization of bare NWs in air dropped to 85% of the initial magnetization compared to 98% and 92.5% of nanocomposites exposed to air and water, respectively. NWs inside the nanocomposite oxidized slowly, due to the fact that PDMS is permeable to water molecules allowing minor interaction between iron and oxygen.

The reduction rate in sensitivity was −0.175 (Ω/mN)/°C, and the reduction in the signal measured at 15 mN (15 kPa) was −2.6 Ω/°C. The operating range was increasing at higher temperatures, which was attributed to the slight stiffening in the nanocomposite cilia. An expected increase in the signal variation was obtained, due to the increase of thermal noise, which as a result reduced the resolution.

CONCLUSIONS

A NWs-based permanent magnetic and highly elastic nanocomposite cilia tactile sensor has been developed for operation at elevated temperatures. Spin-valves detect the change of the stray field created by the permanent magnetic cilia when deflected due to an external force. The concept does not require an external magnetic field; hence the sensors can be highly integrated and employed in different applications without the added bulkiness of electro or permanent magnets.

The tactile sensor performance and operational abilities at high temperature were investigated. The effects of high temperatures on the nanocomposite are a small reduction in the magnetization that can be explained by Curie's law and by a slight oxidation of the NWs. Compared to previous findings for the oxidation of bare NWs, the result shows that the PDMS considerably slows down the oxidation of NWs but does not completely avoid it. This issue could be addressed by using single crystalline iron NWs, in which only the shell oxidizes and the iron core is retained up to 150 °C [24]. The spin-valves provide a stable MR performance up to around 200 °C, at which point the magnetic order of pinned and reference layer are lost.

Overall, the performance of the tactile sensor remains viable at high temperatures and operation to at least 140 °C is possible with the potential to go up to 200 °C. The sensor operation at elevated temperatures comes with a trade off in the sensitivity and signal amplitude. At 140 °C the sensitivity is reduced to 54% of the one at room temperature. This is caused by the reduction in magnetization of NWs and the stiffening of the polymer.

ACKNOWLEDGMENTS

Research reported in this publication is supported by the King Abdullah University of Science and Technology (KAUST). INESC-MN thanks FCT under EXCL/CTM-NAN/0441/2012 Project and Pest-OE/CTM/LA0024/2011. D. C. Leitao acknowledges FCT Grant SFRH/BPD/72359/2010.

REFERENCES

1. Kawasaki, H.; Komatsu, T.; Uchiyama, K. Dexterous anthropomorphic robot hand with distributed tactile sensor: Gifu hand II. *IEEE ASME Trans. Mechatron.* 2002, *7*, 296–303.
2. Sadao, O.; Terunuma, Y. New tactile sensor like the human hand and its applications. *Sens. Actuators A Phys.* 1992, *35*, 9–15.
3. Mukai, T.; Onishi, M.; Odashima, T.; Hirano, S.; Luo, Z. Development of the Tactile Sensor System of a Human-Interactive Robot "RI-MAN". *IEEE Trans. Robot.* 2008, *24*, 505–512.
4. Kerpa, O.; Weiss, K.; Worn, H. Development of a flexible tactile sensor system for a humanoid robot. In Proceedings of the 2003 IEEE/RSJ International Conference on Intelligent Robots and Systems, Las Vegas, NV, USA, 27–31 October 2003; Volume 1, pp. 1–6.
5. Dargahi, J.; Parameswaran, M.; Payandeh, S. A micromachined piezoelectric tactile sensor for an endoscopic grasper-theory, fabrication and experiments. *J. Microelectromech. Syst.* 2000, *9*, 329–335.
6. Hosseini, S.M.; Kashani, S.M.T.; Najarian, S.; Panahi, F.; Naeini, S.M.M.; Mojra, A. A medical tactile sensing instrument for detecting embedded objects, with specific application for breast examination. *Int. J. Med. Robot. Comput. Assist. Surg.* 2010, *6*, 73–82.
7. Yang, Y.; Zhang, H.; Lin, Z.H.; Zhou, Y.S.; Jing, Q.; Su, Y.; Yang, J.; Chen, J.; Hu, C.; Wang, Z.L. Human skin based triboelectric nanogenerators for harvesting biomechanical energy and as self-powered active tactile sensor system. *ACS Nano* 2013, *7*, 9213–9222.
8. Eason, E.V.; Hawkes, E.W.; Windheim, M.; Christensen, D.L.; Libby, T.; Cutkosky, M.R. Stress distribution and contact area measurements of a gecko toe using a high-resolution tactile sensor. *Bioinspir. Biomim.* 2015, *10*.
9. Park, M.; Park, Y.J.; Chen, X.; Park, Y.K.; Kim, M.S.; Ahn, J.H. MoS2-Based Tactile Sensor for Electronic Skin Applications. *Adv. Mater.* 2016, *28*, 2556–2562.
10. Schroeder, P.; Schotter, J.; Shoshi, A.; Eggeling, M.; Bethge, O.; Hütten, A.; Brückl, H. Artificial cilia of magnetically tagged polymer nanowires for biomimetic mechanosensing. *Bioinspir. Biomim.* 2011, *6*.
11. Fonseca, M.A.; English, J.M.; Von Arx, M.; Allen, M.G. Wireless micromachined ceramic pressure sensor for high-temperature applications. *J. Microelectromech. Syst.* 2002, *11*, 337–343.
12. Young, D.J.; Du, J.; Zorman, C.A.; Ko, W.H. High-temperature single-crystal 3C-SiC capacitive pressure sensor.*IEEE Sens. J.* 2004, *4*, 464–470.
13. Liu, G.D.; Cui, W.P.; Hu, H.; Zhang, F.S.; Zhang, Y.X.; Gao, C.C.; Hao, Y.L. April High temperature pressure sensor using a thermostable electrode. In Proceedings of the IEEE 10th International Conference on Nano/Micro Engineered and Molecular Systems (NEMS), Xi'an, China, 7–11 April 2015; pp. 201–204.
14. Leigh, S.J.; Bradley, R.J.; Purssell, C.P.; Billson, D.R.; Hutchins, D.A. A simple, low-cost conductive composite material for 3D printing of electronic sensors. *PLoS ONE* 2012, *7*.
15. Kim, K.N.; Chun, J.; Chae, S.A.; Ahn, C.W.; Kim, I.W.; Kim, S.W.; Wang, Z.L.; Baik, J.M. Silk fibroin-based biodegradable piezoelectric composite nanogenerators using lead-free ferroelectric nanoparticles. *Nano Energy*2015, *14*, 87–94.
16. Taghavi, M.; Mattoli, V.; Sadeghi, A.; Mazzolai, B.; Beccai, L. A Novel Soft Metal-Polymer Composite

for Multidirectional Pressure Energy Harvesting. *Adv. Energy Mater.* 2014, *4*.

17. Alfadhel, A.; Kosel, J. Magnetic Nanocomposite Cilia Tactile Sensor. *Adv. Mater.* 2015, *27*, 7888–7892.
18. Alfadhel, A.; Kosel, J. Magnetic micropillar sensors for force sensing. In Proceedings of the 2015 IEEE Sensors Applications Symposium (SAS), Zadar, Croatia, 13–15 April 2015; pp. 1–4.
19. Khan, M.A.; Alfadhel, A.; Kosel, J. Magnetic Nanocomposite Cilia Energy Harvester. *IEEE Trans. Magn.* 2016,*PP*.
20. Alfadhel, A.; Li, B.; Zaher, A.; Yassine, O.; Kosel, J. A magnetic nanocomposite for biomimetic flow sensing. *Lab Chip* 2014, *14*, 4362–4369.
21. Alnassar, M.; Ivanov, I.P.; Kosel, J. Flexible Magnetoelectric Nanocomposites with Tunable Properties. *Adv. Electron. Mater.* 2016, *2*.
22. Yi, Y.; Zaher, A.; Yassine, O.; Kosel, J.; Foulds, I.G. A remotely operated drug delivery system with an electrolytic pump and a thermoresponsive valve. *Biomicrofluidics* 2015, *9*.
23. Zaher, A.; Li, S.; Yassine, O.; Khashab, N.; Pirmoradi, N.; Lin, L.; Kosel, J. Osmotically driven drug delivery through remote-controlled magnetic nanocomposite membranes. *Biomicrofluidics* 2015, *9*.
24. Ivanov, Y.P.; Alfadhel, A.; Alnassar, M.; Perez, J.; Vázquez, M.; Chuvilin, A.; Kosel, J. Tunable magnetic nanowires for biomedical and harsh environment applications. *Sci. Rep.* 2016, *6*.
25. Parkin, S.P. Giant magnetoresistance in magnetic nanostructures. *Annu. Rev. Mater. Sci.* 1995, *25*, 357–388.
26. Melzer, M.; Kaltenbrunner, M.; Makarov, D.; Karnaushenko, D.; Karnaushenko, D.; Sekitani, T.; Someya, T.; Schmidt, O.G. Imperceptible magnetoelectronics. *Nat. Commun.* 2015, *6*.
27. Pannetier, M.; Fermon, C.; Goff, G.; Simola, J.; Kerr, J.E. Femtotesla magnetic field measurement with magnetoresistive sensors. *Science* 2004, *304*, 1648–1650.
28. Timoshenko, S. *History of Strength of Materials*; McGraw-Hill: New York, NY, USA, 1953.
29. Nielsch, K.; Müller, F.; Li, A.P.; Gösele, U. Uniform Nickel Deposition into Ordered Alumina Pores by Pulsed Electrodeposition. *Adv. Mater.* 2000, *12*, 582–586.
30. Freitas, P.P.; Ferreira, R.; Cardoso, S.; Cardoso, F. Magnetoresistive sensors. *J. Phys. Condens. Matter* 2007, *19*.
31. Freitas, P.P.; Cardoso, S.; Ferreira, R.; Martins, V.R.; Guedes, A.; Cardoso, F.; Loureiro, J.; Macedo, R.; Chaves, R.; Amaral, J.P. Optimization and integration of magnetoresistive sensors. *SPIN* 2011, *1*, 71–91.
32. Silva, A.; Leitao, D.; Amaral, J.; Valadeiro, J.; Cardoso, S.; Freitas, P.P. Linearization strategies for high sensitivity magnetoresistive sensors. *Eur. Phys. J. Appl. Phys.* 2015, *72*.
33. Tang, K.C.; Liao, E.; Ong, W.L.; Wong, J.D.S.; Agarwal, A.; Nagarajan, R.; Yobas, L. Evaluation of bonding between oxygen plasma treated polydimethyl siloxane and passivated silicon. *J. Phys. Conf. Ser.* 2006, *34*, 155–161.
34. Coey, J.M.D. *Magnetism and Magnetic Materials*; Cambridge University Press: Cambridge, UK, 2009.
35. Gehanno, V.; Freitas, P.P.; Veloso, A.; Ferreira, J.; Almeida, B.; Sousa, J.B.; Soares, J.C.; daSilva, M.F. Ion beam deposition of Mn-Ir spin valves. *IEEE Trans. Magn.* 1999, *35*, 4361–4367.
36. Lederman, T. Performance of metallic antiferromagnets for use in spin-valve read sensors. *IEEE Trans. Magn.*1999, *35*, 794–799.
37. Liu, M.; Sun, J.; Chen, Q. Influences ofheating temperature on mechanical properties of polydimethylsiloxane.*Sens. Actuators A Phys.* 2009, *151*, 42–45.

INDEX

G

H

I

K

L

M

N

O

P

R

S